The Literature of Crop Science

Literature of the Agricultural Sciences

Wallace C. Olsen, series editor

Agricultural Economics and Rural Sociology: The Contemporary Core Literature
By Wallace C. Olsen

The Literature of Agricultural Engineering
Edited by Carl W. Hall and Wallace C. Olsen

The Literature of Animal Science and Health
Edited by Wallace C. Olsen

The Literature of Soil Science
Edited by Peter McDonald

The Contemporary and Historical Literature of Food Science and Human Nutrition
Edited by Jennie Brogdon and Wallace C. Olsen

The Literature of Crop Science
Edited by Wallace C. Olsen

THE LITERATURE OF CROP SCIENCE

EDITED BY

Wallace C. Olsen

Cornell University Press

ITHACA AND LONDON

This book was typeset from disks supplied by the staff of the Core Agricultural Literature Project, Albert R. Mann Library, Cornell University. Nicole Kasmer Kresock prepared the machine-readable text, and corrections were made by Sharon Van De Mark. The research was financially supported by the Cornell Agricultural Experiment Station; the National Agricultural Library, the United States Department of Agriculture; and the Rockefeller Foundation.

First published 1995 by Cornell University Press.

Printed in the United States of America

Library of Congress Cataloging-in-Publication Data

The literature of crop science / edited by Wallace C. Olsen.
p. cm. — (The Literature of the agricultural sciences)
Includes bibliographical references and index.
ISBN 0-8014-3138-7 (cloth : alk. paper)
1. Crop science literature. I. Olsen, Wallace C. II. Series.
SB45.65.L58 1995
630—dc20 95-20394

♾ The paper in this book meets the minimum requirements of the American National Standard for Information Sciences—Permanence of Paper for Printed Library Materials, ANSI Z39.48–1984.

Dedicated to Jan R. Olsen, a remarkable administrator and an untiring supporter of access to agricultural literature

Contents

Preface

The Core Agricultural Literature Project at the Albert R. Mann Library, Cornell University, began in 1988 to identify the most useful current literature in the field of agriculture for university-level instruction and research. Portions of this book are conclusions from that work. The Project was funded primarily by the Rockefeller Foundation, which wished to identify this literature for the Third World and then transfer the full texts onto compact disks for easy distribution and use in remote areas. The Project staff worked with Steering Committees for each of the seven subject areas which were studied intensively and reported in this series of books. The volumes published are identified opposite the title page. A final volume will follow: *Forestry and Agroforestry*. The Core Agricultural Literature Project took four years to survey, evaluate, and write the results of these deliberations. The procedures included evaluation and counsel by scientists in all seven disciplines to the level of ranking of titles in lengthy monograph lists. These 600 participants are identified in the volumes; without their deep interest and commitment to scholarship, the core literature could not have been identified with such certainty. An historical literature component was added beginning with the second volume; its purpose was to identify the most valuable American literature for immediate preservation.

This sixth book in the series *The Literature of the Agricultural Sciences* is the broadest in scope and probably the most significant. Food crops are fundamental to the survival of humans. Crop advancements came from a knowledge of natural history and from scientific inquiries. Scientific advancement has been the result of methodical and painstaking effort, usually beginning with the recording of small bits of information which are eventually put together as new knowledge. Botany, soil science, genetics, entomology and numerous related areas of study have all contributed. The first three chapters outline this growth and interdependence and identify the people, literature, and advancements on which today's crop science stands.

This thorough assessment of the literature of crop improvement and protection is valuable at this time because of the increasing globalization of agriculture. Books and journals originating in developed countries are vital sources of information for education and research in Third World countries. Increasingly, Third World countries are creating and discovering much of importance to the rest of the world. Scholars and literature collectors throughout the world will find extensive assistance in this book for evaluating the strength of their literature collections and determining the merits of their journal and report holdings.

The Steering Committee members for the crop science effort reported in this volume were: W. Ronnie Coffman, Cornell University; Gary Heichel, University of Illinois; David J. Hume, University of Guelph; W. Laux, Biologische Bundesanstalt, Berlin; David Pimentel, Cornell University; and H. David Thurston, Cornell University.

WALLACE C. OLSEN
Series Editor

Ithaca, New York

The Literature of Crop Science

1. How Crop Improvement Developed

JOHN M. POEHLMAN
University of Missouri

Plant improvement is the art and science of changing the heredity of plants for either economic or esthetic reasons. The natural process of hereditary change has progressed slowly in nature for thousands of years. Humans eventually began to adapt and change some plants in order to protect crops and to provide better food. Modern crop biotechnology often overlooks the fact that crop improvements would not be possible except for numerous basic contributions to knowledge which slowly emerged after humans began critically to examine plants and crops. The purpose of this chapter is to provide a chronological overview of some of the major contributions to the knowledge of plant reproduction and heredity, and the resulting improvement of plants and crops made possible by the application of this knowledge.

Crop improvement began when people started gathering and planting seeds for the production of food. In time, they learned to select and harvest variant plants with desirable traits. Thus, selection became the earliest form of crop improvement. The plants selected may have had greater vigor, produced larger seeds, been more easily harvested, yielded fruits more pleasing to the taste, or possessed other desirable features that attracted attention. How the variant plants arose, or whether the new traits would be observed again if the seeds were planted, was unknown; indeed, such ideas surely didn't occur to primitive people. But slowly, over several thousand years, the habit of selecting plants with useful features forcefully shaped the evolution of many species that are cultivated today, leaving an indelible legacy for plant breeders in our time.

Wheat, for example, was domesticated in the Near East region of western Asia about ten thousand years ago.[1] The early forms of wild wheats had a brittle rachis that caused the seed-bearing spikelets to break apart and fall to

1. (a) Dettweiler, "Aryan Agriculture," *Journal of Heredity* 5 (1914): 473–481. (b) N. W. Simmonds, *Evolution of Crop Plants* (London: Longman, 1976).

the ground; hulls covering the seeds adhered tightly and were difficult to remove so that the seeds could be eaten. Eventually, mutant forms of wheat appeared without the fragile rachis and with free-threshing seeds. Wheats that did not shatter were easier to gather, and seeds that threshed out cleanly were easier to prepare for food, so these mutant forms were gathered by early humans. Over several thousand years they evolved into the cultivated wheats grown today.

A. Sexual Reproductive Systems in Crop Plants

The physical basis for crop improvement began when human beings first recognized and understood the presence of sex in plants. The crop improvement procedures that emerged differed according to the crop's reproductive system, whether the plant is normally self-fertilized or normally cross-fertilized. Before the fact of sex in plants was understood, systematic improvement procedures were not possible. Archeological records show, for example, that "sterile" (male) and "fruit-bearing" (female) trees of the date palm were familiar to the Assyrians and the Babylonians, and that they were hand pollinating the "fruit-bearing" trees as early as the eighth century B.C. But, while they may have recognized the similarity of the flowering habit in the date palm to sexual behavior in animals, clearly, it was not understood. Centuries passed before the function of pollen in fertilization and seed production was understood.

Sex in plants was discovered in 1694 by Rudolph J. Camerarius, a professor of natural philosophy in the University of Tübingen, Germany.[2] Camerarius removed the staminate flowers from isolated dioecious plants of spinach and hemp and a monoecious plant of maize, and demonstrated that without pollen the plants did not bear seeds. Three-quarters of a century passed before anyone recognized and acted on the significance of this discovery. Joseph G. Köelreuter, a professor of natural history in the University of Karlsruhe, Germany, came upon a letter in 1771, in which Camerarius had described his observations on the function of the stamens in fertilization. Köelreuter was so inspired by this information that he began to cross-pollinate plants. His activity in plant hybridization was so extensive that he became known as the "great hybridizer."[3] During the course of his

2. (a) H. F. Roberts, *Plant Hybridization before Mendel* (Princeton, N.J.: Princeton University Press, 1929). (b) J. von Sachs, *History of Botany (1530–1860)*, translation by H. E. F. Garnsey and I. B. Balfour (London: Oxford Press, 1890). (c) C. Zirkle, *The Beginnings of Plant Hybridization* (Philadelphia: University of Pennsylvania Press, 1935).

3. Roberts, *Plant Hybridization before Mendel*.

hybridizing, Köelreuter made some significant observations: the hybrid plants increased in vigor and possessed characteristics intermediate between the parents, proving that a contribution had been received from the male parent. He also observed insect pollination in plants, an event confirmed by Anton Sprengel in 1793.

Studies clarifying the sexual reproductive process in higher plants were continued throughout the nineteenth century. Some contributions of special importance for the development of crop improvement procedures are listed here:[4]

(1) 1739. Demonstration that pollen is essential for fertilization and seed production in Indian corn or maize (Governor James Logan, Pennsylvania, United States).
(2) 1824. Observation of pollen tubes entering the ovule (G. B. Amici, Italy).
(3) 1875. Description of the chromosomes (E. Strassburger, Germany).
(4) 1875. Clarification of the union of the sperm and the ovum at fertilization (O. Hertwig, Germany).
(5) 1885–87. Identification of two types of cell division, mitosis and reductional division, during fertilization (A. Weismann, Germany).
(6) 1898. Double fertilization observed in higher plants (S. G. Navashin, Russia).

Each of these contributions brought systematic crop improvement one step closer to reality. Plant breeders take this common knowledge for granted today, but before it and other information related to plant reproductive processes became known, crop improvement could not progress much beyond that practiced by early man.

The reproductive systems of particular crop species are altered by mechanisms that reduce self-fertility and impose restrictions on plant breeding. Awareness of the presence of these systems by plant breeders is essential to avoid adverse effects on normal reproductive functions or, alternately, to devise breeding systems that will utilize these mechanisms.

Incompatibility Systems. Inbreeding may be restricted and out-crossing favored by incompatibility systems, even though they produce normal pollen and ovules that mature at the same time. A gametophytic system of incompatibility in tobacco was described by E. M. East and A. J. Mangelsdorf in which the rate of pollen tube growth in stylar tissue is genetically controlled by multiple incompatibility (S) alleles.[5] If the (S) alleles in

4. (a) R. Cook, "A Chronology of Genetics," in *USDA Yearbook* (Washington, D.C.: U.S. Gov. Print. Office, 1937), pp. 1457–1477. (b) R. Frankel and E. Galun, *Pollination Mechanisms, Reproduction, and Plant Breeding* (Berlin: Springer-Verlag, 1977). (c) D. S. Johnson, "Sexuality in Plants," *Journal of Heredity* 6 (1915): 3–16. (d) Sachs, *History of Botany*.

5. E. M. East and A. J. Mangelsdorf, "A New Interpretation of the Hereditary Behavior of Self-Sterile Plants," *Proceedings of the National Academy of Sciences* 11 (1925): 166–171.

the pollen tube nucleus and the stylar tissue are identical, the rate of the pollen tube growth is reduced to the extent that the floral organs may have withered and died before the pollen nucleus reaches the ovule. In some species of grasses, two incompatibility alleles are present, S and Z, that require complementary identity in pollen and style.[6] Self-incompatibility of the gametophytic type in some cross-pollinating crops may be overcome by planting two or more cross-compatible varieties in alternate rows. In a sporophytic system, incompatibility to pollen germination and pollen tube growth is located in the surface of the stigma, with dominance determined by the genotype of the pollen parent. The sporophytic system is utilized in *Brassica* species to manage pollinations for the production of hybrid seed.[7] Homomorphic incompatibility systems based on heterostyly are present in some plant species.

Apomixis. Apomixis, common in forage grass species, is a vegetative form of reproduction in which seeds are formed without union of an egg and a sperm; gene recombination does not occur. In certain forage grasses, genetic control of the apomictic mechanism has been utilized by breeders to produce completely uniform populations.[8]

Male Sterility. Male sterility results from the failure of the plant to produce functional anthers or pollen. Genetic male sterility is normally conditioned by recessive genes, with the dominant alleles restoring the production of fertile anthers and pollen. Genetic male sterility was reported in the sweet pea (*Lathyrus odoratus*) in 1908.[9] It has since been reported in barley, corn, cotton, flax, rice, sorghum, tomato, tobacco, and other field and vegetable crop species. Plant breeders utilize genetic male sterility to eliminate laborous emasculation procedures in self-pollinated species.[10] Cytoplasmic male sterility was identified in flax by W. Bateson and A. E. Gairdner and explained by R. J. Chittenden.[11] Slightly later it was discov-

6. A. Lundqvist, "Self-Incompatibility and the Breeding of Herbage Grasses," in E. Åkerberg, A. Hagberg, G. Olsson, and O. Tedin, eds., *Recent Plant Breeding Research, Svalöf, 1946–1961* (Stockholm: Almqvist and Wiksell, 1948), pp. 193–202.

7. M. L. Odland and C. J. Noll, "The Utilization of Cross-Compatibility and Self-Incompatibility in the Production of F_1 Hybrid Cabbage," *Proceedings of the American Society of Horticultural Science* 55 (1950): 391–402.

8. E. C. Bashaw, "Apomixis and Its Application in Crop Improvement," in W. R. Fehr and H. H. Hadley, eds., *Hybridization of Crop Plants* (Madison, Wis.: ASA and CSSA, 1980), pp. 45–63.

9. W. Bateson, E. R. Saunders, and R. C.Punnett, "Male Sterility in *Lathyrus odoratus*," *Report on Evolution, Royal Society of London* 4 (1908): 16.

10. C. A. Brim and C. W. Stuber, "Application of Genetic Male Sterility to Recurrent Selection Schemes in Soybean," *Crop Science* 13 (1973): 528–530.

11. (a) W. Bateson and A. E. Gairdner, "Male-Sterility in Flax, Subject to Two Types of Segregation," *Journal of Genetics* 11(1921): 269–275. (b) R. J. Chittenden, "Cytoplasmic Inheritance in Flax," *Journal of Heredity* 18 (1927): 337–343.

ered in maize by M. M. Rhoades.[12] Cytoplasmic male sterility is transmitted through the maternal cytoplasm, with normal pollen production being restored by dominant fertility-restoring genes. A procedure for utilization of cytoplasmic male sterility in onion hybrid-seed production was proposed by H. A. Jones and S. L. Emsweller, and in sorghum hybrid-seed production by J. C. Stephens and R. F. Holland.[13] Cytoplasmic male sterility has since been utilized in hybrid seed production of carrot, corn, millet, pepper, rice, sugarbeet, wheat, and other field and vegetable crops.

B. Plant Hybridization before Mendel

It is difficult to know precisely when the idea originated of breeding cultivated plants to improve their performance. Köelreuter's experiments in plant hybridization stimulated extensive efforts in cross-pollination of plants throughout the nineteenth century. Early practitioners of the art were often referred to as "hybridizers." Many early species crosses were made inadvertently because the distinction between species and varieties was often vague, and knowledge of what could or could not be crossed was not always clear. When novel hybrids occasionally arose from the crosses, they were sometimes propagated and marketed. The early hybridizers reported many interesting observations that are considered commonplace in a plant breeding program today. But before the work of Mendel there were no guiding principles to coordinate the new knowledge and to explain the significance of what was observed. Some contributions of significance during this period include the work of:

(1) Thomas Andrew Knight (England, 1795–1858). Produced thirty-two hybrids in fruits and vegetables, and reported dominance of color in hybrids of currant and pea.
(2) Augustin Sageret (France, 1826). Observed contrasting characters of parents in crosses of muskmelon and cantaloupe; characters in hybrids were identical to those of one parent.
(3) William Herbert (England, 1819–1847). Conducted 155 hybridization experiments to improve vegetables and flowers.
(4) Carl Friedrich von Gärtner (Germany, 1835–1860). A prolific hybridizer who attempted 1,332 hybrid experiments involving 107 species.

12. M. M. Rhoades, "Cytoplasmic Inheritance of Male Sterility in *Zea mays*," *Science* 73 (1931): 340–341.

13. (a) H. A. Jones and S. L. Emsweller, "A Male-Sterile Onion," *Proceedings of the American Society of Horticultural Science* 434 (1937): 582–585. (b) J. C. Stephens and R. F. Holland, "Cytoplasmic Male-Sterility for Hybrid Sorghum Seed Production," *Agronomy Journal* 46 (1954): 20–23.

(5) S. M. Schindel (Maryland, United States, 1886). Crossed the Fultz variety of wheat with the Lancaster variety.[14] A selection from the cross named Fulcaster was still popular in the United States as late as 1930.
(6) Gregor Mendel (Austria, 1866). Published the famous *Experiments in Plant Hybridization*, in which he reported his studies of inheritance in garden peas. The publication did not receive recognition until its rediscovery in 1900. The "Laws of Inheritance" that Mendel described marked the beginning of the science of genetics and provided a scientific basis for plant improvement.[15]

Detailed discussions of the work of the hybridizers have been reported by H. de Vries, H. F. Roberts, C. Zirkle, R. Cook, and others.[16]

C. The Improvement of Self-Pollinated Crops by Selection

Improvement in wheat, a self-pollinated crop, began in the early part of the nineteenth century.[17] John Le Couteur (Isle of Jersey) and Patrick Shirreff (Scotland) each selected variant plants from landraces of wheat in the search for improved varieties. Le Couteur and Shirreff multiplied the seed from each plant separately without further selection. For them, the initial choice was critical to finding an improved variety. F. F. Hallett (England, 1857) followed a different procedure, reselecting each year the best seed from the best spike on the best plant and multiplying each culture anew. Wheat breeders in Germany also followed a system of continuous selection. Their method differed from Hallett's as they selected representative plants from landraces and other sources and planted a mixture of seeds from the selected plants. The population was reselected in subsequent years in a similar manner and planted without progeny evaluation of the selections, a procedure known as "mass selection."

This was the state of the art for selection in wheat breeding when in 1888 Hjalmar Nilsson became a wheat breeder at the newly organized Swedish Seed Association in Svalöf.[18] At first Nilsson favored the German method

14. J. A. Clark, "Improvement in Wheat," in *USDA Yearbook* (Washington, D.C.: U.S. Gov. Print. Office, 1936), pp. 207–302.
15. J. A. Peters, *Classic Papers in Genetics* (Englewood Cliffs, N.J.: Prentice-Hall, 1959).
16. (a) H. de Vries, *Plant Breeding* (Chicago: Open-Court Publishing, 1907). (b) H. F. Roberts, "The Contribution of Carl Friedrich von Gärtner to the History of Plant Hybridization," *American Naturalist* 53 (1919): 431–445. (c) Roberts, *Plant Hybridization before Mendel*. (d) Zirkle, *Beginnings of Plant Hybridization*. (e) Cook, "Chronology of Genetics."
17. de Vries, *Plant Breeding*.
18. (a) Å. Åkerman and J. MacKey, "The Breeding of Self-Fertilized Plants by Crossing," in Å. Åkerman, O. Tedin, K. Fröier, and R. O. White, eds., *Svalöf, 1886–1946* (Lund: Carl Bloms Boktryckeri, A.B., 1948), pp. 46–71. (b) L. H. Newnan, *Plant Breeding in Scandinavia* (Ottawa: Canadian Seed Growers Association, 1912). (c) de Vries, *Plant Breeding*.

of continuous or mass selection, but early results were unfavorable. Nilsson observed that the only uniform progenies came from single plants. His observations led him to state two principles for selection in self-pollinated crops: (1) the plant, not an individual seed from a plant, is the correct unit of selection from a mixed population of plants, and (2) a single plant selection only should be made, reselection being unnecessary.

Nilsson's observations were confirmed by the research of W. Johannsen, a Danish botanist who, in 1903, through a selection experiment in garden beans, clarified the distinction between phenotype and genotype as follows: A mixed lot of seed of a self-fertilized species can be separated by selection into different pure lines; after the pure lines have been established, reselection within the pure line is ineffective.[19] Similar conclusions were reached independently by W. M. Hays in Minnesota.[20]

The Progeny Test. During the period when the selection procedure was being defined, an important principle in plant improvement emerged: evaluating a plant selection by the average performance of its progeny. The progeny test was utilized by the Vilmorins in France in the development of the sugarbeet from a weedy species to its cultivated form.[21] Nilsson in Sweden and Hays in the United States were utilizing the progeny test prior to 1900. Progeny evaluation was an essential element in Mendel's experiments.

D. The Improvement of Self-Pollinated Crops by Crossing

Selection procedures for mixed populations of self-pollinated crops were being developed by Nilsson, Hays, and others when Mendel's Laws of Inheritance were rediscovered in 1900 and when Johannsen clarified the distinction between genotype and phenotype in 1903. It became apparent that pure line selection could soon exhaust the variability in landraces of the self-pollinated crops, such as wheat or oats, and breeders began to explore hybridization among varieties as a method of improvement. Hybridization had been practiced in abundance before Mendel, but, as R. F. Biffen observed before Mendel and Johannsen, there was no known method for handling the progenies.[22]

19. Peters, *Classic Papers in Genetics.*

20. W. M. Hays, *Plant Breeding* (USDA, Division of Vegetable Physiology and Pathology, 1901 [*Bulletin* no. 29]).

21. (a) Cook, "Chronology of Genetics." (b) M. J. L. de Vilmorin, *L'Hérédité chez la Betterave Cultivée* (Paris: Gauthier-Villars, 1923).

22. R. H. Biffen, "Mendel's Laws of Inheritance and Wheat Breeding," *Journal of Agricultural Science*, (Cambridge) 1 (1905): 4–48.

W. J. Spillman, a wheat breeder in Washington State, had already concluded that the most common defects in wheat—shattering, lodging, disease susceptibility—would not be solved by selection alone and undertook to correct them by crossing. Spillman, like Mendel, recorded the proportion of plants with each parent characteristic in the hybrid generation and came close to arriving at conclusions similar to those of Mendel.[23]

Biffen pointed out that Mendel had focused "not on the plant as a whole but on its single characters."[24] Biffen listed numerous traits in wheat that exhibit segregation in subsequent hybrid generations and, like Spillman, suggested that breeders should utilize hybridization to combine genes for superior characters from different parents. Hybridization to combine genes for useful parent traits then became the basic method of breeding in self-pollinated crops, as it continues to be.

As knowledge of genetics expanded, several important facts about the structure of hybrid populations following a cross-pollination between two self-pollinated plants became clear: (1) plants are genetically identical in the first generation following a cross between pure line varieties, (2) maximum heterozygosity occurs in the second hybrid generation, and (3) a practical state of homozygosity is not reached until the sixth or later generation.

Efficient selection procedures were needed by breeders to identify the superior genotypes while advancing the hybrid generations through the segregating generations. Three basic selection procedures have emerged, although each may be modified in various respects by different breeders. In *pedigree selection,* plant selections are made in the F_2 generation and repeated in successive generations until satisfactory homozygosis is reached in about the F_5 or F_6 generation. This system was used by Nilsson and H. Nilsson-Ehle at Svalöf before 1910.[25] In *Population or bulk selection*, the progeny of the cross is grown as a population without selection until satisfactory homozygosis is reached, after which plant selections are made. This method was proposed by Nilsson-Ehle at Svalöf in 1908.[26] In *single seed descent*, the progeny of the F_2 plants are advanced as single seeds through several generations before progeny evaluation begins. This procedure was proposed by C. H. Goulden and refined by C. A. Brim for use in soybeans.[27]

23. W. J. Spillman, "Quantitative Studies on the Transmission of Parental Characters to Hybrid Offspring," Proceedings of the 15th Annual Conference of the Association of the American Agricultural Colleges and Experimental Stations (Washington, D.C.: U.S. Govt. Print. Office, 1902 [*USDA, Office of Experiment Stations Bulletin* no. 15]).

24. Biffen, "Mendel's Laws of Inheritance," p. 7.

25. Åkerman, "The Breeding of Self-Fertilized Plants."

26. Ibid.

27. (a) C. H. Goulden, "Problems in Plant Selection," in *Proceedings of the 7th International Genetics Congress*, Edinburgh, Aug. 1939 (1941), pp. 132–133. (b) C. A. Brim, "A Modified Pedigree Method of Selection in Soybean," *Crop Science* 6 (1966): 220–221.

Other hybridizing procedures have been developed to accomplish specific goals in self-pollinated species. *Backcross* is a procedure for adding a gene from a donor parent to a recurring parent by a succession of crosses, described by H. V. Harlan and M. N. Pope.[28] The backcross is also used to transfer "sterile" cytoplasm among genotypes. Backcrossing may be employed in cross-pollinated species if multiple crosses are made in each backcross generation to sample the recurrent parent population. *Composite crossing* is a method for combining genes from a group of parents by crossing parents and F_1s in pairs until all parents enter into a common progeny with equal frequency, described by Harlan, M. L. Martini, and H. Stevens.[29] *Multiline varieties* is a method proposed by N. E. Borlaug to combat a broad spectrum of races of disease producing pathogens, and refined by J. A. Browning and K. J. Frey to manage host genes for control of the crown rust pathogen (*Puccinia coronata*) in oats.[30] A multiline variety is a composite of genetically similar isolines, each isoline possessing a different gene for resistance to a particular disease pathogen, or other trait.

E. Qualitative and Quantitative Inheritance

The traits in garden peas studied by Mendel were conditioned by single genes; the phenotypes could be grouped into easily distinguished, discrete classes, a characteristic of qualitative inheritance. Mendel observed, but did not understand, quantitative inheritance when he stated that flowering time of hybrids "stands almost exactly between those of seed and pollen parents."[31] An explanation of quantitative inheritance was given by Nilsson-Ehle at Svalöf in 1908.[32] Nilsson-Ehle identified three genes for red seed coat color in wheat, each gene contributing to the intensity of the color. Traits affected in an additive manner by genes contributing small effects are designated multigenic or polygenic traits. Similar conclusions regarding the inheritance of quantitative traits were reached independently by Edward M. East from studies of endosperm color and number of rows of kernels in

28. H. V. Harlan and M. N. Pope, "The Use and Value of Back-Crosses in Small-Grain Breeding," *Journal of Heredity* 13 (1922): 319–322.

29. H. V. Harlan, M. L. Martini, and H. Stevens, *A Study of Methods in Barley Breeding* (Washington, D.C.: U.S. Dept. of Agriculture, 1940 [*USDA Technical Bulletin* no. 720]).

30. (a) N. E. Borlaug, "The Use of Multilineal or Composite Varieties to Control Airborne Epidemic Diseases of Self-Pollinated Crop Plants," *Proceedings of the 1st International Wheat Genetics Symposium*, Winnipeg, Canada, 1958 (Winnipeg: University of Manitoba, 1959). (b) J. A. Browning and K. J. Frey, "Multiline Cultivars as a Means of Disease Control," *Annual Review on Phytopathology* 7 (1969): 355–382.

31. Peters, *Classic Papers in Genetics*, p. 14.

32. A. Müntzing, *Genetics: Basic and Applied* (Stockholm: Lts Förlag, 1961).

maize.[33] These pioneer experiments also demonstrated how plants may arise with quantitative traits more extreme than those of the parents. The phenomenon, known as transgressive segregation, is utilized extensively in crop improvement.

Many plant traits important in crop improvement are quantitatively inherited. In quantitative inheritance, the expression of the trait is continuous; discrete differences are not expressed. This has given rise to the field of biometrical genetics, pioneered by R. A. Fisher, S. Wright, and J. B. Haldane, in which means, variances, and covariances are calculated to express genetic effects, replacing classical genetic symbols and ratios. Biometrical genetics is too extensive to review here, except for two concepts that are utilized importantly in crop improvement: (1) The phenotypic variability observed in a quantitative character may be due to its heredity, expressed as genetic variance; to the environment, expressed as environmental variance; or to genotype x environment interactions, expressed as interaction variance. (2) The genetic variance may be further divided into an additive portion, a dominance component, and an epistatic component.[34]

F. Population Improvement in Cross-Pollinated Crops

The basic procedures for the improvement of self-pollinated crops were developed in the early part of the twentieth century according to fundamental principles established by Mendel, Johannsen, Vilmorin, Hays, Nilsson, Nilsson-Ehle, Biffen, and others. Due to the extensive production and importance of wheat in Europe and the northern United States, much of the early research leading to the development of breeding procedures in self-pollinated crops was conducted on wheat. In central and southern United States, corn (maize), a cross-pollinated crop, had become the principal cereal crop and the procedures developed for the breeding of the self-pollinated crops could not be applied to the breeding of corn. In the self-pollinated crops, the breeding focus is on the selection of plants homozygous for qualitative traits that can be propagated in the progeny. In corn and other crops with extensive cross-pollination, individual plants are highly heterozygous, and the exact genotype of a selected plant is not reproduced in its progeny. In breeding, the focus is on quantitative traits with the goal of

33. E. M. East, "A Mendelian Interpretation of Variation That Is Apparently Continuous," *American Naturalist* 44 (1910): 65–82.

34. R. A. Fisher, "The Correlation between Relatives on Supposition of Mendelian Inheritance," *Transactions of the Royal Society of Edinburgh* 52 (1918): 399–433.

improving the average performance of the population rather than the performance of the individual plant.

With corn, population improvement procedures were utilized first to improve open-pollinated varieties, and later to improve germ plasm sources for inbred line development. The basic population improvement procedures are mass selection, ear-to-row selection, and recurrent selection. In cross-pollinated forage crops, such as grasses and legumes, where the stamens and the pistil are present in the same flower, cross-pollinations are not manipulated with the same ease as in corn. In these crops, a procedure for production of synthetic varieties has been developed.

Mass Selection. Mass selection in corn is a system of breeding in which seed is selected by the appearance of the plant or ear, or both, and mixed for planting without progeny evaluation. The procedure was used by farmer/breeders when they selected ears of corn to plant their next crop. Repeated selection for particular traits slowly increases the gene frequency for those traits, although the specifics were unknown to the early practitioners of the art.

Ear-to-Row Selection. In ear-to-row selection, progenies from selected ears are evaluated for a particular trait. Seeds from the superior progenies are bulked, or they may be systematically crossed, to grow the next cycle. The system originated with C. G. Hopkins, at the Illinois Agricultural Experiment Station, to improve the protein and oil composition of the corn kernel.[35]

Recurrent Selection. Recurrent selection is a system of breeding designed to increase the frequency of favorable genes for a quantitative trait by repeated cycles of crossing and selection. The cycles may be repeated as long as improvement in the trait is obtained. The recurrent selection method was proposed by H. K. Hayes and R. J. Garber and by East and D. F. Jones as a way to improve the protein content in corn through successive cycles of selection.[36] M. T. Jenkins described the procedure, and F. H. Hull named it recurrent selection.[37] Recurrent selection has many applications in crop improvement but is utilized most extensively with cross-pollinated crops.

35. C. G. Hopkins, "Improvement in the Chemical Composition of the Corn Kernel," *Illinois Agricultural Experiment Station Bulletin* 55 (1899): 205–240.

36. (a) H. K. Hayes and R. J. Garber, "Synthetic Production of High-Protein Corn in Relation to Breeding," *Journal of the American Society of Agronomy* 11 (1919): 309–318. (b) E. M. East and D. F. Jones, "Genetic Studies on the Protein Content of Maize," *Genetics* 5 (1920): 543–610.

37. (a) M. T. Jenkins, "The Segregation of Genes Affecting Yield of Grain in Maize," *Journal of the American Society of Agronomy* 32 (1940): 55–63. (b) F. H. Hull, "Recurrent Selection for Specific Combining Ability in Corn," *Journal of the American Society of Agronomy* 37 (1945): 134–145.

Synthetic Varieties. Hayes and Garber used the term "synthetic" in their proposal to improve the protein content of corn through repeated cycles of selection, and Jenkins described a procedure for "synthetic varieties" of corn by intercrossing selfed lines or inbreds.[38] Neither fits the procedure utilized in producing a "synthetic variety" of a forage crop. T. J. Jenkin of the University College of Wales, first proposed a procedure, called "strain building," for combining seeds from individual plants of cross-pollinated forage species.[39] The procedure currently used was developed in the alfalfa breeding program at the University of Nebraska, Lincoln, by merging two concepts:[40] (1) genotypes in alfalfa could be perpetuated through clones, and (2) high-yielding hybrids could be produced by combining non-inbred clones. The clones entering into the synthetic are evaluated for their combining ability which is done by harvesting seed from a polycross nursery with open-pollination and growing the polycross seed in performance trials.[41]

G. Utilization of Heterosis or Hybrid Vigor

A new era in plant improvement began in 1909 when G. H. Shull outlined a method for producing seed of hybrid corn. W. J. Beal had proposed the production of variety hybrids in corn thirty years earlier.[42] Beal would plant two varieties in alternate rows, pull out the tassels on one row so that the seeds would be produced from pollen from the other row, and harvest the crossed seeds—a procedure like that used to produce single-cross hybrid seed corn today. At the time, the procedure was too sophisticated for the farmers, who saved their own seed and were not ready to accept the practice of buying seed anew each year.

Shull began inbreeding corn in 1904. The inbred lines declined in vigor with successive generations of inbreeding, but vigor was restored upon crossing the inbreds. Shull focused on the restoration of vigor by crossing rather than on the loss of vigor by inbreeding. He viewed an open-pollinated field of corn as a mixture of many complex hybrids, which inbreeding would reduce to pure lines as in self-pollinated crops. He proposed that the

38. (a) Hayes and Garber, "Synthetic Production of High-Protein Corn." (b) Jenkins, "The Segregation of Genes."

39. T. J. Jenkin, *The Method and Technique of Selection. Breeding and Strain-building in Grasses* (Imperial Bureau on Plant Genetics, 1931 [*Herbage Plants Bulletin* no. 3]).

40. H. M. Tysdal, T. A. Kiesselbach, and H. L. Westover, *Alfalfa Breeding* (Nebraska Agricultural Experiment Station, 1942 (*Research Bulletin* no. 124]).

41. H. M. Tysdal, "History and Development of the Polycross Technique in Alfalfa Breeding," *Report of the 11th Alfalfa Improvement Conference*, Lincoln, Nebr., 1948, pp. 36–39.

42. W. J. Beal, "Indian Corn," in *Report of the Michigan Board of Agriculture* (Lansing, Mich., 1880), pp. 279–289.

breeder should find and maintain the highest yielding hybrid combination; the following year he outlined a procedure for making single-cross hybrid seed corn by crossing inbred lines.[43] East also observed the deteriorating effects from inbreeding corn and the restoration of vigor by crossing inbred lines, but he did not have Shull's clear vision for a practical procedure to utilize the increased vigor.[44] Yet, when Shull became interested in other lines of research and discontinued research in corn, it was East and two of his famous students, D. F. Jones and H. K. Hayes, who became the principal champions for this new method of corn breeding. At first it appeared that the cost of hybrid seed produced on weak inbred lines would prohibit use of the hybrid method for corn improvement. This objection was overcome when Jones proposed that two single-cross hybrids be crossed to produce a double-cross hybrid.[45] In the double-cross hybrid procedure, the seed is produced in abundance on vigorous single-cross plants. The use of hybrid corn expanded rapidly in the 1930s and 1940s. The double-cross hybrid was standard for hybrid corn production in the United States until it was replaced by single-cross hybrids in the 1960s.

Heterosis. Heterosis refers to the increase in size or vigor of a hybrid over the mean of its parents. The term heterosis, coined by Shull, may be used interchangeably with hybrid vigor. A.B. Bruce, F. Keeble and C. Pellew introduced the concept that heterosis results from an accumulation of dominant alleles.[46] East had a different explanation, describing heterosis on the basis of the interaction of alleles at a single locus, a concept later designated overdominance by Hull.[47] Both hypotheses are supported by experimental evidence, but the dominant gene explanation is most generally favored.

Evaluation of Inbred Lines in Single- and Double-Crosses. As the use of hybrid corn expanded in the United States, hybrid corn breeders began systematically to evaluate new inbred lines in single- and double-cross combinations. This proved to be an ardous task since ten inbred lines could be expanded into 45 single-cross and 630 double-cross combinations. More

43. (a) G. H. Shull, "The Composition of a Field of Maize," *Report of the American Breeders' Association* 4 (1908): 296–301. (b) G. H. Shull, "A Pure-Line Method in Corn Breeding," *Report of the American Breeders' Association* 5 (1909): 51–59.

44. E. M. East, "The Distinction between Development and Heredity in Inbreeding," *American Naturalist* 43 (1909): 173–181.

45. D. F. Jones, *The Effects of Inbreeding and Crossbreeding Upon Development* (New Haven: Connecticut Agricultural Experiment Station, 1918 [*Connecticut Agricultural Experiment Station Bulletin* no. 207]).

46. (a) A. B. Bruce, "The Mendelian Theory of Heredity and the Augmentation of Vigor," *Science* 32 (1910): 627–628. (b) F. Keeble and C. Pellew, "The Mode of Inheritance of Stature and the Time of Flowering in Peas (*Pisum sativum*)," *Journal of Genetics* 1 (1910): 47–56.

47. Hull, "Recurrent Selection for Specific Combining."

efficient breeding procedures were needed and over the next dozen years several key concepts emerged that greatly increased efficiency in hybrid corn breeding.

(1) Recognition that the higher yielding double crosses contained inbred lines of diverse origin.[48]
(2) Preliminary evaluation of inbred lines through the use of inbred-variety crosses, called top crosses, thereby reducing the number of single-crosses that it was necessary to make and evaluate.[49]
(3) Estimating the performance of possible double-crosses from the single-cross yields.[50] Only the double-crosses with the highest estimated yield would then be made and evaluated.
(4) Recognition that the top cross performance trials measure the additive gene effects or general combining ability of inbred lines, in contrast to the single-cross performance trials that measure dominance and epistatic gene effects or specific combining ability.[51]

With the change from production of double-cross to single-cross hybrids of corn in the 1960s, evaluation of inbreds was greatly simplified. In current corn breeding programs new inbreds are often designed to replace a specific inbred in a particular single-cross. The new inbred line is evaluated by its substitution in hybrid combinations for the inbred line that it will replace.

Germ Plasm Improvement in Corn. The early inbred lines in corn were generated by isolation from open-pollinated varieties, by selection from single-cross progenies, or by adding desirable genes through backcrosses. These methods were inadequate to improve complex quantitative traits affecting many aspects of crop performance. Experimental results demonstrated that inbred lines extracted from corn populations that had been improved through recurrent selection procedures would have a superior combining ability for quantitative traits affecting performance.[52] In the recurrent selection experiments conducted to improve the germ plasm, contributions were made to quantitative genetic theory, mating designs, and statistical procedures for data analyses. The population improvement procedures that

48. S. K. Wu, "The Relationship between the Origin of Selfed Lines of Corn and Their Value in Hybrid Combinations," *Journal of the American Society of Agronomy* 31 (1939): 131–140.

49. M. T. Jenkins and A. M. Brunson, "Methods of Testing Inbred Lines of Maize in Crossbred Combinations," *Journal of the American Society of Agronomy* 24 (1932): 523–530.

50. M. T. Jenkins, "Methods of Estimating the Performance of Double Crosses in Corn," *Journal of the American Society of Agronomy* 26 (1934): 199–204.

51. G. F. Sprague and L. A. Tatum, "General vs. Specific Combining Ability in Single Crosses of Corn," *Journal of the American Society of Agronomy* 34 (1942): 923–932.

52. J. H. Lonnquist, "Recurrent Selection as a Means of Modifying Combining Ability in Corn," *Agronomy Journal* 43 (1951): 311–315.

were utilized included mass selection, ear-to-row selection, half-sib and full-sib selection, S_1 and S_2 selection, and reciprocal recurrent selection.

Cytoplasmic Male Sterility and Fertility Restoration in Hybrid Seed Production. The utilization of cytoplasmic male sterility (CMS) and fertility restoring (R_f) genes in hybrid seed production was pioneered in the onion.[53] The system is applicable in other cross-pollinated crops in which usable forms of CMS and R_f genes are present. A male sterile (S) female line with recessive restorer genes and a normal (N) male fertile line are planted in alternate rows. Hybrid seed from the female line will have been pollinated from the male fertile line. Utilization of CMS eliminates the detasseling procedure in hybrid seed production in corn and permits production of hybrid seed in crops like sorghum and wheat with perfect flowers. Several sources of CMS have been identified in corn, but the type of CMS most widely used originated from an inbred line in Texas and became known as the T- or Texas-type. Two genes, Rf_1 and Rf_2, and some additional modifier genes are required to restore fertility to the T-type cytoplasm. In 1970, corn containing the T-type cytoplasm was injured by a strain of the pathogen causing southern leaf blight, so use of CMS in corn for the production of hybrid seed was discontinued. In sorghum, CMS was obtained by introducing kafir chromosomes into milo cytoplasm.[54] Fertility was restored by a single dominant gene, Ms_c. In wheat, CMS was obtained by introducing bread wheat chromosomes into cytoplasm of *Triticum timopheevi*.[55]

H. Polyploidy and Plant Breeding

Polyploidy has been studied since 1909 when a "gigas" strain of *Oenothera* was discovered to be a polyploid with the double chromosome number of twenty-eight instead of the normal fourteen. Exploitation of polyploidy was possible after discovery that the alkaloid colchicine would promote polyploidy. Research workers at Svälof, Sweden, were among the first to utilize polyploidy as a means of crop improvement. Polyploidy breeding was initially introduced by doubling the chromosomes to produce autopolyploids in species with a low chromosome number and in forest trees or flowers where increase in size contributes to its economic value. Allopolyploids are polyploids in which the chromosome content of two or

53. H. A. Jones, and A. E. Clark, "Inheritance of Male Sterility in the Onion and the Production of Hybrid Seed," *Proceedings of the American Society of Horticultural Science* 43 (1943): 189–194.
54. Stephens and Holland, "Cytoplasmic Male-Sterility."
55. J. A. Wilson and W. M. Ross, "Male Sterility Interaction of the *Triticum aestivum* Nucleus and *Triticum timopheevii* Cytoplasm," *Wheat Information Service* 14 (1962): 29–30.

more species is combined. Many crop species, including bread wheat (*Triticum aestivum*), are natural allopolyploids that originated in nature with a combination of seven pairs of chromosomes from each of three diploid species. A man-made allopolyploid, triticale (*X Triticosecale*), was produced by combining chromosomes sets from wheat and rye. A classic study of the aneuploids (plants with an irregular number of chromosomes) of common wheat, led to the identification of each of wheat's twenty-one chromosomes.[56] The same study facilitated development of methods for transfer or substitution of individual chromosomes or pairs of chromosomes from one variety to another in a polyploid species.

I. Mutations in Plant Breeding

Hugo de Vries in 1901–03 explained the origin of small variations in natural populations of a plant species as due to mutation. L. J. Stadler reported that mutations in barley are induced by treatments with X-rays and radium.[57] Mutation research expanded with the introduction of new radiation agents such as gamma rays from radioisotopes and neutrons from nuclear reactors, and chemical agents such as the alkalating mutagens. Extensive research has been devoted to utilization of induced mutations in plant breeding. While contributing extensively to genetic information, contributions to plant breeding were often disappointing due to deleterious effects in the mutant plants. The most beneficial results were obtained when mutant genes could be transferred to enhance otherwise desirable genotypes, or when plants with mutant genes could be vegetatively propagated.

J. Transgenic Plants

Advances in plant improvement have largely revolved around two basic breeding practices, selection and hybridization. Selection is limited to isolation of plants from native or hybrid populations, and hybridization is limited to crossing plants of the same species or closly related species. Genetic engineering has given rise to powerful new techniques allowing the creation of new genotypes by transferring DNA from unrelated organisms. The ge-

56. E. R. Sears, *The Aneuploids of Common Wheat*. (Columbia: Missouri Agricultural Experiment Station, 1954 [*Missouri Agricultural Experiment Station Bulletin* no. 572]).
57. L. J. Stadler, "Mutation in Barley Induced by X-Rays and Radium," *Science* 68 (1928): 186–187.

notypes of these new "transgenic" plants contain small functional segments of foreign DNA.

Transgenic plants may be created by introducing foreign DNA into the plant genome through a plant pathogen, *Agrobacterium tumefaciens*, or by direct insertion with a particle gun. In the *Agrobacterium* technique, a segment of foreign DNA is inserted into a break in a bacterial plasmid DNA vector. The foreign segment then becomes incorporated into the plant chromosome.[58] In the particle gun technique, microscopic metal particles are coated with DNA and propelled into plant cells by explosive force, where the DNA segment becomes incorporated into the plant chromosome.[59] The transgenic plant is regenerated from an infected cell. The DNA inserted into the plant cell must replicate and express its unique trait in the new transgenic plant.

Plant traits that are controlled by one to three genes are subject to DNA transfer.[60] Present DNA transfer techniques are not applicable to quantitative traits controlled by multiple loci. Traits currently receiving major attention in DNA transfer research include resistance to infection by plant disease pathogens, resistance to insects, resistance to herbicides, and improved product quality. The new genetic engineering technique is analogous to the backcross technique, in that plant performance of the recipient genotype is altered by addition of new genetic material. The important difference is that the origin of the new DNA is not limited to the same species or a species closely related to the recipient plant, as in the backcross technique, but may come from almost limitless sources. In genetic engineering as with backcrossing, the successful utilization of transgenic plant research for crop improvement depends on the continued availability of appropriate recipient varieties, developed through traditional selection and hybridization procedures employed for the improvement of the whole plant.

K. Experimental Design and Data Interpretation

An event of major importance in plant improvement occurred in 1925 when R. A. Fisher, Chief Statistician at the Rothamsted Experiment Station in England introduced the analysis of variance.[61] The analysis of variance

58. M. D. Chilton, "A Vector for Introducing New Genes into Plants," *Scientific American* 248 (6) (1983):51–59.

59. C. S. Gasser and R. T. Fraley, "Transgenic Crops," *Scientific American* 266 (6) (1992):62–69.

60. Ibid.

61. R. A. Fisher, *Statistical Methods for Research Workers* (Edinburgh: Oliver and Boyd, 1925).

provides a simple mathematical procedure for measuring the relative importance of two or more groups of factors that cause variation in an experiment. Its applications in experimental design, interpretation of data, and tests of significance were further treated in Fisher's publication on the design of experiments,[62] as well as in the many editions of Snedecor's *Statistical Methods*. The analysis of variance is the standard tool for data analysis utilized by all plant breeders and researchers on quantitative inheritance.

62. R. A. Fisher, *The Design of Experiments* (London: Oliver and Boyd, 1937).

2. Plant Pathology Development and Current Trends

George N. Agrios
University of Florida

Plant diseases were a scourge until the mid- to late 1800s when they began to be studied and the first effective chemicals for their control were used. Plant diseases caused many seeds to rot before germination, many seedlings to die before they grew enough to produce fruit, and many plants to have lesions on their roots, stems, leaves, flowers, and fruit, or to become blighted, wilt, and die. Many fruits, vegetables, seeds, bulbs, and other produce rotted while still in the field and were never harvested. But even those that were harvested often continued to rot while stored or transported to market, up to the moment of use. Consumers were forced to eat fruit and other food of low quality (spotted, scabby, partially rotten), or they had to sort it, pare it, and throw away rotten parts, thereby reducing the portion of the food available for use. In some cases, consumption of contaminated or partially rotten seeds and fruits caused poisoning of humans and domestic animals, as happens in the diseases known as ergotism (=St. Anthony's fire), mycotoxicoses, and others.

There is little published information on produce losses caused by diseases in earlier centuries. Losses from plant diseases in a particular field may vary from slight to 100%, depending on the plant, the pathogen, the locality, the environment, and more recently, on the control measures practiced. In 1965, the United States Department of Agriculture published *Losses in Agriculture*, in which it was estimated that, in spite of all crop protection measures and pesticides used, 11.5% of United States agricultural plant production was lost to diseases, 9.7% was lost to insects, and 7.2% was lost to weeds.[1] In 1990 prices, these losses translate into approximately $9.2 billion dollars lost to diseases, $7.9 billion lost to insects and $6.0 billion lost to weeds. In 1967, H. H. Cramer estimated that 11.8% of

1. U.S. Dept. of Agriculture, *Losses in Agriculture* (Washington, D.C.: USDA, 1965 [*USDA Agriculture Handbook*, no. 291]).

the entire world's crop production is lost annually to preharvest diseases, 12.2% is lost to insects, and 9.7% is lost to weeds, for a combined crop loss of 33.7%.[2]

Worldwide crop losses to diseases, insects, and weeds become more depressing when considering the losses are much greater in poorer, overpopulated, and still developing countries than they are in advanced countries.[3] For example, in Europe 25% of the produce was lost to disease, insects, and weeds, while 29% was lost in North and Central America, 33% in South America, 42% in Africa, and 43% in Asia. In developing countries, diseases alone destroy 23% of cereal production vs. 6% lost in developed countries, 42% vs. 20% of potatoes, 39% vs. 12% of other root crops, 35% vs. 18% of sugarcane, 16% vs. 11% of vegetables, and 20% vs. 12% of fruits.

To the above losses must be added the losses caused by diseases after harvesting, which again are much greater in developing, mostly tropical countries than in developed countries. Major factors for developing countries include high temperature and humidity, which favor disease development, and grossly inadequate disease control measures which allow diseases to affect products prior to harvest, and a general lack of refrigeration during storage and transportation. As a result, much of the produce rots rapidly and is discarded after harvest. *Postharvest Food Losses in Developing Countries* estimated the minimum postharvest food losses from all causes in developing countries to be approximately 10% for grain or dry legume crops and 20% for perishable root, vegetable, or fruit crops, the range varying from 2 to 50% for grains and legumes and from 5 to 100% for perishable fleshy produce.[4] Plant diseases are usually a minor factor in postharvest losses of grains and legumes, but rotting diseases caused by fungi and bacteria are the most important factor in postharvest losses of perishable root, vegetable, and fruit crops.

These losses, of course, do not include the cost of pesticides, equipment, and labor used to control diseases and pests, nor do they include the danger to humans and the environment from the use of billions of pounds of pesticides for controlling plant diseases and other pests (1.4 billion pounds annually in the United States alone).

Tremendous progress towards understanding, managing and controlling plant diseases has been made in the last one hundred years. This progress

2. H. H. Cramer, *Plant Protection and World Crop Production*, translated by J. H. Edwards (Leverkusen: Frabenfabriken Bayer AG, 1967 [*Pflanzenschutz Nachrichten Bayer* 20, no. 1]).

3. Ibid.

4. National Research Council, *Postharvest Food Losses in Developing Countries* (Washington, D.C.: National Academy of Sciences, 1978.)

has helped better to feed human populations everywhere. Progress began as occasional trickles of observations which, after many centuries, joined into discernible, yet still meandering, tributaries of experiments and, in the last fifty years, into a well-defined, rolling river of knowledge. The next several pages describe the evolution and development of our knowledge about plant diseases and of our ability to protect our food from them.

A. First Observation and Determination of the Nature of Plant Diseases

Plant Diseases as the Wrath of God

It is likely that plant diseases, and their ability to reduce or destroy plants and food, predate the shift of human evolution from nomadic foragers to rudimentary farmers surviving on small plots of land. There is no doubt that, in some years, the fruits and leaves that groups of nomads used for food were destroyed by disease, and these humans had to look for alternative fruits and plants or to move away. It is probable that crop losses, reduced food, and famines became more common and serious as humans became agriculturists and began to live in villages and to grow many plants of the same kind next to each other. It is not surprising, that plant diseases, such as blights and mildew of cereals and vine crops, are mentioned in several of the earlier Hebrew books of the Old Testament (c. 750 B.C.).[5] When plant diseases struck, they destroyed much of the people's food. Plant diseases were therefore considered to be a curse and God's punishment for people's wrongs and sins. By inference, plant diseases could be avoided if people would abstain from sin.

The Greek philosopher Theophrastus (c. 300 B.C.) was the first to study and write about diseases of trees, cereals, and legumes.[6] Among other things, he noted that plant diseases were generally more severe in the lowlands than on the hillsides and more common on some crops than on others. Although no differential sinning by the farmers in the lowlands and the hillsides could explain the observed differences in disease occurrence or control, Theophrastus still believed that God controlled the weather that "brought about" the disease and that the control of the disease therefore depended on that same superpower. The Romans not only attributed plant diseases and their control to God, they actually created a god, Robigo, who was both responsible for the dreaded rust diseases of grain crops and pro-

5. H. H. Whetzel, *An Outline of the History of Phytopathology* (Philadelphia: Saunders, 1918).
6. G. W. Keitt, "History of Plant Pathology," *Plant Pathology; An Advanced Treatise*, vol. 1, ed. by J. G. Horsfall and A. E. Dimond (New York: Academic, 1959). Pp. 62–97.

tected people from them. Each spring, just before the rusts appeared, the Romans celebrated the Robigalia, a special holiday during which they offered sacrifices of red dogs and sheep in an attempt to appease the god and keep him from sending the rusts to their crops.[7]

For almost two thousand years, people continued to associate plant diseases with sin and God, live fatalistically with repeated losses of food, and endure the hunger and famines that followed.[8] Whatever they observed on diseased plants or on produce they considered to be the product or the result of the disease rather than its cause. Little by little, people, even scientists, came to believe in the spontaneous generation of disease, that is, they came to believe that mildews, decay or other symptoms were diseases that originated naturally on their own, rather than being the causes or effects of disease.

The Theory of Spontaneous Generation

During the long period of fatalism, investigators made a few important observations on the causes and control of certain diseases, but their contemporaries did not believe them, and the generations that followed ignored them. For example, about A.D. 1200, Albertus Magnus of Ballstadt proposed that the mistletoe plant is a parasite obtaining its food from the host plant which it made sick, and that the sick plant could be cured by pruning out the mistletoe plant. Nobody, however, followed up on this important observation.

In England in 1667, R. Hooke used a recently invented tool, the compound microscope, and saw the spores of the rust fungus for the first time. In France in 1705, Joseph Pitton de Tournefort wrote a treatise on plant diseases in which he distinguished between diseases resulting from external (environmental) causes and those resulting from internal ones. In 1660, wheat farmers in Rouen, France, and in 1725, independently, wheat farmers in Connecticut both observed that wheat rust was worst near barberry bushes and concluded that the barberry fathered the rust, which then moved to the wheat. As a result, they asked the legislature to pass a law forcing towns to eradicate barberry bushes and thereby protect the wheat plants from rust. At about the same time, Micheli in Italy described many new genera of fungi, illustrated their reproductive structures, and noted that when he placed them on freshly cut slices of melon these structures gener-

7. G. B. Orlob, "The Concepts of Etiology in the History of Plant Pathology," *Pflanzenschutz Nachrichten Bayer* 17 (1964): 185–268.

8. G. L. Carefoot and E. R. Sprott, *Famine in the Wind* (Chicago: Rand McNally, 1967).

ally reproduced the same kind of fungus that produced them.[9] He proposed that fungi arose from their own spores rather than spontaneously, but nobody believed him.

For some time, farmers used to add common salt brine to wheat seed to protect it from smut or bunt disease. Pluchet in France reported in 1746 that substituting copper sulfate for sodium chloride considerably improved the control of wheat smut. In 1755, M. Tillet in France showed that he could increase the number of wheat plants developing smut by adding smut dust or spores to the wheat seed before planting, and that he could reduce smut occurrence if he pretreated the seed with copper sulfate.[10] Tillet, however, believed that the spore dust contained a poisonous substance that caused the disease rather than that the disease was caused by living microorganisms. In 1797, the Italians Giovanni Targioni-Tozzetti and Felice Fontana independently suggested that wheat rust was an entity (fungus) distinct from the plant, but in his 1773 book on plant diseases, the German Johann B. Zallinger continued to propose that fungi found in association with diseased areas in plants were abnormal structures of the diseased plant.

Finally in 1807 Benedict Prevost in France proved, in a way that now appears conclusive, that wheat bunt (smut) is caused by a fungus.[11] He repeated Tillet's inoculation and copper sulfate treatment experiments and, in addition, he studied under the microscope the production and germination of its spores, as well as the inhibition of spore germination with the addition of a drop of copper sulfate. He reasoned that the fungus spores caused the smut in wheat and that reduced smut observed in treated seeds resulted from the fact that copper sulfate inhibits spores from germinating and growing. Prevost's work and conclusions were not accepted by the French Académie des Sciences because its scientists believed that microorganisms and their spores were the result rather than the cause of disease. The beliefs of the French scientists were shared and expounded for at least another forty years by scientists in Italy, Austria, and Germany.

9. P. A. Micheli, *Nova Plantarum Genera* (Florence, 1729).

10. M. Tillet, *Dissertation sur la Cause qui Corrompt et Noircit les Grains de Bled Dans les Épis; Et sur les Moyens de Prevenir ces Accidents = Dissertation on the Cause of the Corruption and Smutting of Wheat in the Head*, English translation by H. B. Humphrey (Ithaca, N.Y.: American Phytopathological Society, 1937 [*Phytopathological Classics* no. 5]). (Original published, Bordeaux, 1755.)

11. B. Prévost, *Mémoire sur la Cause Immédiate de la Carie ou Charbon des Blés, et de Plusieurs Autres Maladies des Plantes, et sur les Préservatifs de la Carie = Memoir on the Immediate Cause of Bunt or Smut of Wheat, and of Several Other Diseases of Plants, and on Preventitives of Bunt*, English translation by G. W. Keitt (Menasha, Wis.: American Phytopathological Society, 1939 [*Phytopathological Classics* no. 6]). (Original published, Paris: Bernard, Quai des Augustins, 1807.)

Plant Diseases and the Proof of Pathogenicity: The Germ Theory

The devastating epidemics of late blight of potato in Northern Europe, particularly Ireland, in the 1840s, both dramatized the effect of plant diseases on human survival and suffering and greatly stimulated interest in their causes and control. Late blight of potato caused severe potato losses in Northern Europe, but it literally destroyed the potato crop in Ireland in 1845 and 1846. The ensuing widespread famine resulted in the death of hundreds of thousands of people and the immigration of more than one and a half million people from Ireland to the United States.[12]

Several scientists described various aspects of the disease and of the pathogen, and some of them suggested that the fungus was the primary cause of the blight and tuber rot of potato. It was the German A. de Bary who gave the final proof.[13] de Bary had earlier studied the microscopic nature and development of many smut and rust fungi and their relationships to the tissues of the diseased plant, but later he switched to the study of the downy mildews. In 1861, he finally established experimentally beyond criticism that a fungus (*Phytophthora infestans*) was the causal agent of late blight of potato. It was during those years (between 1860 and 1863) that Louis Pasteur proposed and finally established irrefutably that microorganisms arise only from pre-existing microorganisms and that fermentation is a biological phenomenon, not just a chemical one. Pasteur's conclusions, however, were not generally accepted for many years. The proof for involvement of microorganisms (germs) in fermentation and disease provided the basis for the germ theory of disease.

Later, de Bary showed that some rust diseases require two alternate host plants and that the fungus *Sclerotinia* induces rotting of vegetables through substances which diffuse into plant tissues in advance of the fungus. In the meantime, Julius Kuhn published his famous book, *Die Krankheiten der Kulturgewächse, ihre Urachen und ihre Verhütung* (*Diseases of Cultivated Crops, Their Causes and Their Control*) in which he recognized that plant diseases are caused by either unfavorable environment or by parasitic organisms such as insects, fungi, and parasitic plants.[14] Kuhn was the first to demonstrate penetration of the bunt fungus into wheat seedlings and to follow its development in the host from the seedling to the new wheat kernels;

12. C. Woodham-Smith, *The Great Hunger, Ireland 1845–1849* (New York: Harper and Row, 1962).

13. A. de Bary, *Die Gegenwartig Herrschende Kartoffelkrankheit, ihre Ursache und ihre Verhutung* (Leipzig, A. Förstnersche Buchhandlung, 1861).

14. Julius Kuhn, *Die Krankheiten der Kulturgewächse, ihre Urachen und ihre Verhütung* (Berlin: G. Bosselmann, 1858.)

he also promoted development and application of control measures, particularly seed treatments for cereals.

Other books appeared soon after Kuhn's: *Feinde des Waldes* (*Enemies of Trees*) by Heinrich Willkomm, *Phytopathologie, Die Krankheiten der Culturgewächse* (*Diseases of Cultivated Plants*) by Ernst Hallier, *Handbuch der Pflanzenkrankheiten* (*Handbook of Plant Diseases*) by Paul Sorauer, *Important Diseases of Forest Trees* by Robert Hartig, and *Text-Book of the Diseases of Trees* also by Hartig.[15] Most of these followed the same approach as did Kuhn, with descriptions of parasitic diseases steadily gaining in acceptance and slowly increasing in numbers. Up to that point, however, all plant diseases were believed to be caused by fungi.

Discovery of Other Pathogens as Causes of Plant Disease

While numerous botanists, naturalists, physicians and other scientists studied several types of plant diseases caused by fungi, a few scientists published observations on other pathogens, such as nematodes, or on other diseases which later proved to be caused by viruses.

Nematodes. The first report of nematodes associated with plants was made in England by T. Needham in 1743. He observed nematodes within wheat galls (kernels), although he did not show or suggest that they were the cause of the disease. It was not until 1855 that a second plant parasitic nematode, the root knot nematode, was observed in cucumber root galls. In the next four years two other plant parasitic nematodes, the bulb or stem nematode and the sugar beat cyst nematode, were reported from infected plant parts.

Bacteria. The first animal disease proved to be caused by a bacterium was anthrax, as demonstrated by Pasteur and Koch in France in 1876. In 1878, Thomas J. Burrill in Illinois showed that fire blight of pear and apple was also caused by a bacterium. Particularly through the work of E. F. Smith of the U.S. Department of Agriculture, several other plant diseases were shown shortly afterward to be caused by bacteria. As with fungal plant pathogens, however, acceptance of bacteria as causes of disease in plants

15. (a) Heinrich M. Willkomm, ed., *Die Mikroskopischen Feinde des Waldes. Vol. I. Feinde des Waldes* (Dresden: G. Schonfeld's Buchhandlung, 1866–67). (b) Ernst Hallier, ed., *Phytopathologie, Die Krankeiten der Culturgewächse* (Leipzig: W. Engelmann, 1868). (c) Paul Sorauer, ed., *Handbuch der Pflanzenkrankheiten* (Berlin: Wiegandt, Hempel and Parrey, 1874). (d) Robert Hartig, *Important Diseases of Forest Trees: Contributions to Mycology and Phytopathology for Botanists and Foresters*, Translated by William Merrill, David H. Lambert, and Walter Liese (St. Paul, Minn.: American Phytopathological Society, 1975 [*Phytopathological Classics* no. 12]). (Original published, Berlin: Springer, 1874.) (e) Robert Hartig, *Text-Book of the Diseases of Trees*, Rev. and ed., translated by William Somerville (London and New York: Macmillan, 1894).

was slow. For example, as late as 1899, Alfred Fisher, a prominent German botanist, rejected the claim made by Smith and others that they had seen bacteria in plant cells. In the early 1980s Smith was also among the first to study the crown gall disease, a disease he considered similar to cancerous tumors of humans and animals, and which he showed to be caused by bacteria. Not until the late 1970s was crown gall shown to be the result of excessive hormones produced by the uncontrolled expression of certain genes present in a piece of plasmid DNA introduced by the bacterium into the plant genome.

Viruses. Although not recognized as such, viral diseases of plants have been noticed and portrayed in paintings since the early 1600s. An unwitting experimental transmission of a plant virus was reported in 1714 by T. Lawrence. In 1886 in the Netherlands, A. Mayer, showed that he could produce the "tobacco mosaic" disease by injecting juice from infected tobacco plants into healthy plants.[16] Since no fungi were present on the diseased plant or in the filtered juice, Mayer concluded that tobacco mosaic was probably caused by a bacterium. D. I. Ivanowski showed in 1892 that the cause of tobacco mosaic could pass through a filter that retains bacteria and suggested that the disease was caused by a toxin or by very small bacteria that passed through the pores of the filter. It was Beijerinck in 1898, who, after repeating Ivanowski's experiments, concluded that tobacco mosaic was caused not by a microorganism but by a "contagious living fluid," which he called a virus.[17]

The nature, size, and shape of viruses remained unknown for several more years. In 1935, W. M. Stanley in California, observing that an infectious crystalline protein precipitated when juice from infected tobacco was treated with ammonium sulfate, concluded that the virus was an autocatalytic protein that could multiply in living cells.[18] The following year, F. C. Bawden and his colleagues in England demonstrated that the crystalline preparations of the virus observed by Stanley consisted of protein and ribonucleic acid.[19] In 1939, G. A. Kausche and his colleagues saw virus

16. A. Mayer, "Ueber die Mosaikkrankheit des Tabaks" = *Concerning the Mosaic Disesase of Tobacco*, English translation by J. Johnson (Ithaca, N.Y.: American Phytopathological Society, 1942 [*Phytopathological Classics* no. 7]). (Original published in *Landwirtschaftlichen Versuchs-Stationen* 32 (1886): 451–467.)

17. M. W. Beijerinck, "Ueber ein Contagium Vivum Fluidum als Ursache der Fleckenkrankheit der Tabaksblätter" = *Concerning a Contagium Vivum Fluidum as a Cause of the Spot-Disease of Tobacco Leaves*, English translation by J. Johnson (Ithaca, N.Y.: American Phytopathological Society, 1942 [*Phytopathological Classics* no. 7]). (Original published in *Verhandelingen der Koninklyke Akademie van Wettenschappen te Amsterdam* 65 (2) (1898): 3–21.)

18. W. M. Stanley, "Isolation of a Crystalline Protein Possessing the Properties of Tobacco-Mosaic Virus," *Science* 81 (1935): 644–645.

19. F. C. Bawden, N. W. Pirie, J. D. Bernal, and I. Fankuchen, "Liquid Crystalline Substances from Virus-Infected Plants," *Nature* 138 (1936): 1051–1052.

particles for the first time with the electron microscope. In 1956, Gierrer and Schramm showed that it was the nucleic acid of the virus that carried all the genetic information of the virus and that the nucleic acid alone could cause infection and could reproduce the complete virus.[20]

T. O. Diener demonstrated in 1971 that the potato spindle tuber disease was caused by a small, single-stranded, circular molecule of infectious RNA (ribonucleic acid) which he called a viroid. Several other viroids causing disease in plants were subsequently discovered.[21]

Protozoa. Although flagellate protozoa were reported by F. Lafont in 1909 to be present in the latex-bearing cells of laticiferous plants, he did not consider them a cause of disease in these plants. In 1931, G. Stahel reported that flagellates were present in the phloem of wilting coffee trees and were responsible for abnormal phloem development and wilt. Several diseases of coconut and oil palm trees are now known to also be caused by protozoa.[22]

Recent taxonomic studies of certain groups of so-called "lower" fungi, such as the *Plasmodiophoromycetes,* indicate that these organisms belong in the kingdom *Protozoa* rather than the kingdom *Eumycota*, which contains the true fungi. Under the proposed new classification scheme, the *Plasmodiophoromycete* organism has been known to cause disease in plants since 1878 and is one of the first known microorganisms to cause disease in plants.[23]

Mollicutes. For several years after viruses were discovered, many plant diseases were described which showed yellows or symptoms resembling witches'-broom. These were thought to be caused by viruses, but no viruses could be isolated from such plants, nor could they be found with the electron microscope. In 1967, Doi and his colleagues in Japan observed mollicutes (wall-less mycoplasmalike bodies) in the phloem of plants exhibiting yellows and witches'-broom symptoms.[24] That same year this group showed

20. A. Gierrer and B. Schramm, "Infectivity of Ribonucleic Acid from Tobacco Mosaic Virus," *Nature* 177 (1956): 702–703.

21. (a) T. O. Diener, *Viroids and Viroid Diseases* (New York: Wiley, 1979). (b) J. S. Semancik, *Viroids and Viroid-Like Pathogens* (Boca Raton, Fla.: CRC Press, 1987).

22. M. Dollet, "Plant Diseases Caused by Flagellate Protozoa (Phytomonas)," *Annual Review of Phytopathology* 22 (1984): 115–132.

23. M. S. Woronin, "*Plasmodiophora Brassicae, Urheber der Kohlpflanzen-Hernie = Plasmodiophora Brassicae, the Cause of Cabbage Hernia*, English translation by C. Chupp (Ithaca, N.Y.: American Phytopathological Society, 1934 [*Phytopathological Classics* no. 4]). (Original published in *Jahrbücher für Wissenschaftliche Botanik* 11 (1878): 548–574.)

24. Y. Doi, M. Teranaka, K, Yora, and H. Asuyama, "Mycoplasma or PLT Group-Like Microorganisms Found in the Phloem Elements of Plants Infected with Mulberry Dwarf, Potato Witches Broom, Aster Yellows, or Paulownia Witches Broom," *Annals of the Phytopathology Society of Japan* 33 (1967): 259–266.

that the mycoplasmalike bodies and the symptoms disappeared temporarily when the plants were treated with tetracycline antibiotics.

B. The Descriptive Phase of Plant Pathology

As botanists, naturalists, agriculturists, and other scientists such as physicians became familiar with the causes of plant disease, numerous written reports describing diseases affecting a variety of agricultural and ornamental plants appeared in scientific, popular, or semi-popular journals. Improvements and availability of magnifying lenses, and later of compound microscopes made possible the detection and description of numerous fungi, some nematodes, and bacteria associated with diseased plants. Development and introduction of modern techniques for growing microorganisms (fungi and bacteria) in pure culture by Oskar Brefeld, Robert Koch, Julius Petri, and others contributed greatly to plant pathology between 1875 and 1912. Many of the plant pathogenic microorganisms, however, could not be grown on culture media. These included several groups of fungi causing many important plant diseases (rusts, downy mildews, and powdery mildews), all plant parasitic nematodes, protozoa, viruses and mollicutes, and some fastidious bacteria. On the other hand, many non-pathogenic saprophytic fungi and bacteria were often or always present on, and were cultured from, diseased plant tissues, and were often misdiagnosed as the causes of the disease. Finally in 1876 R. Koch proposed a set of rules, known as "Koch's postulates," that must be satisfied to verify that an isolated microorganism is the cause of a disease.[25] Satisfaction of Koch's rules put an end to otherwise interminable speculation and discussion regarding the cause of a disease. Even today, Koch's rules must be satisfied with every new disease that is discovered. The inability to culture many of the plant pathogenic microorganisms presents additional difficulties in proving their pathogenicity. In these cases, isolation of the microorganism in a nearly pure condition, and transmission of it mechanically or by grafting to healthy plants so as to cause disease in them in the absence of any other microorganism, are taken as proof of the microorganism being the cause of the disease.

At the same time, further improvements in compound microcopes and in plant tissue staining techniques allowed cytological and histopathological studies that revealed the location of pathogens (mostly fungi, nematodes,

25. R. Koch, "Die Aetiologie der Milzbrand-Krankheit, Bauf die Entwicklungsgeschichte des *Bacillus Anthracis*," (1876) Vol. 1, *Gesammelte Werke von Robert Koch* (Leipzig, G. Thieme, 1912).

and bacteria) in infected plant cells and tissues. After 1940 the electron microscope made it possible for viruses, viroids, and mollicutes to be visualized and described, and for cytological details and structural characteristics of the various pathogens to be studied.

During the descriptive phase of plant pathology, scientists reported many observations about the biology of microorganisms involved in disease. Such observations included the types of spores produced by the fungal pathogens, the means of spread of the pathogens, the location of their survival during winter, the types of host plants infected, etc. In many cases, such observations were correlated with the prevailing environmental conditions, such as rain and temperature, and with differences in disease severity among different hosts. Different types of control practices, mostly cultural but some chemical, were tried for various diseases. The discovery of the efficacy of Bordeaux Mixture for the control of downy mildew of grape encouraged trials with Bordeaux and some other compounds for the control of many diseases on almost all crops.

In this period, the first plant pathology course in the United States was taught by M. A. Farlow at Harvard University in 1875. The U.S. Department of Agriculture was established in 1862, and the Section of Vegetable Pathology was formed in 1887, the same year that the Hatch Act established the State Agricultural Experiment Station system. Several plant pathologists were added to the U.S. Department of Agriculture laboratories and to the state agricultural experiment stations. Textbooks of plant pathology were published in France (E. E. Prillieux, 1895), in England (G. E. Massee, 1899) and in the United States (B. M. Duggar, 1906).[26]

C. The Flourishing of Experimental Plant Pathology

Following the recognition of the importance of plant diseases and of the new discipline and profession of plant pathology, plant pathologists joined the various USDA and state experiment stations in the late 1800s. These plant pathologists began to experiment in all areas of plant pathology. Although new diseases and pathogens continued to be discovered and described, plant pathologists began to ask, and to design experiments to answer questions regarding the means of pathogen entry into the plant, pathogen multiplication and spread within the plant, mechanisms of host plant cell death and breakdown, pathogen sporulation, spore dispersal, overwintering and germination, vector involvement, effect of environment

26. See S. A. J. Tarr, *The Principles of Plant Pathology* (New York: Winchester Press, 1972).

on disease development, and other areas. They also began tracking variability among plant populations in disease expression and loss. As knowledge accumulated, experimentation found new ways to control diseases and to avoid or reduce the losses from them.

This increased awareness and study of plant diseases necessitated university departments of plant pathology. The first was established at Cornell University in 1907. Soon after, in 1909, a small group of plant pathologists organized into the American Phytopathological Society. They began *Phytopathology*, a journal of plant pathology, in 1911, which is a major journal today. Prior to this and for many years, much information on plant diseases and their control was published as technical or extension bulletins of the USDA and of individual state agricultural experiment stations, as well as in several botanical and agricultural journals. Similar journals were later established in Europe, and much later in Asia, particularly Japan and India.

The Etiological Phase of Plant Pathology

This phase of plant pathology involved observations and experimentation aimed at proving the causes (etiology) of specific plant diseases. Although they began with the proof of pathogenicity of the late blight fungus on potatoes and of the rust and smut fungi on cereals, etiological studies gained momentum with the development of techniques for pure culture of microorganisms and with the necessity to satisfy "Koch's postulates." Numerous publications in the late 1800s and in the first third of the twentieth century described the symptoms of thousands of mostly fungal plant diseases on all types of hosts, the efforts to isolate and culture the suspected pathogen, and the subsequent experiments to prove the pathogenicity of the isolated "pathogen." Many of these papers included information on the losses estimated to be caused by the disease and even information on ways that could control the disease. Several excellent books of plant pathology appeared during this period, particularly Butler's *Fungi and Diseases in Plants*, Stevens and Hall's *Diseases of Economic Plants*, Heald's *Manual of Plant Diseases*, and his *Introduction to Plant Pathology*.[27]

The etiological phase continued and accelerated as new types of pathogens, such as viruses, mycoplasmalike organisms (mollicutes), fastidious bacteria, protozoa, and viroids were discovered. Although the methodologies had to be adapted to the size and properties of each type of pathogen,

27. (a) E. J. Butler, *Fungi and Diseases in Plants* (Calcutta and Simla: Thacker, Spink and Co., 1918). (b) F. L. Stevens and J. G. Hall, *Diseases of Economic Plants* (New York: MacMillan, 1921). (c) F. D. Heald, *Manual of Plant Diseases* (New York: McGraw-Hill, 1926). (d) F. D. Heald, *Introduction to Plant Pathology* (New York: McGraw-Hill, 1937).

the goal and the result remained the determination of the etiology of the disease. The etiological phase often depended on and benefitted from improvements in methodology and instrumentation, such as the availability of the electron microscope, of density gradient centrifugation, and of specialized nutrient media. A number of important books appeared describing various groups of diseases, their causes, and their control: E. C. Large's *The Advance of the Fungi*; the U.S. Department of Agriculture's *Plant Diseases: The Yearbook of Agriculture*; J. C. Walker's *Plant Pathology*; E. C. Stakman and J. G. Harrar's *Principles of Plant Pathology*; Elliott's *Manual of Bacterial Plant Pathogens*; W. R. Jenkins and D. P. Taylor's *Plant Nematology*; F. C. Bawden's *Plant Viruses and Virus Diseases*; K. M. Smith's *A Textbook of Plant Virus Diseases*; Diener's *Viroids and Viroid Diseases*; and K. Maramorosch and S. P. Raychaudhuri's *Mycoplasma Diseases of Trees and Shrubs*.[28] During the same period, J. Kuijt's book *The Biology of Parasitic Flowering Plants* described the characteristics and biology of the many higher plants that cause diseases on other higher plants.[29] In 1964, the British Commonwealth Mycological Institute began the serial publications describing individual plant pathogen fungi, bacteria, and viruses.[30] Four classics in mycology also appeared during that period: E. A. Bessey's *Morphology and Taxonomy of Fungi*; C. J. Alexopoulos' *Introductory Mycology*; F. E. Clements and C. L. Shear's *The Genera of Fungi*; and G. C. Ainsworth's monumental four-volume work: *The Fungi: An Advanced Treatise*.[31]

In 1965, the USDA published *Losses in Agriculture* which sensitized people to the extent of crop losses caused by plant diseases and other pests in the United States.[32] In 1967, Cramer's *Plant Protection and World Crop*

28. (a) E. C. Large, *The Advance of the Fungi* (London: Jonathan Cape, 1940). (b) *Plant Diseases: The Yearbook of Agriculture* (Washington, D.C.: U.S. Department of Agriculture, 1953). (c) J. C. Walker, *Plant Pathology* (New York: McGraw-Hill, 1950). (d) E. C. Stakman and J. G. Harrar, *Principles of Plant Pathology* (New York: The Ronald Press, 1957). (e) C. Elliott, *Manual of Bacterial Plant Pathogens* (London: Bailliere, Tindall and Cox, 1930). (f) W. R. Jenkins and D. P. Taylor, *Plant Nematology* (New York: Van Nostrand-Reinhold, 1967). (g) F. C. Bawden, *Plant Viruses and Virus Diseases* (New York: Ronald Press, 1964). (h) K. M. Smith, *A Textbook of Plant Virus Diseases* (London: Churchill, 1937). (i) Diener, *Viroids and Viroid Diseases*. (j) K. Maramorosch and S. P. Raychaudhuri, *Mycoplasma Diseases of Trees and Shrubs* (New York: Academic Press, 1981).

29. J. Kuijt, *The Biology of Parasitic Flowering Plants* (Berkeley: University of California Press, 1969).

30. Commonwealth Mycological Institute (Great Britain), *CMI Descriptions of Pathogenic Fungi and Bacteria* (Kew, England: Commonwealth Mycological Institute, 1964+).

31. (a) E. A. Bessey, *Morphology and Taxonomy of Fungi* (London: Constable, 1950). (b) C. J. Alexopoulos, *Introductory Mycology* (New York: Wiley, 1950) (Reissued in 1979 with C. W. Mims as a second author). (c) F. E. Clemens and C. L. Shear, *The Genera of Fungi* (New York: Hafner, 1957). (d) G. C. Ainsworth, F. K. Sparrow, and A. S. Sussman, eds., *The Fungi: An Advanced Treatise*, 4 vols. (New York: Academic Press, 1965–1973).

32. U.S. Dept. of Agriculture, *Losses in Agriculture*.

Production shocked people with the severity of crop losses caused by diseases and pests worldwide, but particularly in poor countries everywhere and in the tropics especially.[33] These books clearly focused attention on the magnitude of the problem of food losses to diseases and pests and stimulated research for their control.

Control of Plant Diseases

Reports on efforts to control plant diseases go back to antiquity. About 1000 B.C. Homer mentioned the therapeutic properties of sulfur, and in 470 B.C. Democritus recommended sprinkling plants with pure emulsion of olives to control blight. Most ancient reports, however, dealt with festivals and sacrifices offered to thank, please, or appease a god, and to keep the god from sending the dreaded rusts, mildews, blasts, or other crop scourges. No other information on controlling plant diseases seems to exist in written records for almost two thousand years.

Theophrastus noted in the third century B.C. that some crops, such as legumes, were affected by the rust disease much less than other crops, such as cereals. However, it was not until the mid-1600s that the first report was made that one species or variety, in this case the crabapple, can be more resistant to canker disease than another related species or variety, the common apple. It is likely, however, that in spite of the absence of written reports, growers, knowingly or unknowingly, continued to use differential resistance as a control of plant diseases. This is likely to have occurred because seeds from resistant plants looked bigger and better than those from infected ones, or because in severe disease outbreaks resistant plants were the only ones surviving and their seeds the only ones available for planting.

In the late 1600s, some farmers in southern England planted wheat seed that had been salvaged from a shipwreck and noticed that the wheat plants produced from such seed had considerably less smut (bunt) than plants from other seed. This led some farmers to treat wheat seed with brine (sodium chloride solution) to control bunt. In the mid-1700s, copper sulfate was substituted for sodium chloride, and bunt control improved significantly. This treatment is still used in most parts of the world, although in advanced countries copper sulfate has been replaced by other more effective fungicides.

Diseases of trees were sometimes too obvious to ignore although their causes were unknown. Several cures, many of them worthless, were proposed. As already mentioned, around A.D. 1200, it was noted that a tree

33. Cramer, *Plant Protection and World Crop Production*.

can be cured from mistletoe infections if the mistletoe is pruned out. In the mid-1750s, excision of cankers followed by applications of grafting wax was recommended for control. Some practitioners incorrectly advocated the use of vinegar to prevent canker on trees or the use of worthless mixtures of cow dung, lime rubbish from old buildings, wood ashes, and river sand to cure disease, defects, and injuries of plants. In 1803, however, William Forsyth introduced lime-sulphur for the control of mildew on fruit trees, and in 1821 John Robertson recommended aqueous suspension of sulfur as the only specific remedy for the treatment of mildew of peaches.

Controlling Plant Diseases with Chemicals

In the late 1870s the introduction into Europe of a fungus from America causing the aggressive downy mildew disease of grape stimulated a search by several investigators, especially in France, for chemicals that could control the disease. In 1882, P. M. A. Millardet noticed that vines growing near the road which had been sprayed with a blueish-white mixture of copper sulfate and lime to deter pilferers, retained their leaves throughout the season, whereas the leaves of untreated vines died. After trying several combinations of copper sulfate and lime concentrations in numerous spraying experiments, Millardet concluded in 1885 that a mixture of copper sulfate and hydrated lime could effectively control the downy mildew of grape.[34] This mixture became known as "Bordeaux mixture," and it was soon discovered to be equally effective against the late blight of potato, the other downy mildews, and many other leaf spots and blights. For the next fifty years, Bordeaux mixture was used more than any other fungicide against a wide variety of diseases in all parts of the world. Even today, Bordeaux mixture is one of the most widely used fungicides around the world. Its discovery proved that plant diseases could be controlled chemically and greatly encouraged the study of plant diseases and their control.

In 1913, E. Riehm introduced seed treatment with organic mercury compounds, and such treatments became routine until the 1960s, when all mercury-containing pesticides were removed from the market because of their general toxicity.[35] In the meantime, in 1934, W. H. Tisdale and I. Williams discovered the first dithiocarbamate fungicide (thiram), which led to the

34. Pierre M. A. Millardet, "Sur l'Histoire du Traitement du Mildiou par le Sulfate de Cuivre" = *The Discovery of Bordeaux Mixture*, translated by Felix J. Scheiderhan (Ithaca, N.Y.: American Phytopathological Society, 1933 [*Phytopathological Classics* no. 3]). (Original published in *Journal d'Agriculture Pratique* 2 (1885): 801–805.)

35. E. Riehm, "Prüfung Einiger Mittel zur Bekämpfung des Steinbrandes," *Mitteilungen aus der Kaiserlichen Biologischen Anstalt für Land- und Forstwirtschaft* 14 (1913): 8–9.

development of a series of effective and widely used fungicides, including ferbam, zineb, and maneb.[36] In 1945 and again in 1956, J. G. Horsfall published his *Principles of Fungicidal Action,* which summarized the information available to that point and further increased interest in new fungicides.[37] Many other important protective fungicides followed and in 1965 the first systemic fungicide, carboxin, was introduced. This was followed by several other excellent systemic fungicides, such as benomyl. E. G. Sharvelle's *Chemical Control of Plant Diseases* provided much useful information on chemicals, equipment, and their use to control plant diseases.[38]

Antibiotics, primarily streptomycin, were first used to control bacterial plant diseases in 1950. Soon after, the antibiotic actidione was shown to be effective against several plant pathogenic fungi. In 1967, tetracycline antibiotics were shown to control diseases caused by mollicutes (mycoplasma-like organisms), and in 1972 tetracycline was shown to control fastidious bacteria inhabiting the xylem of their host plants.

In 1954 a few strains of bacterial plant pathogens were noticed to be resistant to certain antibiotics, and in 1963 strains of fungal plant pathogens were found to be resistant to certain protective fungicides. It was in the 1970s, however, when the use of systemic fungicides became widespread, that many new races of fungal pathogens appeared, ones that were resistant to previously effective fungicides.[39] This observation prompted the development of new strategies in controlling plant diseases with fungicides and bactericides. Such strategies included the use of mixtures of fungicides, alternating compounds in successive sprays, and spraying with a systemic compound in the early stages of the disease and with a broad spectrum compound in the later stages.

Following the publication in 1962 of Rachel Carson's *Silent Spring*, which vividly described the dangers of polluting the environment with poisonous chemicals, considerable governmental regulations and restrictions were imposed on the production, testing, and use of pesticides.[40] Some pesticides, such as those containing mercury, were banned outright, while many others came under intense scrutiny. Several pesticides that were found to resist breakdown and to accumulate in the food chain, thereby becoming toxic when they reached higher concentrations, were banned outright. Others were shown to be chemicals that are mutagenic on microorgan-

36. W. H. Tisdale and I. Williams, *Disinfectant* (U.S. Patent no. 1, 972, 961).
37. J. G. Horsfall, *Principles of Fungicidal Action* (Waltham, Mass.: Chronica Botanica Co., 1945, 1956).
38. E. G. Sharvelle, *Chemical Control of Plant Diseases* (College Station, Texas: University Publishing, 1969).
39. C. J. Delp, *Fungicide Resistance in North America* (St. Paul, Minn.: APS Press, 1988).
40. R. Carson, *Silent Spring* (Boston: Houghton Mifflin, 1962).

isms or carcinogenic on experimental animals. Many of them, primarily nematicides and fungicides, were banned by the U.S. governmnent, withdrawn by the manufacturer, or were subjected to stricter regulation and use. The search for and use of effective fungicides against specific fungi intensified, but this accelerated the appearance of resistant pathogen races. For a fuller discussion of pesticide use and abuse see Chapter 3.

Alternative Controls for Plant Diseases

Renewed emphasis has been placed in the last two decades on reexamining and improving many of the old cultural practices for use in controlling plant diseases, and on developing new ones.[41] These practices include sanitation of plant debris and infected plant parts, use of pathogen-free seed, crop rotation with plant species that are immune to the pathogens that affect the other rotating crops, soil fallow, reduced or no tillage, destruction of weeds, use of appropriate forms of fertilizer, appropriate irrigation, adjustment of the time and rate of sowing and of the date of harvest, and minimizing the influx of pathogen vectors through border plants. Modification of cultural practices and use of resistant varieties, as well as monitoring of plant disease epidemics for reduced use of pesticides, have become the basis of integrated management of plant diseases.

Early in the century came reports on soils which, through the antagonistic microorganisms they harbor or through other means, suppress the development of certain diseases caused by soilborne pathogens. Following the observation by A. Fleming in 1928 that certain fungi, such as *Penicillium*, are antagonistic to other fungi and to bacteria, efforts were made to find nonpathogenic microorganisms that, when applied to plants before or after infection with other microorganisms, would antagonize the pathogenic microorganism, and thereby protect the plant from it.[42] Numerous other microorganisms, primarily fungi but also some bacteria, have been shown to antagonize various plant pathogenic fungi, bacteria, and nematodes, and in a few cases protect the host plant from infection by the pathogen.[43] It was later shown that, even with fungi and bacteria in some cases, control of the plant pathogen was obtained by treating the plant with an avirulent or hypovirulent strain of the same species.

In the early 1930s it was shown that infection of a plant with a mild

41. J. Palti, *Cultural Practices and Infectious Crop Diseases* (Berlin: Springer-Verlag, 1981).
42. R. J. Cook and K. F. Baker, *The Nature and Practice of Biological Control of Plant Pathogens* (St. Paul, Minn.: American Phytopathological Society, 1983).
43. D. Hornby, ed., *Biological Control of Soil-Borne Plant Pathogens* (Wallingford, U.K.: CAB International, 1990).

strain of a virus prevented or delayed infection of the plant by a severe strain. This phenomenon is known as cross protection. The first practical biological control of a plant disease with a microorganism was shown in 1963, when inoculation of the surface of freshly cut pine stumps with spores of the nonpathogenic fungus *Peniophora gigantea* protected the stumps from infection by the root and butt rot–causing fungus *Heterobasidion annosum*. In 1972, the crown gall bacterium was reported to be biologically controlled by pre-inoculating seeds or roots of transplants with a related but nonpathogenic bacterium that produces a bacteriocin, an antibiotic specific against related bacteria. That same year, also, tomato seedlings in commercial fields were protected from infection with the tobacco mosaic virus by inoculating them with a nonpathogenic strain of the virus obtained by artificial mutation. Experimentally, biological control has been obtained against many foliage- and root-infecting fungi and bacteria, against some fungi causing postharvest diseases, and against some root-infecting nematodes, but field applications are still mostly ineffective.[44] The use of cross protection to control viral diseases on a large scale, however, was proven against the tristeza disease of citrus and is now used to control several other virus infections. This approach has been further improved by introducing into plant hosts, through genetic engineering, the viral genes for the virus coat protein and some other proteins. When expressed in the plant, these genes prevent or delay infection of the plant by the virus.[45]

Genetic Resistant Varieties

In 1894, Jacob Eriksson in Sweden showed that the cereal rust fungus *Puccinia graminis* consists of different biologic races (subspecies) which are morphologically indistinguishable but are specialized in their pathogenicity to their cereal host (wheat, barley, oats, rye, etc.).[46] In 1905, R. H. Biffen reported that the resistance of two wheat varieties and their progeny to a rust fungus was inherited in a Mendelian fashion. Four years later, W. A. Orton, working with the fusarium wilts of cotton, watermelon, and cowpea, distinguished among disease resistance, disease escape, and disease endurance (tolerance). M. F. Barrus then showed that there is genetic

44. K. G. Mukerji and K. L. Garg, eds., *Biocontrol of Plant Diseases*, 2 vols. (Boca Raton, Fla.: CRC Press, 1988).

45. T. M. A. Wilson and J. W. Davies, *Genetic Engineering with Plant Viruses* (Boca Raton, Fla.: CRC Press, 1992).

46. Jakob Eriksson and E. Henning, *Die Getreideroste, ihre Geschichte und Natur, sowie Massregeln gegen Dieselben*. Translated into German by C. O. Nordgren. (Stockholm: P. A. Norstedt & Soner, 1896). (First published in Swedish in *Meddelanden fran Kongl. Landtbruks-Akademiens Experimentalfält* 38 (1894).

variability within a pathogen species, that is, that different pathogen races infect different varieties of a host species.[47] Soon thereafter, beginning in 1914, E. C. Stakman and his colleagues not only established that morphologically indistinguishable races of a pathogen exist within a pathogen species but also that they differ in their ability to infect different varieties of a host species.[48] Their work helped explain why a variety that was resistant in one geographic area was susceptible in another, why resistance changed from year to year, and why resistant varieties suddenly became susceptible. In each of these cases a different physiological race of the pathogen was involved.

The actual nature of resistance of a plant to a pathogen was considered at first to be due to the presence of a toxic substance in the plant. In 1946, however, H. H. Flor, working with the rust disease of flax, showed that for each gene of resistance in the host there was a corresponding gene for avirulence in the pathogen (a gene-for-gene relationship).[49] J. E. Vanderplank suggested that there are two kinds of resistance: one, controlled by few "major" genes, is strong but is specific only for one or a few races of the pathogen (vertical resistance); the other, determined by many "minor" genes, is weaker but is effective against all races (horizontal resistance).[50] It is assumed that each "major" or "minor" gene for resistance usually operates in conjunction with several other genes; together they enable the plant to produce certain types of plant cell structures and substances that interfere with, or inhibit, the growth, mulitplication, or survival of the attacking pathogen, thereby arresting or inhibiting disease. Some of the plant cell defense structures and substances exist before the plant comes into contact with the pathogen, but in most cases they are produced in response to attack by the pathogen.

In 1946, E. Gaumann proposed that in many host-pathogen combinations, plants remain resistant because they are actually hypersensitive to the pathogen; that is, the attacked plant cells are so sensitive to the pathogen that they and some adjacent cells die immediately and thereby isolate or cause the death of the pathogen.[51] K. O. Muller proposed in 1956, and in 1963 I. A. M. Cruickshank confirmed, that disease resistance is often brought about by phytoalexins, that is, antimicrobial plant substances that

47. Arthur W. Gilbert, Mortier F. Barrus, and Daniel Dean, *The Potato* (New York: Macmillan, 1917).

48. Stakman and Harrar, *Principles of Plant Pathology*.

49. H. H. Flor, "Genetics of Pathogenicity in *Melampsora Lini*," *Journal of Agricultural Research* 73 (1946): 335–357.

50. J. E. Vanderplank, *Disease Resistance in Plants* (Orlando, Fla.: Academic Press, 1968 and 1984).

51. E. Gaumann, *Principles of Plant Infection* (London: Crosby Lockwook, 1950).

are either absent or present at nondetectable levels in healthy plants, but accumulate to high levels as a result of some pathological stimulus.[52]

Plant breeding for resistance to disease has been going on throughout most of this century, having started before the genetics of disease resistance was clearly understood.[53] In the last few decades, however, the development and use of plant varieties resistant to several important diseases has received high priority in most breeding programs.[54] Resistant varieties, when available, have provided the cheapest, easiest, safest, and most effective control of many types of plant diseases.[55] Resistant varieties have been particularly useful against fungal rust, smut, powdery mildew, and vascular wilt diseases and against viral diseases, all of which are difficult or impossible to control by other means. As public consciousness about adding pesticides to the environment has increased, the effort and success in replacing pesticides with host resistance to disease have also increased. Unfortunately, new pathogen races that can attack previously resistant varieties appear with dependable regularity within a few years after such varieties are released, and this necessitates the constant development of new varieties to replace those whose resistance has "broken down."[56] The development of genetic engineering methodologies and technologies promises to improve and accelerate our ability to produce resistant varieties of greater stability.[57]

Interest in the Physiology of Plant Disease

Once it became apparent that plant diseases were caused by microorganisms, efforts began to elucidate the mechanisms by which microorganisms cause disease.[58] In 1886 Anton de Bary, working with the *Sclerotinia* rot disease of vegetables, noted that plant cells were killed in advance of the invading mycelium of the fungus. He also noted that juice from rotted tissue could break down healthy host tissue, but not if the juice was boiled first. de Bary concluded that the pathogen produces enzymes and toxins which move into the host tissue ahead of the pathogen and degrade and kill

52. J. A. Bailey and J. W. Mansfield, eds., *Phytoalexins* (New York: Wiley, 1983).
53. P. R. Day, *Genetics of Host-Parasite Interaction* (San Francisco: Freeman, 1974).
54. R. R. Nelson, ed., *Breeding Plants for Disease Resistance: Concepts and Applications* (University Park: Pennsylvania State University Press, 1973).
55. P. R. Day, ed., *The Genetic Basis of Epidemics in Agriculture* (New York: N.Y. Academy of Sciences, 1977 [*Annals of the N.Y. Academy of Sciences* no. 287]).
56. National Academy of Sciences, *Genetic Vulnerability of Major Crops* (Washington, D.C.: National Academy of Sciences, 1972).
57. T. Kosuge, C. P. Meredith, and A. Hollaender, eds., *Genetic Engineering of Plants* (New York: Plenum, 1983).
58. R. Heitefuss and P. H. Williams, eds., *Physiological Plant Pathology* (Berlin and New York: Springer-Verlag, 1976 [*Encyclopedia of Plant Physiology*, new series, no. 4]).

plant cells. The pathogen then utilizes the degradation products as its nutrients. In 1905, it was reported that cytolytic enzymes were also involved in soft rot diseases of vegetables caused by bacteria, and in 1915, it was recognized that pectic enzymes play a role in the ability of some fungi to cause disease in plants. It was in the 1940s, however, that cellulases were first implicated in the development of plant disease.

Soon after de Bary, many researchers attempted to show that most plant diseases, especially vascular wilts and leaf spots, were caused by toxins secreted by the pathogens; their claims, however, could not be confirmed. In 1933, it was first suggested, and confirmed in the early 1950s, that a toxin was involved in the black spot disease of pears caused by the fungus *Alternaria*. Following a suggestion made in 1925, it was confirmed in 1934 that the bacterium that causes the wildfire disease of tobacco produces a toxin that is responsible for the chlorotic zone surrounding the necrotic spots on the leaves. The wildfire toxin was the first toxin to be isolated in pure form in 1952. In 1947 F. Meehan and H. C. Murphy showed that the fungus that causes blight on Victoria oats produces the toxin victorin, which can induce the symptoms of the disease only on susceptible varieties but has no effect on varieties resistant to the fungus. Numerous new toxins produced by plant pathogenic fungi and bacteria were subsequently detected and identified.[59] Some detailed biochemical studies were carried out to elucidate the mechanisms by which these toxins affect or kill cells or by which cells of resistant plants inactivate them. From the mid-1950s till about 1980 a great many studies were carried out on the effect of infection on host cell respiration and on the role of the latter in plant defenses and resistance to infection. Similarly, numerous studies were carried out on the types of host cell enzymes activated upon infection, the types and amounts of other metabolites accumulating following infection, and, particularly, the types and amounts of phenolic compounds and phenoloxidases produced following infection. These studies provided a wealth of information on many of the biochemical reactions that go on in plant cells following infection, but they did not entirely explain the mechanisms by which plants defend themselves against pathogens.[60]

It had long been observed that in many diseases the affected plants show stunting while in others they show excessive growth, tumors, and other abnormalities. An imbalance in growth regulators in the diseased plant was suspected. In 1926, it was observed that the excessive growth of rice seed-

59. R. D. Durbin, ed., *Toxins in Plant Disease* (New York: Academic Press, 1981).

60. (a) J. G. Horsfall and A. E. Dimond, eds., *Plant Pathology: An Advanced Treatise*, 3 vols. (New York: Academic Press, 1959–1960). (b) J. G. Horsfall and E. B. Cowling, eds., *Plant Disease*, 5 vols. (New York: Academic Press, 1977–1980).

lings infected with the fungus *Giberella* could be reproduced by treating healthy seedlings with culture filtrates of the fungus. In 1939, the growth regulator produced by the fungus was identified and named gibberellin. By the late 1950s numerous plant pathogenic fungi and bacteria were shown to produce the plant hormone indoleacetic acid (IAA), and in the mid-1960s a cytokinin was shown to be produced by several plant pathogenic bacteria and fungi.[61]

Advent of the Epidemiology of Plant Disease

Some epidemiological observations concerning the increase of disease within plant populations and the association of such increase with environmental factors were recorded with most plant diseases early on. However, little effort was made to correlate and utilize this information in controlling plant diseases. In 1944 W. D. Mills working in New York state, developed a table through which he could predict, from knowing the temperature and rain duration at a particular time, whether infection of apple buds, leaves, and fruit by the apple scab fungus would take place and whether control measures (fungicides) should therefore be applied. It was in 1963, with Vanderplank's *Plant Diseases: Epidemics and Control*, that epidemiology was established as an important and interesting area of plant pathology.[62] Vanderplank pointed out the difference in the development and control of monocyclic and polycyclic pathogens and described the general structure and patterns of epidemics. In the late 1960s, researchers attempted to predict the course of epidemics through the modeling of plant diseases, using information collected at various points in time, and under varying environmental conditions, about the host, the pathogen, and their interactions. In 1969 the first computer simulation program was published, developed from modeling each stage of the life cycle of the pathogen as a function of the environment and designed to simulate the fungus-caused early blight epidemics of tomato and potato. Disease modeling and computer simulation of epidemics were developed for many diseases in the last two decades. These, along with newly developed disease monitoring instrumentation, have been used extensively in plant disease forecasting systems.[63] Disease forecasting has become an integral part of integrated pest management and

61. (a) Horsfall and Cowling, *Plant Disease*. (b) R. N. Goodman, Z. Kiraly, and K. R. Wood, *The Biochemistry and Physiology of Plant Disease* (Columbia: University of Missouri Press, 1986).
62. J. E. Vanderplank, *Plant Diseases: Epidemics and Control* (New York: Academic Press, 1963).
63. William E. Fry, *Principles of Plant Disease Management* (New York: Academic Press, 1982).

has helped reduce the amounts of pesticides applied to crops without reducing yields.[64]

The Molecular Phase of Plant Pathology

Since 1980, most physiological-biochemical studies in plant pathology have been replaced by molecular studies in which the emphasis has been on determining the genetic connection of any substance involved in disease development.[65] Because of their small size and the ample information available on viruses and bacteria, molecular studies have been carried out more extensively and more effectively with plant pathogenic viruses, viroids, and bacteria, and considerably less with fungi, mollicutes, and nematodes.[66] As a result, the number, location, size, sequence, and function of most or all genes of many viruses are now known in detail, and many of them have been excised from the virus and transferred either to host plants, to which they often convey resistance, or into experimental bacteria, in which they are expressed and the proteins they code for are isolated and studied.[67] The same transfer has been done with certain pathogenicity-related genes from plant pathogenic bacteria and some fungi.

The molecular phase of plant pathology began with the discovery that the soilborne plant pathogenic bacterium *Agrobacterium tumefaciens* induces galls in plants by introducing into their cells a circular piece of DNA (plasmid), part of which (known as T-DNA) is incorporated into the chromosomes of the infected plant cells.[68] The incorporated T-DNA carries genes that code for excess plant growth regulators that make the cells divide and enlarge uncontrollably and produce tumors (galls). It was noticed that the genes of the bacterium coding for the growth regulators can be removed and replaced with genes from other bacteria or from viruses, and even animals. These genes are then expressed by the plant cells, i.e., the plant cells produce whatever protein these genes code for. It was later discovered that foreign DNA can also be introduced into plant cells by first incorporating it into certain plant viruses, by bombarding plant cells with it and even by growing isolated plant cells in the presence of foreign DNA.[69]

64. (a) C. L. Campbell and L. V. Madden, *Introduction to Plant Disease Epidemiology* (New York: Wiley, 1990). (b) D. Pimentel, ed., *CRC Handbook of Pest Management in Agriculture*, 2d ed. 3 vols. (Boca Raton, Fla.: CRC Press, 1991).

65. T. Kosuge and E. W. Nester, eds., *Plant-Microbe Interactions: Molecular and Genetic Perspectives*, vol. 1 (New York: Macmillan, 1984).

66. R. E. F. Mathews, *Plant Virology* (New York: Academic Press, 1991).

67. S. S. Patil, S. Ouchi, D. Mills, and C. Vance, eds., *Molecular Strategies of Pathogens and Host Plants* (New York: Springer-Verlag, 1991).

68. G. N. Agrios, *Plant Pathology* (San Diego: Academic, 1988).

69. Kosuge et al., *Genetic Engineering of Plants*.

In the meantime, it was discovered that by using enzymes that dissolve the pectin and cellulose components of the plant cell wall, one can get intact living protoplasts of plant cells. Some protoplasts can then be induced to regenerate their cell wall, divide, and form calluses. Subsequently, some of the callus cells differentiate into whole new plants. When isolated cells or protoplasts are treated with foreign DNA in one of the ways mentioned above, for example with a gene that makes a plant resistant to a specific disease, the DNA can become incorporated in the chromosome(s) of the plant cell or protoplast. The callus cells and all cells of the new plants produced from them continue to carry the foreign DNA and to express it, thereby becoming resistant to that disease. Such transformed plants, known as transgenic plants because they carry genes from other plants or other organisms, can then be propagated asexually or sexually, carrying along the new genes.[70]

Rapid propagation and genetic manipulation of normal or transgenic plants has been greatly facilitated by advances in protoplast culture, in the manipulation of their genetic material (DNA), and in the regeneration of whole plants from such protoplasts. Protoplasts can be transformed with the addition of foreign genetic material into them, or by fusion with other protoplasts, the latter process known as somatic hybridization. Many somaclones, that is, clonal plants derived from fused protoplasts, possess new, different, and often useful characteristics, including increased resistance to disease.

In parallel with the developments in plant molecular biology and plant tissue culture, recent advances were made in the molecular biology of plant pathogenic viruses, bacteria, fungi, etc. By use of specific restriction endonucleases, that is, enzymes which cut nucleic acids (DNA and RNA) at certain specific nucleotide sequences, the DNA of each microorganism can be cut into pieces of constant number and length for each kind of enzyme. These DNA pieces, known as RFLPs (random fragment length polymorphisms), each containing one or more genes, can be separated as bands on polyacrylamide gels through electrophoresis. Each RFLP size then can be incorporated into specific plasmids (circular DNAs) which can shuttle the RFLP into certain bacteria. As the bacteria multiply rapidly they also produce many copies or clones of each RFLP. The latter are then isolated and sequenced, revealing the sequence of their nucleotides. At the same time the cloning bacteria and the surrounding media are scrutinized for the presence of new proteins possibly coded for by the gene(s) comprising the

70. D. A. Evans, W. R. Sharp, P. V. Ammirato, and Y. Yamada, eds., *Handbook of Plant Cell Culture*, 5 vols. (New York: Macmillan, 1983–88).

RFLP. In this way the sequence and function of many specific genes have been identified. It has also been possible to distinguish closely related but different microorganisms within a species, e.g., pathogenic bacteria, by comparing their RFLP profiles. Furthermore, RFLPs have been used to detect identical or related DNAs. This is done by attaching a radioisotope or a chromogenic compound to an RFLP and using the tagged RFLP as a DNA probe to search for identical (complementary) DNAs in DNA preparations of microorganisms or other cells tested for having such related DNAs.[71]

Another major development in molecular plant pathology has been the discovery of monoclonal antibodies. Plant pathologists have been using antisera, i.e., fluids containing antibodies (=specific reactive proteins) produced in warm-blooded animals after injection with a microorganism, to detect and identify plant pathogens (viruses, bacteria, etc.) infecting plants. Each microorganism, and actually each protein, has many antigenic sites on its surface and therefore can incite production of many different kinds of antibodies to it. However, each different kind of antibody is produced by a different spleen cell. Because many microorganisms or proteins share identical antigenic sites, antibodies to such sites can react with all the microorganisms or proteins that have such common antigenic sites. Other antigenic sites, however, are unique to a single type of microorganism or protein molecule. A method was discovered—fusion of antibody-producing single spleen cells and autonomously multiplying cancer cells—to obtain hybrid cells (hybridomas) that could produce unlimited amounts of the one kind of (monoclonal) antibody produced by that one spleen cell.[72] Since monoclonal antibodies are specific for individual antigenic sites on a protein or microorganism, they have been used extensively for detection of plant pathogenic microorganisms on infected plants, in healthy-looking seeds and transplants, in soil, in insect and other vectors, in mixtures of microorganisms, etc. Monoclonal antibodies have also been used to detect within cells the specific proteins that elicit them. The latter use has been particularly helpful in studying the location of pathogen infection and multiplication within plant cells and tissues and the location of pathogens such as viruses in their insect or nematode vectors.[73]

Several important practical applications of developments in molecular plant pathology are already in place and many more are in the developmental or experimental stages. The first applications have been the use of spe-

71. B. R. Glick and J. E. Thompson, *Methods in Plant Molecular Biology and Biotechnology* (Boca Raton, Fla.: CRC Press, 1993).

72. J. T. Barrett, *Textbook of Immunology* (St. Louis: C. V. Mosby Co., 1988).

73. E. L. Halk and S. H. DeBoer, "Monoclonal Antibodies in Plant Disease Research," *Annual Review of Phytopathology* 23 (1985): 321–350.

cific DNA probes and monoclonal antibodies to detect and identify plant pathogens and their strains in plant tissues or in mixture with other organisms. Another series of applications has been the incorporation into plant chromosomes of antisense viral nucleic acid and of mutated viral genes that would normally code for proteins that allow the virus to survive, to multiply, to move from cell to cell, or to be spread by its insect vector. For reasons still not well understood, transgenic plants transformed with such mutated genes become resistant to infection by the virus providing the genes. Transgenic plants transformed with fungal genes coding for enzymes that break down the cell wall of fungal mycelium have also been shown to be resistant to infection by certain fungi.

It is expected that many more such useful applications of molecular plant pathology will become available in the near future. One of the biggest prizes is expected to be the ability to locate and isolate plant genes for resistance to disease from any kind of resistant plant and subsequently to modify, combine, introduce, and express them in horticulturally desirable but susceptible plants, thereby making them resistant to disease. Such an accomplishment will protect the productivity of the crop, will reduce costs for controlling the disease, and will reduce or eliminate the use of chemicals for disease control, thereby avoiding contamination of our food and water supplies and of the total environment.

Trends in the Diagnosis of Plant Diseases

The early diagnosis of plant diseases was through characteristic symptoms produced by the infected plant (for example, rust, powdery mildew, root knot). In many cases, this type of diagnosis is still applicable. The development of magnifying lenses and microscopes made it possible to observe the pathogen (fungus, nematode, bacteria) on, in, or outside its host. The use of culture media made it possible to isolate, grow, observe, and distinguish culturable plant pathogenic fungi and bacteria. The development of serological tests since the 1920s made possible the distinction of different viruses in sap from their host plants.[74] In the 1930s, the electron microscope made it possible to see and distinguish viruses for the first time;[75] also, several viruses were distinguished by the kind of insect vector that could transmit them from diseased to healthy host plants. Since the 1950s, suc-

74. R. Hampton, E. Ball, and S. DeBoer, *Serological Methods for Detection and Identification of Viral and Bacterial Plant Pathogens: A Laboratory Manual* (St. Paul, Minn.: APS Press, 1990).
75. R. I. B. Francki, R. G. Milne, and R. Hatta, *Atlas of Plant Viruses* (Boca Raton, Fla.: CRC Press, 1985).

cessful efforts have developed selective culture media for various plant pathogenic fungi and bacteria.[76]

Woody plant diseases caused by viruses, mollicutes, fastidious bacteria, and protozoa were tested and symptoms were observed by tissue transfers (grafting) for several decades, as well as by microscopy, heat treatment, and treatment with antibiotics. Viruses of most herbaceous hosts were also identified by transfer of sap and production of symptoms on selected healthy hosts, and by their tolerance to heating to a certain temperature. In 1977, the Enzyme-Linked Immunosorbent Assay (ELISA) serological technique was introduced to plant virology and quickly became a useful and popular technique for detection and diagnosis of viruses, viral proteins, and other proteins produced by many microorganisms and by specific genes.[77]

In the last fifteen years a number of methods have been developed and used for the detection, identification and differentiation of closely related microorganisms: analysis of isozymes produced by closely related microorganisms, analysis of their fatty acid profiles, analysis of restriction fragment length polymorphisms (RFLPs) of their nucleic acids, analysis of the degree of their nucleic acid hybridization, DNA fingerprinting, and analysis of the nucleotide sequence in the nucleic acids. Since the mid-1980s, radioactive or chromogenic DNA probes complementary to specific segments of the nucleic acid of the microorganism have been developed and are being used extensively for the detection and identification of plant pathogenic microorganisms.

D. Plant Pathology as a Discipline and a Profession

As mentioned earlier, the first people who practiced plant pathology were botanists, naturalists, or physicians. Some had students who became interested in the study of plant diseases, and these became the first plant pathologists. They were hired by state agricultural experiment stations, by the United States Department of Agriculture, and by universities. In the early 1900s, departments of plant pathology began to be established in some of the larger universities, often as departments of botany and plant pathology.

The organization of the American Phytopathological Society (APS) in 1909, and the publication of *Phytopathology* in 1911, gave new visibility and impetus to the development of the profession. Numerous other national

76. A. Y. Rossman, M. E. Palm, and L. J. Spielman, *A Literature Guide for the Identification of Plant Pathogenic Fungi* (St. Paul, Minn.: APS Press, 1987).

77. See Hampton, et al., *Serological Methods*.

or regional phytopathological societies were formed.[78] In the late 1960s, as a result of the international nature of plant diseases, and the extensive cooperation of plant pathologists in different countries, the International Society of Plant Pathology was established. In the last twenty years, several state phytopathological societies have been formed. Many societies soon began to publish their own journals: the British journal *Plant Pathology*, the German *Phytopathologische Zeitschrift,* the French *Annales des Epiphytes,* the *Annals of the Phytopathological Society of Japan*, *Indian Phytopathology*, the *Canadian Journal of Plant Pathology*, etc. The *Review of Applied Mycology* (later the *Review of Plant Pathology*) provided abstracts and reviews of plant pathological papers and, since 1963, the *Annual Review of Phytopathology* has been providing in-depth reviews of current topics of interest.

All societies, through their annual, biennial, or other meetings and through their publications in the form of newsletters, journals, or books, have contributed greatly to promoting the science and the profession of plant pathology. See Chapter 6 for a detailed discussion.

While early plant pathologists were mostly generalists, for the last several years plant pathologists at universities have become specialists in plant pathogenic fungi, plant virology, plant bacteriology, epidemiology, disease physiology, and biological control. Almost all courses cover just one of these specialized areas. Some plant pathologists specialize in diseases of certain crops, e.g. vegetables, potatoes, citrus, etc. A large number of recently graduated plant pathologists and current graduate students specialize in areas of molecular plant pathology where they use the tools of molecular biology to solve hitherto intractable problems.

Each state in the United States usually has only one or two extension plant pathologists whose duty is to transfer to the growers the information developed by research plant pathologists. These extension plant pathologists often work through the extension county agents who interact directly with the growers. In some states, plant pathologists work as private consultants to growers or perform diagnostic services for them. R. S. Cox in Florida has been a pioneer private practitioner plant pathologist.[79] The number of plant pathologists who work as private consultants is still relatively small but is expected to grow considerably in the future. There are already several thousand agronomists, weed scientists, and entomologists who work as pri-

78. U. S. Singh, H. S. Chaube, J. Kumar, and A. N. Mukhopadhyay, *Plant Diseases of International Importance* (Somerville, N.J.: Prentice-Hall, 1992).

79. (a) R. S. Cox, *The Private Practitioner in Agriculture* (Lake Worth, Fla.: Solo Publications, 1971). (b) R. S. Cox, *The Agricultural Consultant* (Crockett, Texas: Publications Development Co., 1982).

vate consultants. These professionals have had, for several years, professional certification programs that guarantee their expertise. On the other hand, practicing plant pathologists only recently developed their Professional Plant Pathologist Certification Program. This program was established in 1991 by the American Phytopathological Society at the request of the plant pathologists serving growers. This program will not only guarantee these practitioners' expertise as plant pathologists but, in time, it will likely make it illegal for persons not certified as professional plant pathologists to operate as such. Because it is not clear-cut, at least at the beginning, whether specific plant health problems in the field or in the greenhouse are caused by a disease, insects, unfavorable soil, or other environmental conditions, it is possible that specialized private consultants may be too narrowly trained to provide the broad-spectrum service needed. It is, therefore, expected that such specialized private consultants will be replaced by private consultants trained in diagnosing and controlling all types of pathogens, pests, and environmental factors that affect plant health.

In early 1993, the first proposal to establish an interdisciplinary Doctor of Plant Medicine degree in a university was submitted at the University of Florida by the plant protection departments (Plant Pathology, Entomology and Nematology) and supported by the plant production departments (Agronomy, Horticultural Sciences, Soil Science). The Doctor of Plant Medicine program is expected to produce professional plant doctors who will be trained to diagnose and offer control recommendations on all types of plant problems: diseases caused by fungi, bacteria, mollicutes, viruses, nematodes and protozoa, as well as by insects, mites, weeds, nutritional deficiencies, soil conditions, pollutants, and vertebrate pests such as field mice and birds. It is expected that graduates of this program will serve as general practitioners for plant health problems, the way veterinarians and family doctors serve for animal and human health problems. The plant doctors will provide diagnostic and plant health management expertise to growers, while the research scientists will continue developing new information, materials, methodologies, and technologies for use by the plant doctors and the growers.

It is expected that plant doctors will work as single-county or multi-county agents, as regional or state county agents, as specialists in plant protection, as specialists for state plant disease clinics, as entry-point interception specialists for the U.S. Plant Quarantine Service, as plant protection specialists for agribusinesses, large estates or cities, as sales representatives for pesticide companies, as proprietors and operators of private plant disease clinics, and, of course, as private consultants to growers. Plant doctors may be particularly useful as extension agents for developing countries, which usually cannot afford to hire the many specialists required to provide the same service. Availability of well and broadly trained plant doctors will help produce more food of better quality. By recommending correct and timely controls with cultural practices, plant doctors will also help reduce the amounts of pesticides used by growers. They will also help avoid the use of excessive or inappropriate pesticides for particular pest problems, thereby saving resources and protecting people, animals, and the environment from unnecessary toxic chemicals.

3. Protecting Crops

David Pimentel
Cornell University

Before 1940, farmers relied on nonchemical techniques for controlling pests. These controls included using crop rotation for control of insects, diseases, and weeds of crops. Both insect and disease control also depended heavily on host-plant resistance, with the farmers automatically selecting the seed for the next year's planting from those plants that appeared healthy and productive. Farmers therefore were acting as plant breeders, and over many centuries this approach has been highly successful. Natural enemies also were highly effective in controlling many insects injurious to crops.

Weed control was generally carried out by hand weeding, with hoes, or with cultivators drawn behind draft animals. All these nonchemical controls remain in use today, as well as numerous other techniques that have been developed.[1]

During World War II, DDT and a few other synthetic pesticides were in use primarily for protecting public health. In 1946 the agriculture community started using DDT and related hydrocarbon insecticides. Parathion, developed as a war gas, was also found to have insecticidal properties and its use quickly spread in world agriculture. Synthetic pesticides are heavily used with crop production today.

A. Past and Current Crop Losses to Pests

An estimated 67,000 different pest species attack agricultural crops worldwide. Included are approximately 9,000 species of insects and mites, 50,000 species of plant pathogens, and 8,000 weeds.[2] In general, less than

1. D. Pimentel, et al., "Environmental and Economic Impacts of Reducing U.S. Agricultural Pesticide Use," in D. Pimentel, ed., *CRC Handbook of Pest Management in Agriculture*, vol. 1 (Boca Raton, Fla.: CRC Press, 1991), pp. 679–718.

2. (a) U.S. Department of Agriculture, *Index of Plant Diseases in the United States* (Washington, D.C.: USDA, Crops Research Division, ARS, 1960). (b) M. A. Ross and C. A. Lembi, *Applied Weed Science* (Minneapolis: Burgess Publishing Co., 1985).

5% are considered serious pests. From 30% to 80% of the pests in any geographic region are native to that region. In most instances the pests have moved from feeding on native vegetation to feeding on crops which were introduced into the region.[3]

Despite the yearly application of an estimated 2.5 million tons of pesticides worldwide, along with the use of biological controls and other nonchemical controls, about 35% of all agricultural crop production is lost to pests.[4] Insect pests cause an estimated 13% crop loss, plant pathogens 12%, and weeds 10%. In the United States, yearly crop losses to pests are estimated to reach 37% of production (13% to insects, 12% to plant pathogens, and 12% to weeds).[5] Indeed, pests are destroying an enormous amount of food and fiber despite all our efforts to control them with pesticides and other methods.

Although pesticide use has increased over the past four decades, crop losses have not shown a concurrent decline. According to survey data collected from 1942 to the present, losses from weeds have fluctuated but slowly declined from 13.8% to 12%.[6] A combination of improved chemical, mechanical, and cultural weed control practices are responsible for this. Over that same period, losses from plant pathogens, including nematodes, have increased slightly, from 10.5% to about 12%. This has resulted, in part because crop rotations were abandoned, field sanitation was reduced, and higher cosmetic standards have been legalized.

The share of crops lost to insects has nearly doubled during the last forty years, despite a more than tenfold increase in both the amount and toxicity of synthetic insecticides used.[7] The increased insect losses have so far been offset by increased crop yields, obtained through higher yielding varieties and greater use of fertilizers and other energy-based inputs.[8]

This increase in crop losses despite intensified insecticide use followed

3. (a) D. Pimentel, "Herbivore Population Feeding Pressure on Plant Hosts: Feedback Evolution and Host Conservation," *Oikos* 53 (1988): 185–238. (b) H. M. T. Hokkanen and D. Pimentel, "New Associations in Biological Control: Theory and Practice," *Canadian Entomologist* 121 (1989): 828–840.

4. D. Pimentel, "Diversification of Biological Control Strategies in Agriculture," *Crop Protection* 10 (1991): 243–253.

5. Pimentel, et al. "Environmental and Economic Impacts."

6. Ibid.

7. (a) L. G. Arrington, *World Survey of Pest Control Products* (Washington, D.C.: U.S. Government Printing Office, 1956). (b) U.S. Bureau of the Census, *Statistical Abstract of the United States* (Washington, D.C.: U.S. Department of Commerce, Bureau of the Census, 1971). (c) U.S. Bureau of the Census, *Statistical Abstract of the United States* (Washington, D.C.: U.S. Department of Commerce, Bureau of the Census, 1991).

8. D. Pimentel and Wen Dazhong, "Technological Changes in Energy Use in U.S. Agricultural Production," in C. R. Carroll, J. H. Vandermeer, and P. M. Rosset, eds., *Agroecology* (New York: McGraw-Hill, 1990), pp. 147–164.

from several major changes in United States agricultural practices. These include the planting of crop varieties that are more susceptible to insect pests; the destruction of natural enemies of certain pests by insecticides, thereby creating the need for additional pesticide treatments;[9] the pests becoming resistant to pesticides;[10] the reduction in crop rotations (which check the growth of pest populations); the increase in monocultures and reduced crop diversity;[11] the lowering of the U.S. Food and Drug Administration tolerance for insects and insect parts in foods, and the enforcement of more stringent "cosmetic standards" by fruit and vegetable processors and retailers;[12] the increased use of aircraft application technology; the reduction in field sanitation, including less attention to the destruction of infected fruit and crop residues;[13] reduced tillage, with more crop residues left on the land surface; the growing of crops in climatic regions where they are more susceptible to insect attack; and the use of herbicides that alter the physiology of crop plants, making them more vulnerable to insect attack.[14] The impact of these factors will be explored further in the discussion of biological and cultural controls as alternatives to pesticide controls.

B. World Pesticide Use

Since the advent of DDT for crop protection in the 1940s, the growth of synthetic chemical agents has been dramatic. The first herbicide developed was 2,4-D; it and other herbicides came onto the market for use in crop production in the late 1940s.

The growth in the use of pesticides in world agriculture since 1945 has been phenomenal (Figure 3.1). From 1945 to date there has been a fortytwofold increase in the use of pesticides. Today, approximately 2.5 million metric tons of pesticides are used worldwide (Table 3.1). About 60% of the

9. R. van den Bosch and P. S. Messenger, *Biological Control* (New York: Intext Educational, 1973).

10. R. T. Roush and J. A. McKenzie, "Ecological Genetics of Insecticide and Acaracide Resistance," *Annual Review of Entomology* 32 (1987): 361–380.

11. (a) D. Pimentel, "Species Diversity and Insect Population Outbreaks," *Annual of the Entomological Society of America* 54 (1961): 76–86. (b) D. Pimentel, "Ecological Basis of Insect Pest, Pathogen and Weed Problems," in J. M. Cherrett and G. R. Sagar, eds., *The Origins of Pest, Parasite, Disease and Weed Problems* (Oxford: Blackwell Scientific Pub., 1977), pp. 3–31.

12. Pimentel, et al., "Pesticides, Insects in Foods, and Cosmetic Standards," *Bioscience* 27 (1977): 178–185.

13. D. Pimentel, "Agroecology and Economics," in M. Kogan, ed., *Ecological Theory and Integrated Pest Management Practice* (New York: John Wiley and Sons, 1986), pp. 299–319.

14. I. N. Oka and D. Pimentel, "Herbicide (2,4-D) Increases Insect and Pathogen Pests on Corn," *Science* 193 (1976): 239–240.

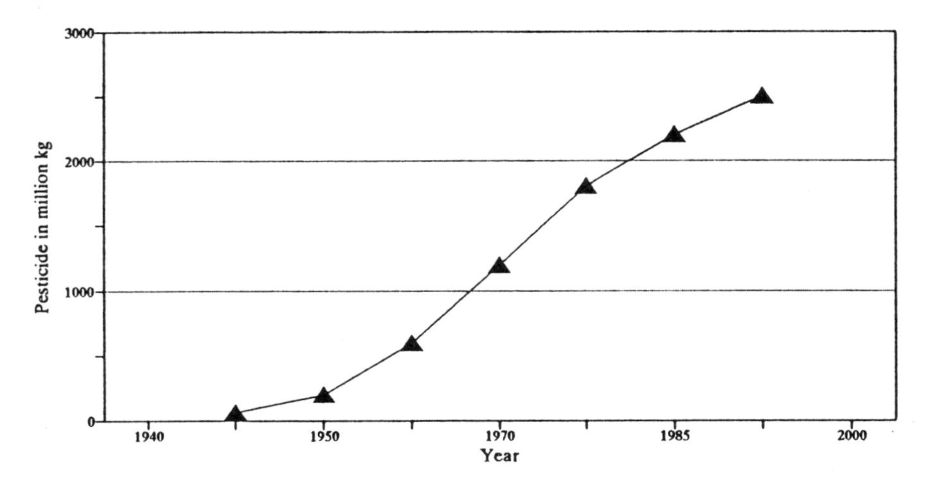

Figure 3.1. Trend in annual worldwide pesticide use. Pesticide use in the years indicated is estimated from the following sources: (a) I. N. Oka and D. Pimentel, "Herbicide (2,4-D) Increases Insect and Pathogen Pests on Corn," *Science* 193 (1976): 239–240; (b) CGIAR, *Facts and Figures—International Agricultural Research* (New York: The Rockefeller Foundation, and Washington, D.C.: IFPRI, 1990); (c) "Farm Chemicals, A Look at World Pesticide Markets," *Farm Chemicals* 148 (9) (1985): 26–34.

pesticide applied in world agriculture is herbicide, 30% insecticide, and 10% fungicide.

In the United States the use of synthetic pesticides has grown thirty-threefold since 1945. In the United States about 0.5 million metric tons of

Table 3.1. Estimated annual pesticide use

Country/Region	Pesticide use (10^6 metric tons)
United States	0.5
Canada	0.1
Europe	0.8
Other developed	0.5
Asia developing	0.3
Latin America	0.2
Africa	0.1
TOTAL	2.5

Source: Modified from CGIAR, *Facts and Figures—International Agricultural Research* (New York: The Rockefeller Foundation and Washington, D.C.: IFPRI, 1990).

pesticide are currently used, of which 68% is herbicide, 23% insecticide, and 9.7% fungicide.[15]

Note that the increase in pesticide use is larger than it appears, since the toxicity and biological effectiveness of these pesticides have increased at least tenfold.[16] For example, in 1945 DDT was applied at a rate of about 2 kg/ha. With more potent insecticides today, comparable effective insect control is achieved with pyrethroids and aldicarb applied at 0.1 kg/ha and 0.05 kg/ha, respectively.

World pesticide use is not evenly distributed amoung countries and regions of the world (Table 3.1). The developed countries account for only 25% of the world population but use nearly 80% of the total of 2.5 million tons of pesticide. The United States alone applies 20% of the total world pesticides; this is almost the same quantity as used in all developing countries together. These data suggest that nonchemical controls are still the primary means of pest control in developing nations.

C. Integrated Pest Management Development

Although integrated pest management (IPM) entails the combined use of a wide array of pest control practices including the use of pesticides,[17] in practice IPM has come to mean monitoring pest populations to determine "when-to-treat." Clearly, this strategy has been a tremendous advancement, because previously most pesticide treatments were made on a routine basis whether the pesticide was needed or not. Worldwide, however, including the United States, most pesticide is still applied with little attention to monitoring the status of pest and natural enemy populations. In part, this is due to a lack of information on pest/economic thresholds and pest monitoring technologies.

The development of criteria for sound economic thresholds in order to determine when-to-treat is highly complex. To establish guidelines for an effective economic threshold for a single pest the following interacting factors must be taken into account: (1) the density of the pest; (2) the densities of its parasites and predators; (3) the temperature and moisture levels and their impact on the crop, the pest, and its natural enemies; (4) the level of soil nutrients available to the crop; (5) the growth characteristics of the

15. Pimental, et al. "Environmental and Economic Impacts"
16. Ibid.
17. (a) D. Pimentel, "Perspectives of Integrated Pest Management," *Crop Protection* 1 (1982): 5–26. (b) Pimentel, "Agroecology and Economics."

particular crop variety; and (6) the crop(s) grown on the land the previous year.

Another important aspect of reducing pesticide use in IPM is enhancing the methods of applying pesticides: targeting them better and minimizing their spread in the environment. Considering that on average less than 0.01% of applied pesticide actually reaches the target pests, it is obvious that enormous quantities of pesticides are not only being wasted but being dispersed into the environment.[18]

The amount of pesticide reaching the target area varies with the application technique. Under ideal conditions only about half of the pesticide applied by aircraft reaches the target hectare.[19] The newest, ultralow volume (ULV) application technology now being adopted by many aerial applicators places only 25% of the pesticide on the target hectare.[20] Several other application techniques are available that will place more pesticide in the target area and on the target pests. For example, a carefully adjusted boom sprayer will place up to 90% of the pesticide in the target pests. Granular pesticides, where appropriate for use, place nearly 99% in the target area. Also, the use of the rope-wick applicator for herbicides is successful in placing 90% of the pesticide on the target weeds.[21]

Another important methodology that reduces the amount of pesticide applied to a crop is "spot treatment." Here, the farmer treats only those areas of the fields where the pest problem is serious and does not treat those areas where there are few or no pests on the crop.

There is a definite trend in the United States and the world for chemical companies and professional pesticide specialists to claim that they are following IPM procedures in order to justify the increased use and application of pesticides. These claims have prompted some specialists to abandon the use of the term IPM and use more specific terms when talking about a particular control technology. That is, nonchemical control technologies are specifically stated as such, and judicious pesticide use is referred to as

18. Pimentel, et al., "Environmental and Economic Impacts."

19. (a) ICAITI, *An Environmental and Economic Study of the Consequences of Pesticide Use in Central American Cotton Production* (Guatemala City, Guatemala: Central American Research Institute for Industry, United Nations Environment Programme, 1977). (b) N. B. Akesson and W. E. Yates, "Physical Parameters Affecting Aircraft Spray Application," in W. Y. Garner and J. Harvey, eds., *Chemical and Biological Controls in Forestry* (Washington, D.C.: American Chemical Society, 1984 [*American Chemical Society Series* no. 238]), pp. 95–111. (c) D. Pimentel and L. Levitan, "Pesticides: Amounts Applied and Amounts Reaching Pests," *BioScience* 36 (1986): 86–91.

20. F. Mazariegos, "The Use of Pesticides in the Cultivation of Cotton in Central America," *UNEP Industry and Environment* (1985): 5–8.

21. J. E. Dale, "Ropewick Applicator—Tool with a Future," *Weeds Today* 11 (2) (1980): 3–4.

pesticide applied after carefully monitoring pest and natural enemy populations.

D. Nonchemical Alternatives for Insect and Weed Control

Most pests worldwide are controlled nonchemically rather than with pesticides. There are numerous alternative technologies for insect and weed control.[22] No single one of these technologies is the "silver bullet." The prime success of the alternative controls is that they can be employed in various combinations depending on the crop, the environment, and the specific insect pests, weeds, and plant pathogens attacking the crop. In the following discussion, the focus is on technologies for insect and weed control.

Biological Control

Biological control involves the manipulation of natural enemies to help control pests.[23] Insect, weed, and pathogen populations in nature and crops are often controlled by natural enemies (predators and parasites). Complexes of natural enemies are present in agricultural systems, and these have been estimated to consume about 50% of pest populations. The common phenomenon of "resurgence" in which chemical pesticides remove natural enemies and exacerbate rather than relieve pest problems illustrates the importance of preserving natural enemies present in agroecosystems.[24]

Biological control focuses on ways to enhance the effectiveness of naturally occurring enemy species by enriching their habitat ("conservation"), restoring an existing enemy-pest relationship in a new area ("classical biological control"), or pairing enemies in new associations with pests ("new associations").

Conservation in this context involves improving the structural or biolog-

22. (a) U.S. Office of Technology Assessment (OTA), *Pest Management Strategies* (Washington, D.C.: Congress of the United States, OTA, 1979). (b) Pimentel, *CRC Handbook*. (c) President's Science Advisory Committee, *Restoring the Quality of Our Environment* (Washington, D.C.: Report of the Environmental Pollution Panel, President's Advisory Committee, Nov. 1965).

23. (a) P. Debach and D. Rosen, *Biological Control by Natural Enemies* (New York: Cambridge University Press, 1991). (b) E. F. Knipling, *Principles of Insect Parasitism Analyzed from New Perspectives* (Washington, D.C.: USDA, Agricultural Research Service, 1992).

24. D. Pimentel, et al., "Assessment of Environmental and Economic Impacts of Pesticide Use," in D. Pimentel and H. Lehman, eds., *The Pesticide Question: Environment, Economics and Ethics* (New York: Chapman and Hall, 1993).

ical diversity of agroecosystems in ways designed to attract and keep natural enemies in the system. An example is the recent use of "mini-hedgerows," raised strips of vegetation planted parallel to rows of cereal crops.[25] These hedgerows provide shelter and food for natural enemies.

Classical biological control involves the restoration of the interaction between a pest and its natural enemies following the accidental introduction of the pest into a new habitat. To do this ecologists search the pest's original habitat for its natural enemy complex, and then introduce those enemy species that are sufficiently specific for the range of prey or hosts they attack.[26]

A new approach is to find a natural enemy that attacks a species that is closely related to the target pest. This "new association" approach expands the potential area of search to all climatically suitable areas where relatives of the pest occur.[27]

Effective biological control has many advantages over the use of chemical pesticides. First and foremost, carefully selected biological control agents have a restricted diet and, with a few exceptions, attack only the target pests. In general, biological control agents maintain and even improve the overall quality of the environment, thus conserving biological diversity.

Sex Attractants

Sex attractants from some insects have been used effectively to trap pest insects for control and for monitoring pest populations to determine when a treatment should be made.[28] For example, the sex pheromone has been used for the effective control of the grape berry moth without the use of toxic insecticides.[29] In other cases the sex attractant can be used to bring pests to an insecticide.[30]

25. M. B. Thomas, S. D. Wratten, and N. W. Sotherton, "Creation of Island Habitats in Farmland to Manipulate Populations of Beneficial Arthropods: Predator Densities and Emigration," *Journal of Applied Ecology* 28 (1991): 906–917.

26. (a) P. H. DeBach, *Biological Control of Insect Pests and Weeds* (New York: Reinhold, 1964). (b) Debach and Rosen, *Biological Control by Natural Enemies.*

27. Hokkanen and Pimentel, "New Associations."

28. H. H. Shorey, "The Use of Chemical Attractants in Insect Control," in Pimentel, *CRC Handbook.* pp. 289–298.

29. T. J. Dennehy, L. G. Clark, and J. S. Kamas, "Pheromonal Control of the Grape Berry Moth: An Effective Alternative to Conventional Insecticides," *N.Y. Food and Life Science Bulletin* 135 (1991): 6.

30. Shorey, "Use of Chemical Attractants."

Crop Rotations

One of the main causes of serious insect, weed, and plant pathogen outbreaks in world agriculture has been the tendency to plant the same crop year after year on the same land. This provides the pest population with a continuous supply of attractive food and the pest population continues to increase each season.[31] An example of this is when the planting of continuous corn (maize) replaces the planting of corn in rotation with soybeans, wheat, and other non-corn hosts. Major problems with insects, such as the corn rootworm complex, and major weed problems have resulted.[32] Pest problems have escalated because of reduced rotations in corn production. Today the corn crop receives more insecticide and more herbicide than any other crop in the United States.[33] If United States corn were grown in rotation with other nonhost crops, the amount of insecticide and herbicide could be substantially reduced and corn yields could be increased significantly.[34]

Host Plant Resistance

Breeding crops for resistance to insects, plant pathogens, and weeds has been practiced for as long as agriculture has been practiced. When early farmers selected seed from the largest, most productive crop plant in their field, they were unintentionally practicing crop breeding. Breeding crops for resistance to insect pests was first formalized by the research of R. H. Painter.[35] Today host plant resistance is one of the first lines of defense in insect pest, plant pathogen, and weed control.[36] The significance of the level and effectiveness of past resistance in crop plants is well illustrated with pea aphids (*Acyrthosiphon pisum*) associated with alfalfa. Five young pea aphids placed on a common crop alfalfa produced a total of 290 offspring in ten days, whereas the same number of aphids placed on a resistant

31. G. E. Brust and B. R. Stinner, "Crop Rotation for Insect, Plant Pathogen, and Weed Control," in Pimentel, *CRC Handbook*, pp. 217–236.

32. (a) National Academy of Sciences, *Pest Control: An Assessment of Present and Alternative Technologies* (Washington, D.C.: National Academy of Sciences, 1975). (b) Pimentel, *CRC Handbook*.

33. Pimentel, et al., "Environmental and Economic Impacts," pp. 679–718.

34. (a) G. A. Helmers, M. R. Langemeir, and J. Atwood, "An Economic Analysis of Alternative Cropping Systems for East-Central Nebraska," *American Journal of Alternative Agriculture* 4 (1986): 153–158. (b) Pimentel, *CRC Handbook*.

35. R. H. Painter, *Insect Resistance in Crop Plants* (Lawrence, Kan.: University Press of Kansas, 1951).

36. (a) Pimentel, "Herbivore Population Feeding Pressure." (b) G. Shaner, "Genetic Resistance for Control of Plant Disease," in Pimentel, *CRC Handbook*, pp. 495–540. (c) W. M. Tingey and J. C. Steffens, "The Environmental Control of Insects Using Plant Resistance," in Pimentel, *CRC Handbook*, pp. 131–156.

variety of alfalfa produced a total of only two offspring in the same period.[37] Obviously, a pest population that has a 145-fold higher greater rate of increase on a host plant will inflict significantly greater damage than one with an extremely low rate of increase.

Genetic engineering will help in breeding plants for resistance to insects. First, genetic engineering will allow scientists to move genes from almost any organism into another organism. Care will have to be taken not to introduce toxins harmful to humans and/or domestic livestock.[38]

Crop Polyculture

Most often in modern agriculture, especially in developed nations, crops are planted in large monocultures. This practice tends to encourage pest outbreaks because the pest has no trouble in finding another host plant.[39] Selecting the right combination of crops and planting a diverse combination of crops can greatly reduce pest attack on crops. The pest then has trouble finding another suitable host plant, and various predators and parasites survive on the alternate host plants.[40]

Planting Time

Crop planting times can be manipulated to minimize insect pest attack. Delaying the planting of a crop until after the pest has emerged and died can control certain pests. For example, delayed seeding of wheat until after its Hessian fly pest has emerged and died has provided effective control of the fall brood of this pest.[41]

Crop Sanitation

Destroying crop residues that harbor insect pests and plant pathogens until the subsequent growing season is another effective pest control strat-

37. R. G. Dahms and R. H. Painter, "Rate of Reproduction of the Pea Aphid on Different Alfalfa Plants," *Journal of Economic Entomology* 33 (1940): 482–485.

38. (a) See S. Krimsky, *Biotechnics and Society: The Rise of Industrial Genetics* (New York: Praeger, 1991). (b) Pimentel, et al., "Benefits and Risks of Genetic Engineering in Agriculture," *BioScience* 39 (1989): 606–614.

39. Pimentel, "Ecological Basis."

40. (a) D. A. Andow, "Control of Arthropods Using Crop Diversity," in Pimentel, *CRC Handbook*, pp. 257–284. (b) W. J. Cromartie, "The Environmental Control of Insects Using Crop Diversity," in Pimentel, *CRC Handbook*, pp. 183–216. (c) Pimentel, "Species Diversity."

41. (a) National Academy of Sciences, *Pest Control*. (b) U.S. Office of Technology Assessment, *Pest Management Strategies*.

egy.[42] For example, shredding cotton stalks and other cotton plant debris helps provide control of the boll weevil in United States cotton.[43]

Fertilizer Nutrients

All organisms, including crop pests, have specific nutrient requirements for their survival.[44] Altering the nutrient level in the soil and, subsequently, the nutrients in the host plant can influence the density of pests feeding on it. For example, L. Haseman reported that the grain aphid pest produced significantly more progeny when feeding on wheat and other small grain plants with high nitrogen content than those on plants with low levels of nitrogen.[45] This was substantiated by J. S. Barker and O. E. Tauber who observed similar associations with pea aphids feeding on the garden pea.[46] Research on the potential of nutrient management for insect control needs to be expanded.

Soil and Water Management

Insects, weeds, and plant pathogens can be controlled by the appropriate manipulation of soil and water resources. The simple procedure of burying crop residues that harbor pests (like the corn borer) diminishes the damage to the next crop.[47]

Water management, that is, the manipulation of water levels including flooding, has proven to be an effective means of controlling insects, diseases, and weeds in rice culture.[48]

No-Till Culture for Erosion Control

Sometimes a soil management technique can have the unwanted effect of encouraging pests. No-till culture of corn and other crops was developed for

42. Pimentel, "Environmental and Economic Impacts."

43. K. M. El-Zik and R. E. Frisbie, "Integrated Crop Management Systems for Pest Control," in Pimentel, *CRC Handbook*, pp. 3–106.

44. Painter, *Insect Resistance in Crop Plants*.

45. L. Haseman, "Influence of Soil Minerals on Insects," *Journal of Economic Entomology* 39 (1946): 8–11.

46. J. S. Barker and O. E. Tauber, "Fecundity of and Plant Injury by the Pea Aphid as Influenced by Nutritional Changes in the Garden Pea," *Journal of Economic Entomology* 44 (1951): 1010–1012.

47. Pimentel, *CRC Handbook*.

48. See the following articles in Pimentel, *CRC Handbook*: E. F. Eastin, "Weed Management in Southern U.S. Rice," pp. 329–340; T. W. Mew, "Disease Managment in Rice," pp. 279–300; K. Moody, "Weed Management in Rice," pp. 301–328; B. M. Shepard, et al., "Management of Insect Pests of Rice in Asia," pp. 255–278.

erosion control.[49] Although the technique is highly effective in controlling soil erosion, it creates serious pest problems because the pests remaining in the crop residue invade subsequently planted crops. The amount of pesticide required for no-till crop culture is two to four times that for conventional culture.[50] No-till culture requires less tractor fuel, but because so much more pesticide is used, more total energy is required for crop production than in conventional culture.[51]

E. Environmental and Public Health Impacts of Pesticides

In both developed and developing nations, there is growing concern about the environmental and public health impacts of pesticides. Human pesticide poisonings and illnesses are clearly the highest price paid for pesticide use. A recent World Health Organization and United Nations Environmental Programme report estimated that one million human pesticide poisonings and about 20,000 deaths occur worldwide each year.[52] In the United States, pesticide poisonings reported by the American Association of Poison Control Centers total about 67,000 each year.[53] J. Blondell has indicated that because of demographic gaps, this figure represents only 73% of the total cases and that the number of accidental fatalities is about twenty-seven per year.[54] In addition to human pesticide poisonings, some pesticides cause cancer, sterilize humans, or alter normal behavior.[55]

In cultivated and wild areas, naturally present predators and parasites help keep pest species in check. When pesticides destroy both pests and beneficial natural enemies, frequently other pests reach outbreak levels. For example, in United States cotton and apple crops, pesticide destruction of

49. R. F. Follett and B. A. Stewart, *Soil Erosion and Crop Productivity* (Madison, Wis.: American Society of Agronomy, Crop Science Society of America, 1985).

50. (a) M. G. Boosalis, B. Doupnik, and G. N. Odvody, "Conservation Tillage in Relation to Plant Diseases," in Pimentel, *CRC Handbook*, pp. 541–568. (b) Pimentel, et al., "Environmental and Economic Impacts."

51. (a) D. Pimentel, ed. *Handbook of Energy Utilization in Agriculture* (Boca Raton, Fla.: CRC Press, 1980). (b) D. Pimentel and G. Heichel, "Energy Efficiency and Sustainability of Farming Systems," in R. Lal and F. J. Pierce, eds., *Soil Managment for Sustainability* (Ankeny, Iowa: Soil and Water Conservation Society, 1991), pp. 113–123.

52. WHO/UNEP, *Public Health Impact of Pesticides Used in Agriculture* (Geneva: World Health Organization/United Nations Environment Programme, 1989).

53. T. L. Litovitz, B. F. Schmitz, and K. M. Bailey, 1989 Annual Report of the American Association of Poison Control Centers National Data Collection System," *American Journal of Emergency Medicine* 8 (1990): 394–442.

54. J. Blondell (EPA, 1990). Personal communication.

55. Pimentel, "Diversification of Biological Control."

natural enemies results in the outbreaks of numerous pests, including cotton bollworm, tobacco budworm, cotton aphid, cotton loopers, European red mite, red-banded leafroller, San Jose scale, and rosy apple aphid.[56] The additional pesticide applications required to control these pests, plus the increased crop losses they cause, are estimated to cost the United States about $520 million per year.[57]

Other vital insects that pesticides frequently kill are honeybees and wild bees, essential for the annual pollination of about $30 billion in fruits and vegetables in the United States.[58] The losses incurred as a result of the destruction of honeybees are conservatively estimated to be $320 million yearly.[59]

Another serious and costly side effect of heavy pesticide use has been the development of pesticide resistance in pest populations of insects, plant pathogens, and weeds. At present some 900 species exhibit resistance to commonly applied pesticides.[60] When resistance occurs, farmers must increase pesticide applications to save their crops. Even so, crop losses frequently are higher than normal. This resistance problem is estimated to cost the United States $1.4 billion each year in increased pesticide costs and reduced crop yields.[61]

Basically, pesticides are applied to protect crops from pests and to increase yields, yet at times the crops themselves are damaged by pesticide treatments. This occurs when the recommended dosages suppress crop growth, development, and yield, when pesticides drift from the targeted crop to damage adjacent valuable crops, and when residual herbicides either prevent chemical-sensitive crops from being planted in rotation or inhibit the growth of such crops. In addition, excessive pesticide residues may accumulate on crops, necessitating the destruction of the harvest. When crop seizures and insurance costs are added to the direct costs of crop losses caused by pesticides, the total yearly loss in the United States is conservatively estimated to be nearly $1 billion.[62]

Ground and surface waters frequently are contaminated by applied pesti-

56. (a) U.S. Office of Technology Assessment, *Pest Management Strategies*. (b) B. A. Croft, *Arthropod Biological Control Agents and Pesticides* (New York: John Wiley and Sons, 1990).

57. D. Pimentel, et al., "Assessment of Environmental and Economic Impacts."

58. W. E. Robinson, R. Nowogrodzki, and R. A. Morse, "The Value of Honey Bees as Pollinators of U.S. Crops," *American Bee Journal* 129 (1989): 477–487.

59. Pimentel, et al., "Assessment of Environmental and Economic Impacts."

60. G. P. Georghiou, "Overview of Insecticide Resistance," in M. B. Green, H. M. LeBaron, and W. K. Moberg, eds., *Managing Resistance to Agrochemicals: From Fundamental Research to Practical Strategies* (Washington, D.C.: American Chemical Society, 1990), pp. 18–41.

61. Pimentel, et al., "Assessment of Environmental and Economic Impacts."

62. Ibid.

cides. Birds, mammals, and other wildlife also are killed, although the full extent of their destruction is difficult to determine because these animals are often hidden from view, highly mobile, and live in protected habitats.

The known costs of human and animal health hazards, including the costs of diverse environmental impacts associated with United States pesticide use, total approximately $8 billion each year.[63] Thus, based on a strictly cost/benefit basis, pesticide use remains beneficial. Decisions about future pesticide use should be based not only on the benefits but on the risks they create. Perhaps in this way an equitable balance can be achieved. No estimate exists for the total environmental and public health costs of using pesticides world wide; but our estimate is that it costs at least $100 billion each year.

F. Growing Replacement of Pesticides with Nonchemical Controls

Because of the environmental and public health impacts of pesticides, there is a growing trend to find ways to reduce the use of pesticides and increase the use of nonchemical alternatives for chemical pest control. Such changes may either increase or decrease the costs of pest control.[64] Three major crops in United States agriculture illustrate the opportunities for reducing insecticide and herbicide use.

Corn and cotton account for approximately 25% of the total insecticide use in United States agriculture. Controlling pests on these two crops through nonchemical alternatives would thus contribute significantly to a reduction in insecticide use. During the early 1940s, little or no insecticide was applied to corn, and losses to insects were only 3.5%.[65] Since then, insecticide use on corn has grown more than 1000-fold, whereas crop losses due to insects have increased to 12%.[66] This increase in insecticide use and the 3.4-fold increase in corn losses to insects are primarily due to the abandonment of crop rotation.[67] Today approximately 40% of United States corn is grown continuously, with 11 million kg of insecticide applied annually.[68] Rotating corn with soybeans or a similar high-value crop generally in-

63. Ibid.
64. Pimentel, et al., "Environmental and Economic Impacts."
65. U.S. Department of Agriculture, *Losses in Agriculture* (Washington, D.C.: Agricultural Research Service, 1954 [*ARS* 20–1]).
66. R. Ridgway, "Assessing Agricultural Crop Losses Caused by Insects," in *Crop Loss Assessment: Proceedings of the E. C. Stakman Commemorative Symposium* (St. Paul: University of Minnesota, 1980), pp. 229–233.
67. Pimentel, et al., "Environmental and Economic Impacts."
68. Ibid.

creases yields and net profits, although rotating corn with wheat or other low-value crops reduces net profit per hectare.[69] From a more comprehensive perspective, however, the rotation of corn with other crops has several advantages, including reducing weed and plant pathogen losses and decreasing soil erosion and rapid water-runoff problems.[70] By combining crop rotations with the planting of corn resistant to the corn borer and chinch bug, it would be possible to avoid 80% of insecticide use on corn while concurrently reducing insect losses to these insects.[71] Such a move is estimated to increase the cost of corn production by $10 per hectare above the current costs of corn grown continuously.[72] A new approach, in which an attractant is combined with insecticides to fight rootworm, has been reported to reduce insecticide requirements 99%.[73]

The potential for reducing pesticide use in United States agriculture is well illustrated by changes in insecticide use in Texas cotton production. Since 1966, insecticide use in Texas cotton has been reduced by almost 90%.[74] The technologies adopted to reduce insecticide use were monitoring pest and natural enemy populations to determine when to treat, biological control, host-plant resistance, stalk destruction (sanitation), uniform planting date, water management, fertilizer management, rotations, clean seed, and changed tillage practices.[75] Currently, a total of twenty-nine million kg of insecticide is applied to cotton, and it is estimated that this amount could be reduced by approximately 40% through the use of readily available technologies.[76] By effectively using a monitoring program, one might reduce insecticide use by approximately 20%. Through the use of pest-resistant cotton varieties and the alteration of planting dates in most growing regions, insecticide use could be reduced by another 3%.[77]

Insecticide use on cotton might be reduced through the implementation of

69. Helmers, Langemeir, and Atwood, "Economic Analysis of Alternative Cropping Systems."
70. Ibid.
71. J. M. Schalk and R. H. Ratcliffe, "Evaluation of the United States Department of Agriculture Program on Alternative Methods of Insect Control: Host Plant Resistance to Insects," *FAO Plant Protection Bulletin* 25 (1977): 9–14.
72. Pimentel, et al., "Environmental and Economic Impacts."
73. J. Paul, "Getting Tricky with Rootworms," *Agrichemical Age* 33 (3) (1989): 6–7.
74. U.S. Office of Technology Assessment, *Pest Management Strategies*.
75. (a) E. G. King, J. R. Phillips, and R. B. Head, "Thirty-Ninth Annual Conference Report on Cotton Insect Research and Control," in *Proceedings of the Beltwide Cotton Production Research Conference* (Memphis, Tenn.: National Cotton Council, 1986), pp. 126–135. (b) U.S. Office of Technology Assessment, *Pest Management Strategies*.
76. Pimentel, et al., "Environmental and Economic Impacts."
77. (a) R. Frans, "A Summary of Research Achievements in Cotton," in R. E. Frisbie and P. L. Adkisson, eds., *Integrated Pest Management on Major Agricultural Systems* (College Station, Texas: Texas Agricultural Experiment Station, 1985), pp. 53–61. (b) R. Frisbie, "Regional Implementation of Cotton IPM," in Frisbie and Adkisson, *Integrated Pest Management*, pp. 638–651.

other pest-control techniques, including cultivation of short-season cotton, fertilizer and water management, improved sanitation, crop rotations, the use of crop seed cleaned of weed seeds during culture and processing, and altered tillage practices.[78] Depending on the particular environment, insecticide use on cotton might even be reduced much more. For example, Shaunak and others have reported that insecticide use in the Lower Rio Grande Valley of Texas could be reduced by 97% through the planting of short-season cotton under dryland conditions.[79] This practice also resulted in a twofold increase in net profits over conventional methods.

Corn and soybeans account for approximately 70% of the total herbicide applied in agriculture; half (53%) of it is applied to corn.[80] More than 3 kg of herbicide are applied per hectare of corn, and more than 90% of the corn hectarage planted is treated. By not requiring total elimination of weeds, in some cases herbicide use can be reduced by 75%.[81] At present, 91% of the corn land is also cultivated to help control weeds.[82]

The average costs and returns per hectare to no-till, reduced-till, and conventional-till culture have actually been found to be quite similar. For example, added labor, fuel, and machinery costs for conventional-till practices for corn were approximately $24/ha higher than those for no-till. However, the costs for the added fertilizers, pesticides, and seeds in the no-till system were $22/ha higher than for conventional-till.[83]

The second largest amount of herbicides is applied to soybeans, with approximately 96% of the soybean hectarage receiving treatment for weed control; 96% of the hectarage also receives some tillage and mechanical cultivation for weed control.[84] Several techniques have been developed that increase the efficiency of chemical applications. The rope-wick applicator has been used in soybeans to reduce herbicide use approximately 90%, and this applicator was found to increase soybean yields 51% over conventional treatments.[85] Also, a new model of recirculating sprayers saves 70–90% of

78. (a) D. W. Grimes, "Cultural Techniques for Management of Pests in Cotton," in: Frisbie, *Integrated Pest Management . . .*, pp. 365–382. (b) U.S. Office of Technology Assessment, *Pest Management Strategies*.

79. R. K. Shaunak, R. D. Lacewell, and J. Norman, *Economic Implications of Alternative Cotton Production Strategies in the Lower Rio Grande Valley of Texas, 1923–1978* (College Station, Texas: Texas Agricultural Experiment Station, 1982 [*Publ.* B-1420]).

80. Pimentel and Levitan, "Pesticides: Amounts Applied."

81. E. E. Schweizer, "Weed Free Fields Not Key to Highest Profits," *Agricultural Research* (1989): 14–15.

82. M. Duffy, *Pesticide Use and Practices* (Washington, D.C.: U.S. Department of Agriculture, Economic Research Service, 1982 [*USDA Agricultural Information Bulletin* no. 462]).

83. M. Duffy and M. Hanthorn, *Returns to Corn and Soybean Tillage Practices* (Washington, D.C.: U.S. Department of Agriculture, 1984 [*Economic Research Service, Agricultural Economic Report* no. 508]).

84. Ibid.

85. Dale, "Ropewick Applicator."

the spray emitted that is not trapped by the weeds.[86] Spot treatments are a third method of decreasing unnecessary pesticide applications. Alternative techniques are available to reduce the need for herbicides, including ridge-till, tillage, mechanical cultivation, row spacing, planting date, tolerant varieties, crop rotations, spot treatments, and reduced dosages.[87] Employing several of these alternative techniques in combination might reduce herbicide use in soybeans by approximately 60%. Despite Tew's report indicating no added control costs for the alternatives, we estimate that the costs of weed control would increase $10/ha.[88]

Farmers worldwide support the trend to reduce pesticide use for three reasons: pesticides are expensive and farmers would like to reduce their costs; pesticides are highly toxic and farmers and their families receive the greatest exposure; and farmers also are concerned about the environment and would like to reduce pesticide impacts.[89]

G. National Goals to Reduce Pesticide Use

Worldwide public demand for reduced pesticide use and farmer demand for more effective pest control have prompted several governments to develop and implement programs to reduce pesticide use. Denmark, for example, was the first nation to develop an action plan to reduce pesticides by 50%.[90] Sweden soon followed, implementing a program to reduce pesticides by 50% within five years, by 1992.[91] Sweden plans another 50% reduction during the next five years. The Province of Ontario, Canada, also developed a program to reduce pesticide use by 50%, and they are well on their way to achieving their goal.[92] The Netherlands is also implementing a program to reduce pesticide use by 50%.[93] Some developing nations are also

86. G. A. Matthews, "Application from the Ground," in P. T. Haskell, ed., *Pesticide Application: Principles and Practice* (Oxford, U.K.: Clarendon Press, 1985), pp. 93–117.

87. (a) F. Forcella and M. J. Lindstrom, "Movement and Germination of Weed Seeds in Ridge-Till Crop Production Systems," *Weed Science* 36 (1988): 56–59. (b) Helmers, Longmeir, and Atwood, "Economic Analysis of Alternative Cropping Systems." (c) J. Russnogle and D. Smith, "More Dead Weeds for Your Dollar," *Farm Journal* 112 (2) (1988): 9–11. (d) B. V. Tew, et al., *Economics of Selected Integrated Pest Management Production Systems in Georgia* (Athens: University of Georgia, College Agricultural Experiment Station, 1982 [*Research Report* no. 395]).

88. (a) Tew, et al., *Economics of Selected Integrated Pest*. (b) Pimentel, et al., "Environmental and Economic Impacts."

89. G. A. Surgeoner and W. Roberts, "Reducing Pesticide Use by 50% in the Province of Ontario: Challenges and Progress," in Pimentel and Lehman, *The Pesticide Question*.

90. P. Hurst, *Pesticide Reduction Programmes in Denmark, the Netherlands and Sweden* (Gland, Switzerland: WWF, 1992).

91. O. Pettersson, "Swedish Pesticide Policy in a Changing Environment," in Pimentel and Lehman, *The Pesticide Question*.

92. Surgeoner and Roberts, "Reducing Pesticide Use."

93. Hurst, *Pesticide Reduction Programmes*.

reducing pesticide use. For example, it was reported that the reason for increased losses of rice due to insects in Indonesia was excessive use of insecticides.[94] When, following the advice of Oka, insecticide use in rice was reduced by 65%, rice yields increased 40%.

Despite the worldwide use of 2.5 million tons of pesticides and numerous nonchemical alternative pest controls, insect pests, weeds, and plant pathogens destroy approximately 35% of potential crop production each year. If pesticide controls were no longer available, crop losses would probably increase to about 45%, a 10% increase over current loss levels. If the nonchemical alternative pest controls were no longer practiced, crop losses would probably increase to 55%.

Because of the multidimensional nature of crops, pests, physical habitats, and natural biota, the factors that cause pest outbreaks are diverse. This knowledge has prompted pest control specialists to expand the use of nonchemical controls and reduce pesticide use in developed and developing nations. In part, this has been stimulated by the enormous environmental and public health problems caused by pesticides.

Research is underway to enlarge and expand the kinds of nonchemical technologies used to control the 67,000 pest species in the world. In crops worldwide, the most important nonchemical controls include biological control with natural enemies, host-plant resistance, crop rotations, and crop polycultures. Combining these nonchemical control technologies with other appropriate nonchemical controls for a particular agroecosystem would enhance pest-control programs and lessen the need for pesticides.

94. I. N. Oka, "Success and Challenges of the Indonesian National Integrated Pest Management Program in the Rice-Based Cropping System," *Crop Protection* 10 (1991): 163–165.

4. Literature Characteristics, Patterns, and Trends in Crop Science

WALLACE C. OLSEN
Cornell University

PETER McDONALD
New York State Agricultural Experiment Station, Geneva

The first task in this study of literature was determining the scope of crop science. This was a relatively involved task solved by a two-step process. First, classifications of knowledge and of crop science were examined and the points of difficulty identified. The second step was to place these peripheral or questionable areas before a Steering Committee of the Core Agricultural Literature Project, Mann Library, for consultation. The committee consisted of scientists from plant and crop breeding, phytopathology, crop production, and entomology. Their determination of what subjects should be included was related to previous subjects studied by this Project, such as agricultural engineering where crop storage questions were partially handled. The final scope is represented in the first three chapters of this book. Some adjunct areas were explored but their influence was minor.

A major aim of the Core Agricultural Literature Project was to identify the current core literature of greatest value to instruction and research in the agricultural sciences. Crop improvement and protection is but one of seven subjects identified and studied. The Project was further interested in determining the differences in this current, core literature between Third World countries and developed countries. Therefore, work progressed on two tracks, and these differences and similarities served as a major influence on the reports in this book. Publishing patterns in crop science, their subject concentrations, and formats are examined along with selected other characteristics.

A recent and very useful book resulted from a workshop held at the CAB International offices in 1989. In this book, crop protection information in its numerous formats, the validity of data, and shortcoming of information are discussed from a worldwide perspective as well as for select specific countries.[1] Two recent literature guides and information source books are recommended for background and details in pest management and entomology. Both have C. J. Hamilton, Library Services Manager of CAB International, as an author.[2] For recent literature guides to all subjects related to crop science see "Reference Update" (Chapter 10).

A. Current Literature Data in Crop Science

Several potential sources for data were examined and discarded as being too expensive of time and money or inadequate in coverage. These included specialized databases, national agricultural bibliographies, and commodity compilations. The literature about databases and the nature of crop literature is proliferating. These often are surveys or data concerning commodities or specialized documentation.[3] These growing literature studies and databases have been used and noted when citation numbers and trends differ significantly. The conclusion, as with the other investigations of the Core Agricultural Literature Project, was that the large, bibliographic, agricultural databases provided the most reliable and accessible sources of data. The reader should also be aware of a major study of the use of bibliometrics for research evaluation in the life sciences by H. F. Moed.[4]

Three large databases concentrate on agriculture: *AGRICOLA* (produced by the U.S. National Agricultural Library), *AGRIS* (produced by the Food and Agriculture Organization), and *CAB Abstracts* (produced by CAB International). All database citations appear on digitized tapes or compact disks,

1. K. M. Harris and P. R. Scott, eds. *Crop Protection Information: An International Perspective; Proceedings of the International Crop Protection Workshop* (Wallingford, U.K.; CAB International, 1989).

2. (a) Pamela Gilbert and Chris J. Hamilton, *Entomology: A Guide to Information Sources*, 2d ed. (London and New York: Mansell, 1990). (b) C. J. Hamilton, compiler, *Pest Management: A Directory of Information Sources; Vol. 1: Crop Protection* (Wallingford, U.K.: CAB International, 1991).

3. Five illustrations in the field of tropical agriculture: (a) "CIRAD Provides Access to Information on Crop Protection" an announcement by the Centre de Coopération Internationale en Recherche Agronomique pour le Développement, *IAALD Quarterly Bulletin* 37 (4) (1992): 238. (b) V. K. Datta, "Coverage of Literature on Mycotoxins by Computer Databases and Comparison of This Coverage with the TDRI In-House Index Facility," *IAALD Quarterly Bulletin* 33 (2) (1988): 61–78. (c) G. Hartmann, "Une Base de Données Concernant les Fruits Tropicaux et Subtropicaux: FAIREC (Fruits Agro-Industrie Régions Chaudes)," *IAALD Quarterly Bulletin* 28 (4) (1983): 209–213. (d) Stephen M. Lawani, "Grain Legume Documentation and Information: The Contributions of the International Institute of Tropical Agriculture," *IAALD Quarterly Bulletin* 27 (2) (1982): 47–53. (e) B.S. Maheswarappa, "Bibliographical Phenomena of Phytomorphology Literature: A Citation Analysis" in *Annals of Library Science and Documentation* 30 (1) (1983):22–30.

4. H. F. Moed, *The Use of Bibliometric Indicators for the Assessment of Research Performance in the Natural and Life Sciences; Aspects of Data Collection, Reliability, Validity and Applicability* (Leiden: DSWO Press, 1989).

as well as in printed abstracting or indexing publications. Prior to 1985, all three databases aimed to cover the world literature of the agricultural sciences in all subject areas. In 1985–1986, *AGRICOLA* altered its coverage and began to concentrate on U.S.-published journals and monographs. "Many foreign journals previously indexed are no longer indexed because of cooperation with the International Information System for the Agricultural Sciences and Technology (*AGRIS*)."[5] *AGRICOLA* and *CAB Abstracts* are selective of citations, choosing them on the basis of the nature, subject, and value of the individual literature pieces or articles. *AGRIS* adds over 130,000 citations a year, *CAB Abstracts* added 153,000 in 1993,[6] while *AGRICOLA* provided fewer entries each year between 1985 and 1988, and in 1989 entered only 69,720. *AGRICOLA* records increased steadily thereafter and hit 102,398 in 1993.[7] *AGRIS*, an international cooperative effort accepting citations from food and agricultural organizations around the world, makes few judgments on the value of the literature indexed. It is somewhat more representative of Third World literature than the other two files, although prior to 1985 the differences were not as great as in 1993. Earlier studies of these three major bibliographic databases indicate that about 20% of each database is comprised of items not in either of the others.[8]

For a picture of the total numbers of articles, books, and proceedings titles in the agricultural sciences, there is no one readily accessible source with complete coverage. These bibliographic databases must be used together to identify the universe of publishing for agriculture and its subdisciplines. The head of the *AGRIS* Coordinating Centre reported that 40.5% of that database was made up of crop science citations in 1979, in the two broad categories of " plant production" and " plant protection." This contrasts with the animal science literature, which comprised 23.1%.[9] This points out the immense importance of crops in the food chain. These large numbers of references in the bibliographic databases are continuing, and the percentage of literature on crops remains close to 41% of all agricultural literature.

5. *List of Journals Indexed in AGRICOLA, 1991* (Beltsville, Md.: National Agricultural Library, 1991), p. v.

6. Personal correspondence, January 1994 from Stella G. Dexter Clarke, CABI.

7. *National Agricultural Library Annual Report for 1993,* (Beltsville, Md.: U.S. Department of Agriculture, 1994). p. 50.

8. Norbert Deselaers, "The Necessity for Closer Cooperation among Secondary Agricultural Information Services: An Analysis of *AGRICOLA*, *AGRIS*, and CAB," *Quarterly Bulletin of the International Association of Agricultural Librarians and Documentalists* 31 (1) (1986): 19–26.

9. Abe Lebowitz, "*AGRIS* since the First Technical Consultation," *Agricultural Libraries Information Notes* 6 (9/10) (1980): 5–7.

Correlating data from the major agricultural databases proved difficult in part because the subject of crop science is of immense scope, and because concise delineations or clear comparisons between databases are difficult. Analyses by the Core Agricultural Literature Project sought to match the subject categories and counts from *AGRICOLA* with those of *CAB Abstracts*. Further analysis of *AGRIS* gave comparative data of publishing trends in smaller developing countries.

In all three databases, citations are assigned one or more subject category codes at the time of indexing. An important function of category coding is that it permits the extraction of data on narrow, specified subjects. The coding in *AGRICOLA* tends to emphasize production aspects of crop science, while *CAB Abstracts* indexes more heavily in areas of plant biology and related disciplines. In a field as large as crop science, coding overlap within databases will necessarily be extensive and must be accounted for at the outset. The software used with *AGRICOLA* and *CAB Abstracts* preclude the screening of the secondary codes assigned, making difficult the determination of indexing overlap.

Analysis of double subject coding conducted by the Core Agricultural Literature Project revealed that some codes, notably in the areas of plant breeding and crop biology, had higher duplication rates than others, for example CABI's code OT, used for its *Helminthological Abstracts, Series B*. In the case of CABI's *Plant Breeding Abstracts* file (code 0P), double coding with other files stood at 17.6% in the sample, but this percentage could be higher because there is considerable duplication between categories within the OP file itself. Conversely, CABI files dealing with crop pathology and protection, such as Pathology (code OM) and Weeds (code W), had less double coding, at 9.6% and 6.5% respectively. These calculations were made on the basis of examination of hundreds of printed records in each major subject.

Double coding in *AGRICOLA* is less problematic, in part because indexing by the National Agricultural Library (NAL) tends to group crop-related subjects under broader headings than in *CAB Abstracts,* thus obviating the need for multiple category entries. Few of the files exceed 4–5% duplicate coding with the heaviest duplication between code F120 (Plant Production/Field Crops) and F300 (Plant Ecology) at 6.2% based on an analysis of 250 citations to material published in 1985. These and other data skews are caveats when approaching bibliographic analyses such as this.

B. Subject Concentration

Tables 4.1 and 4.2 show numerical counts of crop science citations in *AGRICOLA* over a ten-year period divided by subject categories. The appropriate subject categories have been broadly divided into "Crop Improvement" and "Crop Protection and Pathology." Similar divisions for *CAB Abstracts* have been made for Tables 4.3 through 4.9. Tables 4.10 and 4.11 strive to match citation data by grouping subject categories in the two databases.

Twenty-six subject categories in *AGRICOLA* comprise the coding schema for crop science, twelve under Crop Improvement and fourteen under Crop Pathology and Protection. The decade 1980–1989 served as the time base. The counts steadily decrease in both Tables 4.1 and 4.2 from 1980 to 1989. Recent administrative policies at NAL allowing the database coverage to decline probably account for this ten-year drop.[10] The codes F000–F700, which comprise crop improvement in *AGRICOLA*, account for 31.7% of the entire *AGRICOLA* database, 1980–1989 (Table 4.1).

Given the current NAL schema, it is not possible to extract a specific crop from the categories in *AGRICOLA* such as F110 (Plant Production - Horticulture) or F120 (Plant Production - Field Crops). F120 for instance,

Table 4.1. AGRICOLA category code counts for crop improvement, 1980–1989

Category code	Subject	1980	1983	1986	1989	1980–1989
F000	Plant Science	429	322	211	195	2,747
F100	Plant Production	679	548	342	230	4,891
F110	Plant Production - Horticulture	5,475	4,401	2,684	2,189	36,623
F120	Plant Production - Field Crops	4,731	3,824	2,620	1,309	32,629
F130	Plant Production - Range & Pasture	1,101	997	577	303	7,406
F140	Plant Production - Misc.	503	430	286	156	3,570
F200	Plant Breeding & Genetics	8,064	8,677	6,426	5,916	72,986
F300	Plant Ecology	2,922	2,516	1,749	1,131	21,238
F400	Plant Structure	2,661	2,351	1,910	1,822	21,564
F500	Plant Nutrition	1,532	1,662	1,325	954	13,932
F600	Plant Physiology	12,287	13,176	11,533	8,848	116,437
F700	Plant Taxonomy	4,535	5,049	3,742	2,553	39,654
	TOTALS	44,919	43,953	33,405	25,606	373,677

10. (a) Wallace Olsen, "Characteristics of Agricultural Engineering Literature," in Carl W. Hall and Wallace Olsen, eds., *The Literature of Agricultural Engineering*, (Ithaca, N.Y.: Cornell University Press, 1993), p. 221. (b) Wendy Simmons, "The Development of *AGRIS*: A Review of the United States Response," *Quarterly Bulletin of the International Association of Agricultural Librarians and Documentalists* 31 (1) (1986): 11–18.

Table 4.2. AGRICOLA category code counts for crop pathology and protection, 1980–1989

Category Code	Subject	1980	1983	1986	1989	1980–1989
F800	Protection of Plants	711	622	464	237	5,199
F820	Animal Pests	328	249	209	181	2,546
F821	Insect Pests	5,610	4,997	4,301	2,966	45,538
F822	Nematodes	808	684	602	392	6,376
F830	Plant Disease: General	526	439	335	218	3,864
F831	Plant Disease: Fungal	4,614	3,883	2,973	1,643	32,981
F832	Plant Disease: Bacterial	714	677	539	334	6,172
F833	Plant Disease: Viral	1,371	1,079	885	552	9,809
F840	Plant Disease: Physiological	404	287	258	169	2,858
F841	Miscellaneous Disorders	1,340	1,054	866	635	9,404
F850	Protection of Products	378	353	210	114	2,758
F851	Protection of Products - Insects	387	409	255	218	3,181
F900	Weeds	2,305	1,971	1,592	1,515	18,341
H000	Pesticides	2,635	3,068	1,993	1,854	23,745
	TOTALS	22,131	19,772	15,482	11,028	172,772

covers an immense range of plants: legumes, nine types of marketable wheat, forage crops including perennial and annual grasses, alfalfa, cotton, tobacco, clover, sugar crops, and fourteen types of oil seeds. Short of using complex subject searching strategies incorporating Boolean operators connecting strings of genus/species names, it is not practicable to extract reliable data on specific crops in *AGRICOLA*. Hence, individual crop comparisons with *CAB Abstracts* are exceedingly difficult where most crop types are given individual codes.

In Table 4.2, code F821 and other pest-related categories need clarification, especially since *AGRICOLA*'s counts for insects and nematodes so poorly match those in *CAB Abstracts*. In keeping with its emphasis on crop production, the Insect Pests category in *AGRICOLA* (F821) almost exclusively indexes records pertaining to insect infestation and damage in crops. Less emphasis is placed on insect physiology, biology, or taxonomy. For aspects of the biology of insect pests, one must combine F821 with L200–L700, and L001 (Entomology-related Animal Science). Here again, it is not clear how much of the L001 category not double coded with F821 deals with the biology of crop insect pests as opposed to animal or forest pests. Since no logical strategies allow for these distinctions to be sorted, L001 is not listed in Table 4.2, although many of its 50,000+ citations (1980–1989) undoubtedly deal with plant pests. Citations on the pollination of certain crops by bees and other insects, while not numerically significant, demand either the additional code L001 (Animal Production) or L002

Table 4.3. Citations in *Plant Breeding Abstracts* of *CAB Abstracts*

Sequencing Code	Subject	1980	1983	1986	1989	1980–1989
0P0101	Breeding	N/D	141	230	428	
0P0111	Molecular Genetics	N/D	0	0	6	
0P0112–13/116	Genetics	N/D	100	136	141	
0P11	Plant Genetics	117	0	N/D	N/D	
0P13 (0P013)	Cytology	63	78	74	61	616
0P14 (0P014)	Reproduction	14	73	101	94	740
0P15	Botany	74	301	N/D	N/D	
0P17/19	Evolution & Taxonomy	80	107	N/D	N/D	
0P2/31 (0P019/1)	Grain Crops	3,471	2,508	4,240	4,086	33,538
0P32–36 (0P252)	Herbage	143	154	30	27	906
0P3	Grasses	198	203	261	302	2,410
0P41–44	Legumes	390	478	408	502	4,127
0P45–47 (0P46)	Root Crops	412	492	479	458	4,484
0P48	Fibre Crops	360	259	368	319	2,985
0P49	Sugar Crops	210	201	244	235	2,079
0P51	Stimulants	280	309	332	357	3,082
0P52–53 (0P52)	Oil Crops	352	388	535	464	4,518
0P54–55 (0P54)	Misc. Crops	159	104	179	234	1,575
0P56–57/64–65 (0P55–65)	Fruits	660	753	1,197	1,083	8,656
0P68–69/7 (0P67–69/7)	Vegetable Crops	1,870	2,036	2,295	2,344	20,226
	TOTALS	8,853	8,685	11,109	11,141	89,942

(Apiculture-related), depending on the citation being indexed. These subject divisions make comparisons nearly impossible with the heavy emphasis on invertebrate biology in CABI's Entomology subfile (code 0E). Nor do the taxonomic or cytological aspects of nematodes and other plant pests figure as prominently in the *AGRICOLA* counts of category code F822 (Nematodes) or even F820 (Animal Pests, e.g., rodents, birds, mammals) as they do in the corresponding files in *CAB Abstracts*. For this reason the 6,300 citation total for the *AGRICOLA* category code F822 (Nematodes) is half of CABI's *Helminthological Abstracts* count for the same period. CABI's more inclusive indexing policies also account for a portion of the difference.

Another category in *AGRICOLA*, H000 (Pesticides), poses additional comparison problems with *CAB Abstracts*. The policy at NAL is to index general considerations of pesticides in the HOOO category. These include records on fungicides, nematicides, bactericides, some antibiotics, insecticides, rodenticides, acaricides and herbicides. Where specific hosts, pests, or weeds are given in the article, the records are coded in either F820–F833 or F850–F851. In *CAB Abstracts* there is no clear equivalent of H000. Pesticides are coded separately in the various subfiles of *CAB Abstracts* (e.g., 0M05, 0W8, 0T62); these codes also cover environmental toxicology and soil impacts, which in *AGRICOLA* are coded H000. Here again, clear comparisons are hampered by the vagaries of the two indexing systems. The H000 code with codes F800 through F900, comprising crop pathology and protection, account for 14.6% of the *AGRICOLA* database, 1980–1989.

Tables 4.3 through 4.9 cover the discipline of crop science in *CAB Abstracts*, where coverage is more extensive both in numbers of citations and subject specificity. The three subfiles given in Tables 4.3 to 4.5 deal with crop improvement; those in Tables 4.6 through 4.9 deal with aspects of crop pathology, pests, and protection.

By dividing crop improvement into separate subfiles, with eight categories apiece in the field crop and horticultural files, and twenty codes in plant breeding, *CAB Abstracts* has made it easier to use its sequencing codes to categorize specific crops. In the Plant Breeding subfile (Table 4.3), the general headings, codes 0P0101–0P19, are followed by specific crop categories, codes 0P2–0P7. A similar organization is used for the two other subfiles. The Plant Breeding subfile accounts for 6.8% of the total *CAB Abstracts* database, 1980–1989.

The *Horticultural Abstracts* file, code 0C (Table 4.4), is of interest because it is the only crop file which separately indexes tropical plants, notably in codes 0C4, 0C6, 0C7 and 0C8. Combining these four codes with the publishing years 1985, and searching the first hundred records in each category, revealed that upwards of 31% of the records deal with aspects of tropical horticulture. Since these four categories comprise 54% of the total

Table 4.4. Citations in *Horticultural Abstracts* of *CAB Abstracts*

Sequencing Code	Subject	1980	1983	1986	1989	1980–1989
0C1	General Aspects	886	547	609	1,043	6,916
0C2	Temperate Tree Fruits and Nuts	1,339	1,402	1,508	1,643	13,730
0C3	Small Fruits	767	959	908	1,008	8,526
0C4	Vegetables, Temperate, Tropical & Greenhouse	1,577	2,157	2,348	2,895	21,841
0C5	Ornamental Plants	1,445	1,682	1,895	2,178	16,911
0C6	Minor, Temperate and Tropical Industrial Crops	1,378	1,594	1,866	2,249	17,203
0C7	Subtropical Fruit & Plantation Crops	651	867	678	704	6,772
0C8	Tropical Fruit and Plantation Crops	934	936	891	1,269	9,475
	TOTALS	8,977	10,144	10,703	12,989	101,374

0C file, a rough estimate of the total tropical input in the OC file as a whole stands at about 16%. Coffee, sugarcane, bananas, papayas, avocados, and similar low-latitude or tropical fruits and plantation crops dominate.

Field crops (Table 4.5) account for 7.6% of the total records of *CAB Abstracts* for the decade 1980–1989. In both *AGRICOLA* and *CAB Abstracts,* cereals are clearly the crop type with the most records indexed. Cereals are by far the most important crop group by any standard. Some of the more extensive literature concerns rice, which seems not to be covered as well by standard bibliographic tools, particularly Japanese language literature, according to a study by K. Morooka.[11] The wheat literature is extensive, and many studies report on the increase in maize or corn literature, in developing countries in particular, as exemplified in an article by B. O. Ikhizama.[12]

As noted, overlap throughout the *CAB Abstracts* crop subfiles is extensive. Actual records are not duplicated in *CAB Abstracts* but there are multiple codings for each record for different *Abstract* uses. This is especially true of the 0Q file. For example, between Field Crop (code 0Q) and Plant Breeding (code 0P) there is extensive overlap, as high as 15%, notably with cereals/grains (codes 0Q00 and 0P2) and legumes (codes 0Q05–9 and 0P41–4). When other overlapping subfiles are added, the overlap may be as high as 45% in some areas of crop science. Within the 0Q subfile itself, the code 0Q3 (Crop Botany) is often double-coded with other 0Q categories. When the Core Agricultural Literature Project wrote to CABI for an esti-

Table 4.5. Citations in *Field Crop Abstracts* of *CAB Abstracts*

Sequencing Code	Subject	1980	1983	1986	1989	1980–1989
0Q00/02–04	Cereals	3,797	3,708	3,784	4,469	38,382
0Q05–9	Legumes	2,168	2,184	1,849	2,405	20,819
0Q10/12–14	Root Crops	839	856	843	869	8,493
0Q15/19	Fibre Crops	401	364	320	401	3,485
0Q20–21/24	Oilseed Crops	409	493	419	627	4,753
0Q25–27	Misc. Crops	276	263	261	415	2,852
0Q29	Field Crops, General	95	115	126	197	1,268
0Q3	Crop Botany	585	561	323	441	4,763
	TOTALS	8,570	8,544	7,925	9,824	84,815

11. Kuzuko Morooka, "Analysis of Bibliographic Data Relating to the *International Bibliography of Rice Research*," *IAALD Quarterly Bulletin* 30 (4) (1985): 91–100.

12. B. O. Ikhizama, "The Development of Maize Literature in Nigeria," *IAALD Quarterly Bulletin* 27 (4) (1982): 122–128.

mate on primary and secondary coding and problems of duplication, CABI provided no numerical answers.[13]

CAB Abstracts crop improvement files clearly emphasize biological aspects of crop production, and they do an excellent job of coding at the genus level. Indeed, in the Plant Breeding file, the general codes on breeding, genetics, reproduction, and taxonomy are rarely used, compared to the crop type categories 0P2–0P7. In total, the three major crop improvement files in *CAB Abstracts* account for over 20% of the database. Combined with the crop pathology and protection files, the total for all of crop science is just over 41% of the entire database.

Four subfiles of *CAB Abstracts* concern crop pests and protection, and these four are also published as abstracting journals. Tables 4.6 through 4.9

Table 4.6. Citations in *Review of Plant Pathology* of *CAB Abstracts*

Sequencing Code	Subject	1980	1983	1986	1989	1980–1989
0M03	Bacteria	594	588	647	643	6,146
0M05	Fungicides	133	120	160	258	1,633
0M06	Antibiotics	14	18	13	31	169
0M07	Regulations	17	18	29	35	241
0M08	Mycorrhizas	162	176	289	341	2,372
0M09	Fungi Physiology	193	129	147	246	1,577
0M13	Viruses	1,025	1,094	1,317	1,083	10,746
0M14	Crops, General	144	126	144	124	1,269
0M15	Cocoa	23	31	21	26	275
0M16	Cereals	1,019	1,050	1,282	1,145	11,250
0M17	Citrus	89	134	123	103	1,105
0M18	Coffee	36	37	23	33	290
0M19	Cotton	80	55	75	69	694
0M20	Other Fibre Crops	32	27	14	14	235
0M21	Ornamental Crops	213	249	216	195	2,292
0M23	Fruit Crops (but not citrus)	514	454	596	480	4,938
0M24	Viticulture	119	86	121	92	1,044
0M25	Hops	15	20	8	9	139
0M27	Palms	44	41	33	29	340
0M28	Potatoes	295	288	264	251	2,660
0M29	Rubber	14	18	28	16	181
0M30	Sugarcane	83	71	80	64	679
0M31	Tea	8	12	9	15	99
0M32	Tobacco	83	126	99	152	1,038
0M33	Tomatoes	132	148	184	187	1,592
0M35	Vegetables	1,020	1,149	1,317	1,169	11,813
	TOTALS	5,076	5,171	5,922	5,727	64,817

13. Personal correspondence with C. Ogbourne, CABI, 1993.

Table 4.7. Citations in *Weeds Abstracts* of *CAB Abstracts*

Sequencing Code	Subject	1980	1983	1986	1989	1980–1989
0W0	General	68	90	34	59	543
0W1	Annual Field Crops	973	1,101	986	702	8,528
0W2	Grassland & Herbage Crops	201	203	192	129	1,589
0W31	Vegetable Crops	178	177	153	109	1,377
0W32	Ornamental Plants	83	96	84	59	715
0W33	Fruit Crops	111	107	88	102	952
0W4	Plantation Crops	117	51	47	52	524
0W7	Weed Biology	1,294	1,336	1,578	2,023	14,788
0W8	Herbicides	990	1,094	864	1,202	9,562
	TOTALS	4,015	4,255	4,026	4,437	38,578

are the *CAB Abstracts* equivalents of the *AGRICOLA* Table 4.2. One major *CAB Abstracts* subfile is Plant Pathology (code 0M), given in Table 4.6. Duplication of codes for different abstract use in the 0M file is about 10%. Vegetables (18.2%), cereals (17.3%), and fruits (7.6%) are the most heavily represented commodities and together comprise 43.1% of the OM total.

Typical tropical or subtropical crops such as cocoa, coffee, cotton, palms, rubber, sugarcane, and tea comprise only 4.9% of the file. *CAB Abstracts* certainly does a better job than *AGRICOLA* in covering crop production in the Third World, but clearly the emphasis in both is on crop pathology in the developed world. However, this is probably reflective of

Table 4.8. Citations in *Helminthological Abstracts, Series B; Plant Nematology*[a] of *CAB Abstracts*

Sequencing Code	Subject	1980	1983	1986	1989	1980–1989
0T52	Arthropod Hosts	99	61	98	127	1,046
0T53	Associations	68	44	77	101	712
0T55	Morphology & Taxonomy	225	216	187	193	1,992
0T59	Biology	415	335	466	418	4,037
0T60	Ecology	154	98	126	96	1,090
0T61	Distribution	7	3	33	42	221
0T62	Control	316	283	313	331	3,038
0T63	Techniques	65	42	58	71	572
0T64	Cytology & Evolution	48	38	49	89	612
	TOTALS	1,397	1,120	1,407	1,468	13,320

[a]Title changed to *Nematological Abstracts* in 1990.

Table 4.9. Citations in *Review of Agricultural Entomology* of *CAB Abstracts*

Sequencing Code	Subject	1980	1983	1986	1989	1980–1989
0E01	Insect Anatomy	167	218	267	322	2,419
0E02	Insect Reproduction	223	218	182	272	2,042
0E03	Insect Physiology	689	790	791	1,121	8,341
0E04	Insect Genetics	152	181	172	291	1,883
0E05	Insect Ecology	1,187	1,929	2,596	2,665	21,030
0E08	Agricultural Arthropods & Their Host Plants	4,700	6,237	6,604	6,447	60,203
0E10	Agricultural Diseases & Disorders	567	700	762	938	7,311
0E11	Biological Controls	1,948	2,557	2,881	2,737	25,190
	TOTALS	9,633	12,830	14,255	14,793	128,419

the percentages of literature published on temperate and tropical crops. Viruses (0M13), 16.6% of the OM file, are the most intensely written about pathogens. Because the breeding of crops to resist virus attack is the best method of preventing crop loss due to viral infestation, the 0M13 category overlaps heavily with the Plant Breeding file. File 0M accounts for 4.9% of the *CAB Abstracts* total for the decade 1980–1989.

After insects, weeds are the most bothersome and widespread problem in crop production. In Table 4.7 categories 0W5 (forest weeds) and 0W6 (aquatic weeds) have been removed, since these fall outside the scope of crop science in this volume. Weed Biology (0W7) contains 38.3% of the records in the 0W file. There is no analogous category in *AGRICOLA* covering weed biology, anatomy, physiology, and ecology. Of the total *CAB Abstracts* database, 2.9% of the records pertain to non-aquatic, non-forest weeds.

Nematodes are a worldwide crop pest, and account for 1% of the citations in *CAB Abstracts* for the years 1980 through 1989 (Table 4.8). Both *AGRICOLA* (code F822) and *CAB Abstracts* (subfile 0T) seem to aim for comparable helminthological subject coverage, including nematode biology; yet *CAB Abstracts* has more than twice the records and appears to do a far better job of indexing Third World conferences on this subject as well as non-English publications.

The 0E (Entomology) subfile in *CAB Abstracts* is enormous and accounts for 9.7% of the total CABI database (Table 4.9). Insect infestation of agricultural crops is clearly of great economic importance worldwide. It is not clear from the *CAB Abstracts Online Manual*[14] whether all records on crop insecticides are intended to be indexed in the 0E subfile. Clearly, many are not. Of all the "insecticide" citations in *CAB Abstracts* for 1987–1989,

14. CAB International, *CAB Abstracts Online Manual* (Wallingford, U.K.: CABI, 1989).

26.4% are not in the 0E subfile. Some records dealing with specific applications of pesticides on field and horticultural crops are found in those subfiles only. The remaining records deal heavily with human and animal insect pests and with toxicology. Table 4.9 gives the decade totals for agricultural entomology.

Comparisons of the 0E file with *AGRICOLA's* code F821 (Insect Pests) clearly show the numerical discrepancies between the two databases. The 0E file is almost three times larger than F821. This is partly explained by the fact that a great deal of invertebrate biology in *AGRICOLA* is coded under animal science, notably in codes L001 (Entomology-related), L200 through L700, and L100 (Entomology-related Animal Culture). As noted, however, many of these records deal with insect infestation of livestock, not of plants. There is no easy way to extract purely plant-related data from the L001 category in *AGRICOLA*. The Core Agricultural Literature Project analysis showed with *CAB Abstracts*, that almost 4% of the 0E file deals with forestry and should be discounted. Also of note is the fact that 0E11 (Biological Controls) is the second largest group in entomology, reflecting the growth of this emerging field. There is no corresponding code for biological controls in crop protection in *AGRICOLA*.

In all the agricultural disciplines analyzed by the Core Agricultural Literature Project, an effort was made to correlate citation data between *AGRICOLA* and *CAB Abstracts*. The foregoing discussion of the category codes for crop science should alert the reader that comparisons in this area are difficult. Tables 4.10 through 4.11 show combined codes for both databases in order to measure comparative subject sets. The methodology is imprecise because the match is uneven. Nevertheless, a comparison seems worthwhile.

Table 4.10 points up the problems of comparisons. The *AGRICOLA* cate-

Table 4.10. Comparative data on crop production and improvement in *CAB Abstracts* and *AGRICOLA*, 1980–1989

AGRICOLA			*CAB Abstracts*		
F110	Plant Production-Horticulture	36,623	101,374	0C	*Horticultural Abstracts*
F120	Plant Production-Field Crops	32,629	84,815	0Q	*Field Crop Abstracts*
PLANT PRODUCTION SUBTOTAL		69,252	186,189	—	—
F300	Plant Ecology	21,238	—	—	—
F400	Plant Structure	21,564	—	—	—
F500	Plant Nutrition	13,932	—	—	—
F600	Plant Physiology	116,437	—	—	—
—	Plant Taxonomy	39,654	—	—	—
	TOTAL	282,007	186,189	TOTAL	

Table 4.11. Comparative data on crop pathology and protection in *AGRICOLA* and *CAB Abstracts*, 1980–1989

	AGRICOLA		*CAB Abstracts*		
F822	Nematodes	6,376	13,320	0T	*Helminthological Abstracts*
F821, F851	Insects	48,719	128,419	0E	*R. Agricultural Entomology*
F830–F840	Plant Diseases	55,684	64,817	0M	*R. Plant Pathology*
F900	Weeds	18,341	38,578	0W	*Weed Abstracts*
F820	Animal Pests	2,546	—	—	—
F800	Plant Protection	23,745	—	—	—
F850	Product Protection	5,199	—	—	—
H000	Pesticides	2,758	—	—	—
	TOTAL	163,368	244,864	TOTAL	

gories F110 and F120 deal exclusively with crop production, whereas the *CAB Abstracts* files 0C and 0Q include citations on physiology, ecology, taxonomy and nutrition in addition to production. Both the subtotals and the totals in the two columns therefore represent non-similar counts. Otherwise, how could *AGRICOLA*, with its steady decline in indexed items for ten years, exceed *CAB Abstracts* by 100,000 citations? Yet if codes F300–F700 are not added to match meaningful subject parameters, *AGRICOLA* falls far behind. The crop production (codes F110–F120) citations in *AGRICOLA* are 24.5% of all the crop improvement total (all F codes).

The comparison of crop breeding is less problematic. The F200 category in *AGRICOLA* contains 72,986 records for the decade 1980–1989. In *CAB Abstracts*, the entire *Plant Breeding Abstracts* (code 0P) for the same period contains 89,942 records, a close match when *AGRICOLA's* declining counts are factored in. This is an immensely important field, and both databases cover it extensively.

Turning to crop protection and pathology, Table 4.11 compares the category codes F800–H000 in *AGRICOLA* with the subfiles 0M, 0T, 0W, and 0E in *CAB Abstracts*. As noted earlier, the last four *AGRICOLA* categories are concentrated in the four *CAB Abstracts* codes.

Over the past decade *CAB Abstracts* has indexed 11% more records in its entire database than *AGRICOLA*. If this average is applied to all categories, it still does not explain the overwhelmingly higher number of records in *CAB Abstracts* for crop pathology and protection. Plant diseases (F830–F840 or 0M) is the category that comes closest to the 11% figure (at 14%). But the difference in the number of records for entomology is 2.6 times greater in *CAB Abstracts* than in *AGRICOLA*, and similar figures hold for both nematodes and weeds.

The coverage of the biological and zoological aspects of pests is greater in *CAB Abstracts* than *AGRICOLA*. Furthermore, CABI indexing remains strong with Third World publications; it is also willing to index non-English publications, particularly if English abstracts are included with the article.[15] While imprecise, the comparisons given in Tables 4.10 and 4.11 serve to illustrate ranges of bibliographic activity.

Another view of the CABI indexing and publication effort is seen in their abstracting journals, where citations from the database are entered as deemed appropriate for the subject of any abstracting publication. Entries then may appear in two or more abstracting journals. The numbers and averages for seven dispersed years are shown in Table 4.12, which demonstrates subjects with growth or added coverage by CABI.

A comparative analysis of the *AGRIS* database was also made, since its coverage of Third World publications is more extensive than *AGRICOLA* and nearly the same as that of *CAB Abstracts*, especially with respect to report series and Third World journals. *AGRIS* and *AGRICOLA* reworked their category codes so they matched in the 1970s. In 1985–1986, these codes were readjusted and today are not in exact agreement. The *AGRIS* change resulted in half as many codes as *AGRICOLA*, thereby allowing less specificity when assigning them. In late 1993 CAB International announced that it would change its category and journal codes in 1994 to match those

Table 4.12. Citations in printed CABI abstracting journals in crop science

Year	*Field Crop Abstracts*	*Horticultural Abstracts*	*Plant Breeding Abstracts*	*Review of Plant Pathology*	*Nematological Abstracts*[a]	*Review of Agricultural Entomology*[b]	*Weed Abstracts*
1992	8,919	10,584	11,511	8,091	1,703	11,300	4,777
1991	9,127	11,686	11,932	8,336	1,761	12,048	4,549
1990	9,218	10,456	12,851	8,447	2,009	12,036	4,513
1989	10,104	10,511	10,938	6,059	1,732	10,129	4,089
1986	9,944	10,226	11,293	6,295	1,202	6,180	4,463
1983	11,389	9,152	9,842	5,066	1,699	8,208	3,646
1980	10,789	9,725	11,131	6,079	1,918	6,737	4,322
Total	69,490	72,340	79,498	48,373	12,024	66,638	30,359
Yearly average	9,927	10,334	11,357	6,910	1,718	9,520	4,337

[a]Title prior to 1990: *Helminthological Abstracts, Series B*. [b]Title prior to 1991: *Review of Applied Entomology, Series A*.

15. For a discussion of CABI non-English language indexing see: E. Stage, "ELFIS — An Agricultural Bibliographic Database of Forestry-Related Information," in Alois Kempf and R. Louis, eds. *Information Systems for Forestry-Related Subjects: Access, Search Techniques and User Needs* (Birmensdorf, GDR: International Union of Forestry Research Organizations, 1988), pp. 63–70.

of *AGRICOLA* more closely.[16] This was done by CABI and the 1995 online and compact disk versions reflect this realignment. Duplicate entries have also been removed in recent releases of tapes, online and on compact disks.

There are two version of the *AGRIS* database readily available in the United States, online and on CD-ROM. The Food and Agriculture Organization which manages *AGRIS* accepts indexing and cataloging input from centers around the world; the United States input is provided by the National Agricultural Library. Therefore, the U.S. publications indexed for *AGRICOLA* are the same as those sent to *AGRIS*. The National Agricultural Library and *AGRIS* have an agreement that the U.S. portion of *AGRIS* will not be placed online in the United States, because it is already online there in *AGRICOLA*. The *AGRIS* database online elsewhere in the world includes the U.S. citations. The CD-ROM version of *AGRIS*, which is purchased worldwide, also includes the U.S. input. Therefore, in the United States the online *AGRIS* does not include U.S. citations, but the CD-ROM version of *AGRIS* does. The CD-ROM version of *AGRIS* has approximately 25% more citations than the online version in the U.S. But the online version offers a unique window for analysis of Third World and non-U.S. literature in agri-

Table 4.13. AGRIS crop improvement and protection online in the United States but without U.S. imprints

Category Code	Subject	1980	1985	1989	1980–1989
F00	Plant Production	8,563	795	0	15,203
F01	Crop Husbandry	N/D	5,180	6,134	33,932
F02	Plant Propagation	N/D	754	1,320	5,903
F08	Cropping Patterns	N/D	889	1,133	6,169
F30	Plant Breeding	4,977	4,202	2,285	48,090
F40	Plant Ecology	1,253	886	485	10,507
F50	Plant Structure	1,114	729	392	8,110
F60	Plant Physiology	7,923	2,877	1,303	49,215
F61	Plant Nutrition	N/D	971	1,387	6,282
F62	Plant Growth	N/D	1,385	2,009	9,827
F63	Plant Reproduction	N/D	313	393	2,093
F70	Plant Taxonomy	668	401	377	7,204
H00	Plant & Production Protection	1,376	197	0	6,450
H01	Plant & General Aspects	N/D	352	288	2,696
H10	Pests of Plants	4,353	3,872	1,927	44,787
H20	Plant Diseases	4,752	4,007	2,237	46,883
H50	Misc. Plant Disorders	610	995	461	9,095
H60	Weeds	1,521	1,105	549	14,702
	TOTAL				327,148

16. *CAB International Database News*, no. 18 (Dec. 1993), p. 2.

culture. It is by no means a complete picture of world agriculture, since the omission of publications from the United States (the primary scientific publisher worldwide) slants the citation counts dramatically. This caveat should be kept in mind when examining Table 4.13, which gives the numbers for the online version only, which does not include U.S. imprints. Whereas *AGRICOLA* has twenty-six category codes, *AGRIS* now has fifteen, up from seven in the early 1980s. All of the plant protection citations have been moved from the F file into H01–H60. Table 4.13 gives the online version in the U.S. of the decade totals in *AGRIS*.

Five subjects dominate the counts: crop husbandry which is the same as *AGRICOLA*'s production (F01), plant breeding (F30), plant physiology (F60), plant pests (H10), and plant diseases (H20). These match similarly high counts in both *AGRICOLA* and *CAB Abstracts*. Crop improvement accounts for 24.1% of the *AGRIS* database and crop protection 16.3%, for a total of 40.4%.

C. Language Concentrations

The worldwide scope and subject inclusiveness of the agricultural bibliographic databases provide an excellent source for information on the languages most heavily used in crop science. The *AGRICOLA* file has concentrated on United States imprints since 1985, when it began systematically to reduce its non-English language coverage. Therefore, it is likely that data on language distribution taken from *CAB Abstracts* are more representative of worldwide publishing. In *CAB Abstracts* English has steadily increased over the past nine or so years from about 67.5% in 1984 to about 74% currently.[17] The *CAB Abstracts* searching manual reports 61% of all documents in the database as far back as 1970 are in English.[18] This shows a decided shift to English in recent years. In 1990, crop science literature written in English ranged from 71.2% in plant breeding to a high of 82.3% in nematology (Table 4.14 and Figure 4.1).

Table 4.14 provides details on seven subject areas for three recent time periods as represented in *CAB Abstracts*. The percentage of indexed materials in English increased in each of the seven subjects from 1980 to 1990. Fig. 4.1 illustrates this clear trend throughout world agricultural literature.

Within *CAB Abstracts*, Russian literature is second-ranked in six of the seven crop areas in 1990, the highest concentration being in plant breeding (13.0%). The pattern also reflects the British knowledge of the value of

17. Personal communication from a CABI reviewer, July 1992.
18. *CAB Abstracts Online Manual*, 1989, p. 8.

Table 4.14. Languages in *CAB Abstracts* by year and crop science subjects

Category subfile	English	German	French	Spanish	Russian
Entomology (0E)					
1980	72.7%	3.7%	5.1%	2.1%	6.9%
1984–1986	72.2	2.7	3.2	2.7	5.6
1990	78.9	3.4	2.6	1.9	3.5
Plant Pathology (0M)					
1980	71.3	4.7	3.7	1.3	6.2
1984–1986	70.7	4.3	3.0	2.2	4.2
1990	79.0	4.7	2.0	1.3	5.0
Nematology (0T)					
1980	68.9	2.1	2.3	3.1	11.4
1984–1986	71.8	4.1	3.5	2.0	9.6
1990	82.3	2.0	2.0	2.7	2.3
Weeds (0W)					
1980	74.9	4.4	3.0	1.5	4.8
1984–1986	72.9	3.8	3.3	3.1	4.4
1990	80.6	4.8	2.4	2.3	3.4
Horticulture (0C)					
1980	65.1	6.3	4.5	1.8	6.1
1984–1986	60.5	5.5	4.6	2.7	7.8
1990	74.5	3.2	3.4	1.3	3.6
Field Crops (0Q)					
1980	67.0	3.2	2.4	1.8	10.7
1984–1986	64.8	3.1	2.5	2.7	9.3
1990	74.0	2.0	1.5	1.3	4.5
Plant Breeding (0P)					
1980	55.6	3.9	2.0	1.6	23.8
1984–1986	59.8	2.4	1.8	2.2	17.4
1990	71.2	1.5	1.8	1.0	13.0

Russian plant breeding and its literature. Such trends, if carefully read, provide a map of research and literature concentrations.

The non-English languages shown in Table 4.14 are graphically represented in Figure 4.2 as a percentage of each of the seven crop subjects. The changes in the languages of crop literature in *CAB Abstracts* are clearly demonstrated. In the top six graphs in Figure 4.2 different non-English languages range from 1 to 12% of each subject file. The seventh graph, on plant breeding, is on a scale of 1 to 24%. The upswing of English since 1984–1986 has lowered the representation of other languages shown in Figure 4.2. Of these languages, only two show upswings in two or three subjects, none of them greater than 1% over the last six year period.

These language percentages for all of crop science have variations by commodities which do not influence the overall percentages. For example, the languages of the tropical commodities literature vary a great deal from

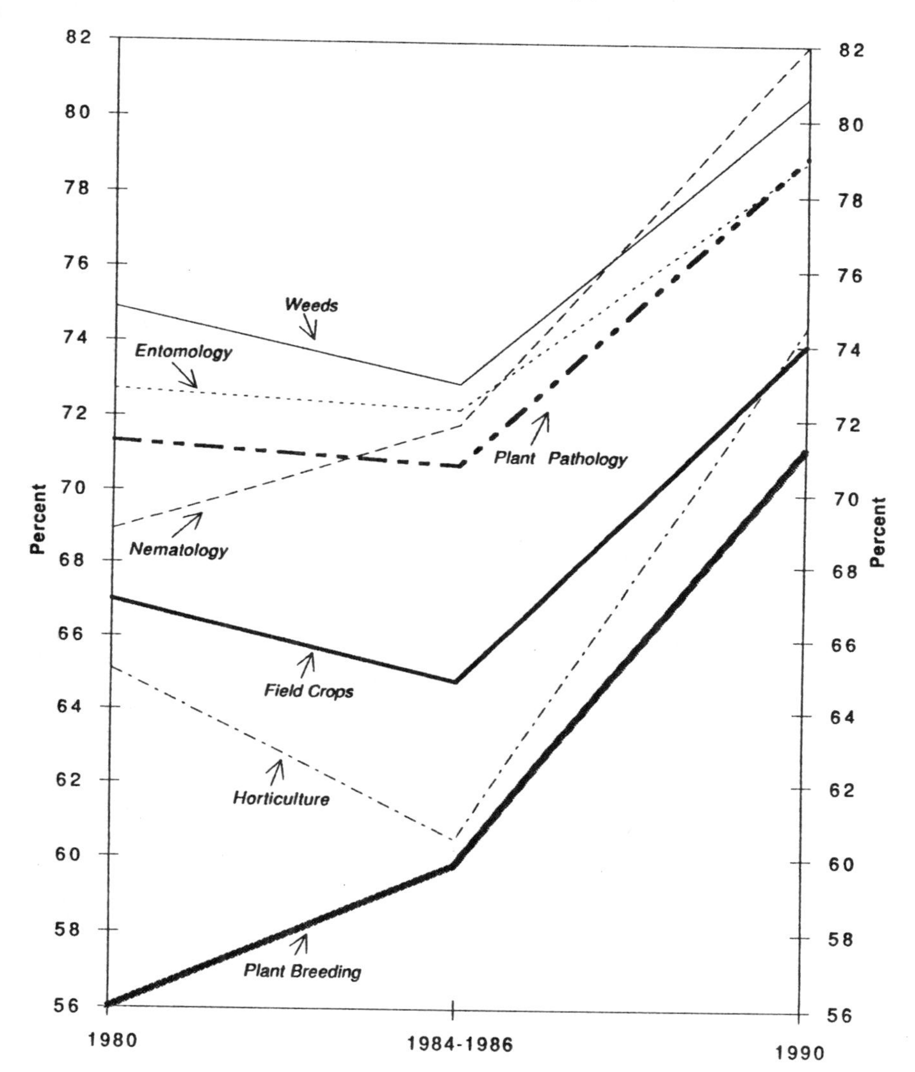

Figure 4.1. English-language percentage of *CAB Abstracts* on crop subjects. Subjects and category codes are the same as those of Tables 4.3–4.9 and 4.14.

these norms. Such crops as cassava have a higher percentage of literature in Spanish and Portuguese. Many years ago, the literature of tropical and subtropical agriculture in French was reported as being of second rank behind English.[19] This has changed in the past two decades. A major difference in

19. S. M. Lawani, "Periodical Literature of Tropical Agriculture," *UNESCO Bulletin for Libraries* 26 (2) (1972): 88–93.

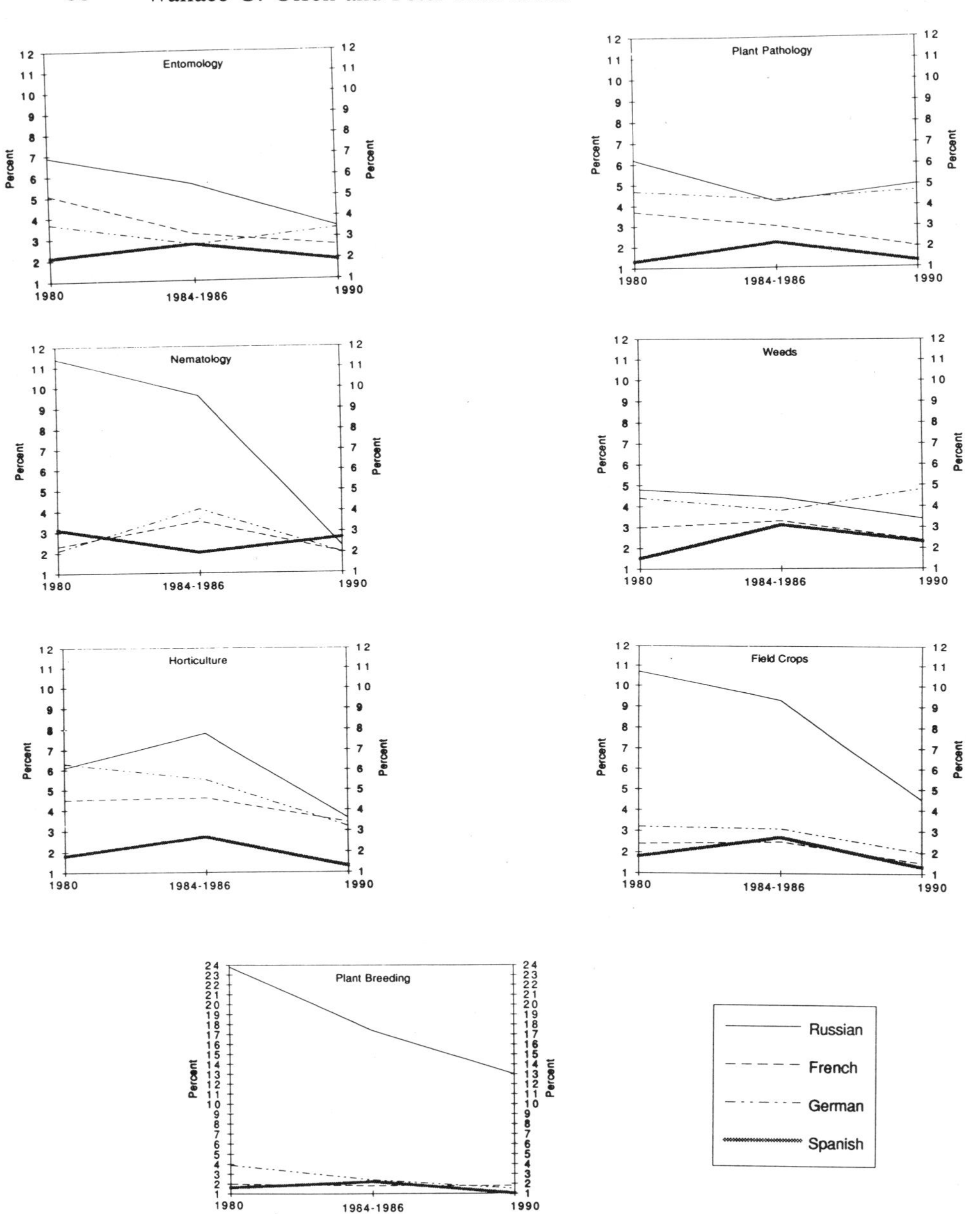

Figure 4.2. Non-English language percentages of *CAB Abstracts* within seven crop subjects for three recent time periods.

language percentages was reported by K. Morooka on rice for the publishing year 1983: this literature was "written in twenty-two of the world's major languages . . . with 65% in English and 16% in Japanese."[20] He further indicated that for the same year 21% of the rice literature was in Asian languages but that English publications were increasing at a greater rate. Ten years earlier, S. M. Lawani and T. A. B. Seriki reported that 41% of the world's literature concerning rice was in Japanese.[21]

D. Types of Publications

Both the *AGRICOLA* and *CAB Abstracts* databases are encoded with the type of document from which each bibliographic entry emanated, although this is not always precisely done. *CAB Abstracts* has four format codes: Numbered Parts (journal articles), Numbered Wholes (mostly reports in series), Unnumbered Parts, and Unnumbered Wholes. These designations changed in 1994. The last three categories match the definitions of monographs and reports, or parts thereof, as used in the studies in following chapters. Although the *AGRICOLA* file uses different descriptions, the divisions may be grouped similarly. Table 4.15 compares these two sources as well as the data from the citation analysis of the Core Agricultural Literature Project detailed in Chapter 8.

Greater format details on the subjects of crop science are provided in Table 4.16. The subject categories are the same as those used in section B above.

We acknowledge the extensive assistance of Peter Wightman, CAB International, for suggestions on the text of this chapter as well as for clarifying

Table 4.15. Formats of crop science literature

Source	Journals	Monographs and reports	Dissertations
CAB Abstracts[a]	86.4%	13.6%	
AGRICOLA[a]	90.1	9.9	
Citation analysis data (Chapter 8)	67.7	31.5	0.8%

[a]Averaged data from online files, 1980–1990.

20. K. Morooka, "Analysis of Bibliographic Data," p. 97.
21. S. M. Lawani and T. A. B. Seriki, "Some Characteristics of the World Literature on Rice," *International Rice Commission Newsletter* 23 (1) (1974): 1–15.

Table 4.16. Format of publication by categories

CAB Abstracts, 1980–1990			*AGRICOLA*, 1980–1990		
Sequencing (code)	Journals	Monographs and reports	*Category*	Journals	Monographs and reports
Entomology (OE)	89.4%	10.6%	Entomology	96.4%	3.6%
Pathology (OM)	88.3	11.7	Pathology	95.9	4.1
Nematology (OT)	82.0	18.0	Nematology	96.1	3.9
Weeds (OW)	80.8	19.2	Weeds	93.7	6.3
Horticulture (OC)	90.3	9.7[a]	Crop Improvement	93.4	6.6
Field Crops (OQ)	88.0	12.0			
Plant Breeding (OP)	86.1	13.9			

[a]This low figure is accounted for in part by the publication of much horticultural conference literature in *Acta Horticulturae*; the conference papers are cited here as journal articles, not reports.

comments on the numbers. Similar assistance with *AGRICOLA* was received from John Forbes and Caroline Early of the National Agricultural Library. The draft chapter was sent to Abraham Lebowitz of *AGRIS* for comment, but no reply was received.

5. Trends in the *PHYTOMED* Bibliographic Database

WOLFRUDOLF LAUX

Biologische Bundesanstalt für Land- und Forstwirtschaft (Federal Biological Research Center for Agriculture and Forestry)

In the second half of the nineteenth century plant protection developed into a science which not only described plant diseases, pests, and their damage symptoms but also searched for the reasons for their occurrence. The actual effects of damage to plants were investigated in order to create a basis for the purposeful control of diseases and pests.

A state-owned research center, the Biologische Abteilung am Kaiserlichen Gesundheitsamt (Biological Division at the Imperial Health Office), was founded in 1898. In 1905 it became independent as the Imperial Biological Research Center; later it became the Biological Research Center for Agriculture and Forestry of the Reich; and today it is the Federal Biological Research Center for Agriculture and Forestry (Biologische Bundesanstalt für Land- und Forstwirtschaft). It is entrusted with research in the field of phytomedicine, plant diseases, and plant protection, as well as with organizational functions such as testing of pesticides.

From its beginning in 1898, the research center was authorized to gather international literature and to carry out a bibliographic description.[1]

The Biological Research Center began its documentation work after World War I. Hermann Morstatt founded the *Bibliographie der Pflanzenschutzliteratur* in 1921, which summarized the international literature on plant protection from 1914 to 1919. Morstatt continued to produce this series in annual volumes until 1939. After World War II Bärner worked on annual volumes for 1940 to 1956 and 1958. At this point, it became clear that traditional methods in documentation as well as in plant protection and plant diseases could no longer be continued.

1. W. Laux, "Contributions to the Development of Phytomedicine in Berlin," *Englera* 7 (1987): 51–84.

A. New Challenges and Methods

In the early 1960s agricultural documentation in Germany developed a new organizational structure, a decentralized but centrally coordinated system of specialized documentation institutions. These documentation institutions agreed on uniform technology, standards for collecting literature data, and the definition and scope of special fields. The most important task of this organization, known today as Fachinformationssystem für Ernährung, Landwirtschaft und Forsten (Special Field Information System for Food, Agriculture and Forestry), is the provision of German agricultural scientific literature citations to the *AGRIS* bibliographic system of the FAO.[2]

For phytomedicine (plant protection and phytopathology) an automated database was deemed necessary to make detailed information rapidly accessible. In 1965 a reorganization of the documentation of plant protection and its indexing began with the application of modern filing methods and some automated equipment. More than 120,000 citations were placed on peek-a-boo cards, which allow a limited degree of information access. These advancements were new and very useful.[3] Computers were introduced in 1970 and the data recorded since 1965 on paper tapes were then input. This was the beginning of the database *PHYTOMED*, which reached 418,000 citations by the end of 1993.

The technical developments permitted easier setting and editing of the *Bibliography of Plant Protection*, which changed from annual volumes to a quarterly.[4] The old subject arrangement developed by Morstatt was extended to modern aspects of plant protection, e.g., biological control, toxicology of pesticides, etc. Coding was restructured using an eight-figure classification system. At present, the *Bibliographie der Pflanzenschutzliteratur = Bibliography of Plant Protection* has a title and a table of contents in four languages (German, English, French, Spanish), a citation portion divided into four main chapters, an index of authors, an index of descriptors, and an "English-German Reference List" for those descriptors which

2. W. Laux and E. Stage, "*AGRIS* Input from the Federal Republic of Germany: An Example of Cooperation Between Decentralized Agricultural Documentation Centers," *IAALD Quarterly Bulletin* 35 (1990): 53–60.

3. W. Laux and W. Sicker, "Kombinierte Anwendung von Sichtlochkarten und Kerblochkarten zur Erfassung Biologischer Objekte," *Nachrichten für Dokumentation* 18 (1967): 185–187.

4. W. Laux, "Die Bibliographie der Pflanzenschutzliteratur: Ein Traditionsreiches Werk, Erstellt mit Modernen Dokumentationsmethoden," in *Acquisition et Exploitation de l'Information dans le Domain de l'Agriculture et de l'Alimentation, Techniques Modernes et Cooperation Internationale = Acquisition and Exploitation of Information in the Field of Food and Agriculture, Modern Technics and International Cooperation*. [Title also in German and Spanish]. Proceedings of the 4th World Congress of Agricultural Librarians and Documentalists, Paris, 1970. (Paris: Institut National de la Recherche Agronomique, 1971), pp. 267–270.

are available only in the German language (in contrast to the internationally understandable scientific names of biological objects or chemicals). Each volume of the *Bibliography of Plant Protection* comprises four numbers. The last number contains the cumulated contents and indexes for the respective year of indexing. The total citation database was recently published in thirty-five volumes as the *International Bibliography of Plant Protection, 1965–1987*, with cumulated literature citations, contents, and indexes for twenty-two years.[5]

Before 1984 the *Bibliography of Plant Protection* entered titles in German, English, or French without translation because scientists in those days were sufficiently capable of using these languages. Since then all non-English citations have included a translated English title.

The classification system of the *Bibliography of Plant Protection* was transplanted into *PHYTOMED* and is used to print or recall chapters of the *Bibliography of Plant Protection*. Being classified in four languages facilitates international access. In 1994 volume 30 of the *Bibliography of Plant Protection* was published.

B. The *PHYTOMED* Database

PHYTOMED is the most important information product of the Federal Biological Research Center for Agriculture and Forestry. Since 1965 it has produced worldwide coverage on plant diseases and plant protection, phytopathology, phytomedicine, weed research, virology, entomology, stored product protection, nematology, and related subjects. It is accessible online at the host Deutsches Institut für Medizinische Dokumentation und -information (DIMDI) in Cologne where it is available for international use under the responsibility of Zentralstelle für Agrardokumentation und Information in Bonn.

Descriptors in the *PHYTOMED Thesaurus* in English and German provide subject access.[6] Full implementation of English descriptors in the database is in preparation.

The greatest number of descriptors are scientific names of biological objects (plants, pests, agents of diseases), chemical terms, and the names of pesticides. Use of these descriptors is possible without knowledge of the German language. Indexing facilitates special access to agents of diseases

5. W. Laux, ed., *International Bibliography of Plant Protection, 1965–1987* (Munich: Saur, 1989).

6. D. Blumenbach and W. Laux, *PHYTOMED Thesaurus* (Berlin-Dahlem: Biologische Bundesanstalt für Land- und Forstwirtschaft, 1986).

and pests as well as to all scientific names, both at the species or genus level and at the level of higher systematic units such as orders (for example, insects). It is possible to obtain all literature on the insect order "Orthoptera" as well as particular literature on the migratory locusts *Schistocerca gregaria* or *Locusta*. These can also be joined with terms for crops, for example "cereal," "vegetable," "fruit." A free-text search is possible in the database, too, because English translations of titles have been entered since 1977. For 1965 to 1977, more than 80% of the indexed publications have an English title.

A further access point in *PHYTOMED* searching is an eight-figure code in four languages. These classification codes can be combined with descriptors for a more refined search.

PHYTOMED is updated with more than 4,000 citations quarterly, but a shorter updating period is planned. All publications in the database are available in the libraries of the Federal Biological Research Center for Agriculture and Forestry.

PHYTOMED contains only a small number of abstracts, mostly for German publications which are also supplied to the *AGRIS* system. The combination of English titles, descriptors, and classification codes offers excellent searching possibilities (see Figure 5.1). Because of geographic proximity and linguistic knowledge of the staff, *PHYTOMED* has recorded literature from Eastern Europe to a very great extent. A number of further improvements have been made or are planned: duplicate-check methods for combining searches with other agricultural scientific databases (available since 1991); possible production of a CD-ROM edition of *PHYTOMED*; a shorter updating frequency to improve topicality; support of an online host to improve access to the database outside Europe; introduction of CAS numbers

6.00/000004 ZADI: -PHYTOMED /COPYRIGHT BBA
AU: Pflueger W; Grau R
TI: Abschaetzung und Beurteilung der Gefaehrdung landlebender Wirbeltiere durc Pflanzenschutzmittel
Assessment and evaluation of pesticide risks for terrestrial vertebrates
SO: Mitteilungen aus der Biologischen Bundesanstalt fuer Land- und Forstwirtschaft Berlin-Dahlem (Germany, F.R.); ISSN 0067-5849; (1990); (no.264) p. 32–41
(De)/3 ill.; 3 ref. Summaries (De, En)
CT: VERTEBRATE; PESTIZID; TOXIKOLOGIE; NEBENWIRKUNG; METHODIK; MAMMALIA; D
UT: Letaldosis

Figure 5.1. Citation printout from *PHYTOMED*.

for pesticides; and adoption of the projected "Unified Agricultural Thesaurus" to replace the *PHYTOMED Thesaurus*, provided that the years since 1965 remain accessible.

PHYTOMED's 418,000 literature citations (end of 1993) on plant diseases and plant protection, including extensive information on the biology of pests and agents of diseases, should be of economic importance for scientists, governments, and biological or agricultural libraries.

C. Changes in Contents in *PHYTOMED* throughout Twenty-Seven Years

PHYTOMED is based on the holdings of the libraries of the Federal Biological Research Center for Agriculture and Forestry which includes the central libraries in Berlin, Brunswick, and Kleinmachnow along with some of the holdings of specialized institutes. Such holdings, subject to little change, include the important international professional journals. Proceedings of special field congresses, the most important monographs, and relevant German dissertations are all obtained. *PHYTOMED* reflects continuity and change in phytomedicine and its literature.

Since the initial 6,000 citations in 1965, between 14,000 and 17,000 citations have been added to the database annually. Four time snapshots at six- to ten-year intervals were taken for analysis: 1965 and 1966 = 17,000 citations; 1975 = 17,000; 1985 = 16,000; 1991 = 16,000. The results are calculated on the basis of 16,000 citations per time period.

Considering first the important crop plants *Triticum*, *Oryza*, and *Zea* (Table 5.1) it turns out that their frequency in phytomedicine literature has remained nearly the same or increased. The strong show of citations on wheat (*Triticum*) is surprising. The same pattern shows up in sorghum, a crop of growing interest and a limited growing range but of special importance to the Third World.

Figures concerning the potato (*Solanum*) indicate not an increase but a

Table 5.1. Citation changes for five crops

Subject	1965/66	1975	1985	1991
Triticum sp.	337	590	767	1,034
Oryza sativa	243	302	505	433
Zea mays	384	475	576	601
Sorghum bicolor	89	95	177	199
Solanum tuberosum	477	657	578	575

stagnation. This might be due to reduced consumption and research activities in Middle Europe.

Viruses and mycoplasmae are agents of diseases that have provoked the interest of phytopathologists since their discovery. Table 5.2 shows the continuing interest of scientists over the years. The slight augmentation of citations in the nineties is probably due to the intensified activities on the decoding of nucleic acid sequences in viruses as well as biotechnological issues. Citations concerning viruses in leguminoses have decreased; here the most important research lies before the 1965 input period. Viruses on fruit have also declined in importance. With regard to mycoplasmae an increase in the seventies is revealed. Once they were identified as agents of plant diseases, mycoplasmae were a preferred subject of research for some years until it was determined that they were of limited significance as agents of plant diseases in the temperate zones, despite their great importance for tropical crop plants.

The use of pesticides is increasingly rejected particularly in industrial countries. Biological and integrated control of plant pests and diseases is increasing worldwide. Table 5.3 indicates a considerable increase of citations in this area in the past few years. It must be noted that publications dealing only with occurrence of *Hymenoptera* as parasites of *Lepidoptera* without mention of biological control are not counted here. An especially important beneficial insect, the oophagous parasite *Trichogramma*, still attracts much interest even though its large-scale cultivation and application have reached a plateau. The growing number of publications dealing with biological weed control is good evidence of the effort to find alternatives to herbicide use. This trend is witnessed as well in citations on ecology and environment, and indicates the same public interest in these questions.

Data on DDT and Atrazine, two pesticides once heavily used and now in disfavor, show contrasting trends in the literature (Table 5.4). Prohibition of DDT in many countries has reduced the number of publications, at least in plant protection. More recent publications mainly deal with questions of residues of this strongly persistent insecticide. As for Atrazine, because of its massive application in maize cultivation and its unexpected contamination of groundwater, this herbicide has a growing number of citations.

Table 5.5 indicates the growing scientific literature on important agents of widespread diseases in the field of bacteria and fungi. *Erwinia*, the agent of fire blight on fruit, shows an increase undoubtedly resulting from spread of the disease in Europe as well as the numerous efforts to contain it.

Literature trends regarding animal pests are indicated in Table 5.6. *Lepidoptera/Heliothis* and *Spodoptera* show a stagnation or slight increase in the number of citations, as do the nematodes. The opposite is true for *Laspeyresia/Carpocapsa*, which decreased exceptionally in 1991 for reasons

Table 5.2. Viruses and mycoplasma citation changes

Subject	1965/66	1975	1985	1991
virus	1,063	1,124	1,116	1,228
virus + legumes	96	91	89	57
virus + fruits	141	133	107	98
mycoplasma	80	175	96	68

Table 5.3. Citation changes on control methods

Subject	1965/66	1975	1985	1991
Biological control	505	368	379	884
Trichogramma	28	44	37	83
Biological weed control	42	58	47	12
Ecology	509	426	533	785

Table 5.4. Citation changes with DDT and Atrazine

Subject	1965/66	1975	1985	1991
DDT	129	200	147	91
Atrazine	47	54	79	106

Table 5.5. Citation changes with select bacteria and fungi

Subject	1965/66	1975	1985	1991
Erwinia spp.	41	118	137	205
Fusarium spp.	198	388	472	488
Phytophthora spp.	213	291	324	341

Table 5.6. Citation changes with select animal pests

Subject	1965/66	1975	1985	1991
Heliothis spp.	52	91	136	205
Spodoptera spp.	19	88	137	131
Carpocapsa spp.	58	96	158	31
Orthoptera	157	211	223	203
Nematodes	635	490	853	801
Aphis spp.	28	85	85	130

still unknown. The number of citations to *Orthoptera* is stagnant, although it is still counted as the main pest worldwide.

The examples used in Tables 5.1 to 5.6 are based on larger subject groups which were searched by one descriptor or one classification code, or by a simple combination of descriptors. The examination of more restricted or specific facts or circumstances would be even more interesting. Such a study is beyond the scope of this publication, however. A "living" database adapts to changes of science and terminology, which means limited correlations on subjects when comparing over extended periods.

The *PHYTOMED* retrieval base of 400,500 citations was searched from 1965 to 1992 on select major subjects (Table 5.7). The literature represented in this search included some basic concepts in plant pathology. The last six listings in Table 5.7 demonstrate the importance of topics that pertain specifically to the United States. In the case of two subjects, 32 to 62% is literature related to the United States.

Table 5.7. Cumulative citations on select subjects (1965 to 1992)

Subject	Number
Herbicides	23,412
Insecticides	22,289
Viruses (except TMV and Cucumber Mosaic Virus)	30,101
Tobacco Mosaic Virus (TMV)	2,307
Biological control	10,953
Human health with regard to plant protection	2,393
Biological weed control	1,354
Glycine	5,858
Glycine (U.S.A. only)	3,624
Arachis	2,487
Arachis (U.S.A. only)	933
Conifers	10,406
Conifers (U.S.A. only)	3,352

D. Language in *PHYTOMED*

Language is an important criterion in scientific publications because language barriers constrain the international flow of information.[7] The full *PHYTOMED* database has thirty-five languages represented. Twenty-eight languages were represented in 1965/66 and 1975, twenty-seven in 1985, and twenty-six in 1991 (Figure 5.2). It is not known whether the slight

7. W. Laux, "Zur Sprachenverteilung Phytomedizinischer Literatur," *Nachrichtenblatt des Deutschen Pflanzenschutzdienstes* 38 (1986): 75–79.

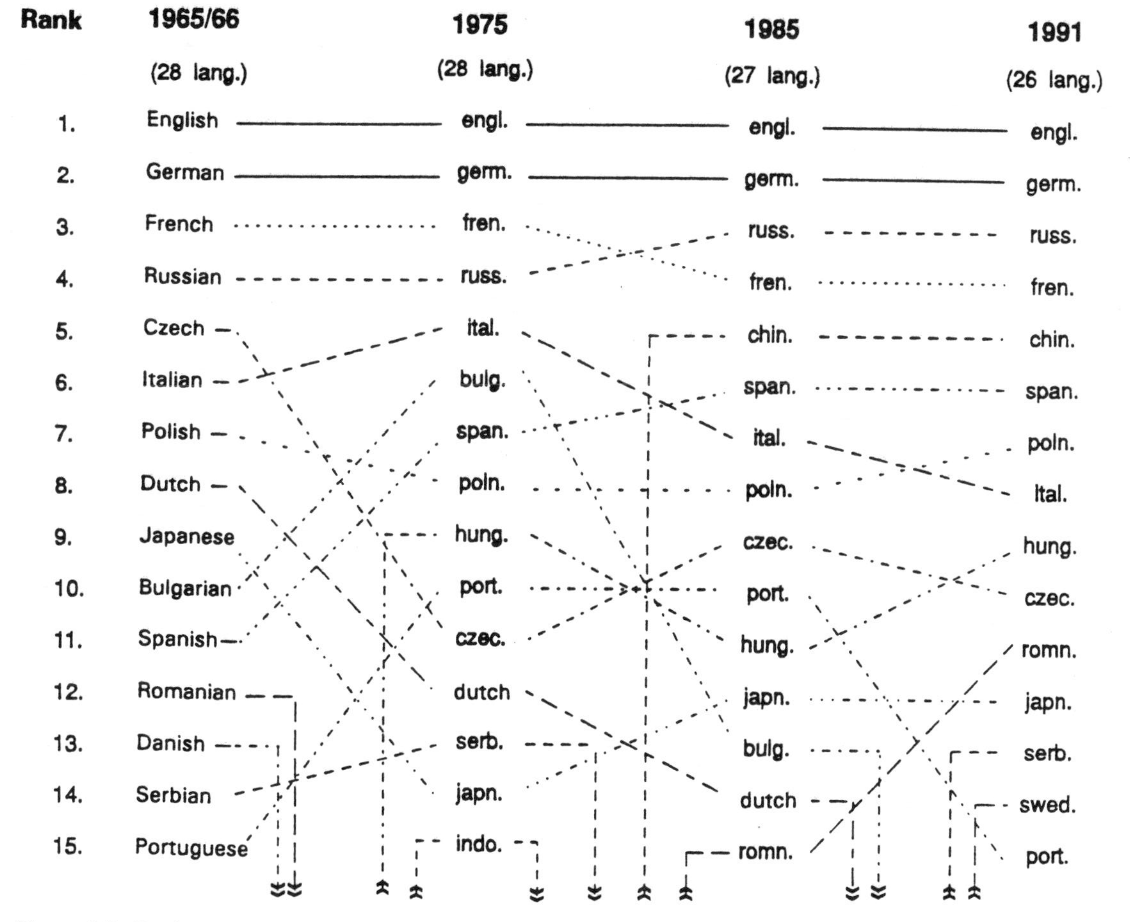

Figure 5.2. Rank changes in languages in *Phytomed*.

decrease to twenty-six represents a trend. Figure 5.3 also shows the fifteen most important languages during the four periods. Latin or Cyrillic alphabets or other languages with translated titles are included in the database. Chinese or Arabic publications without summaries or translations of their titles into other languages are not entered.

The English language is the most frequently used, as with all scientific literature in the past fifty years. The great number of German citations certainly represents the documentation center's bias toward local literature. Much of the German literature cited in *PHYTOMED* is not available in other agricultural databases. The number of citations in East-European languages is relatively high. An intensive literature exchange with these states and a staff well versed in East-European languages have made possible extensive coverage and indexing for *PHYTOMED*. The high number of Bulgarian, Romanian, Czechoslovakian, Polish, and Hungarian citations represented in the database reflects the agricultural scientific research and practice in these countries. Much of this literature is also missing in other major agricultural bibliographic databases.

Over time, the English and German languages have remained relatively stable, as have French and Russian. This is also true for Spanish, which advanced from eleventh to sixth place. The occurrence of citations in Chinese is interesting and logical. The Chinese language's sixteenth place in 1965/66 and 1975 corresponds to the political isolation of the country during and after the Cultural Revolution. Then the language advanced to a secure fifth place. If the data also included Chinese literature not accessible by English translations of titles or English abstracts, Chinese would move to an even higher rank.

Low priority or small numbers in the less common languages do not represent the quantity of publications in these countries. A typical example is the low showing of Dutch and Japanese citations. The scientific publications of Japan and the Netherlands increasingly appear in English, a trend also true for Danish, Swedish, and Finnish, and which will probably also continue in the future with the East European languages. The continuing decrease of the German language probably demonstrates the same move to English (see Table 5.8).

Figure 5.3 presents the same data as Figure 5.2 but with selected absolute numbers calculated on 16,000 titles in each of four groupings. These figures demonstrate the great distance between the first three to five places and languages of low priority. The line for 85% of all citations includes English, German, and Russian; the line at 93.75% includes seven languages in 1991. The fact that in 1975 ten languages lay within the 93.75% range

	Rank	1965/66 (28 lang.)		1975 (28 lang.)		1985 (27 lang.)		1991 (26 lang.)	
	1.	English	9603	engl.	9344	engl.	10177	engl.	10737
	2.	German	2681	germ.	2653	germ.	2367	germ.	2204
	3.	French	709	fren.	875	russ.		russ.	
85 %	4.	Russian		russ.		fren.	676	fren.	469
	5.	Czech		ital.	256	chin.		chin.	
	6.	Italian	313	bulg.		span.		span.	
	7.	Polish		span.		ital.	257	poln.	
	8.	Dutch		poln.		poln.		ital.	214
	9.	Japanese		hung.		czec.		hung.	
93,75 %	10.	Bulgarian	261	port.	172	port.	200	czec.	107
	11.	Spanish		czec.		hung.		romn.	
	12.	Romanian		duth.		japn.		japn.	
	13.	Danish		serb.		bulg.		serb.	
	14.	Serbian		japn.		duth.		swed.	
	15.	Portguese	83	indo.	92	romn.	57	port.	50

Figure 5.3. Growth and rank shifts of language concentrations.

Table 5.8. Citation trends for English and German

Citations	1965/66	1975	1981	1991
English	9,603	9,344	10,177	10,737
German	2,661	2,653	2,367	2,204

underlines the tendency to publish primary scientific phytomedicine literature in an increasingly limited number of languages.

This trend toward fewer languages is a positive one for international research and science, since it means that the language barrier will become increasingly less influential. We must not forget that literature relating to practice and local use in a country follow the language of that country. This is particularly true in an applications field such as phytomedicine, where research results must be placed in the local language for effective use. It is important to note that much agricultural science research deals with topics that are important on a supraregional basis and are reported in a language with wide readership. In contrast, local, regional, or developing countries' studies often do not have wide application and do not need to be published in any language except that of the country.

6. Publishing Influence of Crop Improvement and Protection Societies

C. LEE CAMPBELL AND PAUL D. PETERSON, JR.
North Carolina State University

RAYMOND J. TARLETON
The American Phytopathological Society

A. Communication in Science

For scientific knowledge to be of value the research results in the laboratory, the work in the field, and the creative process of the scientist must be communicated. Seven different functions of scientific communication have been identified: "1. providing answers to specific questions, 2. helping researchers stay abreast of new developments in a chosen field, 3. helping researchers acquire an understanding of a new field of research, 4. identifying the major trends in a selected field and the relative importance of that field to the broader discipline out of which it comes, 5. verifying the reliability of information by additional evidence, 6. redirecting or broadening a researcher's span of interest and attention, 7. obtaining critical response to a researcher's own work."[1] Among the most compelling roles of scientific communication is that of a record of science. This record serves as a yardstick to measure current research and as a guidepost to extend knowledge. In many ways, communication is the glue that binds together the cumulative archive of learning and permits the advancement of science.

Two major forms of scientific communication have developed. Some mode of verbal communication has always been linked to scientific enterprise. Today, verbal communication in various applications has become "an increasingly strong competitor" for the attention of the busy scientist.[2] The seminar, conference, symposia, telephone, FAX, and electronic mail transmit information rapidly. A good speaker also can convey the zeal and pas-

1. R. D. Walker and C.D. Hurt, *Scientific and Technical Literature: An Introduction to Forms of Communication* (Chicago: American Library Association, 1990), pp. xiii–xiv.
2. P. H. Abelson, "Scientific Communication," *Science* 209 (1980): 61.

sion of scientific investigations. Yet there are weaknesses. Verbal communication excludes some scientists and "lacks the permanence of the written word." Scientists need a better method to communicate current information and to record scientific and technical developments. As a result, "scientific literature has and will have a continuing important role" in the interchange of scientific ideas and results.[3]

Since the seventeenth century, scientific literature has played an increasingly important role in the exchange and recording of scientific information. The groundsel of this literature has been the journal and its scientific paper. Many scientists consider the journal the vehicle of choice for communication. Walker and Hurt found that journals have developed four key roles: (1) to communicate the results of original research and ideas, (2) to function as a repository or guardian of accumulated knowledge; (3) to assign priority to scientific discovery, (4) to preserve scholarly integrity through a system of peer review and editing.[4] The major function of the journal has been and remains the communication of original research. But as the above listing suggests, journals have assumed more responsibility. Indeed, they now include excerpts and abstracts of presentations and demonstrations, advertisements of products, book reviews, and society and organization news.

B. Role of Scientific Societies in Communication

The origins of modern scientific communication are linked permanently with the rise of scientific societies in the seventeenth century. Not only bolstering experimental science by the creation of laboratories and observatories, learned societies, like London's Royal Society created in 1662, "undertook to publish, periodically, news of the work done under their auspices, and often the work of other learned men, in order to make it known as quickly and as widely as possible."[5] Before societies, men of science communicated their ideas by an assortment of calendars, almanacs, newspapers, book catalogs, books published as serials, and personal correspondence. Societies, however, had been created with the goal of coordinating scientific information and reaching a wider audience. As a result of societies, access postal systems, and improved transportation, the scientific journal, much as we recognize it, came into existence.[6]

3. Ibid.
4. Walker and Hurt, *Scientific and Technical Literature*
5. M. O. Bronfenbrenner, *The Role of Scientific Societies in the Seventeenth Century* (Chicago: University of Chicago Press, 1928), p. 198. (Reprint ed. New York: Arno Press, 1975).
6. (a) Ibid. (b) D. Kronick, *A History of Scientific and Technical Periodicals: The Origins and Development of the Scientific and Technological Press, 1665–1790* (New York: Scarecrow Press, 1962). (c) A. J. Meadows, *Communication in Science* (London: Butterworths, 1974).

These learned societies were broad in mission, encompassing practically the entire known field of learning. To study science meant investigating all the natural world. Before long, a recognition of common purposes led groups to organize along more specialized lines. In agriculture, a general movement began in the late eighteenth and early nineteenth centuries that promoted the creation of state and regional agricultural societies. As members of these societies, gentlemen agriculturists and amateur scientists published their own society journals championing the cause of agricultural improvement.[7]

Agricultural societies were a reasonable stage in the growth and development of agricultural science. During the middle decades of the nineteenth century, however, encouraged by the promises of chemistry, economic entomology, and botany, the societies were soon replaced by colleges and universities, experiment stations, and government bureaus. These institutions became the new homes for the teaching, research, and communication of agricultural science. The professional scientist also replaced the amateur.

This activity was most visible in the United States where a "significant linkage of agriculture, science, and government" in the closing decades of the nineteenth century brought remarkable expansion and organization to the field.[8] Reflective of the growing specialization in other professions, between 1880 and 1920 approximately eleven professional societies in different disciplines were established in the United States to encourage agricultural science. The one driving force behind the creation of professional societies was communication. Indeed, "the chief synthesizing force for a new organization was usually its journal. . . . It took on the role of a symbol or banner for the field, serving to identify it, make it visible to others, and define its subject matter and methodology."[9] In addition to university training and full-time employment, societies also strengthened a scientist's professional standing.

C. Early Publishers of Crop Improvement and Protection Journals

Agricultural societies gradually replaced the sundry forms of publishing agricultural information in the late eighteenth century and became the cen-

7. (a) J. E. McClellan, *Science Reorganized: Scientific Societies in the Eighteenth Century* (New York: Columbia University Press, 1985). (b) M. W. Rossiter, "The Organization of Agricultural Improvement in the United States, 1785–1865," in *The Pursuit of Knowledge in the Early American Republic: American Scientific and Learned Societies from Colonial Times to the Civil War*, edited by A. Oleson and S. Brown (Baltimore: The Johns Hopkins University Press, 1976).

8. M. W. Rossiter, "The Organization of the Agricultural Sciences," in *The Organization of Knowledge in Modern America, 1860–1920*, edited by A. Oleson and J. Voss (Baltimore: The Johns Hopkins University Press, 1979), p. 212.

9. Ibid., p. 219.

ters of publication until the middle of the next century. They aided the diffusion of agricultural knowledge through the publication of society journals in the forms of proceedings, transactions, and memoirs. Society members often published separate treatises on specific subjects of interest. The societies also played an important role in furnishing information to privately owned agricultural periodicals when these appeared in substantial numbers during the first half of the nineteenth century.[10]

With the rise of agricultural science, professional scientists, and government sponsored teaching and research in the second half of the nineteenth century, government reports and bulletins joined the list of agricultural publications. In the United States, the reports and bulletins of the U.S. Department of Agriculture and the state agricultural experiment stations became important sources of information for agricultural scientists. Before long, the new scientist in this rapidly expanding and specializing field found these publications of limited use because they did not include adequate detail. The need for better avenues of communication was directly related to the enormous growth of specialized societies and their professional journals between 1880 and 1920.[11]

D. Societies for Crop Improvement and Protection and Their Publications

The publications of a scientific society, in essence, *are* the society.[12] They are the primary tangible evidence of the society's existence, its vigor, and vitality. Other activities certainly occur, such as annual meetings, committee activities, and cooperative efforts with other organizations. The publications of a scientific society, in most instances, set the standard by which the society and its members are evaluated and determine the degree of prestige the society has in the scientific world.[13]

The improvement and protection of crops has long been the goal of many learned and professional societies. Certainly this was a primary goal of most of the general agricultural societies that developed in the United

10. (a) D. B. Marti, "Agricultural Journalism and the Diffusion of Knowledge: The First Half-Century in America," *Agricultural History* 54 (1980): 28–37. (b) Rossiter, "Organization of Agricultural Improvement."

11. Rossiter, "Organization of the Agricultural Sciences."

12. (a) G. M. Browning, "The Role of Professional Societies in Resource Development," *Journal of Soil and Water Conservation* 17 (1962): 59–61. (b) J. F. Lutz, "History of the Soil Science Society of America," *Soil Science Society of America Journal* 41 (1977): 152–173.

13. H. H. Laude, et al., "History of the American Society of Agronomy, 1907–1957," *Agronomy Journal* 54 (1962): 57–69.

States. As agricultural sciences became more the province of professionals than amateurs in the late nineteenth and early twentieth centuries, societies became more specialized and agricultural scientists narrowed the focus of their endeavors.

The disciplines of agronomy, entomology, and plant pathology emerged as distinct sciences during the era of specialization between 1860 and 1920.[14] Crop science and soil science were recognized as separate sections of agronomy as early as 1924, both products of further maturation of the agricultural sciences.[15] Weed science had its genesis in the 1940s after the discovery and use of (2,4-dichlorophenoxy) acetic acid (2,4-D).[16] Entomology and plant pathology had an earlier start but also experienced rapid growth during the initial stages of the "chemical era" of pest control during the 1950s and 1960s. Nematology has developed as a separate academic department at several universities in the United States, but it largely remains associated with departments of plant pathology or entomology and is often seen as a component of these disciplines.

To illustrate the publishing influence, we will emphasize primarily societies of crop science, agronomy, horticulture, entomology, nematology, plant pathology, and weed science. The societies associated with these disciplines provide excellent examples of a range of publishing activities. Our emphasis also is primarily on United States societies for crop improvement and protection, because they have, in general, been far more active in the publishing arena than their counterparts in other countries.

We examine a case study of the development and specialization of society journals and publications based upon the activities of the American Phytopathological Society (APS). The APS has made a very successful entry into the publishing arena and has more extensive activities in this area than many professional societies. The activities of the APS Press demonstrate the spectrum of influence a society for crop protection can have within a discipline. We then briefly examine the origins and publishing activities of the American Society of Agronomy, the American Society for Horticultural Science, the Entomological Society of America, the Society of Nematologists, and the Weed Science Society of America.

14. Rossiter, "The Organization of the Agricultural Sciences."

15. W. E. Larson, "Soil Science Societies and Their Publishing Influence," in *The Literature of Soil Science*, edited by Peter McDonald (Ithaca, N.Y.: Cornell University Press, 1994), pp. 123–142.

16. F. L. Timmons, "A History of Weed Control in the United Sates and Canada," *Weed Science* 18 (1969): 294–307.

E. Development and Specialization of Society Journals: A Case Study of the American Phytopathological Society

The American Phytopathological Society (APS) was founded in 1908. However, it was not until 1911 that the membership was sufficiently large to justify publishing a journal. By the end of 1991, APS had 4,500 members in eighty-four countries and had established an active publishing program with an international reputation. The APS is one of the largest nonprofit publishers in the field of plant pathology/plant protection. It publishes three research journals; a monthly newsletter for its members; a series of monographs on specific plant diseases; a series of plant pathology classics covering translations of published works in non-English journals which have proven to be key research contributions; a series of disease compendia covering all major commodity crops, such as wheat, corn, barley, rice, and soybean, ornamentals, turf, and significant vegetable crops; a symposia series; an annual *Biological and Cultural Tests* series; and a collection of color slide sets designed for teaching purposes. The Society publishes an average of ten books per year and has an inventory of over 100 titles in addition to its journal publications.

Prior to 1973, all APS titles were typeset and printed by commercial printing companies. However, as commercial typesetting costs continued to increase, the Society decided to embark on a new approach to maintain quality in typesetting and graphics, but with lower costs. In mid-1972, the Society launched an experimental in-house typesetting project and began producing its newsletter internally except for the actual printing. The success of this project encouraged the Society to broaden the scope of publications produced internally, and, by late 1973, all journals and book titles published by the APS were edited, typeset, and prepared for commercial printing at the Society's headquarters.

To augment this production program, a marketing department was established to promote the sale of APS publications. The governing body of the Society (the Council) also established an oversight committee to supervise all Society journals and books. The Publications Coordinating Committee supervised existing publications and their individual editorial committees, and launched new publishing ventures, the first of which was the *Compendia* series.

In early 1973, faculty members at the University of Illinois completed a manuscript on the diseases of corn, a project sponsored by the U.S. Department of Agriculture. The Society was offered the opportunity to publish this manuscript and to hold the copyright. In late summer of 1973, the APS published its first compendium, *The Compendium of Corn Diseases*. This particular title has been revised twice and has sold more than 40,000 copies since its original issue.

The APS Council determined that its publishing efforts were making significant contributions to its members and to the science of plant pathology. As a non-profit organization, the Society could price its journals and books somewhat lower than commercial firms and, thus, make its publications available to an audience with limited financial resources, especially in the Third World. It also could publish significant titles of limited scope with very limited sales appeal, since it did not have to generate a surplus on every title published.

As publishing activities increased, the APS Council reviewed the procedures and decided to streamline the book publishing and to provide additional oversight on financing. Thus, the APS Council established APS Press and appointed an editor-in-chief and a group of senior editors in 1983. The charge from Council was to expand the book publishing activities of the Society while maintaining close surveillance over the finances and the quality of the editorial content.

The Society's Major Publications

The first journal published by the Society is *Phytopathology*, which appeared in February 1911, and six issues were published by the end of that year; volume one consisted of 204 pages plus index. Volumes grew in size over the years, and in 1987 the journal peaked at 1,687 pages plus index. More specialized journals of the Society grew in size as *Phytopathology* stabilized in content. At its largest circulation, *Phytopathology* was received by over 2,000 libraries in addition to its international membership. Since 1979, *Phytopathology* has carried APS division and annual meeting abstracts and the annual report.

Another important APS publication is *Plant Disease*. In 1917 the U.S. Department of Agriculture began publishing a monthly, *Plant Disease Reporter*, which was distributed to a wide audience of individual scientists and to libraries, mostly free of charge. In the late 1970s, during government publishing budget cuts, the U.S. Department of Agriculture began conversations with the American Phytopathological Society which resulted in a takeover of *PDR* by the Society in 1979. As a result of grants from the U.S. Department of Agriculture, the Society was able to expand the scope of *PDR*, add new features, and publish many of the photographs in color. Color reproductions are especially appreciated by readers since color aids detection of disease symptoms. In addition to the physical upgrading, a peer review system was established to ensure a high quality of published articles.

The decision also was made to continue the volume sequence of *PDR* to connect it with its new name, *Plant Disease*. The first issue of *Plant Disease* was published in late 1978 and dated January 1979, volume sixty-five. *Plant Disease* published its largest volume in 1987 at approximately 1,200 pages.

With the first journal in 1911 the society had a convenient vehicle to communicate with its members. Thus, *Phytopathology* carried a page or two of news of the membership and the organization in each issue. This means of communicating with the membership was used until the end of 1966, when the Council approved a separate vehicle to communicate with the members. In January 1967, *Phytopathology News* was issued for the first time and continues on a monthly basis. *PN* varies from twelve to twenty-four pages. Over the years, it has carried information and abstracts for the annual meeting, and from 1970 to 1973 it was the vehicle chosen by Council to carry the Society's annual report.

Until 1970, concerns of the divisions and annual meeting abstracts appeared in *Phytopathology*. In an attempt to control costs and to limit *Phytopathology* to publishing original research, Council decided to publish the annual report, division abstracts, and annual meeting abstracts in *Phytopathology News*. This decision was reviewed and revised in early 1974, when a new publication was established, *Proceedings of the American Phytopathological Society*. The *Proceedings* contained the annual reports, annual meeting symposia, division and annual meeting abstracts, and information on the awards programs. *Proceedings* proved to be unpopular with the membership, and at the 1978 annual meeting the membership asked that *Proceedings* be discontinued and that the abstracts and annual report be returned to *Phytopathology*. Since mid-1979, *Phytopathology* has carried all division and annual meeting abstracts and reports.

Fungicide and Nematicide Tests has been published annually by an official committee of the Society since 1953. While not an official publication under the authority of the Council, *F & N Tests* publishes field tests of new chemical treatments, and is sold on a break-even basis by the committee.

Molecular Plant-Microbe Interactions began in late 1987 with volume one, dated January 1988. The molecular approach to plant disease control had become an increasingly important area of interest in the 1980s. Younger members of plant pathology departments frequently had training in biochemistry and molecular biology rather than in the traditional areas of plant pathology. Key members of APS began a dialogue which culminated in a new journal published by APS Press and under its direction and control. Editorial responsibilities were shared with scientists attempting to organize a new society in this area.

A key to *MPMI*'s success was rapid publication (three to four months from submission of manuscript) and the absence of page charges. Using the latest technologies of electronic manuscript submission and fax transmission of reviews and correspondence, the journal was able to maintain its publication goal. In 1990, a new scientific society, the International Society of Molecular Plant-Microbe Interactions, was formally organized in Interlaken, Switzerland, and adopted *MPMI* as its official journal. Active members of that group share editorial responsibilities with the APS.

After the APS established its research journal in 1911, the Society launched a unique editorial project—a series of translations from classic phytopathological literature. The first translation was published in 1926, and fifteen titles have been published through 1990. A list of these historically important titles, the *Phytopathological Classics*, is provided in an appendix.

A monograph series was started in 1961 with the publication of Bruehl's *Barley Yellow Dwarf*. This was followed by eight other titles, the last being published in 1986. The original goal was to publish a complete history of the major crop diseases primarily intended for the research scientist. While not discontinued, the series has been largely replaced by the disease compendium series.

Biological and Cultural Tests for Control of Plant Diseases was first issued in 1986 as an annual series designed to improve biological and cultural testing and reporting. Each volume contains reports based on replicated disease control tests involving nonchemical means. Eight volumes have been published through 1992.

A disease compendia series was launched in 1973 with corn as the first crop to be reported. At the end of 1994, the following twenty-eight titles had been published: *Alfalfa*, *Apple and Pear*, *Barley*, *Bean*, *Beet*, *Citrus*, *Corn*, *Cotton*, *Elm*, *Grape*, *Onion and Garlic*, *Ornamental Foliage*, *Pea*, *Peanut*, *Potato*, *Raspberry/Blackberry*, *Rhododendron and Azalea*, *Rice*, *Rose*, *Sorghum*, *Soybean*, *Strawberry*, *Sweet Potato*, *Tobacco*, *Tomato*, *Tropical Fruit*, *Turfgrass*, and *Wheat*. One or two additional titles or new editions are published yearly. The average size of the volumes is near eighty pages. The series is written expressly for the extension agent, commercial grower, and active hobbyist; it reports disease detection, diagnosis, and control. Liberal use of color aids the reader in diagnosis. By late 1992, 260,000 copies of the compendia series had been sold worldwide. Responding to requests from teachers for copies of the many colored plates used in the compendia series, the Society issued a slide set based on each of the major titles. Ten sets had been issued through 1992.

A Plant Disease Video Image Resource, introduced in the summer of

1992, is the latest APS effort to assist researchers and teachers in the subject area of plant pathology. This resource is composed of two parts: a videodisc and a set of computer diskettes. The videodisc contains nearly 10,000 full-color plant disease slide images selected from APS publications and private slide collections. The computer diskettes contain records describing each videodisc image, together with retrieval software to find specific images. This new resource provides subject searching, basic image research, and viewing when used with a personal computer, videodisc player, and monitor. When combined with additional software and/or hardware, more advanced applications are possible. For example, self-guided tutorial programs are available to help the student study specific aspects of plant pathology.

F. Publishing Activities of Other Major Societies in the United States

American Society of Agronomy

Not long after the term "agronomist" had come into use by the U.S. Department of Agriculture and land-grant colleges, the American Society of Agronomy (ASA) was organized on December 31, 1907. Its objective was "the increase and dissemination of knowledge concerning soils and crops and the conditions affecting them."[17] As with other societies forming during this period, communication was a key goal. The first publication of the society was its *Proceedings* in 1910. C. R. Ball, its editor in 1911, insisted: "The need of a suitable medium for the prompt publication of papers relating to American agronomy is becoming increasingly evident. The time is now ripe for our Society to consider founding a high-class journal which shall adequately meet this need."[18] Thus, the *Proceedings* was discontinued after 1912 and replaced with the *Journal of the American Society of Agronomy* in 1913, and the *Agronomy Journal* in 1949.

The annual program of the ASA was divided into topics on soils and crops in 1924, and by 1931 the ASA was formally reorganized into crops and soils sections. The ASA initiated publication of *Soil Science Society of America Proceedings* as a quarterly journal in 1952. A nontechnical magazine, *What's New in Crops and Soils*, first appeared in October 1948. The title was changed to *Crops and Soils* in 1958, and although the publication has not been financially self-sustaining, it has served well as the extension and outreach arm of the ASA, the Soil Science Society of America (SSSA),

17. Laude, et al., "History of the American Society of Agronomy," p. 58.
18. Ibid., p. 59.

and the Crop Science Society of America (CSSA). In 1959 the membership of the ASA approved the publication of a new journal, *Crop Science* (first published in 1961).

Agronomy Monographs, a very successful and ongoing series designed to examine specific subjects in depth, was published initially by Academic Press from 1949–1956. However, the ASA assumed responsibility for publishing the monographs beginning with volume seven in 1957. It was decided in 1962 that the Society should not publish books, which could be published by commercial firms.[19] In 1963, the ASA initiated a series of *Special Publications*, and beginning in 1966, often in conjunction with the SSSA and the CSSA, the ASA published a number of books in this series largely resulting from symposia and technical conferences.

American Society for Horticultural Science

The ASHS was organized in September 1903 and held its first annual meeting in St. Louis in December 1903.[20] The first publication was volume one and two of the *Proceedings of the Society for Horticultural Science*, in 1905, which reported on the first and second annual meetings. This publication, which primarily reported the research presented at the annual meetings, continued on an annual basis until 1942 when it became biannual. In 1966 *HortScience* debuted as a supplement to the *Proceedings*. In 1969 the *Proceedings* were replaced by the *Journal of the American Society for Horticultural Science*, a bimonthly publication which is still published. The *Journal* had a format that differed from the *Proceedings* and represented a time of revitalization for the ASHS. The *Journal* was designed with a versatile format and encouraged publication of brief (two-page) papers.

ASHS recognized in the 1980s the need to expand the Society's publications beyond journals aimed primarily at horticultural researchers, so *HortTechnology: Applied Horticultural Science* was established as a peer-reviewed journal in 1991. The publication was designed as technology-based and aimed at crop consultants, industry representatives, vocational agriculture teachers, extension agents, gardening writers, and other professional horticulturalists. *HortTechnology* represented a new level of outreach commitment on the part of the ASHS toward professional horticulturalists and practitioners.

19. D. C. Smith, "Development of the American Society of Agronomy, 1958–1977," *Agronomy Journal* 72 (1980): 227–246.

20. (a) A. C. True, "Science and Technical Societies Dealing with Agriculture and Related Subjects," *Proceedings of the Association of Land-Grant Colleges and Universities* 42 (1928): 37–58. (b) C. B. Link, "The First Twenty Years," *American Society for Horticultural Science Newsletter* 8 (1992): 6–7.

The ASHS has also prepared special publications at irregular intervals. These have included a *Glossary for Horticultural Crop*, directories such as *Women in Horticulture* and *Vegetable Breeders*, *Classic Papers in Horticultural Science*, histories of the development of horticulture in various areas, and the proceedings of regional meetings and symposia.

American Society of Plant Physiologists

The American Society of Plant Physiologists (ASPP) grew from the Physiological Section of the Botanical Society of America and became a separate entity in 1924.[21] The official journal of the society, *Plant Physiology*, began publication in January 1926 with quarterly issues. The journal continued the efforts begun earlier by *Physiological Researches*, which was founded in 1913, and was edited, managed, and published by Burton E. Livingston, of the Laboratory of Plant Physiology, Johns Hopkins University.[22]

The ASPP also has published books at irregular intervals from symposia, workshops, and other conferences. These books have been published by arrangement with commercial publishers (e.g., Academic Press, Waverly Press) or university presses (e.g., University of California, Riverside; University of Illinois, Urbana).

Entomological Society of America

The progenitors of the current Entomological Society of America were the American Association of Economic Entomologists (AAEE), founded in 1889, and its sister society, the Entomological Society of America (ESA), founded in 1906. The two societies merged in 1953 as the Entomological Society of America. The AAEE, formed shortly after the passage of the Hatch Act, fulfilled the need for an organization to advance the professional objectives of economic entomologists within agriculture.[23] The proceedings and papers of the AAEE annual meetings were published by the U.S. Department of Agriculture in *Insect Life* from 1889–1894 and in twelve bulletins of the USDA Bureau of Entomology from 1895–1906.[24] The Association launched its *Journal of Economic Entomology* in 1908, intended for

21. J. B. Hanson, *History of the American Society of Plant Physiologists* (Rockville, Md.: American Society of Plant Physiologists, 1989), pp. 1–4.

22. Ibid., pp. 5–6, 42–46.

23. E. H. Smith, "The Entomological Society of America: The First Hundred Years, 1889–1989," *Bulletin of the Entomological Society of America* 35 (1989): 10–32.

24. True, "Science and Technical Societies."

"papers dealing with the economic impact of insects on humans, animals, and plants to meet the information needs of economic entomologists."[25] The original ESA, founded to meet the needs of the expanding entomological community, offered in 1908 its first official publication, *Annals of the Entomological Society of America*. The *Annals* was "founded as the first national journal to publish results of studies examining the basic aspects of insect biology."[26]

When the two societies merged to form the current ESA in 1953, a new publication, *Bulletin of the Entomological Society of America*, was mandated to ". . . include items of current and timely interest, such as proceedings of meetings, lists of members, current notes, obituary notices, etc."[27] The *Bulletin* first appeared in 1955 as a quarterly publication and was transformed in 1983 into a journal providing a forum for subjects of interest not only to entomologists but to the broader areas of biology, science, and society. In 1990 the *Bulletin* was renamed the *American Entomologist* with the continued purpose of communicating scientific information and promoting discussion among entomologists and other scientists.

In 1972, the ESA initiated a third journal, *Environmental Entomology*. The journal was created in response to the growing number of manuscript submissions to the other two bimonthly journals and as a focus for insect ecologists and those concerned with environmental insect science.

The fourth bimonthly journal currently published by the ESA, *Journal of Medical Entomology*, was founded in 1964 by the Bishop Museum in Honolulu, Hawaii to publish research articles on medical and veterinary entomology and acarology. The journal was sold to the ESA in a "friendly takeover" in 1986.

Other publications from the ESA include the *Thomas Say Publications in Entomology*, which include the former *Miscellaneous Publications* and the *Thomas Say Monographs*. Currently there are three series within the *Thomas Say Publications in Entomology*. These are *Systematic Monographs of the ESA*, intended primarily for lengthy taxonomic revisions and reviews; *Memoirs of the ESA*, for original comprehensive works outside the field of systematics; and *Symposia of the ESA*. Also, *Insecticide and Acaricide Tests* has been published annually beginning in 1976 as a survey of the results from field, greenhouse, and laboratory tests of experimental and standard products for the control of insects and mites. In 1992 ESA initiated a new series of *Handbooks of Insect Pests* intended to provide information

25. G. M. Chippendale, "Publications of the Entomological Society of America," *Bulletin of the Entomological Society of America* 35 (1989): 167.
26. Ibid., p. 167.
27. Ibid., p. 168.

to entomologists, to the broader scientific community, and to the general public.[28]

Society of Nematologists

Although founded in 1961, the first annual meeting of the Society of Nematologists was held August 1962 on the Oregon State University campus. Previously, nematologists had met as a section of the American Phytopathological Society. The *Journal of Nematology* began in 1969 as a forum for original research in all areas of nematology, not just plant nematology.[29] In 1987 a supplemental publication, *Annals of Applied Nematology*, was initiated to concentrate on application and control.

Weed Science Society of America

Weed science in the United States had its genesis in a series of state and regional weed conferences.[30] Idaho and Utah organized what were probably the first state weed conferences in 1931; the first regional meeting was the Western Weed Control Conference organized in Denver in 1938. With the continued interest in weed science, the Association of Regional Weed Control Conferences was organized in 1949, which sponsored the journal *Weeds* in 1951. The Weed Science Society of America (WSSA) was organized in 1954 and adopted *Weeds* as its official journal and primary publication, renaming it *Weed Science*.

A tabular summary of the foregoing societies and their current journals is provided in Table 6.1.

G. Publishing Activities of Major Societies outside the United States

Most industrialized and developing countries have societies of crop science or agronomy, crop protection, phytopathology, nematology, entomology, or weed science. Some societies are more generalized in purpose than others; for example, crop protection societies may be more general in their

28. L. R. Nault, "The Past, Present, and Future of Publications of the Entomological Society of America," *American Entomologist* 38 (1992): 69–73.

29. W. F. Mai and R. E. Motsinger, "History of the Society of Nematologists," in *Vistas on Nematology: A Commemoration of the Twenty-fifth Anniversary of the Society of Nematologists*, edited by J. A. Veech and D.W. Dickson (Hyattsville, Md.: Society of Nematologists, 1987), pp. 1–6.

30. Timmons, "A History of Weed Control."

Table 6.1. Journals published by several major U.S. societies for crop improvement and protection

Society	Journal	First published
American Phytopathological Society	*Phytopathology*	1911
	Plant Disease	1979
	Molecular Plant-Microbe Interactions	1988
American Society of Agronomy	*Agronomy Journal*	1913
	Journal of Environmental Quality	1972
American Society for Horticultural Science	*Journal of the American Society for Horticultural Science*	1905
	HortScience	1966
	HortTechnology	1991
American Society of Plant Physiologists	*Plant Physiology*	1924
Crop Science Society of America	*Crop Science*	1961
Entomological Society of America	*Annals of the Entomological Society of America*	1908
	Journal of Economic Entomology	1908
	Environmental Entomology	1972
	Journal of Medical Entomology	1964
Society of Nematologists	*Journal of Nematology*	1969
	Annals of Applied Nematology	1987
Weed Science Society of America	*Weed Science*	1951

scope of activities than phytopathological or entomological societies. Most of these societies publish a journal in which members may disseminate the results of their research. Representative lists of the major journals published by or for specific societies in crop protection, phytopathology, nematology, and entomology are provided in Table 6.2.

H. Impact of Society Journals of Crop Improvement and Protection

Professional societies are the publishers of a significant portion of the literature in the areas of crop improvement and crop protection (Tables 6.1 and 6.2). Because of the importance and relatively high cost of production of journals, societies occasionally evaluate the importance of their publications. This is often done by questionnaires to subscribers, particularly professional society members who usually receive a journal as part of their membership fee. Another method of evaluation is a count of the number of subscriptions or copies sold, whether as part of a membership or as a separate subscription. The frequency and page count of a volume of a journal

Table 6.2. Major society journals for crop improvement and protection published outside the United States

Society	Journal
CROP PROTECTION	
Arab Society for Plant Protection	*Arab Journal of Plant Protection*
Inter-African Phytosanitary Council	*African Journal of Plant Protection*
Korean Society of Plant Protection	*Korean Journal of Plant Protection*
Malaysian Plant Protection Society	*Journal of Plant Protection in the Tropics*
Plant Protection Association of India	*Indian Journal of Plant Protection*
Quebec Society for Plant Protection	*Phytoprotection*
Société Française de Phytiatrie et de Phytopharmacie	*Phytiatrie-Phytopharmacie*
PHYTOPATHOLOGY	
Asociacion Colombiana de Fitopatologia y Ciencias Afines	*ASCOLFI Informa*
Asociacion Latinamericana de Fitopatologia	*Fitopatologia*
Association pour les Etudes et les Recherches de Zoologie appliquée et de Phytopathologie	*Parasitica*
Australasian Plant Pathology Society	*Australasian Plant Pathology*
British Society for Plant Pathology	*Plant Pathology*
Canadian Phytopathological Society	*Canadian Journal of Plant Pathology*
Deutsche Akademie der Landwirtschaftswissenschaften	*Archiv für Phytopathologie und Pflanzenschutz*
Egyptian Phytopathological Society	*Egyptian Journal of Phytopathology*
Hungarian Society for Plant Protection	*Acta Phytopathologica et Entomologica Hungarica*
Indian Phytopathological Society	*Indian Phytopathology*
Indian Society of Mycology and Plant Pathology	*Indian Journal of Mycology and Plant Pathology*
Iranian Phytopathological Society	*Iranian Journal of Plant Pathology*
Korean Society of Plant Pathology	*Korean Journal of Plant Protection = Han'guk Singmul Poho Hakhoe Chi*
Netherlands Society of Plant Pathology	*Netherlands Journal of Plant Pathology*
Phytopathological Society of Japan	*Annals of the Phytopathological Society of Japan = Nihon Shokubutsu Byori Gakkaiho*
Sociedade Brasileira de Fitopatologia	*Fitopatologia Brasileira = Brazilian Phytopathology*
Sociedade Brasileira de Fitopatologia, Grupo Paulista de Fitolpatologia	*Summa Phytopathologica*
Sociedad Mexicana de Fitopatologia	*Revista Mexicana de Fitopatologia*
NEMATOLOGY	
Nematological Society of India	*Indian Journal of Nematology*
Organization of Tropical American Nematologists	*Nematropica*
Pakistan Society of Nematologists	*Pakistan Journal of Nematology*
Sociedade Brasileira de Nematologia	*Nematologia Brasileira*
ENTOMOLOGY	
Acarology Society of India	*Indian Journal of Acarology*

Table 6.2. Continued

Society	Journal
Australian Entomological Society	*Journal of the Australian Entomological Society*
Deutsche Entomologische Gesellschaft	*Deutsche entomologische Zeitschrift*
Deutsche Gesellschaft für allgemeine und angewandte Entomologie	*Mitteilungen der Deutschen Gesellschaft für allgemeine und angewandte Entomologie*
Deutsche Gesellschaft für angewandte Entomologie	*Zeitschrift für angewandte Entomologie*
Entomological Society of Australia	*General and Applied Entomology*
Entomological Society of Canada	*Canadian Entomologist*
	Memoirs of the Entomological Society of Canada
Entomological Society of Egypt	*Bulletin of the Entomological Society of Egypt*
Entomological Society of India	*Indian Journal of Entomology*
	Memoirs of the Entomological Society of India
Entomological Society of Iran	*Journal of the Entomological Society of Iran*
Entomological Society of Israel	*Israel Journal of Entomology*
Entomological Society of New Zealand	*The New Zealand Entomologist*
Entomological Society of Southern Africa	*Journal of the Entomological Society of Southern Africa*
Nederlandse Entomologische Vereniging	*Entomologia, Experimentalis et Applicata*
	Tijdschrift voor Entomologie
Philippine Association of Entomologists	*Philippine Entomologist*
Royal Entomological Society of London	*Ecological Entomology*
	Medical and Veterinary Entomology
	Physiological Entomology
	Systematic Entomology
Sociedad Chilena de Entomologia	*Revista Chilena de Entomologia*
Sociedad Colombiana de Entomologia	*Revista Colombiana de Entomologia*
Sociedad Entomologica Argentina	*Revista de la Sociedad Entomologica Argentina*
Sociedad Entomologia del Peru	*Revista Peruana de Entomologia*
Sociedad Mexicana de Entomologia	*Folia Entomologica*
Sociedade Brasileira de Entomologia	*Revista Brasileira de Entomologica Italiana*
Societa Entomologica Italiana	*Memorie della Societa Entomologica Italiana*
Société Entomologique de France	*Annales de la Société Entomologique de France*
Societas Entomologica Scandinavica	*Entomologica Scandinavica*
WEED SCIENCE	
European Weed Research Society	*Weed Research*
Indian Weed Science Society	*Indian Journal of Weed Science*
Korean Weed Science Society	*Korean Journal of Weed Science*
South African Weed Science Society	*Applied Plant Science*
Weed Science Society of the Philippines	*Philippine Journal of Weed Science*

along with the numbers of papers rejected each year are also used as guides to importance or value.

In recent years counts of the numbers of times other journals cite a journal has been used extensively, because the Institute for Scientific Information (ISI) indexes the citations in journal articles and makes this information available. A relative index has been established by ISI by dividing the numbers of articles in a journal during a given year by the number of times the journal's two preceding volumes were cited by other journals. ISI calls this the journal's impact factors. If the impact is 1.0, this means that the number of citations in the current journal is exactly the same as the number of times that journal was cited in other scientific journals in the two preceding years. An impact of 1.0 is a high ranking. Based upon impact factors from the ISI 1989 edition of the *Science Citation Index Journal Citation Report*, journals published by societies are among the most influential in several crop improvement and protection disciplines (Tables 6.3 and 6.4).

These citation evaluations are often quoted as indications of value. For example, in the *Proceedings* of the 1990 Symposium on the Peer-Review Editing Process sponsored by the American Society of Agronomy, the first chapter examines the impact of journals published by the American Society.[31] Such examinations allow members and society officers to judge the effectiveness of journal publications in the peer scientific community.

In crop science and agronomy (Table 6.3), society journals from the American Society of Agronomy and the Crop Science Society of America have the third and sixth highest impact factors, respectively. In plant pathology and nematology (Table 6.4), four of the six journals with the highest impact factors are published by societies, and a fifth title, *Plant Pathology*, is sponsored by the British Society for Plant Pathology. In entomology four of the top fifteen journals are published by the Entomological Society of America, and two others are sponsored by the Royal Entomological Society of London.

I. Retrospective and Prospective Publishing Influence of Crop Societies

The information on crops available to the scientific community in the United States in the late nineteenth and early twentieth centuries came largely from government publications and the newly initiated journals of

31. E. Garfield, "The Effectiveness of American Society of Agronomy Journals: A Citationist's Perspective," in *Research Ethics, Manuscript Review, and Journal Quality*, edited by H. F. Mayland and R. E. Sojka (Madison, Wis.: American Society of Agronomy, 1992), pp. 1–14.

Table 6.3. Leading crop improvement journals based upon 1989 ISI impact factors

Journal	ISI impact factor	Publisher and location
AGRONOMY AND CROP SCIENCE		
Advances in Agronomy	1.00	Academic Press, San Diego
Australian Journal of Agricultural Research	0.74	Commonwealth Scientific and Industrial Research Organization, Melbourne
Agronomy Journal	0.71	American Society of Agronomy, Madison, Wis.
Zeitschrift für Pflanzenzüchtung = Journal of Plant Breeding	0.71	Paul Parey, Berlin
Field Crops Research	0.63	Elsevier Science Publishers, Amsterdam
Crop Science	0.61	Crop Science Society, Madison, Wis.
The Journal of Agricultural Science	0.49	Cambridge University Press, United Kingdom
Acta Agriculturae Scandinavica	0.48	Scandinavian Association of Agricultural Scientists, and the Royal Swedish Academy of Agriculture and Forestry, Stockholm
New Zealand Journal of Agricultural Research	0.43	Sir Publishing, Wellington
American Potato Journal	0.40	Potato Association of America, Orono, Maine
Australian Journal of Experimental Agriculture	0.40	Commonwealth Scientific and Industrial Research Organization, Melbourne
Netherlands Journal of Agricultural Science	0.36	Royal Netherlands Society for Agricultural Science, Wageningen
Potato Research	0.33	European Association for Potato Research, Wageningen
Agriculture, Ecosystems, and Environment	0.27	Elsevier Science Publishers, Amsterdam
Experimental Agriculture	0.25	Cambridge University Press, United Kingdom
Outlook on Agriculture	0.25	CAB International, United Kingdom
Archiv für Acker und Pflanzenbau und Budenkunde = Archives of Agronomy and Soil Science	0.24	Akademie Verlag, Berlin
New Zealand Journal of Crop and Horticultural Science	0.21	Sir Publishing, Wellington
Zeitschrift für Acker und Pflanzenbau	0.16	Paul Parey, Berlin
Tropical Agriculture	0.11	Butterworth-Heinemann, London
HORTICULTURE		
HortScience	0.42	American Society for Horticultural Science, Alexandria, Virginia
Journal of American Society for Horticultural Science	0.35	American Society for Horticultural Science, Alexandria, Virginia

Table 6.4. Leading crop protection journals based upon 1989 ISI impact factors

Journal	ISI impact factor	Publisher and location
PLANT PATHOLOGY AND NEMATOLOGY		
Physiological and Molecular Plant Pathology	1.93	Academic Press, London
Phytopathology	1.51	American Phytopathological Society, St. Paul
Journal of Nematology	0.77	Society of Nematologists, Lake Alfred, Florida
Plant Pathology	0.64	Blackwell Scientific Publications for the British Society for Plant Pathology, Oxford
Netherlands Journal of Plant Pathology	0.55	Nederlandse Planteziektenkundige Vereniging, Wageningen
Plant Disease	0.53	American Phytopathological Society, St. Paul
Journal of Phytopathology = Phytopathologische Zeitschrift	0.52	Paul Parey, Hamburg
Phytoparasitica	0.45	Priel Publishers, Rehovot, Israel
Canadian Journal of Plant Pathology	0.44	Canadian Phytopathological Society, Guelph, Ontario
Zeitschrift für Pflanzenkrankheiten und Pflanzenschutz	0.32	Eugen Ulmer GMBH Co., Stuttgart
Nematropica	0.20	Organization of Tropical American Nematologists, Auburn, Alabama
Nematologica	0.09	E.J. Brill, Leiden, The Netherlands
Archiv für Phytopathologie	0.06	Akademie Verlag, Berlin
Acta Phytopathologica Hungarica	0.01	Akadémiai Kiadó, Budapest
WEED SCIENCE		
Weed Science	0.59	Weed Science Society of America, Champaign, Illinois
Weed Reasearch	0.35	For the European Weed Research Society by Blackwell Scientific Publications, Oxford
Weed Technology: A Journal of the Weed Society of America	0.21	Weed Science Society of America, Champaign, Illinois
ENTOMOLOGY		
Advances in Insect Physiology	3.00	Academic Press, London
Archives of Insect Biochemistry and Physiology	1.49	Wiley-Liss, New York
Journal of Insect Physiology	1.49	Pergamon Press, Devon, United Kingdom
Insect Biochemistry	1.44	Pergamon Press, Elmsford, New York
Physiological Entomology	1.30	For the Royal Entomological Society by Blackwell Scientific Publications, Oxford
Ecological Entomology	0.92	For the Royal Entomological Society by Blackwell Scientific Publications, Oxford

Table 6.4. Continued

Journal	ISI impact factor	Publisher and location
Journal of Medical Entomology	0.87	The Entomological Society of America, Lanham, Maryland
Entomologia, Experimentalis et Applicata	0.84	Kluwer Academic Publishers, Dordrecht, The Netherlands for Nederlandse Entomologische Vereniging
Bulletin of Entomological Research	0.80	CAB International, London
Journal of Economic Entomology	0.74	Entomological Society of America, Lanham, Maryland
Environmental Entomology	0.73	Entomological Society of America, Lanham, Maryland
Medical and Veterinary Entomology	0.60	Blackwell Scientific Publications, Oxford
Journal of the American Mosquito Control Association	0.59	The Association, Lake Charles, Louisiana
Annals of the Entomological Society of America	0.55	Entomological Society of America, College Park, Maryland
International Journal of Insect Morphology and Embryology	0.51	Pergamon Press, Oxford
Entomophaga	0.42	Lavoisier Abonnements, Paris
The Canadian Entomologist	0.41	Entomological Society of Canada, Ottawa
Applied Entomology and Zoology	0.36	Japan Society of Applied Entomology and Zoology, Tokyo
Experimental and Applied Acarology	0.35	Elsevier Science Publisher, Amsterdam and New York
Insectes Sociaux	0.34	Birkhauser Verlag, Basel, Switzerland
Florida Entomologist	0.31	Florida Entomological Society, Gainesville
Zeitschrift für Angewandte Entomologie	0.30	Paul Parey, Hamburg
Systematic Entomology	0.29	Blackwell Scientific Publications, Oxford
Journal of the Kansas Entomological Society	0.29	The Society, Lawrence, Kansas
Journal of Entomological Science	0.29	Georgia Entomological Society, Tifton, Georgia
The Southwestern Entomologist	0.27	Southwestern Entomological Society, College Station, Texas
Sociobiology	0.23	California State University, Chico, Calif.
Journal of the Australian Entomological Society	0.22	The Society, Indooropilly
Journal of the New York Entomological Society	0.20	Allen Press, Lawrence, Kansas
GENERAL		
Crop Protection	0.45	Butterworth-Heinemann, Guildford, United Kingdom
Phytoprotection	0.21	Quebec Society for the Protection of Plants, Recherches Agricoles, Quebec City, Canada

Table 6.4. Continued

Journal	ISI impact factor	Publisher and location
PESTICIDES		
Environmental Toxicology and Chemistry	1.50	For the Society of Environmental Toxicology and Chemistry by Pergamon Press, New York
Pesticide Biochemistry and Physiology	1.34	Academic Press, Orlando, Florida
Journal of Environmental Quality	1.20	American Society of Agronomy, Crop Science Society of America, Soil Science Society of America, Madison, Wis.
Pesticide Science	1.02	For the Society of Chemical Industry by Blackwell Scientific Publications, Oxford
Ecotoxicology and Environmental Safety	0.87	Academic Press, New York

scientific societies. Because journals served as a means of disseminating information from members and as a signpost for the uniqueness and maturity of a discipline, the publication of a journal was the highest priority for the newly formed societies. Journals published by the societies were a highly significant guiding force in the progress of knowledge for crop improvement and protection.

From their beginnings to the present, the publications of scientific societies have served as the foundations for the continuing development of the sciences associated with crop improvement and protection. Journals continue as the chief publication of most societies; however, nearly all societies have one or more series of special book publications which also are significant. Books published by the societies are often of prime importance to a rather specialized audience which may be too small to attract the attention of the large, commercial publishing houses.

The scientific success of APS Press for the American Phytopathological Society demonstrates clearly that societies can serve effectively in the dissemination of information not only to members but to other scientists and professionals as well. The financial success of APS Press illustrates the appropriateness and viability of the role of societies in providing non-journal publications for disciplinary markets. The initiation of a series of *Handbooks of Insect Pests* by the Entomological Society of America in 1992 signifies the growing role of societies as publishers of disciplinary information.

Page charges, which are currently often required to publish scientific papers in society journals, have the potential to alter the publishing influence of crop protection and improvement societies in the future. Page charges are the costs per published page paid by an author to help defray the expense of publication. These charges are split between authors or their institutions and societies, which derive funds for publishing from member dues and other income-generating activities.

Cost of publication may indeed be a determining factor in the future publishing influence of societies. For scientists without institutional support or where such support is declining, the tendency in the future may be to publish more in commercial journals that do not levy page charges.[32] The dilemma is that commercial publishers support their costs of publication through subscriptions to libraries, and the costs of such subscriptions may be quite high. The result has been the proliferation of costly new journals at a time when library budgets are declining. The effect at many universities is that not only are the costly new journals not being added to library subscription lists, but many journals (often expensive commercially published ones) are also being deleted from such lists. Authors must, in some cases, balance the cost of publication of their research papers with the breadth of dissemination of the information.

Another factor in the changing publication influence of societies is the desire of researchers to publish in more specialized journals and in those journals that have the lowest submission-to-publication time. Commercial publishers are often eager to fulfill the wishes of scientists in reaching a narrower, more specialized audience than that of the traditional society journals. Also, some commercial journals have better time-to-publication statistics than society journals. Both of these factors are challenges that must be addressed by societies if society journals are to continue as the main outlet for member papers. That societies are addressing both issues is indicated by the initiation of the journal *Molecular Plant- Microbe Interactions* by the American Phytopathological Society in 1988. The journal is more specialized than other APS journals and has achieved a relatively short (three- to four-month) submission-to-publication time.

In this early era of electronic publication, the new role and influence of societies of crop improvement and protection in publishing is yet to be determined. Service to members is the principal reason that societies publish journals. Many of the challenges of electronic publication of journals are still to be resolved. However, it is incumbent on scientific societies to resolve these challenges to serve their members best. As the cost of printing

32. Nault, "Past, Present, and Future of Publications."

and distribution increase, society and commercial publishers are looking to media other than paper to distribute scientific data.

Online electronic distribution of scientific journals has been available for a number of years. The major fault is the current low resolution of the average computer screen. While there has been significant progress in this area in recent years, normal photographs as well as photomicrographs are still difficult to reproduce on the average computer monitor. In addition, line charges are significant if any volume of material is being reviewed. For this reason, alternate reproduction media have been investigated, and CD-ROM technology appears to solve some of the prevailing problems.

A leader in the study of electronic distribution of data in the agricultural sciences is the National Agricultural Library in Beltsville, Maryland, a branch of the U.S. Department of Agriculture. Recent studies with CD-ROM databases suggest that this technology is suitable for reproducing scientific journals. The NAL has established the National Agricultural Text Digitizing Program and will be seeking publishers willing to license their publications to this program. The first of the major agricultural science societies to take part in the program is the American Society of Agronomy, which has made available its *Agronomy Journal* starting with volumes one through sixteen.

The advantage of CD-ROM technology lies in its vast storage capacity and inexpensive reproduction. One CD-ROM disk can hold up to 680 megabytes of data, the equivalent of thousands of pages of text. While reproduction of photographs depends upon the resolution quality of the computer monitor, downloading to a high resolution laser printer will capture much of the quality of the original image. CD-ROM disks are relatively light in weight and can be mailed at reasonable cost.

Appendix

Prepared by the editorial staff of this book.

The *Phytopathological Classics* were published by the American Phytopathological Society in order to make important, classical works readily available in English. They were published by the Society beginning in 1926 and have become important historical works in themselves.

1. Fabricius, Johann Christian. *Attempt at a Dissertation on the Diseases of Plants.* Translated by Margaret Kolpin Ravn. Lancaster, Pa.: American Phytopathological Society, 1926. 66p. Original published as "Forsog Tilen Afhandling om Planternes Sygdomme," *Det Kongelige Norske Videnskabers Selskabe Skrifter* (1774) 5: 431–492.

2. Fontana, Felice. *Observations on the Rust of Grain*. Translated by Pascal Pompey Pirone. Washington, D.C.: Hayworth Printing, 1932. 40p. Original published as *Osservazioni sopra la Ruggine del Grano*. Lucca, Italy: Jacob Giusti, 1767.
3. Millardet, Pierre Marie Alexis. *The Discovery of Bordeaux Mixture; Three Papers: I. Treatment of Mildew and Rot; II. Treatment of Mildew with Copper Sulphate and Lime Mixture; III. Concerning the History of the Treatment of Mildew and Copper Sulphate*. Translated by Felix John Schneiderhan. Ithaca, N.Y.: Cayuga Press, 1933. 25p. Original published as "Traitement du Mildiou et du Rot." *Journal d'Agriculture Pratique* (1885) 2: 513–516; "Traitement du Mildiou par le Melange de Sulphate de Cuivre et de Chaux." *Journal d'Agriculture Pratique* (1885) 2: 707–710; "Sur l'Histoire du Traitement du Mildiou par le Sulfate de Cuivre." *Journal d'Agriculture Pratique* (1885) 2: 801–805.
4. Woronin, Michael Stephanovitch. *Plasmodiophora Brassicae: The Cause of Cabbage Hernia*. Translated by Charles Chupp. Ithaca, N.Y.; American Phytopathological Society, 1934. 32p. and six plates. Original published as *Jahrbücher für Wissenschaftliche Botanik*, 11 (1878).
5. Tillet, M. *Dissertation on the Cause of the Corruption and Smutting of the Kernels of Wheat in the Head and on the Means of Preventing These Untoward Circumstances*. Translated by Harry Baker Humphrey. Ithaca, N.Y.; American Phytopathological Society, 1937. 191p. Original published as *Dissertation sur La Cause qui Corrompt et Noirict Les Graines de Bled dans Les Epis*. Bordeaux: Pierre Brun, 1755. Title page of original French dissertation is photographically reproduced in the translation.
6. Prévost, Benedict. *Memoir on the Immediate Cause of Bunt or Smut of Wheat, and of Several Other Diseases of Plants, and on Preventives of Bunt*. Translated by George Wannamaker Keitt. Menasha, Wis.; American Phytopathological Society, 1939. 95p. and 3 plates. Original published as *Memoire sur La Cause Immediate de La Carie ou Charbon des Bles . . .* Paris; Bernard, Quai des Augustins, no. 25, 1807. 93p.
7. This Classic includes four papers:
   (1) Mayer, Adolf. *Concerning the Mosaic Disease of Tobacco*. Original published as "Ueber die Mosaikkrankheit des Tabakas." *Die Landwirtschaftlichen Versuchs-Stationer* 32 (1896): 451–67. (2) Ivanowski, Dmitrii. *Concerning the Mosaic Disease of the Tobacco Plant*. Original published as "Ueber die Mosaikkrankheit der Tabakspflanze." *St. Petersbourg Académie Imperiale des Sciences. Bulletin. 35* (ser. 4, v. 3) (Sept. 1892): 67–70. (3) Biejerinck. M. W. *Concerning a Contagium Vivum Fluidum as Cause of the Spot Disease of Tobacco Leaves*. Original published as "Ueber ein contagium vivum fluidum als Ursache der Fleckenkrankheit der Tabaksblätter." *Verhandelingen der Koninklyle akademie van Wettenschappen te Amsterdam*. 65 (2) (1898): 3–21. (4) Baur, Erwin. *On the Etiology of Infectious Variegation*. Original published as "Zur Aetiologie der infectiösen Panachierung." *Berichte der Deutschen Botanischen Gesellschaft* 22 (1904): 453–460. All translated by James Johnson. Ithaca, N.Y.: American Phytopathological Society, 1942. 62p.
8. Berkeley, M. J. *Observations, Botanical and Physiological, on the Potato Murrain . . .* together with selections from Berkeley's *Vegetable Pathology*. East Lansing, Mich.: American Phytopathological Society, 1948. 108p. *Observations . . .* were

published originally in the *Journal of the Horticultural Society of London* (1846) 1: 9–34; *Vegetable Pathology* appeared as a series of 173 articles in *Gardeners' Chronicle* between 7 Jan. 1854 and 3 Oct. 1857.

9. Tozetti, Giovanni Targioni. *True Nature, Causes and Sad Effects of the Rust, the Bunt, the Smut, and Other Maladies of Wheat, and of Oats in the Field;* Part V of *Alimurgia, or Means of Rendering Less Serious the Dearths, Proposed for the Relief of the Poor.* Translated by Leo R. Tehon. Ithaca, N.Y.; American Phytopathological Society, 1952. 139p. Original published in Florence; il Mock, 1767.
10. Bassi, Agostino. *Del Mal del Segno.* Translated by P. J. Yarrow. Baltimore, American Phytopathological Society, 1958. 49p. Includes a facsimile of original title page. Original published as *Del Mal del Segno, Calcinaccio o Moscardino . . .* Lodi: Tipografia Orcesi, 1835–36. Translation of Pt. 1 on theory of silkworm diseases.
11. de Bary, Anton. *Investigations of the Brand Fungi and the Diseases of Plants Caused by Them with Reference to Grain and Other Useful Plants.* Translated by R. M. S. Heffner, D. C. Arny, and J. Duain Moore. Ithaca, N.Y.: American Phytopathological Society, 1974. 93 p. and 8 plates. Original published as *Untersuchungen über die Brandpilze und die durch sie verursachten Krankheiten der Pflanzen.* Berlin: Verlag von G.W.F. Muller, 1853.
12. Hartig, Robert. *Important Diseases of Forest Trees; Contributions to Mycology and Phytopathology for Botanists and Foresters.* Translated by William Merrill, David H. Lambert, and Walter Liese. St. Paul, Minn.: American Phytopathological Society, 1975. 120p. and six plates. Original published as *Wichtige Krankheiten der Waldbäume; Beitrage zur Mycologie und Phytopathologie für Botaniker und Forstmänner.* Berlin: Verlag von Julius Springer, 1874.
13. Fischer, Alfred, and Erwin F. Smith. *The Fischer-Smith Controversy: Are There Bacterial Diseases of Plants?* Translated and prepared by C. Lee Campbell. St. Paul, Minn.: American Phytopathological Society, 1981. 65p. Material is from two sources but primarily from the *Centralblatt für Bakteriologie*, Abteilung II, 1899–1901.

Unnumbered

Campbell, C. Lee, and G. W. Bruehl, eds. *Viruses in Vectors: Transovarial Passage and Retention.* St. Paul, Minn.: APS Press, 1986. 53p. Includes one paper by Hirofaro Ando, two by Harold Haydon Storey, and three by Teikichi Fukushi, originally published between 1910 and 1939.

Holmes, Francis W., and Hans M. Heybroek, eds. *Dutch Elm Disease—The Early Papers; Selected Works of Seven Dutch Women Phytopathologists.* St. Paul, Minn.: APS Press, 1990. 154p. The original papers, translated by the editors, were published between 1921 and 1970.

7. Publishing Patterns of Crop Research Institutions in Select Countries of the Third World

Barbara A. DiSalvo
Wallace C. Olsen
Cornell University

The development of crop research capabilities in parts of the Third World is well known among crops academicians and researchers. Scientists in the Third World have created extensive research, information, and collaboration networks covering such diverse subjects as peanuts, soil management, and agroforestry. A recent publication lists these networks which flourished in the 1980s, most of which were organized by the international agricultural research centers.[1] The book lists sixty-five research and academic networks of which thirty-six (55%) have to do with crops. Four new cooperative networks were created in 1992 and 1993. The African Coffee Research Network was established under the aegis of the Organisation Inter-Africaine du Cafe (OIAC) with the aim of coordinating research in Africa. It was established as a result of a conference in Lisbon in 1992. Cassava Biotechnology Network is a similar group formed to discuss cassava biotechnology issues and to foster research. It is headquartered at CIAT in Colombia and issues *CBN Newsletter*. The third, BioNET-International, has a more elaborate structure and is run by the Consultative Group of CAB International in London. It was conceived with regional or subregional operational partnerships and technical support institutions. This is a network of institutions concerned with the biosystematics of arthropods, nematodes, and micro-organisms. The International Ecological Agriculture Network concerned with sound farming practice has its home in Brussels.[2] Interna-

1. Donald L. Plucknett, Nigel J. H. Smith, and Selcuk Ozgediz, *International Agricultural Research: A Database of Networks* (Washington, D.C.: The World Bank, 1990 [*CGIAR Study Paper* no 26.]). A detailed book on this topic was written by the same authors and published by Cornell University Press in 1990 as *Networking in International Agricultural Research*.

2. These networks were announced in the Dec. 1993 and 1994 issues (nos. 48 and 54) of *SPORE*.

tional cooperative networks of scientists and practitioners represent a growing trend among Third World agriculturalists. The networks provide a broad structure for operation and the potential for cooperative efforts and solutions of problems. Determining which research institutions are progressing quickly is an involved process. One method used extensively is determination of research publishing patterns and their impacts. Three recent publications provide summaries of the methodologies, scientific basis, and some results in agriculture.[3]

As a means of determining which institutions in the Third World are important today in crops research, several actions were taken to frame the parameters of a study. Crop specialists at Cornell University, some of them from the Third World but all with extensive Third World experience, were asked to identify primary institutions with important and substantive work in crop research within the past ten years. These were supplemented with agricultural research information from directories.[4] A substantial list of institutions was gathered for each of three geographic areas: Asia, Africa, and Latin America. A quick check was made of these institutions and their recent publications on crops. The resulting list of institutions grew and was then pared to about thirty, based on institutional activities, size, and the coverage of citations in standard bibliographic databases. Division by continent or region provides a focus on the production, improvement, and protection of particular crops and their pests based on similarities in climate and soils.

Some problems are inherent in utilizing bibliographic databases for determining the of publishing output of an organization or a country. The most important is the coverage of the bibliographic database and how extensively it includes publications from the Third World. Journals recognized as international are fully included in the primary agricultural bibliographic files. International journals are useful to get a measure of the publishing of individuals or organizations. They are also considered to be the most important scientifically because they tend to have the widest application. The next

3. (a) H. F. Moed, *The Use of Bibliometric Indicators for the Assessment of Research Performance in the Natural and Life Sciences: Aspects of Data Collection, Reliability, Validity and Applicability* (Leiden: DSWO Press, 1989), p. 230. (b) Jacques Gaillard, "Use of Publication Lists to Study Scientific Production and Strategies of Scientists in Developing Countries," in *Les Indicateurs de Science pour les Pays en Développement = Science Indicators for Developing Countries*, edited by Rigas Arvanitis and Jacques Gaillard (Paris: Editions de l'ORSTOM, 1992). (c) Wallace C. Olsen, *Agricultural Economics and Rural Sociology: The Contemporary Core Literature* (Ithaca, N.Y.: Cornell University Press, 1991). (See Chapters 2 and 4).

4. (a) *Agricultural Research Centers*, 9th ed. (Essex, U.K.: Longman, 1988). (b) R. Vernon, ed., *Directory of Research Workers in Agricultural and Allied Sciences* (Oxon, U.K.: CAB International, 1989).

level includes regional publications which tend to span a continent or a large geographic region. These, too, are usually well covered by the agricultural bibliographic databases. At the country or more local level, a much lower percentage of journals are included for indexing. Most bibliographic databases survey material with a locus focus but then must decide whether the articles or publications will be valuable to a wider audience. This material usually offers solutions to local farming or food problems and is viewed as site-specific without wide application outside a country or portion of a country. Therefore, little is cited in large agricultural bibliographic databases. *AGRIS*, the computerized bibliographic database created by the Food and Agricultural Organization, includes a larger portion of this type of material than the other databases.

The citation counts in this study primarily reflect research or findings with wide applications or distribution value. *CAB Abstracts* was chosen for searching because it consistently covers the greatest number of national and international publications.[5]

A. Methodology

In this search, crops were broadly defined as including both edible and ornamental crops, and research as covering all levels from molecular studies to the improvement, production, protection, storage, and marketing of specific crops. The author address (AD) field in *CAB Abstracts* provides the home institution of the principal author and is usually where the research was conducted.

A search was made of the entire *CAB Abstracts* database for 1984–91 in compact disk format for 28 of the 30 institutions and their derivative names, yielding 6,750 references. This resulted in an immense bank of organizational names and numbers even after machine editing of unrelated subunits such as veterinary schools. Citations were printed for all of the organizations, and those outside the study's subject scope were removed. Therefore, the entire publishing effort of an organization such as the Kenya Agricultural Research Institute was not assessed, but only the portion related to crops. Coordination of variant organizational titles was necessary along with other idiosyncracies. The final, appropriate citations for the study numbered 5,324.

5. Norbert Deselaers, "The Necessity for Closer Cooperation among Secondary Agricultural Information Services: An Analysis of *AGRICOLA*, *AGRIS*, and *CAB*," *IAALD Quarterly Bulletin* 31 (1) (1986): 20.

These are the organizations which were suggested for investigation:

Asia:

Bangladesh	Bangladesh Agricultural Research Institute
India	Indian Council of Agricultural Research; Punjab Agricultural University
Indonesia	Agency for Agricultural Research and Development
Korea	Office of Rural Development
Malaysia	Malaysian Agricultural Research and Development Institute
Pakistan	Pakistan Agricultural Research Center
Philippines	University of the Philippines at Los Banos
Taiwan	Institute Academia Sinica; Taiwan Agricultural Research
Thailand	Kasetsart University; Chiang Mai University

Africa:

Ghana	Crops Research Institute
Kenya	Kenya Agricultural Research Institute
Malawi	Ministry of Agriculture, Agricultural Research Dept., Experiment Stations
Nigeria	National Horticultural Research Institute
Tanzania	Tanzania Agricultural Research Organization
Uganda	Makerere University
Zimbabwe	Dept. of Research and Specialist Services

Latin America:

Brazil	EMBRAPA; Instituto Agronomico de Campinas
Colombia	Instituto Colombiano Agropecuario
Costa Rica	Ministry of Agriculture
Ecuador	Instituto Nacional de Investigaciones Agropecuarias
Honduras	Fundacion Hondurena de Investigacion Agricola
Mexico	Instituto Nacional de Investigaciones Agricolas; Universidad Autonoma Chapingo
Peru	Instituto Nacional de Investigacion y Promocion Agropecuaria
Trinidad	University of the West Indies, St. Augustine
Venezuela	Centro Nacional de Investigaciones Agropecuarias

EMBRAPA and Instituto Agronomico de Campinas in Brazil were not searched, although they were among those recommended. The number of individual institutions which comprise EMBRAPA makes accurate retrieval of all possible citations questionable. The form of the institutional name varies from full to abbreviated to acronymic and inclusion of the parent organization's name is not consistent in either form or occurrence. For example, the Centro de Pesquisa Agropecuaria do Tropico Humido (Agricultural Research Center of the Humid Tropics) appeared in the AD field in its full form as given here, in an abbreviated form with the prepositions dropped, as an acronym—CPATU, and in combination with EMBRAPA in both full and acronymic forms. This problem occurred consistently with each subordinate EMBRAPA organization searched. Because of these complications with EMBRAPA and entry in the database, all Brazilian citations were excluded from consideration.[6]

Following the decision to exclude Brazil, some thought was given to excluding all countries considered as scientifically peripheral, especially if other bibliometric studies had already been conducted.[7] However, since there was likely to be valuable information in a determination of the types of journals in which the research from each country was being published, other scientifically peripheral countries remained in the scope of the study.[8]

Search statements were formulated using both full and variant abbreviated forms of the names of organizations. In the case of Malawi, the individual names of experiment stations were used. For centers conducting agricultural research in agricultural fields other than crops, terms distinguishing departments, such as dairy and veterinary, were eliminated from retrieved results. Broader terms such as zoology, animal, etc. were not eliminated in order to avoid dropping citations referring to the entomological and fodder aspects of crop science. Citations which had no pertinence to crops were removed from the final count following individual review of results. The problem of duplicate coding of citations appearing in *CAB Abstracts*, described in a similar study by Bennell and Thorpe, was handled by a review of each search result.[9]

Citations were categorized based on the format used in the source (SO) field in the database, with several exceptions. Although formatted as journals, BIOTROP *Special Publications*, FFTC *Book Series*, ACIAR *Proceedings Series*, and *IFDC Special Publications* were counted as monographs since each publication is unique in subject coverage and complete in itself.

6. Léa Velho, "The 'Meaning' of Citation in the Context of a Scientifically Peripheral Country," *Scientometrics* 9 (1–2) (1986): 71–89.

7. (a) Yvon Chatelin and Rigas Arvanitis, *Strategies Scientifiques et Developpement: Sols et Agriculture des Régions Chaudes* (Paris: Editions de l'ORSTOM, 1988), pp. 37–38. (b) K. C. Garg and B. Dutt, "Bibliometrics of Agricultural Research in India," *IAALD Quarterly Bulletin* 37 (3) (1992).

8. Gretchen Whitney, "Access to Third World Science in International Scientific and Technical Bibliographic Databases," in *Les Indicateurs de Science pour les Pays en Développement* (Science Indicators for Developing Countries), edited by Rigas Arvanitis and Jacques Gaillard (Paris: Editions de l'ORSTOM, 1992), p. 402.

9. Paul Bennell and Peter Thorpe, "Crop Science Research in Sub-Saharan Africa: A Bibliometric Overview," *Agricultural Administration & Extension* 25 (1987): 102.

B. Data and Results

The results of the analysis are summarized in Table 7.1. Judging by total number of citations, it is clear that a substantial amount of literature has come from organizations in Asia and Latin America. No determination was made for numbers of citations per scientist within institutions. Within the countries targeted for searching, no citations were retrieved for organizations in Costa Rica, Honduras, and Zimbabwe. Therefore, these three countries were dropped from consideration. Citations were additionally divided into other categories to assess the nature of the material produced at the organizations. Those categories are:

1. Items published within each country by national organizations or publishers which concentrate on national, site-specific problems
2. Articles published by journals within the three regional divisions which have application beyond one country or which concentrate on regional or continental topics
3. Papers published in journals (nearly all peer reviewed) outside the region or continent, or by internationally known organizations; of generally broad subject application and from authoritative sources.
4. Monographic publications, primarily conference proceedings and reports.

From this data some observations can be made.

1. The organizations with greater than 50% percent of their crops literature published in international journals probably have more advanced research programs. Among these are the Crops Research Institute (Ghana), the Kenya Agricultural Research Institute, the Malawi agricultural research stations, and Makerere University in Africa.[10] In Latin America, the University of the West Indies (Trinidad-Tobago) published twenty of its sixty-two articles appearing in international journals in one journal only, *Tropical Agriculture*, which, although published in the United Kingdom, is edited by the UWI Faculty of Agriculture and from its scope of authorship must be categorized as an international publication. The Agency for Agricultural Research (Indonesia), with 36% publishing in international journals, could also be included in the list of more advanced research programs.

   Some additional considerations must be taken into account before drawing rigid conclusions based on numbers alone. The decision to publish in English or in the official language of the country may automatically determine or limit the choice of publication. If the author's native language is English, the choice is likely to be an international journal. If English is not the author's primary lan-

10. Bennell, "Crop Science Research in Sub-Saharan Africa," pp. 121–122.

Table 7.1. Crops research results from *CAB Abstracts*, 1984–1991

Country	Organization[a]	Total no. of citations	In national journals		In regional journals		In international journals		Citations to monographs	
			No.	% of Total	No.	% of Total	No.	% of Total	No.	% of Total
Africa (South of Sahara)										
Ghana	CRI	25	0	0.0	1	4.0	17	68.0	7	18.0
Kenya	KARI	43	4	9.3	0	0.0	31	72.1	8	18.6
Malawi	Ministry Agric. Res. Stations	28	1	3.6	2	7.1	15	53.6	10	35.7
Nigeria	NHRI	59	32	54.3	0	0.0	24	40.7	3	5.0
Tanzania	TARO	31	16	51.6	0	0.0	8	25.8	7	22.6
Uganda	Makerere University	12	0	0.0	2	16.6	9	75.0	1	8.4
Latin America										
Colombia	ICAR	89	55	61.8	7	7.9	17	19.1	10	11.2
Ecuador	INIAP	17	1	5.9	0	0.0	7	41.1	9	53.0
Mexico	Chapingo University	274	156	56.9	18	6.6	84	30.7	16	5.8
	INIA	138	65	47.1	2	1.5	57	41.3	14	10.1
Peru	INIPA	5	1	20.0	0	0.0	2	40.0	2	40.0
Trinidad	UWI	83	1	1.2	0	0.0	62	74.7	20	24.1
Venezuela	CENIAP	158	139	88.0	3	1.9	7	4.4	9	5.7
Asia										
Bangladesh	BARI	165	100	60.6	31	18.8	20	12.1	14	8.5
India	Punjab Agric. University	1,951	1,521	78.0	5	0.2	345	17.7	80	4.1
	ICAR	505	379	75.1	6	1.2	85	16.8	35	6.9
Indonesia	AARD	175	55	31.4	11	6.3	64	36.6	45	25.7
Korea	RDA	331	308	93.1	13	3.9	7	2.1	2	0.6
Malaysia	MARDI	224	152	67.9	16	7.2	15	6.6	41	18.3
Pakistan	PARC	192	120	62.5	8	4.3	46	7.8	18	9.4
Philippines	University, Los Banos	255	165	64.7	19	7.5	35	13.7	36	14.1
Taiwan	Academia Sinica	118	64	54.2	2	1.7	36	30.5	16	13.6
	TARI	341	305	89.4	7	2.1	10	2.9	19	5.6
Thailand	Chiang Mai University	62	13	21.0	8	12.9	17	27.4	24	38.7
	Kasetsart University	71	54	76.1	3	4.2	1	1.4	13	18.3

[a]For abbreviations, see the preceding list of organizations.

guage, this fact may restrict even some wide application research to regional or national journals.[11]

The importance of publishing as a means to advancement in the field rather than merely a means of communication must also be considered, especially when comparing one region with another.[12]

2. The Taiwan Agricultural Research Institute work in Taiwan and all other Asian institutions except two concentrate on local publications. This probably signifies scientific concentration on solving local problems, although there may be other influences as well, such as govermental proprietary rights.
3. None of the institutions searched published heavily in regional journals, in part because the citations which fell into the regional category were mostly from national journals of other countries within a region, which often also fall into the developing country category. No distinction has been made between developing and developed countries with regard to regional publications.[13] Only one truly regional journal title demonstrated any significant impact among the citations retrieved. This was *Pasturas Tropicales*, published by the International Center for Tropical Agricultural (CIAT), with eighteen citations. All others appeared only once or twice each. A number of titles fell into the international category, although they were published in the developing countries. Most contributions to these came from within the region. The most significant of these are *Turrialba, East African Agriculture and Forestry Journal,* the *Proceedings of the Tropical Region* of the American Society for Horticultural Science, and the *Journal of Plant Protection in the Tropics.*
4. Only one organization has monographs in over half of its citations: the Instituto Nacional de Investigaciones Agropecuarias (Ecuador). These monographs represent reports in series and proceedings. This category of publication represents a greater majority of items concerned with local or site-specific questions, but also indicates participation in international conferences.

Continental groupings, as reflected in the *CAB Abstracts*, provide additional information about primary international, regional, and local journals. Two types of journals are of particular interest, the international journals and those which are regionally published and regionally cited.

All of the journals in Table 7.2 were cited a minimum of four times within their continental groups. In the international journal literature there is little unanimity of publishing from Third World authors. Only *Acta Horticulturae, Tropical Pest Management*, and *Tropical Agriculture* are well represented in all three continental areas. *Fertilizer Research* is a middle-

11. *Multilanguage Publication in Agriculture Workshop Report and Description of Participating Agencies* (Manila: International Rice Research Institute, 1984).

12. Peter Thorpe and Philip G. Pardey, "The Generation and Transfer of Agricultural Knowledge: A Bibliometric Study of a Research Network," *Journal of Information Science* 16 (3) (1990): 184.

13. Thorpe and Pardey, "The Generation and Transfer of Agricultural Knowledge." (see Table 5, p. 188)

Table 7.2. Ranked primary international journals by continental groups

	Asia	Latin America	Africa
International Rice Research Newsletter	1		
Acta Horticulturae	2	1	4–5
Tropical Pest Management	3	5–6	1
Fertilizer Research	4–5		6
Plant & Soil	4–5		
Euphytica	6		
Journal of Agricultural Science	7		
Tropical Agriculture	8	4	3
International Pest Control	9		
Sorghum Newsletter		2	
American Potato Journal		3	
Insect Science & Its Application			2
Plant Genetic Resources Newsletter			4–5

scorer in two continental areas. The two highly-ranked newsletters are represented only within one continental group, which raises doubts about their international character. Upon examination of both, it is clear that the term newsletter is a misnomer, used to represent quick publishing of research results. Authorship in both titles goes beyond two continents. The same is true of the plant resources newsletter. The thirteen titles in Table 7.2 represent a small percentage of all international journals cited; the dispersion among journals is great.

In each continental group, an examination was made of the top institution, that is, the one with the greatest number of published papers in international and regional journals. Of these top three, the National Horticultural Research Institute in Nigeria was lowest in citations, with only twenty-four; similarly low figures were true of the whole sub-Sahara group of organizations. Chapingo University had 102 international and regional publications and was highest in the Latin American group. With 350 publications in international or regional journals, Punjab University outdistanced by far all other organizations surveyed in Asia. Only three titles, *Fertilizer Research, Acta Horticulturae,* and *Tropical Agriculture* have a minimum of three citations per institution. All three are published in Europe, by Nijhoff, by the International Society for Horticultural Science in the Hague, and by a commercial press in England, respectively (the third title is edited by the Imperial College of Tropical Agriculture in Trinidad-Tobago). Punjab University authorship covers a great diversity of international publications in its top ten titles, including titles from the Netherlands (four titles), England (three titles), the Philippines, Germany, Italy, and Trinidad-Tobago (one each). If one can judge by its acceptance in peer-reviewed, international journals, it appears that advanced work of quality is coming from Punjab University.

Regional journals require a different display of their respective ranks (Table 7.3). The lines between titles or groups of titles signify a division of rank. In the case of Latin American titles there are only two ranks. The data for Latin America are inconclusive because of low numbers of citations, and the exclusion of Brazilian authors and organizations; however, it is clear that the *Pasturas Tropicales*, published by CIAT, is a strong regional publication. In Asia, there is clear evidence that regional publications are strong; diverse countries, developing as well as developed countries, issue publications.

Recently, a study was made of the publication productivity of the agricultural centers of the Consultative Group on International Agricultural Research, a group not represented in this examination. The study, which used bibliometric methods, is of a specialized group of research centers, but the

Table 7.3. Primary regional journals by continental groups

	Rank	
Africa:		Three journals were cited but lacked a statistically valid level of citations.
Latin America:	1	*Pasturas Tropicales* (CIAT)
(N=24)	2	*Revista Brasileiras de Fisologia Vegetale* *Revista Ceres* (Universidade Federal de Vicosa) *Phyton* (Buenos Aires)
Asia:	1	*Extension Bulletin* (Taiwan)
(N=68)	2	*Economic Affairs* (India)
	3	*Bacterial Wilt Newsletter* (Australia) *Japanese Journal of Tropical Agriculture* *Leucaena Research Reports* (Taiwan) *Thai Journal of Agricultural Science*
	4	*ASEAN Food Journal*
	5	*Annals of the Phytopathological Society of Japan* *Soil Science and Plant Nutrition* (Japan) *Soybean Rust Newsletter* (Taiwan) *Tropical Agriculture Research Series* (Japan) *Tropical Grasslands* (Australia)

details of correlation or influences of budgets, research and support staffs, and publication output is of some interest.[14] A bibliometric study of the crop science literature used in Ghana was also published recently.[15]

14. J. MacLean and Carmella Janagap, "The Publication Productivity of International Agricultural Research Centers," *Scientometrics* 28 (1993): 329–348.
15. K. O. Darko-Ampem, "An Analysis of the Literature of Crop Science in Ghana, 1977–1992," *IAALD Quarterly Bulletin* 38 (4) (1993): 191–196.

8. Crop Science Citation Analysis and Core Lists of Primary Monographs, Post-1950

WALLACE C. OLSEN
Cornell University

Citation counting and analyses have been used extensively in libraries and scholarship to answer questions or to observe patterns and trends in literature use. Beginning in the 1950s, citation analysis grew dramatically as a result of two major thrusts: (1) the implementation of computing storage and speeds in compilation and analysis, and (2) publication of immense citation databases, *Science Citation Index* and the *Social Science Citation Index,* which index the citations in articles in approximately 5,500 journals.[1] Bibliometric techniques have been used in a variety of applications within the publishing, library, and scholarly communities. They have proven useful in measuring research and education productivity, in determining a scholar's output and impact, and as indicators of social and economic growth.

Readers are referred to a discussion of citation analysis and its applications in the agricultural sciences in a recent publication by the author.[2] For a general overview of citation analysis, readers are directed to a comprehensive collection of articles edited by Christine L. Borgman, on bibliometric methods for the study of scholarly communication.[3]

A structured citation database is necessary to obtain correlations and conclusions. The Core Agricultural Literature Project began by examining the Institute for Scientific Information's print and online tools.[4] It was clear upon examination, however, that these publications did not adequately

1. *Science Citation Index (1961–)* and *Social Science Citation Index (1977–)* are products of the Institute for Scientific Information, Philadelphia.
2. Wallace C. Olsen, *Agricultural Economics and Rural Sociology: The Contemporary Core Literature* (Ithaca, N.Y.: Cornell University Press, 1991), chapter 4.
3. Christine L. Borgman, ed., *Scholarly Communication and Bibliometrics* (Newbury Park, Calif., and London: Sage Publications, 1990).
4. *Science Citation Index* and *SCI Journal Citation Report.*

cover crops and the related sciences and, therefore, had limited use. (Some useful data from ISI sources concerning the journal literature are given in the following chapter.) One problem with the ISI database is that it deals almost exclusively with research literature in journals. Citations at the ends of journal research articles are reflective of the point of view and approach of a researcher. They may or may not reflect the literature used by an educator or an applications practitioner, particularly in the agricultural sciences. It was necessary to establish another path for identification of titles and for quantifying what we wanted to know. Findings then had to be related to other studies or databases wherever valid.

A. Purpose and Methods

The primary aim of the study reported here was to determine the core literature of crop improvement and protection of the past forty years that still has impact in academic teaching and research today. A further aim was to determine the relative rank and merit of the titles for the worldwide academic community and especially for the developing or Third World countries. This focus required a careful analysis of the tools and citation analysis methods needed to determine the core literature. As indicated earlier, the research literature analyses are relatively numerous and useful, but they do not measure the literature of greatest value to advanced students and beginning researchers. The aim of this study was to identify the literature in the middle of a continuum from undergraduate college education through advanced and graduate studies to post-doctoral research. There is of course a large overlap between the center of the continuum and the undergraduate and the advanced research literatures.

A variety of literature tools exist to aid in this process such as specialty abstracting tools, like *Plant Breeding Abstracts*.[5] These tools present comparison problems caused by differing subject definitions and groupings, time periods, and formats of coverage as discussed in Chapter 4. Literature studies already published offer some help, but they are few and tend to concern themselves with journals only, whereas the aim here was to examine all formats.

Some qualitative evaluations of crops literature have been done. These are largely literature reviews, but they also include selected readings brought together in book form, reserve readings used in academic departments as adjuncts to the classroom, and landmark crop science mono-

5. *Plant Breeding Abstracts* has been published since 1930 by Commonwealth Agricultural Bureaux and CAB International.

graphs. Overviews of the literature such as these meet the aims of the project; they are a valid means to establish a base and to identify literature use patterns. Overview literature incorporates two desired quality factors: the works are widely accepted as classics in their own right, and they are peer-reviewed. Surveying the references in these monographs, it becomes clear that crop scientists are as closely tied to their literature and citation justification as animal scientists are to theirs.

Citation analysis has pitfalls in methods and applications and must be carefully executed within different disciplines. Provisions must be made for different databases and time periods. These are summarized and discussed in a previous work.[6] Potential problems can be eliminated or minimized through a variety of techniques, some statistical, some empirical. An extensive knowledge of the literature of the field can obviate false starts and mistakes.

An effort was made to locate those monographs and literature reviews which best fit established criteria and lack possible skews. Such works would have adequate coverage of different dates of publication, a subject scope to match crop science in the past forty years, and a suitably wide geographic distribution of publications or authors. Attention was also paid to publishing organizations with a heavy influence and credibility in the discipline. The Core Agricultural Literature Project had a particular interest in determining if the developed country literature was different from that for the Third World. A bias toward Third World literature was therefore allowed for the purposes of comparing these two groups. Provisions had to be made for climate-dependent crops such as rice, tropical fruits, alfalfa, etc.

The Core Agricultural Literature Project created a Steering Committee to guide and counsel these efforts. One of the tasks of this group was to aid in determining the source documents to be used for citation analysis. The Steering Committee members were:

W. Ronnie Coffman
Cornell University
Ithaca, New York

Gary Heichel
University of Illinois
Urbana, Illinois

David J. Hume
University of Guelph
Guelph, Canada

W. Laux
Biologische
Bundesanstalt
Berlin, Germany

David Pimentel
Cornell University
Ithaca, New York

H. David Thurston
Cornell University
Ithaca, New York

Thirty-one monographic works were identified and approved by the Steering Committee for citation analysis, along with four journal literature reviews.

6. Olsen, *Agricultural Economics and Rural Sociology*, chapter 4.

Sources of Citations Analyzed in Crop Science

Asterisked items (*) were not analyzed, but monographs were extracted from these titles for inclusion in developed and Third World lists for evaluation. Items with a D were analyzed for developed countries data; items with a T were analyzed for Third World data.

Monographs

*Adams, C. R., K. M. Bamford and M. P. Early. *Principles of Horticulture*. London; Heinemann, 1984. 197p.

Agrios, George N. *Plant Pathology*. 3d ed. New York; Academic, 1988. 803p. D&T

Akobundu, I. Okezie, ed. *Weeds and Their Control in the Humid and Subhumid Tropics; Proceedings of the Conference of the International Institute of Tropical Agricultural, Ibadan, Nigeria, July 1978*. Nigeria; Weed Science Society of Nigeria. 421p. T

Altieri, Miguel A. and Matt Liebman, eds. *Weed Management in Agroecosystems: Ecological Approaches*. Boca Raton, Fla.; CRC Press, 1988. 339p. T

*Baker, Ralph R. and Peter E. Dunn, eds. *New Directions in Biological Control: Alternatives for Suppressing Agricultural Pests and Diseases*; *Proceedings of a UCLA Colloquium held at Frisco, Colorado, January, 1989*. New York; Alan R. Liss, Inc., 1990. 837p.

Beadle,C.L., S.P. Long, S.K. Imbamba, D.O. Hall, and R.J. Olembo. *Photosynthesis in Relation to Plant Production in Terrestrial Environment*. Oxford, Eng.; Tycooly Pub., Ltd., 1985. 156p. D&T

Beck, D. P. and L. A. Materon, eds. *Nitrogen Fixation by Legumes in Mediterranean Agriculture*; *Proceedings of a workshop on Biological Nitrogen Fixation on Mediterranean-Type Agriculture, ICARDA, Syria, April, 1986*. Dordrecht, Boston, and Lancaster; M. Nijhoff Publishers, 1988. 379p. D&T

Bothe, H., F.J. de Bruijn and W.E. Newton. *Nitrogen Fixation: Hundred Years After*; *Proceedings of the 7th International Congress on Nitrogen Fixation, Cologne, Germany, Mar. 1988*. Stuttgart, Germany and New York; Gustave Fischer, 1988. 878p. D&T

Brown, A.H.D., et al. *The Use of Plant Genetic Resources*. Cambridge and New York; Cambridge University Press, 1989. 382p. D&T

*Cook, R. James and Kenneth F. Baker. *The Nature and Practice of Biological Control of Plant Pathogens*. St. Paul, Minn.; The American Phytopathological Society, 1983. 539p.

Fadeev, IU. N. and K. V. Novozhilov, eds. *Integrated Plant Protection* (Translation of Integriovannaia Zashchita Rastenii). New Delhi, India; Amerind Publ. Co., 1987. 333p. D

Fehr, Walter R., ed. *Principles of Cultivar Development. Volume 1: Theory and Technique, D&T, Volume 2: Crop Species*, D. New York and London; Macmillan, 1987.

Fehr, Walter R. and Henry H. Hadley. *Hybridization of Crop Plants*. Madison, Wis.; American Society of Agronomy and Crop Science Society of America, 1980. 765p. D

*Gair, R., J. E. E. Jenkins, and E. Lester. *Cereal Pests and Diseases*. Ipswich, Suffolk, Eng.; Farming Press Ltd., 1987. 268p.

Grayson, B.T., M.B. Green, and L.G. Copping, eds. *Pest Management in Rice*. London and New York; Elsevier, 1990. 536p. D&T

*Gresshoff, Peter M., L. Evans Roth, Gary Stacey, and William E. Newton, eds. *Nitrogen Fixation: Achievements and Objectives; Proceedings of the 8th International Congress on Nitrogen Fixation, Knoxville, Tenn., May, 1990*. New York and London; Chapman and Hall, 1990. 869p.

Guan-Soon, Lim and Di Yuan-Bo, eds. *Status and Management of Major Vegetable Pests in the Asia-Pacific Region With Special Focus Towards Integrated Pest Management*. Bangkok, Thailand, Food and Agriculture Organization, 1990. 125p. T

Gupta, O. P. and P. S. Lamba. *Modern Weed Science in the Tropics and Subtropics*. New Delhi; Today and Tomorrow Printers and Publishers, 1978. 421p. T

*Harris, K.M. and P.R. Scott, eds. *Crop Protection Information: An International Perspective; Proceedings of the International Crop Protection Information Workshop,* CAB International, Wallingford, Eng., April 1989. Wallingford, Eng.; CAB International, 1989. 321p.

*Heiser, Charles B. *Seed to Civilization*. London and Cambridge, Mass.; Harvard University Press, 1990. 228p.

*Holliday, Paul. *A Dictionary of Plant Pathology*. Cambridge; Cambridge University Press, 1989. 369p.

*Holm, Leroy and James Herberger. *World List of Useful Publications for the Study of Weeds and Their Control*. Madison, Wis.; Dept. of Horticulture, University of Wisconsin, 1971. 609p. (also in PANS Vol. 17 No. 1, March 1971)

Hussein, M.Y. and A.G. Ibrahim, eds. *Biological Control in the Tropics; Proceedings of the 1st Regional Symposium on Biological Control, Universiti Pertanian Malaysia, Serdang, Sept. 1985*. Serdang, Malaysia; Penerbit Universiti Pertanian Malaysia, 1986. 516p. T

International Symposium on Pesticide Use in Developing Countries-Present and Future; Proceedings of a Symposium on Tropical Agriculture Research, Kyoto, Sept. 1982. Japan; Tropical Agriculture Research Center, 1983. 199p.

**International Symposium on Virus Diseases of Rice and Leguminous Crops in the Tropics; Proceedings of a Symposium on Tropical Agriculture Research, Tsukuba, Oct. 1985*. Tsukuba, Japan; Tropical Agriculture Research Center, Ministry of Agriculture, Forestry and Fisheries, 1986. 262p. (*Tropical Agriculture Research Series No. 19*)

Janick, Jules, ed. *Plant Breeding Reviews*. Vol. 10. New York; J. Wiley and Sons, Inc., 1992. 374p. D

Janick, Jules. *Horticultural Science*. 4th ed. New York; W. H. Freeman, 1986. 746p. D

Jayaraj S., ed. *Integrated Pest and Disease Management; Proceedings of the National Seminar . . .* Tamil Nadu Agricultural University, Coimbatore, India, 1985. 416p. T

*Jones, D. Gareth and Brian C. Clifford. *Cereal Diseases: Their Pathology and Control*. 2d ed. Chichester and New York; J. Wiley, 1983. 309p.

Kass, Donald C. L. *Polyculture Cropping Systems: Review and Analysis*. Ithaca, N.Y.; NYS College of Agriculture and Life Sciences, Cornell University, 1978. 69p. (*Cornell International Agriculture Bulletin* no. 32) T

Kranz, J., H. Schmutterer, and W. Koch, eds. *Diseases, Pests and Weeds in Tropical Crops*. Berlin; Verlag Paul Parey, 1977. 666p.; and New York; Wiley-Interscience, 1978. 704p. T

*Kranz, J., H. Schmutterer, and W. Koch, eds. *Krankheiten, Schädlinge und Unkräuter im Tropischen Pflanzenbau*. Berlin and Hamburg, Germany; Verlage P. Parey, 1979. 679p.

*Lorenz, Klaus J. and Karel Kulp, eds. *Handbook of Cereal Science and Technology.* New York; M. Dekker, Inc., 1991. 882p.

Miller, Ross. *Insect Pests of Wheat and Barley in West Africa and North Africa.* Aleppo, Syria; International Center for Agricultural Research in Dry Areas, 1987. 209p. (Technical Manual No. 9) T

Mukerji, K. G. and K. L. Garg, eds. *Biocontrol of Plant Diseases.* Boca Raton, Fla.; CRC Press, Inc., 1988. 2 vols. D&T

Mathur, S. B. and Johs Jørgensen, eds. *Seed Pathology.* Meppel, Holland; Technical Centre for Agricultural and Rural Co-operation, CTA, 1992. 412p. T

Pimentel, David, ed. *CRC Handbook of Pest Management in Agriculture.* 2nd ed. Boca Raton, Fla.; CRC Press, 1991. Vol.1, D&T; Vols. 2–3, D

Poehlman, John M. *Breeding Field Crops.* 3rd ed. Westport, Conn.; AVI Publishing Co., Inc., 1987. 724p. D

*Raychaudhuri, S.P. and J.P. Verma. *Review of Tropical Plant Pathology.* New Delhi, India; Today and Tomorrow's Printers and Publishers, 1986. 2 vols.

Rehm, Sigmund and Gustav Espig. *The Cultivated Plants of the Tropics and Subtropics: Cultivation, Economic Value, Utilization.* Weikersheim, Germany; Margraf, 1991. 552p. T

Röbbelen, Gerhard, R. Keith Downey, Amram Ashri, eds. *Oil Crops of the World: Their Breeding and Utilization.* New York and London; McGraw-Hill, 1989. 553p. T

Russell, G. E., ed. *Progress in Plant Breeding—1.* London and Boston; Butterworths, 1985. 325p. D&T

*Schlösser, Eckart. *Allgemeine Phytopathologie.* Stuttgart; Georg Thieme Verlag, 1983. 257p.

Simmonds, N. W. *Principles of Crop Improvement.* London and New York; Longman, 1979. 399p. D&T

Singhal, G. S., et al., eds. *Photosynthesis: Molecular Biology and Bioenergetics; Proceedings of the International Workshop on the Application of Molecular Biology and Bioenergetics of Photosynthesis.* New Delhi; Springer-Verlag and Narosa Publishing House, 1989. 441p. D&T

Steward, Frederick C. and R. G. S. Bidwell, eds. *Plant Physiology: A Treatise. Vol. VIII: Nitrogen Metabolism.* Orlando, Fla.; Academic Press, 1983. 449p. D

Summerfield, R. J. and A. H. Bunting, eds. *Advances in Legume Science.* Vol. 1 of the *Proceedings of the International Legume Conference, Kew, July–August 1978.* Richmond, Eng.; Royal Botanic Gardens, Kew; and Ministry of Agriculture, Fisheries and Food, 1980. 667p. D

Thurston, H. David. *Sustainable Practices for Plant Disease Management in Traditional Farming Systems.* Boulder, Colo.; Westview Press, 1991. T

Journal Articles and Serial Chapters

Annual Review of Phytopathology

Vol. 26 (1988). Pp. 331–350: Perspectives on Progress in Plant Virology, by Myron K. Brakke. 112 refs.

Vol. 25 (1987). Pp. 67–85: Evolving Concepts of Biological Control of Plant Pathogens, by Kenneth F. Baker. 13 refs.

Vol. 25 (1987). Pp. 87–110: The Impact of Molecular Genetics on Plant Pathology, by Allen Kerr. 151 refs.

CRC Critical Reviews in Plant Sciences
*Vol. 1 (1983–84)
Pp. 49–93: Improvement of Protein Quality in Cereals, by Simon W. J. Bright and Peter R. Shewry.
Pp. 133–181: Current Status of Crop Plant Germplasm, by Garrison Wilkes.
Pp. 183–201: Salt Tolerance of Food Crops: Prospectives for Improvements, by W. J. S. Downton.
Vol. 4 (1986)
Pp. 1–46: Genetic Transformation in Higher Plants, by Robert T. Fraley, Stephen G. Rogers and Robert B. Horsch. 317 refs.
**Plant Breeding Abstracts*, Vol. 60 (1990)
May. Pp. 461–467: Rice Breeding—Accomplishments and Challenges, by Gurdev S. Khush. 6 refs.
July. Pp. 728–731: The Evolution of Agriculturally Significant Legumes, by J. Smartt. 14 refs.

B. Compilation and Citation Analysis

The 37,963 citations in these source publications were systematically analyzed and the following data gathered:

Title of publication
Date of publication
Format of publication (e.g., journal or monograph)
Category of publisher of monographs (e.g., commercial press, university, government)

During this process, the titles of monographs were noted and entered into a computerized list. Each time the same monograph or a chapter in it was cited, a tally was made for that title. Similarly, each time a journal or report series was cited, a tally was made for it. The end result is a systematic count of journals, report series, and monographs cited and the numbers of times. Additional select data were gathered which provide the basis for some analysis in this and the following chapter.

Before we examine the results, some definitions must be understood. Distinctions among journals, serial works, and monographs are necessary. Series issues, short works, and books were treated as monographs when they were cited as distinct works with an author or editor, when a title was distinctive, and when the item was complete in itself. Therefore, a work such as a U.S. Dept. of Agriculture *Technical Bulletin* in which the subject was complete in the issue and distinct from the next issue in the series was counted and identified as a monograph. The same was true of proceedings volumes with distinctive titles which varied from one year to the next. Those without distinctive titles or special subject foci were counted as journals. Journals follow the pattern of having several articles on different and very specific subjects, each usually only a few pages long. Chapters in books were counted as monographic titles. These definitions worked well since they followed the citation patterns of crop scientists.

In the compilation process select materials were excluded:

(1) Very short monographs, fifty pages or fewer, which were cited only once
(2) Country, state, or provincial documents of brief pagination, often highly specialized and site specific
(3) Select, esoteric works when in an uncommon language, or when very limited in geographic scope
(4) Specialized geographic materials when not in a national or international context, which were cited only once
(5) Early background, technique, and statistical working tools (some frequently cited works of this type were retained)
(6) Early editions, which were combined with the latest edition and listed only once

The purpose of an article or book dictates the nature of the literature it cites. Therefore, a journal article reporting the results of research will tend to cite research literature, supporting methodologies, and overviews from earlier works. As mentioned, this study has attempted to identify advanced university instructional literature along with basic research literature. To accomplish this, the choice of the literature had to be systematic. Much of the journal literature analyzed is oriented toward the latest developments and current research. It is logical to expect some citation differences between this journal corpus and that of the monographs. Since the divergence was minor, the two types of literature are not segregated in this summary.

Citations from the source documents were 67.7% to journals, 31.5% to monographs, and .8% to dissertations. This is somewhat different than the literature percentages from citation analysis in the earlier studies of the Core Agricultural Literature Project.[7] Variations are shown in Table 8.1.

Technical report series where each report is on a specific subject and complete in that issue were counted as monographs since these individual items are actually short monographs. They are put in a report series for convenience of publishing, storage, and distribution. Monographs therefore may be subdivided into reports in series and more traditional monographs. Reports amounted to 16.0% of all monographs cited, and only 3.4% of the total 37,963 citations analyzed.

7. (a) Wallace C. Olsen, ed., *The Literature of Animal Science and Health* (Ithaca, N.Y.: Cornell University Press, 1993), p. 79. (b) Peter McDonald, ed., *The Literature of Soil Science* (Ithaca, N.Y.: Cornell University Press, 1994), chapter 9. (c) Jennie Brogdon and Wallace C. Olsen, eds. *The Contemporary and Historical Literature of Food Science and Human Nutrition* (Ithaca, N.Y.: Cornell University Press, 1995), chapter 4.

Table 8.1. Literature formats of citations analyzed in journals and monographs

	% to journals	% to monographs	% to dissertations, patents, standards
Agricultural Economics and Rural Sociology	54.8%	43.5%	1.7%
Agricultural Engineering	58.1	38.2	3.7
Animal Science and Health	75.0	24.0	1.0
Soil Science	72.9	27.0	.1
Food Science and Human Nutrition	75.6	22.7	1.7
Crop Science	67.7	31.5	.8

C. Monograph Peer Evaluation and Tabulations

The monograph list compiled by citation analyses was examined by the Steering Committee members prior to being sent for peer evaluation. Two committee members with extensive experience in developing countries were also asked to look at the list and determine which titles had application for the Third World. Third World scientists did the same.

These processes resulted in:

(1) Breaking the master list into two portions, one for the developed countries and a second for the Third World nations. There was extensive overlap between the two.

(2) A developed countries list that included nearly 2,500 titles. Since the next step was to have these titles ranked by crop scientists, the Steering Committee felt that evaluations would be unrealistic unless the titles were placed into subject groups and distributed to specialists in those subjects. This resulted in six subject lists. The first subject list was sent to all reviewers along with the appropriate list for their subject specialty. Lists were issued in these categories:
   (a) Crop Ecology, Production, and Management; Plant Structure; Nutrition; Taxonomy; Fertilizers, and Growth Regulators; Miscellaneous
   (b) Crop Breeding; Pollination; Resistance; Germplasm
   (c) Crop Physiology; Seed Physiology; Cell Biology; Most Myco- and Aflo-Toxins
   (d) Pests, Insects and Other Animals, and Control
   (e) Plant Diseases and Control; Fungal; Bacterial; Viral; Other
   (f) Weeds and Control

Instructions to the reviewers carefully discussed the aim of identifying monographs of academic and instructional value for Third World institutions, or those which should be represented in a university library in the developed world. These six lists for developed countries went through several iterations. Titles were dropped and added upon the recommendation of reviewers (although recommendations were few) as well as through the citation analysis process. Approximately 2,500 monograph titles were evaluated in the developed countries list, of which 1,208 made the final core compilation. In a similar manner a maximum of 2,000 monographs were evaluated in the Third World listing, which resulted in a core of 1,286. This is a rejection rate of 35.9% for Third World monographs and 51.7% for developed countries titles.

Evaluators were sought with a minimum of fifteen years experience in their profession and with extensive instructional and research experience. People with more than twenty years' experience proved particularly valuable in putting short term theories and practices into perspective.

Rejected titles tended to fall into these categories: (1) the work is outdated, although it may have been very good twenty or thirty years before; (2) the language is too difficult for scientists to handle easily, and the work therefore is less well known or used than others; (3) the subject matter is acceptable but the approach is arcane or badly presented.

Following are the names of reviewers for each type of list.

Third World Crop Science Monograph Reviewers

S. Adisoemarto
Center for Research on Biotechnology
Bogor, Indonesia

J. Lawrence Apple
North Carolina State University

Amarjit S. Basra
Punjab Agricultural University
Ludhiana, India

Malcolm Blackie
Rockefeller Foundation
Lilongue, Malawi

Elkin Bustamante
Centro Agronomico Tropical de Investigacion y Ensenanza
Turrialba, Costa Rica

M. Y. Chiang
Taiwan Agricultural Chemical and Substances Research Institute
Taiching, Taiwan

Desiree Cole
University of Zimbabwe
Harare

Jan deWet
International Crops Research Institute for the Semi-Arid Tropics
Andhra Pradesh, India

Jose Carlos Dianese
Universidad de Brasilia
Brasilia, Brazil

Jack Harlan
New Orleans, La.

E. W. Kitajima
Universidad de Brasilia
Brasilia, Brazil

D. Mbewe
University of Zambia
Lusaka

D. Muraleedharan
University of Kerala
Trivadrum Kerala, India

Mano D. Pathak
Mahanagar Ext.,
Lucknow, India

F.W. Penning de Vries
Center for Agrobiological Research
Wageningen, Netherlands

Vincent W. Saka
Bunda College of Agriculture
Lilongue, Malawi

S. Sastrapradja
Center for Research on Biotechnology
Bogor, Indonesia

C.N. Sun
Tainshui, Taitei
Taiwan

James Teri
Sokoine University of Agriculture
Morogoro, Tanzania

Combined Reviews
National Chung Hsing University,
Taipei, Taiwan
T. T. Chang
S. T. Hsu
T. H. Su
University of Ilorin, Nigeria
J. Adedokun
E.O. Salawu
J.A. Ogunwale
Olofintoye

Developed Countries Crop Science Monograph Reviewers

Pedro J. Aparicio
Centro de Investigaciones Biológicas
Madrid, Spain
A.P. Appleby
Oregon State University
R.B. Austin
Cambridge, U.K.
Y.I. Chu
National Taiwan University
Taipei
M.F. Claridge
University College
Cardiff, U.K.
W. Ronnie Coffman
Cornell University
David Conner
University of Melbourne
Clive A. Edwards
Ohio State University
Lester E. Ehler
University of California, Davis
Gary Heichel
University of Illinois
Charles B. Heiser
Indiana University
Paul C. Holliday
Leichester, U.K.
David J. Hume
University of Guelph, Canada
D. Gareth Jones & Colleagues
University College of Wales
Dyfed, U.K.
Jürgen Kranz
University of Giessen, Germany
W. Laux
Biologische Bundesanstalt
Berlin, Germany
Jill Lenné
International Germplasm Associates, Cumbria
U.K.
Arlene E. Luchsinger
University of Georgia
K.N. Mehrotra
Banaras Hindu University
Varanasi, India
David Pimentel
Cornell University
James Ridsdill-Smith
CSIRO, Wembley
Western Australia
Carmelo Rigano
Università di Napoli, Italy
George A. Schaefers
Agricultural Experiment Station
Geneva, N.Y.
John Scott
CSIRO, Wembley
Western Australia
Fumiki Takahashi
Hiroshima University
H. David Thurston
Cornell University
N.C. Turner
CSIRO, Wembley
Western Australia
Milton Zaitlin
Cornell University
Joint Evaluation; Wageningen Group
Andrei Dusink
Technical Centre for Agricultural and Rural Co-operation
Gerbrand Kingma
Formerly with CIMMYT
Gijsbert A. M. van Marrewijk
Wageningen Agricultural University

Several elements were used to rank the titles. Numeric scores were assigned to these elements and a computation made for each title. Procedures were as follows:

(1) The counts made each time a monograph or a chapter of a monograph was cited in the Sources of Citations listed earlier in this chapter were used. This element was given the weight of 1 per citing.
(2) Rankings by reviewers were coded for each of the two top rankings. These were graded 2 and 1 and multiplied times the number of recommendations in

each category. A statistical equalization on the six developed countries lists was required because the first, a general list, was sent to each reviewer resulting in eight to nine times as many evaluations as for the specialty lists.
(3) If a title was reprinted it was given a score of 1.
(4) If a title went through more than one edition it was given a score of 2.

The equation for the computations is:

$$(\# \text{ Counts} \times 1) + (\# \times 2 + \# \times 1) + (\text{Reprint} \times 1) + (\text{Editions} \times 2)$$

This formula was used for both lists of monographs, which were ranked separately. Using this equation, the peer evaluations accounted for between 78% and 90% of the cumulative scores in the two lists.

Three ranked categories within each list are provided to aid in making decisions for purchase, preservation, or collection assessment. Within each list the scores were broken into three logical ranges. The fewest titles in both lists are in the first rank, the most important category. Percentages of titles in each list by ranking are:

Developed countries list			Third World list	
No. of titles	Percentage		No. of titles	Percentage
152	12.6%	First rank	181	14.1%
582	48.1	Second rank	537	41.7
474	39.3	Third rank	568	44.2
1,208	100.0%		1,286	100.0%

Readers are reminded that all the titles in this core list are valuable monographs for instruction and research today. Those which are less valuable were never entered for consideration in select cases, or were removed by peer-evaluations with a drop rate of 35.9% for Third World countries and 51.7% for developed countries. Therefore, all the titles in the two lists should be viewed as vital literature.

The two lists have been merged in the following compilation. Rankings for a title are within its own listing only, i.e., developed countries or Third World list. Where no ranking is shown, the title is not in that list. Of the 1,663 unique titles in the combined list 831 are common to both, a 49.9% duplication. These 831 constitute the most valuable core titles, since they are highly rated in both groups. Figure 8.1 shows the score on monographs by cumulation.

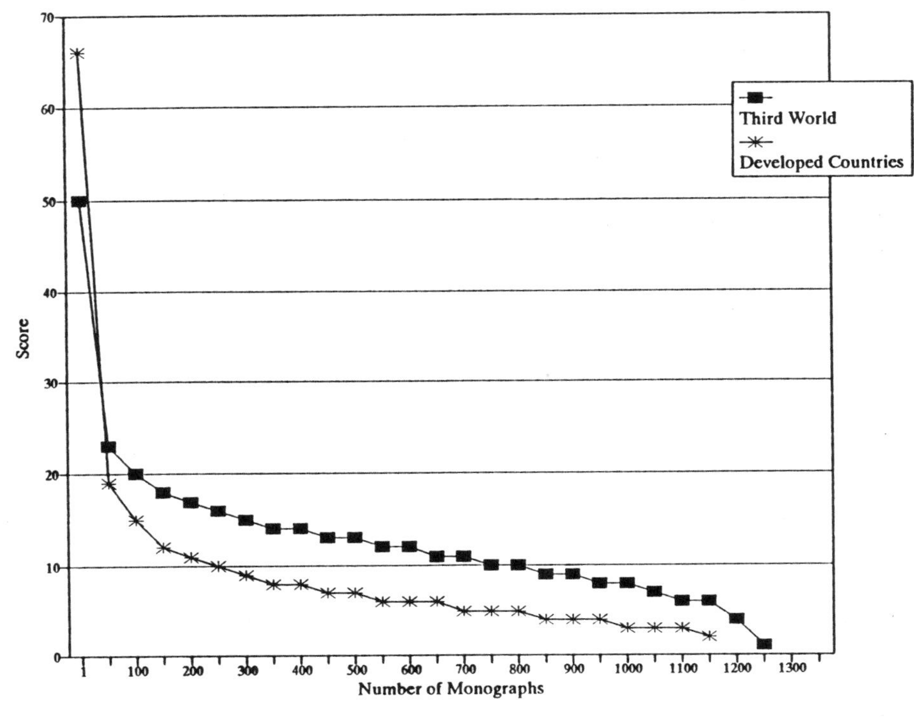

Figure 8.1. Final scoring on monographs.

Core Monographs for Developed and Third World Countries, 1,663 titles

Developed Countries Ranking		Third World Ranking
	Abawi, G. S., and M. A. Pastor-Corrales. Root Rot of Beans in Latin American and Africa: Diagnosis, Research Methodologies, and Management Strategies. Cali, Colombia; CIAT, 1990. 114p.	Third
Second	Abbott, A. J., and R. K. Atkin, eds. Improving Vegetatively Propagated Crops. London, Eng. and San Diego, Calif.; Academic Press, 1987. 416p.	Third
Third	Abeles, Frederick B., Morgan W. Page, and Mikal E. Saltveit. Ethylene in Plant Biology. 2d ed. San Diego, Calif.; Academic Pres, 1992. 414p. (1st ed. by F. B. Abeles, New York; Academic Press, 1973. 302p.)	
	Acland, Julien D. East African Crops: An Introduction to	Third

Developed Countries Ranking		Third World Ranking
	the Production of Field and Plantation Crops in Kenya, Tanzania and Uganda. London; Longman, 1971. 252p. (Reprint, 1975. 252p.)	
	Adams, C. R., K. M. Bamford, and M. P. Early. Principles of Horticulture. London; Heinemann, 1984. 197p. (Also available in Spanish as Principios de Hortofruticultura. Zaragoza; Acribia, 1989. 241p.)	Second
Third	Agarwal, Vijendra K., and James D. Sinclair. Principles of Seed Pathology. Boca Raton, Fla.; CRC Press, 1987. 2 vols.	Third
Second	Agricultural Research Institute (U.S.). Improving On-Target Placement of Pesticides; A conference, Reston, Va., June 1988, sponsored by the Agricultural Research Institute in cooperation with the U.S. Dept. of Agriculture, Science and Education. Bethesda, Md.; Agricultural Research Institute, 1988. 220p.	
First rank	Agrios, George N. Plant Pathology. 3d ed. San Diego; Academic Press, 1988. 803p. (1st ed., New York; Academic Press, 1969. 629p.)	First rank
Second	Ahmad, Sami, ed. Herbivorous Insects: Host-seeking Behavior and Mechanisms. New York; Academic Press, 1983. 257p.	Second
	Ahn, Peter M. West African Agriculture. 3d ed. Vol. 1. West African Soils by P.M. Ahn and Vol. 2. West African Crops by F.R. Irvine. London, Eng.; Oxford University Press, 1970. 2 vols. (1st -2d eds., by F.R. Irvine titled: A Text-book of West African Agriculture, Soils and Crops. 2d ed., 1953. 367p.)	Second
First rank	Ainsworth, G. C., and Alfred S. Sussman, ed. The Fungi: An Advanced Treatise. New York; Academic Press, 1965–73. 4 vols.	First rank
Second	Ainsworth, Geoffrey C. Introduction to the History of Plant Pathology. Cambridge, Eng. and New York; Cambridge University Press, 1981. 315p.	Third
	Aiyer, A. K. Y.N. Field Crops of India: With Special Reference to Karnataka. 7th ed. Bangalore; BAPPCO, 1980. 564p. (3d ed., 1950 titled: . . . Special Reference to Mysore. 686p.)	Second
Second	Akehurst, B. C. Tobacco. 2d ed. London and New York; Longman, 1981. 764p. (1st ed., Harlow; Longmans, 1968. 551p.)	First rank
	Akinola, J. O., P. C. Whiteman, and E. S. Wallis. The Agronomy of Pigeon Pea (Cajanus Cajan). Farnham Royal, Eng.; Commonwealth Agricultural Bureaux, 1975. 57p.	Second

Developed Countries Ranking		Third World Ranking
Third	Akobundu, I. Okezie. Weed Science in the Tropics: Principles and Practices. Chichester, Eng. and New York; Wiley, 1987. 522p.	First rank
	Akobundu, I. O., and A. E. Deutsch, eds. No-Tillage Crop Production in the Tropics; Proceedings of a symposium, Monrovia, Liberia, Aug. 1981, sponsored by the W. African Weed Science Society and the International Weed Science Society in collaboration with the W. African Rice Development Association. Corvallis, Or.; International Plant Protection Center, Oregon State University, 1983. 235p.	Second
Second	Akoyunoglou, George, ed. Photosynthesis; Proceedings of the 5th International Congress of Photosynthesis, Halkidiki, Greece, Sept. 1980. Philadelphia; Balaban International Science Services, 1981. 6 vols.	Third
	Akoyunoglou, George, and Horst Senger, eds. Regulation of Chloroplast Differentiation; Proceedings of an International Meeting, Rhodes, Greece, July 1985. New York; A.R. Liss, 1986. 790p.	
First rank	Aldrich, Richard J. Weed Crop Ecology: Principles in Weed Management. N. Scituate, Mass.; Breton Publishers, 1984. 465p.	First rank
Third	Aldrich, Samuel R., Walter O. Scott, and Robert G. Hoeft. Modern Corn Production. 3d ed. Champaign, Ill.; A & L Publications, 1986. 358p. (1st ed. by Samuel R. Aldrich and Earl R. Leng. Cincinnati, Ohio; Farm Quarterly, 1965. 308p.)	Second
	Alexander, Martin, ed. Biological Nitrogen Fixation: Ecology, Technology, and Physiology; Proceedings of a Training Course, . . . Caracas, Venezuela, January, 1982. New York; Plenum Press, 1984. 247p.	Second
First rank	Alexopoulos, Constantine J. Introductory Mycology. 3d ed. New York; Wiley, 1979. 632p. (1st ed., 1962.) (Also available in Spanish as Introducción a la Micología. 3d ed. Buenos Aires; Eudeba. 614p.)	Second
	Allaby, Michael. World Food Resources: Actual and Potential. London; Applied Science Publishers, 1977. 418p.	Second
First rank	Allard, Robert W. Principles of Plant Breeding. New York; J. Wiley, 1960. 485p. (Also available in Spanish as Principios de la Mejora Genética de las Plantas. Barcelona; Omega. 498p.)	First rank
Second	Allen, D. J. The Pathology of Tropical Food Legumes: Disease Resistance in Crop Improvement. Chichester, Eng. and New York; Wiley, 1983. 413p.	Second

Developed Countries Ranking		Third World Ranking
First rank	Allen, George E., Carolo M. Ignoffo, and Robert P. Jaques, eds. Microbial Control of Insect Pests: Future Strategies in Pest Management Systems; Selected papers from NSF-USDA-University of Florida Workshop, Gainesville, Fla., Jan. 1978, sponsored by National Science Foundation, et al. 1979. 290p.	
Second	Allen, O. N., and Ethel K. Allen. The Leguminosae: A Source Book of Characteristics, Uses, and Nodulation. Madison, Wis.; University of Wisconsin Press, 1981. 812p.	Third
	Allen, Patricia, and Debra Van Dusen, eds. Global Pespectives on Agroecology and Sustainable Agricultural Systems; Proceedings of the 6th International Scientific Conference of the International Federation of Organic Agriculture Movements, University of California, Santa Cruz, Calif., Aug. 1986. Santa Cruz, Calif.; Agroecology Program, University of California, 1988. 2 vols. 730p.	First rank
	Allen, Peter W. Natural Rubber and the Synthetics. London, Eng.; C. Lockwood, 1972. 255p.	Third
	Altieri, Miguel A. Agroecology: The Scientific Basis of Alternative Agriculture. 2d ed. Boulder, Colo.; Westview Press, 1994. 300p. (2d ed., Berkeley, Calif.; M.A. Altieri, 1984. 162p. Berkeley, Calif.; University of California at Berkeley, 1983.)	Second
	Altieri, Miguel A., and Matt Liebman, eds. Weed Management in Agroecosystems: Ecological Approaches. Boca Raton, Fla.; CRC Press, 1988. 354p.	Second
Second	Alvim, Paulo de T., and T. T. Kozlowski, eds. Ecophysiology of Tropical Crops. New York; Academic Press, 1977. 502p.	Second
Second	American Association of Cereal Chemists. Advances in Cereal Science and Technology. St. Paul, Minn.; American Association of Cereal Chemists, 1976–90. 10 vols.	Third
Second	American Chemical Society. Fate of Organic Pesticides in the Aquatic Environment; A symposium sponsored by the Division of Pesticide Chemistry at the 161st meeting of the American Chemical Society, Los Angeles, Calif. March 1971. Washington, D.C.; American Chemical Society, 1972. 280p.	
Second	American Chemical Society. Pesticides in Tropical Agriculture. A collection of papers comprising the Symposium on Pesticides in Tropical Agriculture, presented before the Division of Agricultural and Food Chemistry at the 126th meeting of the American Chemial Society,	Third

Developed Countries Ranking		Third World Ranking
	New York, Sept. 1954. Washington, D.C.; American Chemical Society, 1955. 102p.	
Second	American Oil Chemists' Society. Official and Tentative Methods of the American Oil Chemists' Society. Editor of Analytical Methods, 1945–50, V.C. Mehlenbacher; 1950–58, T.H. Hopper; 1958– E.M. Sallee. 2d ed. Chicago, Ill.; The Society, 1945–. 1 vol.	
	American Peanut Research and Education Association. Peanuts — Culture and Uses; Proceedings of a Symposium . . . Stillwater, Okla.; American Peanut Research and Education Association, 1973. 684p.	Second
Second	Amesz, J., ed. Photosynthesis. Amsterdam and New York; Elsevier, 1987. 355p.	Third
	Ananthakrishnan, T. N., and A. Raman, eds. Dynamics of Insect-Plant Interaction: Recent Advances and Future Trends. New Delhi, India; Oxford and IBH Pub. Co., 1988. 223p.	Third
Third	Anderson, Edgar S. Plants, Man and Life. Rev. ed. Berkeley, Calif.; University of California Press, 1971. 251p. (1st ed., Boston; Little, Brown, 1952. 245p.)	Second
	Anderson, Harry W. Diseases of Fruit Crops. New York; McGraw-Hill, 1956. 501p.	Third
Second	Anderson, J. M., A. D. M. Rayner, and D. W. H. Walton, eds. Invertebrate-Microbial Interactions; Joint symposium of the British Mycological Society and the British Ecological Society, University of Exeter, Sept. 1982. Cambridge, Eng. and New York; Cambridge University Press, 1984. 349p.	
Third	Anderson, Jock R., and Peter B. R. Hazell, eds. Variability in Grain Yields: Implications for Agricultural Research and Policy in Developing Countries. Baltimore, Md.; Johns Hopkins University Press, 1989. 395p.	Second
	Anderson, R. G., ed. Proceedings of the Wheat Triticale and Barley Seminar, El Batan, Mexico, 1973. El Batan, Mexico; Centro Internacional de Mejoramiento de Maiz y Trigo, 1973. 378p.	Second
Second	Anderson, Wood P. Weed Science: Principles. 2d ed. St. Paul, Minn.; West Pub. Co., 1983. 655p. (1st ed., New York; West Publishing, 1977. 598p.)	First rank
	Andrew, C. S., and E. J. Kamprath, eds. Mineral Nutrition of Legumes in Tropical and Subtropical Soils; Proceedings of a workshop, CSIRO Cunningham Laboratory, Brisbane, Australia, Jan. 1978. Melbourne, Australia; Commonwealth Scientific and Industrial Research Organization, 1978. 415p.	First rank
First rank	Andrewartha, Herbert G. Introduction to the Study of Ani-	

Developed Countries Ranking		Third World Ranking
	mal Populations. 2d ed. Chicago, Ill.; University of Chicago Press, 1971. 283p. (1st ed., 1961. 281p.)	
Second	Andrews, Jean. Peppers: The Domesticated Capsicums. Austin, Tex.; University of Texas Press, 1984. 170p.	Second
	Angladette, Andre. Le Riz . . . Paris, France; G.P. Maisonneuve et Larose, 1966. 931p.	Second
	Angladette, Andre, and Louis Deschamps. Problemes et Perspectives de l'Agriculture dans les Pays Tropicaux. Paris, France; G.P. Maisonneuve et Larose, 1974. 770p. (Techniques Agricoles et Productions Tropicales No. 25–27)	Second
Second	Annecke, David P., and F. C. Moran. Insects and Mites of Cultivated Plants in South Africa. Durban and Woburn, Mass.; Butterworths, 1982. 383p.	Second
	Appert, Jean, and Jacques Deuse. Les Ravageurs des Cultures Vivrieres et Maraicheres sous les Tropiques. Paris, France; G.P. Maisonneuve and Larose, Agence de Cooperation Culturel le et Technique, 1982. 420p. (Techniques Agricoles et Productions Tropicales no. 31)	Second
First rank	Apple, J. L., and R. F. Smith. Integrated Pest Management. New York; Plenum, 1976. 200p.	First rank
Third	Araullo, E. V., D. B. De Padua, and Michael Graham, eds. Rice: Postharvest Technology. Ottawa, Canada; International Development Research Centre, 1976. 394p.	Second
	Araullo, E. V., Barry Nestel, and Marilyn Campbell, eds. Cassava Processing and Storage; Proceedings of an interdisciplinary workshop, Pattaya, Thailand, Apr. 1974, sponsored by the International Development Research Centre. Ottawa, Canada; International Development Research Centre, 1974. 125p.	Third
Third	Arber, Werner, et al, eds. Genetic Manipulation: Impact on Man and Society; Papers from the 3d GOGENE Symposium, Cologne, Germany, Apr. 1983, organized with the collaboration of the Institute of Genetics, University of Cologne and the Federation of European Biochemical Societies. Cambridge, Eng. and New York; Cambridge University Press, 1984. 250p.	Third
	Are, L. A., and D. R. G. Gwynne-Jones. Cacao in West Africa. Ibadan, Nigeria; Oxford University Press, 1974. 146p.	Second
	Arnon, Itzhak. Crop Production in Dry Regions, edited by Nicholas Polunin. New York; Barnes and Noble Books, 1972. 2 vols.	Second
	Arnon, Itzhak. Mineral Nutrition of Maize. Bern; International Potash Institute, 1975. 452p.	Third
Third	Arora, D. K., et al, eds. Soil and Plants. New York; M.	Third

Developed Countries Ranking		Third World Ranking
	Dekker, 1991. 720p. (Handbook of Applied Mycology No. 1)	
Second	Arx, Josef A. von. The Genera of Fungi Sporulating in Pure Culture. 3d rev. ed. Vaduz; J. Cramer, 1981. 424p. (1st ed., Lehre, J. Cramer, 1970. 288p.)	Second
Third	Asada, Yasuji, et al, eds. Plant Infection: The Physiological and Biochemical Basis; Proceedings of the U.S.-Japan seminar, Brainerd, Minn., May 1981. Tokyo, Japan; Japan Scientific Societies Press and Berlin, Germany and New York; Springer-Velag, 1982. 362p.	Third
Second	Asher, M. J. C., and P. J. Shipton, eds. Biology and Control of Take-All. London and New York; Academic Press, 1981. 538p.	
	Ashri, A., ed. Sesame and Safflower—Status and Potentials; Proceedings of an expert consultation. Rome; Food and Agriculture Organization, 1985. 223p. (FAO Plant Production and Protection Paper No. 66)	Second
Third	Ashri, A., ed. Sesame: Status and Improvement; Proceedings of expert consultation, Rome, Italy, Dec. 1980. Rome; Food and Agriculture Organization, 1981. 198p. (FAO Plant Production and Protection Paper No. 29)	Third
Second	Ashton, Floyd M., and Alden S. Crafts. Mode of Action of Herbicides. 2d ed. New York and Chichester, Eng.; Wiley, 1981. 525p. (1st ed., 1973. 504p.)	First rank
Second	Ashton, Floyd M., and Thomas J. Monaco. Weed Science: Principles and Practices. 3d ed. New York; Wiley, 1991. 466p. (1st ed. by G.C. Klingman and F.M. Ashton, 1975. 431p. A rev. of Weed Control: As a Science, 1961. 421p.)	Second
	Asian Productivity Organization. Vegetable Production and Marketing in Asia and the Pacific. Tokyo; Asian Productivity Organization, 1989. 497p.	Third
Second	Askew, R. R. Parasitic Insects. New York; American Elsevier, 1971. 316p.	
	Association for the Advancement of Agricultural Sciences in Africa. Food Crisis and Agricultural Production in Africa: Problems, Policies and Solutions; Proceedings of the 3d General Conference of the AAASA, University of Ibadan, Ibadan, Nigeria, Apr. 1978. Addis Ababa, Ethiopia; Association for the Advancement of Agricultural Sciences in Africa, 1981+. 3 vols.	Third
Third	Association of Applied Biologists. Environmental Aspects of Applied Biology; Texts and papers and posters presented at a residential meeting of the Association of Applied Biologists, University of York, Eng., Sept. 1988.	

Developed Countries Ranking		Third World Ranking
	Wellesbourne, Eng.; Association of Applied Biologists, 1988. 2 vols.	
	Association of Applied Biologists. Influence of Environmental Factors on Herbicide Performance and Crop and Weed Biology; Symposium, Dec. 1983, St. Catherine's College, Oxford . . . with European Weed Research Society and Society for Chemical Industry. Wellesbourne, Warwick; The Association, 1983. 552p.	Third
	Association of Japanese Agricultural Scientific Societies (Nihon Nogakkai), ed. Rice in Asia; Selection of translated papers from the Association's symposia on Rice in the World held between 1966 and 1972. Tokyo, Japan; University of Tokyo Press, 1975. 600p.	Second
Third	Association of Offical Seed Certifying Agencies (AOSCA). AOSCA Certification Handbook. Clemson, S.C.; AOSCA, 1971 and later. 1 vol. (looseleaf) (AOSCA Publication No. 23)	
Third	Atherton, J. G., ed. Manipulation of Flowering; Proceedings of the 45th Nottingham Easter School in Agricultural Science, Sutton Bonington, Apr. 1986. London and Boston, Mass.; Butterworths, 1987. 438p.	
Third	Atherton, J. G., and J. Rudich, eds. The Tomato Crop: A Scientific Basis for Improvement. London, Eng. and New York; Chapman and Hall, 1986. 661p.	Third
Third	Atkinson, Burr G., and David B. Walden, eds. Changes in Eukaryotic Gene Expression in Response to Environmental Stress. Orlando, Fla.; Academic Press, 1985. 379p.	
Third	Atkinson, Daniel E. Cellular Energy Metabolism and its Regulation. New York; Academic Press, 1977. 293p.	
Third	Atkinson, John D. Diseases of Tree Fruits in New Zealand. Wellington; Govt. Print. Off., 1971. 406p.	
	Atwal, A. S. Agricultural Pests of India and South-East Asia. 2d ed. New Delhi, India; Kalyani Pub., 1986. 509p. (1st ed., 1976. 502p.)	Third
Third	Aubert, J-P, P. Beguin, and J. Millet, eds. Biochemistry and Genetics of Cellulose Degradation; Proceedings of a symposium, organized by the Federation of the Microbiological Societies and the French Society for Microbiology, Sept. 1987. London, Eng. and New York; Academic Press, 1988. 428p.	
Second	Audus, Leslie J. Herbicides: Physiology, Biochemistry, Ecology. 2d ed. New York; Academic Press, 1976. 2 vols. (1st ed., 1964, titled: The Physiology and Biochemistry of Herbicides. 555p.)	Third

Developed Countries Ranking		Third World Ranking
	Audus, Leslie J. Plant Growth Substances. 3d ed. New York; Barnes and Noble, 1972. 495p. (1st ed., New York; Interscience Publishers, 1953. 465p.)	Second
Second	Auld, B. A., K. M. Menz, and C. A. Tisdell. Weed Control Economics. London and Orlando, Fla.; Academic Press, 1987. 177p.	Third
Second	Austin, R. B., ed. Decision Making in the Practice of Crop Protection; Proceedings, 1982 British Crop Protection Symposium, University of Sussex, April 1982. Croyden, Eng.; BCPC Publications, 1982. 238p.	
	Ayanaba, A., and P. J. Dart, eds. Biological Nitrogen Fixation in Farming Systems of the Tropics. Based on papers presented at a Symposium, International Institute of Tropical Agriculture, Ibadan, Nigeria, Oct. 1975. Chichester, Eng. and New York; Wiley, 1977. 377p.	Second
	Ayensu, E. S., et al, eds. Striga: Biology and Control; Papers presented at a workshop on the Biology and Control of Striga, Dakar, Senegal, Nov. 1983. Miami, Fla.; ICSU Press, 1984. 216p.	Second
	Ayoub, Sadek M. Plant Nematology: An Agricultural Training Aid. Rev. Sacramento, Calif.; NemaAid Publication, 1980. 195p.	Second
Second	Ayres, P. G., and Lynne Boddy, eds. Water, Fungi and Plants; Symposium of the British Mycological Society, University of Lancaster, Apr. 1985. Cambridge, Eng. and New York; Cmabridge University Press, 1986. 413p. (British Mycological Society Symposium No. 11)	Third
Third	Bahadir, M., et al, contrib. Controlled Release, Biochemical Effects of Pesticides, Inhibition of Plant Pathogenic Fungi. Berlin, Germany and New York; Springer-Verlag, 1990. 312p. (Chemistry of Plant Protection No. 6)	
Third	Bailey, John A., ed. Biology and Molecular Biology of Plant-Pathogen Interactions; Proceedings of the NATO Advanced Research Workshop, Dillington College, Ilminster, Eng., Sept. 1985. Berlin, Germany and New York; Springer-Verlag, 1986. 415p.	Third
Second	Bailey, John A., and B. J. Deverall, eds. The Dynamics of Host Defence. Sydney, Australia and New York; Academic Press, 1983. 233p.	Second
Second	Bailey, John A., and John W. Mansfield, eds. Phytoalexins. New York; Wiley, 1982. 334p.	Third
Second	Bailey, Liberty H. Manual of Cultivated Plants Most Commonly Grown in the Continental United States and Canada. Rev. ed. New York; Macmillan Co., 1954. 1116p. (1st ed. entitled Manual of Cultivated Plants, 1924. 851p.)	

Developed Countries Ranking		Third World Ranking
	Bailey, Peter, and Don Swincer, eds. Pest Control: Recent Advances and Future Prospects; Proceedings of the 4th Australian Applied Entomological Research Conference, Adelaide, S. Australia, Sept. 1984. Adelaide, S. Australia; South Australia Dept. of Agriculture, 1984. 520p.	Third
Third	Bajaj, Y. P. S., ed. Crops II. Berlin and New York; Springer-Verlag, 1988. 578p. (Biotechnology in Agriculture and Forestry, 6)	Third
	Bajaj, Y. P. S., ed. Potato. Berlin; Springer-Verlag, 1987. 509p. (Biotechnology in Agriculture and Forestry, 3)	Third
Third	Bajaj, Y. P. S., ed. Rice. Berlin and New York; Springer-Verlag, 1991. 645p. (Biotechnology in Agriculture and Forestry, 14)	Third
Third	Bajaj, Y. P. S., ed. Wheat. Berlin; Springer-Verlag, 1990. 687p. (Biotechnology in Agriculture and Forestry, 13)	Third
	Baker, F. W. G., and P. J. Terry, eds. Tropical Grassy Weeds; Sponsored by CASAFA Workshop on Integrated Control of Grassy Weeds in Tropical Subsistence Farming Systems, International Centre for Insect Physiology and Ecology, Nairobi, Feb. 1990. Wallingfor, Eng.; CAB International, 1991. 203p. (CASAFA Report Series no. 2).	Third
First rank	Baker, Kenneth F., and R. J. Cook. Biological Control of Plant Pathogens. St. Paul, Minn.; American Phytopathological Society, 1982. 433p. (Reprint. Originally published San Francisco; W.H. Freeman, 1974.)	First rank
Second	Baker, M. F., and W. M. Williams, eds. White Clover. Wallingford, Eng.; CAB International, 1987. 534p.	
Second	Baker, N. R., and J. Barber, eds. Chloroplast Biogenesis. Amsterdam and New York; Elsevier, 1984. 379p. (Topics in Photosynthesis No. 5)	
Second	Baker, N. R., and S. P. Long, eds. Photosynthesis in Contrasting Environments. Amsterdam and New York; Elsevier, 1986. 423p. (Topics in Photosynthesis No. 7)	Second
First rank	Baker, Ralph R., and Peter E. Dunn, eds. New Directions in Biological Control: Alternatives for Suppressing Agricultural Pests and Diseases; Proceedings of a UCLA Colloquium, Frisco, Colo., Jan. 1989. New York; Alan R. Liss, Inc., 1990. 837p. (UCLA Symposia on Molecular and Cellular Biology, New Series No. 112)	First rank
Third	Baldev, B., S. Ramanujam, and H. K. Jain, eds. Pulse Crops (Grain Legumes). New Delhi, India; Oxford and IBH Pub. Co., 1988. 626p.	Second
Second	Barber, J., ed. Electron Transport and Photophosphorylation. Amsterdam and New York; Elsevier Biomedical Press, 287p. (Topics in Photosynthesis No. 4)	

Developed Countries Ranking		Third World Ranking
Second	Barber, J., ed. The Light Reactions. Amsterdam and New York; Elsevier, 1987. 595p. (Topics in Photosynthesis No. 8)	Third
Second	Barber, J., ed. Photosynthesis in Relation to Model Systems. Amsterdam and New York; Elsevier, 1979. 434p. (Topics in Photosynthesis No. 3)	
Second	Barber, J., ed. Primary Processes of Photosynthesis. Amsterdam and New York; Elsevier, 1977. 516p. (Topics in Photosynthesis No. 2)	Third
Second	Barber, J., and N. R. Baker, eds. Photosynthetic Mechanisms and the Environment. Amsterdam, Holland and New York; Elsevier, 1985. 565p. (Topics in Photosynthesis No. 6)	Second
Second	Barbosa, Pedro, and Jack C. Schultz, eds. Insect Outbreaks. San Diego, Calif.; Academic Press, 1987. 578p.	Second
Third	Barigozzi, C., ed. The Origin and Domestication of Cultivated Plants; Symposium, Accademia Nazionale dei Lincei, Rome, Nov. 1985, organized by Centro Linceo Interdisciplinare di Scienze Matematiche e loro Applicazioni. Amsterdam and New York; Elsevier, 1986. 218p.	
Second	Barile, M. F., and S. Razin, eds. The Mycoplasmas. New York; Academic Press, 1979–89. 5 vols.	Third
Second	Barker, K. R., C. C. Carter, and J. N. Sasser, eds. An Advanced Treatise on Meloidogyne; Proceedings of the final conference conducted under the International Meloidogyne Project, Raleigh, N.C., April 1983. Raleigh, N.C.; Dept. of Plant Pathology, North Carolina State University, 1985. 2 vols.	First rank
Second	Barksdale, T. H., J. M. Good, and L. L. Danielson. Tomato Diseases and Their Control. Rev. Washington, D.C.; U.S. Dept. of Agriculture, Agricultural Research Service, 1972. 109p. (USDA Agriculture Handbook No. 203) (1st published by S.P. Doolittle, A.L. Taylor, and L.L. Danielson; 1st rev. in 1967 by R.E. Webb, J.M. Good, and L.L. Danielson.)	Second
	Barlow, Colin. The Natural Rubber Industry: Its Development, Technology, and Economy in Malaysia. Kuala Lumpur, Malaysia and New York; Oxford University Press, 1978. 500p.	Second
Second	Barnes, Ervin H. Atlas and Manual of Plant Pathology. New York; Plenum, 1979. 325p. (1st printed, New York; Appleton Century-Crofts, 1968.)	Second
Third	Barnes, Horace F. Gall Midges of Economic Importance. London; C. Lockwood, 1946–69. 8 vols.	

Developed Countries Ranking		Third World Ranking
Second	Barnes, P. J., ed. Lipids in Cereal Technology. New York; Academic Press, 1983. 425p.	
First rank	Barnes, Robert F., Darrell A. Miller and C. Jerry Nelson, eds. Forages. 5th ed. Ames, Iowa; Iowa State University Press, 1995. 1 vol. (4th ed., 1985, edited by Maurice E. Heath, et al.; 1st ed., 1951. 724p., primary author Harold D. Hughes)	Third
First rank	Barnett, Horace L., and Barry B. Hunter. Illustrated Genera of Imperfect Fungi. 4th ed. New York; Macmillan, 1987. 218p. (1st ed., Minneapolis, Minn.; Burgess Pub. Co., 1955. 218p.)	Second
Second	Barr, Barbara A., Carlton S. Koehler, and Ray F. Smith. Crop Losses, Rice: Field Losses to Insects, Diseases, Weeds, and Other Pests. Berkeley, Calif.; UC/AID Pest Management and Related Environmental Protection Project, 1975. 64p.	Second
	Barry, Sheila, and D. R. Houghton, eds. Biodeterioration 6; Papers presented at the 6th International Biodeterioration Symposium, Washington, D.C., Aug. 1984, sponsored by the George Washington, University, Virginia Commonwealth University. Farnham Royal, Eng.; CAB International Mycological Institute, Biodeterioration Society, 1986. 691p.	Third
Second	Bartholomew, William V., and Francis E. Clark, eds. Soil Nitrogen. Madison, Wis.; American Society of Agronomy, 1965. 615p. (Agronomy Mongraph Series No. 10)	First rank
Second	Bassam, N. el, M. Dambroth, and B. C. Loughman, eds. Genetic Aspects of Plant Mineral Nutrition. Boston, Mass. and London; Kluwer Academic Publishers, 1990. 558p.	Third
Second	Bassett, Mark J., ed. Breeding Vegetable Crops. Westport, Conn.; AVI Pub. Co., 1986. 584p.	Third
Second	Bates, David M., Richard W. Robinson, and Charles Jeffrey, eds. Biology and Utilization of the Curcurbitaceae. Ithaca, N.Y. and London; Comstock Publ. Associates, Cornell University Press, 1990. 485p.	Second
	Bates, William N. Mechanization of Tropical Crops. London, Eng.; Temple Press, 1957. 410p.	First rank
Second	Baum, Bernard R. Oats, Wild and Cultivated: A Monograph of the Genus Avena L. (Poaceae). Ottawa; Biosystematics Research Institute, Canada Department of Agriculture, Research Branch, 1977. 463p. (Monograph, Canada Dept. of Agriculture, Research Branch No. 14)	Second
Second	Baur, Fred J., ed. Insect Management for Food Storage	Third

Developed Countries Ranking		Third World Ranking
	and Processing. St. Paul, Minn.; American Association of Cereal Chemists, 1984. 384p.	
	Bavappa, K. V. A., M. K. Nair, and Kumar T. Prem, eds. The Arecanut Palm (Areca Catechu Linn.). Kasaragod, India; Central Plantation Crops Research Institute, 1982. 340p.	Third
Third	Beadle, C. L., et al. Photosynthesis in Relation to Plant Production in Terrestrial Environments. Oxford, Eng.; Tycooly Pub. for the U.N. Environment Programme, 1985. 156p. (Natural Resources and the Environment Series No. 18)	
Second	Beck, Stanley D. Insect Photoperiodism. 2d ed. New York; Academic Press, 1980. 387p. (1st ed., 1968. 288p.)	
	Becker, Walter A. Manual of Quantitative Genetics. 4th ed. Pullman, Wash.; Academic Enterprises, 1984. 190p. (1st ed. publ. as Manual of Procedures in Quantitative Genetics. Pullman, Washington State University, 1964. 70p.)	Third
Second	Beckman, Carl H. The Nature of Wilt Diseases of Plants. St. Paul, Minn.; APS Press, 1987. 175p.	Third
Third	Beets, Willem C. Multiple Cropping and Tropical Farming Systems. Aldershot, Hants., Gower, and Boulder Colo.; Westview Press, 1982. 156p.	First rank
	Bell, E. A., and B. V. Charlwood, eds. Secondary Plant Products. New York; Springer-Verlag, 1980. 674p. (Encyclopedia of Plant Physiology; New Series No. 8)	Third
Second	Bell, William J., and Ring T. Carde, eds. Chemical Ecology of Insects. Sunderland, Mass.; Sinauer Associates, Inc., 1984. 524p.	Third
Second	Bellotti, Anthony, and Aart van Schoonhoven. Cassava Pests and Their Control, edited by Trudy Brekelbaum. Cali, Colombia; Cassava Information Center, Centro Internacional de Agricultura Tropical, 1978. 71p.	First rank
Third	Bennett, J. W., and Linda L. Lasure, eds. Gene Manipulations in Fungi; Based on a conference on Gene Manipulations in the Exploitation and Study of Fungi, South Bend, Ind., May 1983, sponsored by the American Society for Microbiology and Miles Laboratories. Orlando, Fla.; Academic Press, 1985. 558p.	
Third	Berg, D., and M. Plempel, eds. Sterol Biosynthesis Inhibitors: Pharamaceutical and Agrochemial Aspects. Chichester, Eng.; Ellis Horwood; Distributors for U.S.A., New York; VCH Publishers, 1988. 583p.	
	Berger, Josef. The World's Major Fibre Crops: Their Cul-	First rank

Developed Countries Ranking		Third World Ranking
	tivation and Manuring. Zurich, Switzerland; Centre d'Aetude de l'Azote, 1969. 294p. (Summaries in French and Spanish.)	
Third	Bergersen, F. J., ed. Methods for Evaluating Biological Nitrogen Fixation. Chichester, Eng. and New York; J. Wiley, 1980. 702p.	Second
Third	Bergersen, F. J. Root Nodules of Legumes: Structure and Functions. Chichester, Eng. and New York; Research Studies Press, 1982. 164p.	Second
First rank	Bergey's Manual of Systematic Bacteriology. 9th ed. Edited by John G. Holt, Noel R. Krieg, et al. Baltimore, Md.; Williams & Wilkins, 1994. (1923 ed. by D. H. Bergey. 442p. Also published with title: *Bergey's Manual of Systematic Biology*, edited by John G. Holt and Noel R. Krieg. 1984–1989. 4 vols.)	First rank
Third	Bergmann, H., et al, contrib. Degradation of Pesticides, Desiccation, and Defoliation, ACh-receptors of Insects. Berlin, Germany and New York; Springer-Verlag, 1989. 256p. (Chemistry of Plant Protection No. 2)	Third
Third	Berkeley, R. C. W., and M. Goodfellow, eds. The Aerobic Endospore-Forming Bacteria: Classification and Identification; Papers from a Conference organized by the Systematics Group of the Society for General Microbiology, Cambridge University, April 1979. London and New York; Academic Press, 1981. 373p.	
Second	Bernard, M., and S. Bernard, eds. Genetics and Breeding of Triticale; Proceedings of the 3d EUCARPIA Meeting of the Cereal Section on Triticale = Genetique et Selection du Triticale: Comptes-rendus des Conferences presentees au Congres de la Section Ceeles d'EUCARPIA sur le Triticale, INRA, Station d'Amelioration des Plantes, Clermont-Ferrand, France, Juillet 1984. Paris, France; Institut National de la Recherche Agronomique, 1985. 703p.	Second
Third	Bernard, Richard L., Gail A. Juvik, and Randall L. Nelson. USDA Soybean Germplasm Collection Inventory. Urbana, Ill.; International Soybean Program (INSTOY), University of Illinois, 1987–89. 2 vols.	Third
Second	Beroza, M., ed. Pest Management with Insect Sex Attractants and Other Behavior-Controlling Chemicals. Washington, D.C.; American Chemical Society, 1976. 192p.	Second
	Berrie, G. K., A. Berrie, and J. M. O. Eze. Tropical Plant Science. Harlow; Longman Group U.K. Ltd., 1987. 421p.	Second
Second	Bewley, J. Derek, and Michael Black. Physiology and	Third

Developed Countries Ranking		Third World Ranking
	Biochemistry of Seeds in Relation to Germination. Berlin, Germany and New York; Springer-Verlag, 1982–83. 2 vols. (Corrected printing of the 1st ed., 1978. 2 vols.)	
Third	Bewley, J. Derek, and Michael Black. Seeds: Physiology of Development and Germination. 2d ed. New York; Plenum Press, 1994. 445p. (1st ed., 1985. 367p.)	Third
Second	Bezdicek, D. F., et al, eds. Organic Farming: Current Technology and Its Role in a Sustainable Agriculture; Proceedings of a Symposium sponsored by the American Society of Agronomy, the Crop Science Society of America, and the Soil Science Society of America, Nov.–Dec. 1981. Madison, Wis.; The Societies, 1984. 192p. (American Society of Agronomy Special Publication No. 46)	First rank
First rank	Biggins, J., ed. Progress in Photosynthesis; Proceedings of the 7th International Congress on Photosynthesis, Providence, R.I., 1986. Dordrecht and Boston; Nijhoff; Distributor for the U.S. and Canada, Kluwer Academic Publishers, 1987. 4 vols.	
Third	Biggs, Alan R., ed. Handbook of Cytology, Histology, and Histochemistry of Fruit Tree Diseases. Boca Raton, Fla.; CRC Press, 1993. 330p.	
	Biggs, Huntley H., and Ronald L. Tinnermeier, eds. Small Farm Agricultural Development Problems. Ft. Collins, Colo.; Colorado State University, 1974. 168p.	Second
Third	Bilgrami, K. S., ed. Mycotoxins in Food and Feed; Proceedings of all India symposium held at Bhagalpur, Feb. 1983. Bhagalpur, India; Allied Press, 1983. 315p.	Third
Second	Biological Control in Crop Production. Papers presented at an annual symposium organized by BARC Symposium V Committee; sponsored by and held at Beltsville Agricultural Research Center, Md., May 1980. Totowa, N.J. and London, Eng.; Allanheld, Osmun Granada, 1981. 461p.	Second
Second	Birch, M. C., ed. Pheromones. Amsterdam; North-Holland Publishing, 1974. 495p.	Third
Second	Bird, Julio, and Karl Maramorosch, eds. Tropical Diseases of Legumes; Papers presented at a workshop, Rio Piedras Agricultural Station, University of Puerto Rico, June 1974. New York; Academic Press, 1975. 171p.	Second
	Black, Charles A. Soil-Plant Relationships. 2d ed. New York; Wiley, 1968. 792p. (Reprinted in 1984. 1st ed., 1957. 332p.)	Second
Second	Blackburn, Frank. Sugar-Cane. London, Eng. and New York; Longman, 1984. 414p.	First rank

Developed Countries Ranking		Third World Ranking
First rank	Blackman, Roger L., and V. F. Eastop. Aphids on the World's Crops: An Identification Guide. Chichester, Eng. and New York; Wiley, 1984. 466p.	Third
Second	Blakeman, J. P. Microbial Ecology of the Phylloplane; Papers presented at the 3d International Symposium on the Microbiology of Leaf Surfaces, Aberdeen, Sept. 1980. London and New York; Academic Press, 1981. 502p.	
	Blanch, Harvey W., Stephen Drew, and Daniel I. C. Wang, eds. The Practice of Biotechnology—Current Commodity Products. Oxford, Eng. and New York; Pergamon, 1985. 1 vol. (Comprehensive Biotechnology No. 3)	Second
	Bleasdale, J. K. A. Plant Physiology in Relation to Horticulture. 2d ed. London; Macmillan, 1984. 143p. (1st published: Westport, Conn.; AVI Pub. Co., 1977. 144p.)	Second
Second	Blum, Abraham. Plant Breeding for Stress Environments. Boca Raton, Fla.; CRC Press, 1988. 223p.	Second
Second	Blum, Murray S. Chemical Defenses of Arthropods. New York; Academic Press, 1981. 562p.	Third
Second	Blumer, Samuel. Echte Mehltaupilze (Erysiphaceae). Ein Bestimmungsbuch fur die in Europa Vorkommenden Arten. Jena; G. Fischer, 1967. 436p.	
Second	Blumer, Samuel. Rost- und Brandpilze auf Kulturpflanzen: Ein Bestimmungsbuch fur die in Mitteleuropa Vorkommenden Arten. Jena; G. Fischer, 1963. 379p.	
Third	Bodenheimer, Friedrich S. Citrus Entomology in the Middle East: With Special References to Egypt, Iran, Iraq, Palestine, Syria, Turkey. Gravenhage; Junk, 1951. 663p.	Third
Third	Bodenheimer, Friedrich S., and E. Swirski. The Aphidoidea of the Middle East. Jerusalem, Israel; Weizmann Science Press of Israel, 1957. 378p.	
Second	Boethel, D. J., and R. D. Eikenbarry, eds. Interactions of Plant Resistance and Parasitoids and Predators of Insects. Chichester, Ellis Harwood and New York; Halsted Press, 1986. 224p.	Second
Second	Bogdan, A. V. Tropical Pasture and Fodder Plants. London and New York; Longman, 1977. 475p.	First rank
Second	Bokx, J. A. de, and J. P. H. vand der Want, eds. Viruses of Potatoes and Seed-Potato Production. 2d ed. Wageningen, Netherlands; PUDOC, 1987. 259p. (1st ed., 1972. 233p.) (Also available in Spanish as Virosis de la Papa y de la Semilla de Papa. Buenos Aires; Hemisferio Sur., 1980. 303p.)	Second

Developed Countries Ranking		Third World Ranking
Third	Bond, D. A., ed. Vicia Faba: Feeding Value, Processing, and Viruses; Proceedings of a seminar in the EEC Programme of Coordination of Research on the Improvement of the Production of Plant Proteins, Cambridge, Eng., June 1979. Hague, Netherlands and Boston, Mass.; M. Nijhoff; Distributors for the U.S. and Canada, Boston, Mass.; Kluwer, 1980. 422p.	
First rank	Bonner, James F., and Joseph E. Varner, eds. Plant Biochemistry. 3d ed. New York; Academic Press, 1976. 925p. (1st ed., 1950 by James F. Bonner. 537p.)	
Second	Booth, Colin. Fusarium: Laboratory Guide to the Identification of the Major Species. Kew, Eng.; Commonwealth Mycological Institute, 1977. 58p.	First rank
First rank	Booth, Colin. The Genus Fusarium. Kew, Eng.; Commonwealth Mycological Institute, 1971. 237p.	Second
	Bornemisza, Elemer, and Alfredo Alvarado, eds. Soil Management in Tropical America; Proceedings of a Seminar, CIAT, Cali, Colombia, Feb. 1974, co-sponsored by the University Consortium on Soils of the Tropics, Centro Internacional de Agricultura Tropical, U.S. Agency for International Development. Raleigh, N.C.; Soil Science Dept., North Carolina State University, 1975. 565p.	Second
Second	Bos, Luite. Introduction to Plant Virology. London, Eng. and New York; Longman, 1983. 160p.	Second
	Bos, Luite. Symptoms of Virus Diseases in Plants: With Indexes of Names of Symptoms in English, Dutch, German, French, Italian, and Spanish. 3d ed. rev. Wageningen, Netherlands; Research Institute for Plant Protection, 1978. 225p. (1st ed., Wageningen, Netherlands; Centre for Agricultural Publications and Documentation, 1963. 132p.)	Second
	Bose, T. K., and M. G. Som, eds. Vegetable Crops in India. Calcutta; Naya Prokash, 1986. 775p.	Third
	Boserup, Ester. The Conditions of Agricultural Growth: The Economics of Agrarian Change Under Population Pressure. Chicago, Ill.; Aldine Pub. Co., 1965. 124p.	Third
Third	Bostanian, N. J., L. T. Wilson, and T. J. Dennehy, eds. Monitoring and Integrated Management of Arthropod Pests of Small Fruit Crops; an outgrowth of papers presented at a symposium at the 18th International Congress of Entomology, Vancouver, July 1988. Andover, Eng.; Intercept, 1990. 301p.	
	Boswell, K. F., and A. J. Gibbs. Viruses of Legumes	Second

Developed Countries Ranking		Third World Ranking
	1983: Descriptions and Keys from Virus Identification Data Exchange. Canberra, Australia; Australian National University, Research School of Biological Sciences, 1983. 139p.	
First rank	Bothe, H., F. J. de Bruijn, and W. E. Newton, eds. Nitrogen Fixation: Hundred Years After Gustav Fischer; Proceedings of the 7th International Congress on (Triple-bond) Nitrogen Fixation, Cologne, Germany, Mar. 1988. Stuttgart, Germany and New York; G. Fischer, 1988. 878p.	Second
	Boudet, G. Manuel sur les Paturages Tropicaux et les Cultures Fourrageres. 3d ed. Paris, France; Ministere de la Cooperation, Institut d'Elevage et de Medecine Veterinaire des Pays Tropicaux, 1978. 258p. (2d ed., 1975. 254p.)	Third
Second	Boulter, D., and B. Parthier, eds. Nucleic Acids and Proteins in Plants. Berlin, Germany and New York; Springer-Verlag, 1982. 2 vols. (Encyclopedia of Plant Physiology New Series No. 14A)	
	Bouriquet, Gilbert. Le Vanillier et la Vanille dans le Monde. Paris, France; P. Lechevalier, 1954. 748p. (Encyclopedie Biologique No. 46)	Third
Third	Bousquet, Yves. Beetles Associated with Stored Products in Canada: An Identification Guide. Ottawa, Canada; Research Branch, Agriculture Canada, 1990. 220p.	
	Bouwkamp, John C., ed. Sweet Potato Products: A Natural Resource for the Tropics. Boca Raton, Fla.; CRC Press, 1985. 271p.	First rank
	Bradbury, J. Howard, and Warren D. Holloway. Chemistry of Tropical Root Crops: Significance for Nutrition and Agiculture in the Pacific. Canberra, Australia; Australian Centre for International Agricultural Research, 1988. 201p.	Third
Second	Bradbury, J. F. Guide to Plant Pathogenic Bacteria. Farnham Royal, Eng.; CAB International Mycological Institute, 1986. 332p.	First rank
Third	Brajcich, P., W. Pfeiffer, and E. Autrique. Durum Wheat: Names, Parentage, Pedigrees and Origins. Mexico City, Mexico; International Maize and Wheat Improvement Center, 1986. 102p.	Third
Second	Brathwaite, Chelston W. D. An Introduction to Diseases of Tropical Crops. 1st ed. Port-of-Spain, Trinidad and Tobago; C.W.D. Brathwaite, 1985. 184p.	Second
	Brent, K. J., and R. K. Atkin, eds. Rational Pesticide Use;	Third

Developed Countries Ranking		Third World Ranking
	Proceedings of the 9th Long Ashton Symposium, 1984. Cambridge, Eng. and New York; Cambridge University Press, 1987. 348p.	
Third	Brett, C. T., and K. Waldron. Physiology and Biochemistry of Plant Cell Walls. London, Eng. and Boston, Mass.; Unwin Hyman, 1990. 194p.	
Second	Brewbaker, James L. Agricultural Genetics. Englewood Cliffs, N.J.; Prentice-Hall, 1964. 156p.	
	Breymeyer, A. I., and G. M. Van Dyne, eds. Grasslands: Systems Analysis and Man. Cambridge, Eng. and New York; Cambridge University Press, 1980. 950p.	Third
Second	Briggs, Dennis E. Barley. London, Eng.; Chapman and Hall; Distributed in USA by New York; Wiley, 1978. 612p.	Second
Third	Briggs, Fred N., and P. F. Knowles. Introduction to Plant Breeding. New York; Reinhold Pub. Corp., 1967. 426p.	Second
	Bringi, N. V., ed. Non-Traditional Oilseeds and Oils in India. New Delhi, India; Oxford and IBH Pub. Co., 1987. 254p.	Second
First rank	British Crop Protection Conference—Weeds. Proceedings . . . Nottingham, Eng.; Boots co., 1953–91. 3 vols. (1128p.)	
Third	British Crop Protection Council. Opportunities for Chemical Plant Growth Regulation; Proceedings of a symposium, University of Reading, Jan. 1978, jointly organized by the British Crop Protection Council and the British Plant Growth Regulator Group. Malvern, Eng.; British Crop Protection Council, 1978. 222p. (British Crop Protection Council Monograph No. 21)	Third
Second	British Crop Protection Council. Weed Control Handbook: Principles. 8th ed., edited by R.J. Hance and K. Holly. Oxford, Eng. and Boston, Mass.; Blackwell Scientific Publications, 1990. 582p. (1st ed. by British Crop Protection Council, 1958, 245p.; 2d ed. edited by E.K. Woodford, 1960, 264p.; 3d–4th and 6th eds. edited by E.K. Woodford and S.A. Evans, [3d, 1963, 356p.; 4th, 1965, 434p.; 6th, 1968–70, 2 vols. 5th ed. edited by J.D. Fryer and S.A. Evans, 1968, 2 vols.; 7th ed. edited by H.A. Roberts, 1982, 533p. The 1972 edition, edited by J.D. Fryer and R.J. Makepeace, is also called 7th ed.)	Third
First rank	British Weed Control Conference, 1st-12th. Proceedings, . . . organized by the British Crop Protection Council. Nottingham; Boots co., 1953–74. 12 vols.	
	Briton-Jones, Harry R. The Diseases of the Coconut Palm.	Third

Developed Countries Ranking		Third World Ranking
	Revised by Ernest E. Cheesman. London; Bailliere, Tindall and Cox, 1940. 176p.	
Third	Broertjes, C., chair. Induced Variability in Plant Breeding; International Symposium of the Section Mutation and Polyploidy of the Euopean Association for Research on Plant Breeding, EUCARPIA, Wageningen, Netherlands, Aug.–Sept. 1981. Wageningen, Netherlands; PUDOC, 1982. 143p.	Third
Third	Broertjes, C., and A. M. Van Harten. Applied Mutation Breeding for Vegetatively Propagated Crops. Rev. ed. Amsterdam and New York; Elsevier, 1988. 345p. (Developments in Crop Science No. 12) (1st ed. entitled: Application of Mutation Breeding Methods in the Improvement of Vegetatively Propagated Crops: An Interpretive Literature Review. 1978. 316p.)	Second
Second	Brooks, G. T. Chlorinated Insecticides. Cleveland, Ohio; CRC Press, 1974. 2 vols.	
Second	Broughton, W. J., ed. Nitrogen Fixation. Oxford, Eng.; Clarendon Press and New York; Oxford University Press, 1980–1985. 4 vols.	Third
	Browder, John O., ed. Fragile Lands of Latin America: Strategies for Sustainable Development; Collection of rev. papers originally presented at the Symposium, . . . held during the 14th Congress of the Latin American Studies Association, New Orleans, La., Mar. 1988. Boulder, Colo.; Westview Press, 1989. 301p.	Second
Second	Brown, Anthony H. D., et al, eds. Plant Population Genetics, Breeding, and Genetic Resources. Sunderland, Mass.; Sinauer Associates, 1990. 449p.	
Second	Brown, Anthony H. D., et al, eds. The Use of Plant Genetic Resources. Cambridge, Eng. and New York; Cambridge University Press, 1989. 382p.	Third
Second	Brown, Anthony W. A. Ecology of Pesticides. New York; Wiley, 1978. 525p.	Third
Second	Brown, Anthony W. A., and R. Pal. Insecticide Resistance in Arthropods. 2d ed. Geneva, Switzerland; World Health Organization, 1971. 491p. (WHO Monograph Series No. 38) (1st ed., 1958. 240p.)	Third
Third	Brown, Lester R., and Pamela Shaw. Six Steps to a Sustainable Society. Washington, D.C.; Worldwatch Institute, 1982. 63p. (Worldwatch Paper No. 48)	Third
Third	Brown, R. H., and B. R. Kerry, eds. Principles and Practice of Nematode Control in Crops. Sydney, Australia and Orlando, Fla.; Academic Press, 1987. 447p.	Second
Third	Brown, W. L., et al, eds. Conservation of Crop Germ-	Second

Developed Countries Ranking		Third World Ranking
	plasm: An International Perspective; Proceedings of a symposium, Washington, D.C., Aug. 1983, sponsored by the Crop Science Society of America. Madison, Wis.; Crop Science Society of America, 1984. 67p. (CSSA Special Publication No. 8)	
	Browne, Francis G. Pests and Diseases of Forest Plantation Trees: An Annotated List of the Principal Species Occurring in the British Commonwealth. Oxford, Eng.; Clarendon Press, 1968. 1330p.	Second
	Brucher, Heinz. Useful Plants of Neotropical Origin, And Their Wild Relatives. Berlin, Germany and New York; Springer-Verlag, 1989. 296p.	Third
Second	Büchel, K. H., ed. Chemistry of Pesticides. New York; J. Wiley, 1983. 518p. (Translated by Graham Holmwood of: Pflanzenschutz und Schädlingsbekämpfung. Stuttgart, Germany; Thieme, 1977. 247p.)	Second
Second	Buckner, Robert C., and Lowell P. Bush, eds. Tall Fescue. Madison, Wis.; American Society of Agronomy, Crop Science Society of America, and the Soil Science Society of America, 1979. 351p. (Agronomy No. 20)	
Third	Buczacki, S. T., ed. Zoosporic Plant Pathogens: A Modern Perspective. London, Eng. and New York; Academic Press, 1983. 352p.	Third
Second	Bull, David. A Growing Problem: Pesticides and Third World Poor. Oxford, Eng.; Oxfam, 1982. 192p.	Second
Second	Bulla, L. A., ed. Regulation of Insect Populations by Microorganisms. New York; New York Academy of Sciences, 1973. 243p. (New York Academy of Science Annals No. 217)	Third
Second	Bu'Lock, John, and Bjorn Kristiansen, eds. Basic Biotechnology. London, Eng. and Orlando, Fla.; Academic Press, 1987. 561p.	First rank
	Bunting, E. S., ed. Production and Utilization of Protein in Oilseed Crops; Proceedings of a seminar in the EEC Programme of Coordination of Research on the Improvement of the Production of Plant Proteins organized by the Institut fur Pflanzenbau und Pflanzenzuchting at Braunschweig, July 1980, sponsored by the Commission of the European Communities, Directorate-General for Agriculture, Coordination of Agricultural Research. Hague, Netherlands and Boston, Mass.; M. Nijhoff; Distributors for the U.S. and Canada, Kluwer Boston, 1981. 382p.	Second
First rank	Burges, Horace D. Microbial Control of Pests and Plant Diseases 1970–1980. London, Eng. and New York; Academic Press, 1981. 949p.	First rank

Developed Countries Ranking		Third World Ranking
	Burges, Horace D., and N. W. Hussey, eds. Microbial Control of Insects and Mites. London, Eng. and New York; Academic Press, 1971. 861p.	First rank
Third	Burgess, Abraham H. Hops: Botany, Cultivation, and Utilization. London; L. Hill and New York; Interscience, 1964. 300p.	
Second	Burkill, H. M. The Useful Plants of West Tropical Africa. 2d ed. Kew, Eng.; Royal Botanic Gardens, 1985. 1 vol. (1st ed. by J.M. Dalziel. London, Eng.; The Crown Agents for the Colonies, 1937. 612p.)	First rank
	Burkill, Isaac H. A Dictionary of the Economic Products of the Malay Peninsula. 2d ed. Kuala Lumpur, Malaysia; Ministry of Agriculture and Co-operatives, 1966. 2 vols. 2444p. (1st ed., London, Eng.; Crown Agents for the Colonies)	Third
First rank	Burn, A. J., T. H. Coaker, and P. C. Jepson, eds. Integrated Pest Management. London and San Diego, Calif.; Academic Press, 1987. 474p.	First rank
Second	Burnett, John H. Fundamentals of Mycology. 2d ed. London, Eng.; E. Arnold; Distributed in USA, New York; Crane Russak, 1976. 673p. (1st ed., New York; Martin's Press, 1968. 546p.)	Second
Third	Burnham, Charles R. Discussions in Cytogenetics. Minneapolis, Minn.; Burgess Pub. Co., 1962. 375p.	
Third	Burton, William G. Post-Harvest Physiology of Food Crops. London and New York; Longman, 1982. 339p.	Third
First rank	Burton, William G. The Potato. 3d ed. Harlow, Eng.; Longman and New York; Wiley, 1989. 742p. (1st ed. titled: The Potato: A Survey of Its History and of Factors Influencing Its Yield, Nutritive Value and Storage. London; Chapman and Hall, 1948. 319p.)	First rank
First rank	Bushnell, William R., and Alan P. Roelfs, eds. The Cereal Rusts. New York; Academic Press, 1984–85. 2 vols.	First rank
Second	Bushuk, Walter, ed. Rye: Production, Chemistry, and Technology. St. Paul, Minn.; American Association of Cereal Chemists, 1976. 181p. (American Association of Cereal Chemists Monograph Series No. 5)	Third
	Busson, Felix F. Les Plantes Alimentaires de l'ouest Africain: Étude Botanique, Biologique et Chimique. Marseille, France; L'Imprimerie Leconte, 1965. 568p.	Second
Third	Butler, G. W., and R. W. Bailey, eds. Chemistry and Biochemistry of Herbage. London and New York; Academic Press, 1973. 3 vols.	
Third	Buyckx, E. J. E. Precis des Maladies et des Insectes Nuisibles Rencontres sur les Plantes Cultivees au Congo, au	Third

Developed Countries Ranking		Third World Ranking
	Rwanda et au Burundi. Brussels; Institute National pour l'Etude Agronomique du Congo, 1962. 708p.	
Second	Callow, J. A., ed. Biochemical Plant Pathology. Chichester, Eng. and New York; Wiley, 1983. 484p.	
Second	Campbell, C. Lee, and Laurence V. Madden. Introduction to Plant Disease Epidemiology. New York; Wiley, 1990. 532p.	
Third	Campbell, Richard E. Plant Microbiology. London, Eng. and Baltimore, Md.; E. Arnold, 1985. 191p. (1st ed., New York; Wiley, 1977. 2d ed., Blackwell Scientific, 1983.)	Third
Second	Camper, N. D., ed. Research Methods in Weed Science. 3d ed. Campaign, Ill.; Southern Weed Science Society, 1986. 486p.	Third
Third	Canny, M. J. Phloem Translocation. Cambridge, Eng.; University Press, 1973. 301p.	
Second	Cantwell, George E., ed. Insect Diseases. New York; M. Dekker, 1974. 2 vols. 595p.	Third
	Cardarelli, Nate F. Controlled Release Pesticide Formulations. Cleveland, Ohio; CRC Press, 1976. 210p.	Second
	Cardenas, Juan, Carols E. Reyes, and Jerry D. Doll. Tropical Weeds. Malenzas Tropicales, edited by Fernando Pardo. Bogota, Colombia; Instituto Colombiano Agropecuario, 1972. 1 vol.	Second
	Carlile, W. R. Control of Crop Diseases. London, Eng.; E. Arnold, 1988. 100p.	Second
Third	Carlson, Peter S., ed. The Biology of Crop Productivity. New York; Academic Press, 1980. 471p.	Second
Third	Carnegie Institution of Washington. Experimental Studies on the Nature of Species. Washington, D.C.; Carnegie Institution of Washington, 1940–1982. 6 vols. (Carnegie Institution of Washington. Publication No. 520, 564, 581, 615, 628, 636)	
Second	Carroll, C. Ronald, John H. Vandermeer, and Peter Rosset, eds. Agroecology. New York; McGraw-Hill, 1990. 641p.	First rank
First rank	Carson, Rachel. Silent Spring. 1st Ballantine Books ed. New York; Fawcett Crest, 1982. 304p. (1st ed., Boston; Houghton Mifflin, 1962. 368p.) (Also available in Spanish as Primavera Silenciosa. Barcelona; Grijalbo.)	Second
Second	Carter, Jack F., ed. Sunflower Science and Technology. Madison, Wis.; American Society of Agronomy, 1978. 505p.	First rank
Second	Carter, N., A. F. G. Dixon, and R. Rabbinge. Cereal Aphid Populations: Biology, Simulation and Prediction.	

Developed Countries Ranking		Third World Ranking
	Wageningen, Netherlands; Centre for Agricultural Publishing and Documentation, 1982. 91p.	
	Carvajal, Jose F. Cafeto: Cultivo y Fertilizacion. 2d ed. Bern, Switzerland; International Potash Institute, 1984. 254p. (1st ed., Coffee: Its Cultivation and Fertilization, 1972. 141p.)	Third
Second	Casida, John E., ed. Pyrethrum, the Natural Insecticide; Proceedings of the International Symposium on Recent Advances with, . . . Minnesota 1972. New York; Academic Press, 1973. 329p.	Third
Second	Caswell, George H. Agricultural Entomology in the Tropics. London; E. Arnold, 1962. 152p.	Third
	Cavalcante, Paulo B. Frutas Comestiveis da Amazonia. 4th ed. Belem, Para; Museau Paraense Emilio Goeldi, 1988. 279p. (1st ed., 1974.)	Third
Second	Centre for Overseas Pest Research (Great Britain). Pest Control in Rice. 2d ed, comp. rev. London, Eng.; Centre for Overseas Pest Research, 1976. 295p. (PANS Manual No. 3) (1st ed. by S.D. Feakin, 1970.)	Second
Second	Centre for Overseas Pest Research (Great Britain). Pest Control in Tropical Grain Legumes. London, Eng.; Centre for Overseas Pest Research, Overseas Development Administration, 1981. 206p.	Second
Second	Centre for Overseas Pest Research (Great Britain). Pest Control in Tropical Tomatoes. London, Eng.; Centre for Overseas Pest Research, Overseas Development Administration, 1983. 130p.	Second
	Centro de Ensenanza, Investigacion y Capacitacion para el Desarrollo Agropecuario, Forestal y Acuicola del Sureste (CEICADES). Taller de Fitopatologia Tropical. 2d ed. Chapingo, Mexico; CP and CONACYT, Copyright by Colegio de Postgraduados, 1987. 461p.	Third
	Centro Internacional de Mejoramiento de Maiz y Trigo. Barley Yellow Dwarf; Proceedings of a Workshop, CIMMYT, Mexico, Dec. 1983, sponsored by the United Nations Development Programme and CIMMYT. Mexico; Centro Internacional de Mejoramiento de Maiz y Trigo, 1984. 109p.	Third
	Chadha, K. L., G. S. Randhawa, and R. N. Pal, eds. Viticulture in Tropics; Proceedings of the Working Group on Viticulture in Southeast Asia formed by the International Society of Horticultural Science, Bangalore, India, Feb. 1972. Bangalore, India; Horticultural Society of India, 1977. 348p.	Second
Third	Chaleff, R. S. Genetics of Higher Plants: Applications of	

Developed Countries Ranking		Third World Ranking
	Cell Culture. Cambridge, Eng. and New York; Cambridge University Press, 1981. 184p.	
Third	Chan, Harvey T., ed. Handbook of Tropical Foods. New York; M. Dekker, 1983. 639p.	First rank
	Chandler, Robert F. Rice in the Tropics: A Guide to the Development of National Programs. Boulder, Colo.; Westview Press, 1979. 256p.	First rank
Third	Chandra, S., ed. Edible Aroids. Oxford, Eng.; Clarendon Press and New York; Oxford University Press, 1984. 252p.	First rank
Third	Chang, S. T., and W. A. Hayes, eds. The Biology and Cultivation of Edible Mushrooms. New York; Academic Press, 1978. 819p.	Second
Third	Chang, Te-Tzu. Manual on Genetic Conservation of Rice Germ Plasm for Evaluation and Utilization. Los Banos, Philippines; International Rice Research Institute, 1976. 77p.	Third
Second	Charles-Edwards, David A. Physiological Determinants of Crop Growth. Sydney and New York; Academic Press, 1982. 161p.	Third
Third	Charles-Edwards, David A., David Doley, and Glynn M. Rimmington. Modelling Plant Growth and Development. Sydney, Australia and Orlando, Fla.; Academic Press, 1986. 235p.	
First rank	Charudattan, R. Biological Control of Weeds with Plant Pathogens. New York; J. Wiley, 1982. 293p.	First rank
	Cheremisinoff, Paul N., and Robert P. Ouellette, eds. Biotechnology: Applications and Research. Lancaster, Pa.; Technomic Pub. Co., 1985. 699p.	Third
First rank	Cherrett, J. M., and G. R. Sagar, eds. Origins of Pest, Parasite, Disease and Weed Problems; Proceedings of the 18th Symposium of the British Ecological Society, Bangor, April 1976. Oxford, Eng.; Blackwell, 1977. 413p.	Third
Second	Chet, Ilan, ed. Biotechnology in Plant Disease Control. New York; Wiley-Liss, 1993. 373p. (1st ed. with title: Innovative Approaches to Plant Disease Control. New York; Wiley, 1987. 372p.)	Second
Second	Chiarappa, L., ed. Crop Loss Assessment Methods: FAO Manual on the Evaluation and Prevention of Losses by Pests, Disease and Weeds. Farmham Royal, Eng.; Commonwealth Agricultural Bureaux, 1971–77. 1 vol. (Supplements no. 1 and 2 inserted into the manual. Supplement no. 3 issued separately.)	First rank
Second	Child, Reginald. Coconuts. 2d ed. London; Longman, 1974. 335p. (1st ed., 1964. 216p.)	First rank

Developed Countries Ranking		Third World Ranking
	Christensen, Clyde M., and Henry H. Kaufmann. Grain Storage: The Role of Fungi in Quality Loss. Minneapolis, Minn.; University of Minnesota Press, 1969. 153p.	Second
Second	Christiansen, M. N., and Charles Lewis, eds. Breeding Plants for Less Favorable Environments. New York; Wiley, 1982. 459p. (Also available in Spanish as Mejoramientos de las Plantas en Ambientes Poco Favorables. Mexico; Limusa-Noriega, 1987. 531p.)	Second
Third	Christie, B. R., ed. CRC Handbook of Plant Science in Agriculture. Boca Raton, Fla.; CRC Press, 1987. 2 vols.	Second
Third	Christie, William W. Lipid Analysis: Isolation, Separation, Identification, and Structural Analysis of Lipids. 2d ed. Oxford, Eng. and New York; Pergamon Press, 1982. 207p. (1st ed., 1973. 338p.)	
Third	Ciferri, Orio and Leon Dure, eds. Structure and Function of Plant Genomes; Proceedings of a NATO Advanced Study Institute, Porto Portese, Italy, Aug.–Sept. 1982. New York; Plenum Press, 1983. 495p.	
Second	Clark, C. A., and J. W. Moyer. Compendium of Sweet Potato Diseases. St. Paul, Minn.; APS Press, 1988. 74p.	Second
Second	Clark, L. R., P. W. Geier, R. D. Hughes, and R. F. Morris, Collaborators. The Ecology of Insect Populations in Theory and Practice. London, Eng.; Methuen, 1967. 232p.	Third
	Clark, Norman, and Calestous Juma. Biotechnology for Sustainable Development: Policy Options for Developing Countries. Nairobi, Kenya; African Centre for Technology Studies Press, 1991. 117p.	First rank
Third	Clarke, R. J., and R. Macrae, eds. Coffee. London, Eng. and New York; Elsevier, 1985–88. 6 vols.	First rank
First rank	Clausen, Curtis P., ed. Introduced Parasites and Predators of Arthropod Pests and Weeds: A World Review. Washington, D.C.; Agricultural Research Service, U.S. Dept. of Agriculture, 1978. 549p. (USDA Agriculture Handbook No. 480)	Third
Third	Clements, Harry F. Sugarcane Crop Logging and Crop Control: Principles and Practices. Honolulu; University of Hawaii Press, 1980. 520p.	Second
Third	Cleveland, David A., and Daniela Soleri. Food from Dryland Gardens. Tucson, Ariz.; Center for People, Food and Environment with the support of UNICEF, 1991. 387p.	First rank
Second	Clifford, B. C., and E. Lester, eds. Control of Plant Diseases: Costs and Benefits; From a meeting of the British Society for Plant Pathology, University of Manchester,	Second

Developed Countries Ranking		Third World Ranking
	Dec. 1984. Oxford, Eng. and Boston, Mass.; Blackwell Scientific Publications, 1988. 263p.	
Second	Clifford, M. N., and C. K. Willson, eds. Coffee: Botany, Biochemistry, and Production of Beans and Beverage. American ed. London and Westport, Conn.; Croom Helm AVI Pub. Co., 1985. 457p.	First rank
	Coats, Joel R., ed. Insecticide Mode of Action. New York; Academic Press, 1982. 470p.	Third
Second	Cobley, Leslie S. An Introduction to the Botany of Tropical Crops, rev. by W.M. Steele. 2d ed. London, Eng. and New York; Longman, 1976. 371p. (1st ed., 1956. 357p.)	First rank
	Cock, James H. Cassava: New Potential for a Neglected Crop. Boulder, Colo.; Westview Press, 1985. 191p. (Also available in Spanish as Yuca: Nuevo Potencial para un Cultivo Tradicional, 2d ed. Cali; Ciat, 1989. 240p.)	First rank
Second	Coffman, Franklin A. Oat History, Identification, and Classification. Rev. ed. Washington, D.C.; Dept. of Agriculture, Agricultural Research Service, 1977. 356p. (USDA Technical Bulletin No. 1516) (Supersedes Technical Bulletin No. 1100, Oat Identification and Classification by T.R. Stanton.)	
Second	Coffman, Franklin A. Oats and Oat Improvement. Madison, Wis.; American Society of Agronomy, 1961. 650p.	
Second	Cole, J. A., and S. J. Ferguson, eds. The Nitrogen and Sulphur Cycles; 42nd Symposium of the Society for General Microbiology, University of Southampton, Jan. 1988. Cambridge, Eng. and New York; Cambridge University Press, 1988. 490p.	Third
	Cole-Desiree, L., ed. Proceedings of the Workshop for the Southern Africa Region on Basics of Soybean Cultivation and Utilization, University of Zimbabwe, Harare, Zimbabwe, Feb.–Mar. 1986. Harare, Zimbabwe; Crop Science Dept., University of Zimbabwe, 1987. 199p.	Third
Second	Coley-Smith, J. R., K. Verhoeff, and W. R. Jarvis, eds. The Biology of Botrytis. London and New York; Academic Press, 1980. 318p.	Third
Third	Collins, Glenn B., and Joseph G. Petolino, eds. Applications of Genetic Engineering to Crop Improvement. Boston, Mass.; Martinus Nijhoff and W. Junk Publishers, 1984. 604p.	
Third	Collins, Julius L. The Pineapple: Botany, Cultivation, Utilization. New York; Interscience Publishers, 1960. 294p.	Second

Developed Countries Ranking		Third World Ranking
	Colvin, T. S., and J. H. Turner. Applying Pesticides: Management, Application, Safety. Athens, Ga.; American Association for Vocational Instructional Materials, 1976. 96p.	Third
First rank	Commonwealth Agricultural Bureaux. Descriptions of Plant Viruses. Farnham Royal, Eng.; Commonwealth Agricultural Bureaux, 1970–89. 22 vols.	First rank
	Commonwealth Agricultural Bureaux. Perspectives in World Agriculture. Farnham Royal, Eng.; Commonwealth Agricultural Bureaux, 1980. 532p.	Second
	Commonwealth Agricultural Bureaux. A Review of Nitrogen in the Tropics with Particular Reference to Pastures; A Symposium, Brisbane, Feb. 1960. Farnham Royal, Eng.; Commonwealth Agricultural Bureaux, 1962. 185p. (Commonwealth Bureau of Pastures and Field Crops Bulletin No. 46)	Second
Second	Commonwealth Institute of Biological Control (Canada). A Catalogue of the Parasites and Predators of Insect Pests. Belleville, Ontario; The Imperial Parasite Service, 1943–65. 18 vols.	Second
First rank	Commonwealth Mycological Institute (Great Britain). CMI Descriptions of Pathogenic Fungi and Bacteria. Kew, Eng.; Commonwealth Mycological Institute, 1964+. (Set 102 in 1990.)	First rank
Second	Commonwealth Mycological Institute (Great Britain). Distribution Maps of Plant Diseases. Kew, Eng.; Commonwealth Agricultural Bureaux, 1942+. (No. 611 in 1989.)	Second
Third	Commonwealth Science and Industrial Research Organization (CSIRO). Ecology of Photosynthesis in Sun and Shade. Melbourne, Australia; CSIRO, 1988. 358p.	Third
	Compagnon, P. Le Caoutchouc Naturel: Biologie, Culture, Production. Paris, France; G.P. Maisonneuve and Larose, 1986. 595p. (Techniques Agricoles et Productions Tropicales no. 35)	Second
Third	Conger, B. V., ed. Cloning Agricultural Plants via *in Vitro* Techniques. Boca Raton, Fla.; CRC Press, 1981. 273p.	Third
	Conway, Gordon. Agroecosystem Analysis of Research and Development. Bangkok; Winrock International Institute for Agricultural Development, 1986. 111p.	Second
Second	Conway, Gordon R., ed. Pest and Pathogen Control: Strategic, Tactical, and Policy Models. Chichester, Eng. and New York; Wiley, 1984. 488p.	
	Cook, Allyn A. Diseases of Tropical and Subtropical Field, Fiber and Oil Plants. New York and London; Macmillan, 1981. 450p.	First rank

Developed Countries Ranking		Third World Ranking
	Cook, Allyn A. Diseases of Tropical and Subtropical Fruits and Nuts. New York; Hafner, 1975. 317p.	First rank
	Cook, Allyn A. Diseases of Tropical and Subtropical Vegetables and Other Plants. New York; Hafner Press, 1978. 381p.	First rank
First rank	Cook, R. James, and Kenneth F. Baker. The Nature and Practice of Biological Control of Plant Pathogens. St. Paul, Minn.; American Phytopathological Society, 1983. 539p.	First rank
Second	Cook, R. James, and R. J. Veseth. Wheat Health Management. St. Paul, Minn.: American Phytopathological Society, 1991. 152p.	Second
Second	Cooke, G. W., N. W. Pirie, and G. D. H. Bell. Agricultural Efficiency. London; Royal Society, 1977. 227p. (Previously published as The Management of Inputs for Yet Greater Agricultural Yield and Efficiency: A Discussion. Royal Society (Great Britain), Philosophical Transactions, Series B, Biological Sciences, v. 281, no. 980)	Third
	Coombs, J., et al, eds. Techniques in Bioproductivity and Photosynthesis, sponsored by the U.N. Environment Programme. 2d ed. Oxford and New York; Pergamon Press, 1985. 298p. (1st ed. J. Coombs and D. O. Hall, eds., 1982. 171p.)	Second
Second	Cooper, J. P., ed. Photosynthesis and Productivity in Different Environments. Synthesis Meeting on the Functioning of Photosynthetic Systems in Different Environments, Aberystwyth, 1973. Cambridge and New York; Cambridge University Press, 1975. 715p.	
Third	Copeland, Lawrence O., and M. B. McDonald. Principles of Seed Science and Technology. 2d ed. Minneapolis, Minn.; Burgess Pub. Co., 1985. 321p. (1st ed., 1976. 369p.)	Third
Second	Coppel, Harry C., and James W. Mertins. Biological Insect Pest Suppression. Berlin, Germany and New York; Springer-Verlag, 1977. 314p. (Advanced Series in Agricultural Sciences No. 4)	Third
Third	Coppin, L. G., M. B. Green, and R. T. Rees, eds. Pest Management in Soybean. London and New York; Elsevier Applied Science, 1992. 369p.	Third
Second	Corbett, John R., K. Wright, and A. C. Baille. The Biochemical Mode of Action of Pesticides. 2d ed. London, Eng. and Orlando, Fla.; Academic Press, 1984. 382p. (1st ed. 1974. 330p.)	
Third	Corner, Edred J. H. The Natural History of Palms. London, Eng.; Weidenfeld and Nicolson, 1966. 393p.	Third

Developed Countries Ranking		Third World Ranking
Third	Cornish, P. S., and J. E. Pratley, eds. Tillage: New Directions in Australian Agriculture. Melbourne, Australia; Inkata Press, 1987. 448p.	
	Cornuet, Pierre. Elements de Virologie Vegetale. Paris; Institut National de la Recherche Agronomique, 1987. 206p.	Third
	Coste, Rene. Le Cafeier. Paris; G.P. Maisonneuve et Larose, 1968. 311p. (Techniques Agricoles et Productions Tropicales no. 14)	Second
Second	Cotton, Richard T. Pests of Stored Grain and Grain Products. Minneapolis, Minn.; Burgess Publishing Co., 1963. 318p. (1st published in 1941 under title: Insect Pests of Stored Grain and Grain Products.)	Second
Third	Cottrell, Helen J., ed. Pesticides on Plant Surfaces. Chichester, Eng. and New York; Wiley, 1987. 86p.	
Second	Couch, Houston B. Diseases of Turfgrasses. 2d ed. Huntington, N.Y.; R.E. Krieger Pub. Co., 1973. 348p. (1st ed., New York; Reinhold Pub. Corp., 1962. 289p.)	
Second	Coursey, Donald G. Yams: An Account of the Nature, Origins, Cultivation and Utilization of the Useful Members of the Dioscoreaceae. London; Longmans, 1967. 230p.	First rank
Third	Coyier, Duane L., and Martha K. Roane, eds. Compendium of Rhododendron and Azalea Diseases. St. Paul, Minn.; American Phytopathological Society Press, 1986. 65p.	
Second	Crafts, Alden S. Modern Weed Control. Berkeley, Calif.; University of California Press, 1975. 440p.	Third
Second	Cramer, H. H. Plant Protection and World Crop Production. Translated by J.H. Edwards. Leverkusen; Farbenfabriken Bayer A, 1967. 542p.	Second
Second	Crane, Eva, and Penelope Walker. The Impact of Pest Management on Bees and Pollination. London, Eng.; Tropical Development and Research Institute, 1983. 129p.	Second
Third	Cronquist, Arthur. Basic Botany. 2d ed. New York; Harper and Row, 1982. 662p. (Also available in Spanish as Botánica Básica. Mexico; Cecsa, 1978. 587p.)	Third
Third	Cronquist, Arthur. The Evolution and Classification of Flowering Plants. 2d ed. Bronx, NY; New York Botanical Garden, 1988. 555p. (1st ed., Boston, Mass; Houghton Mifflin, 1968. 396p.)	Third
Third	Crop Protection Chemicals Reference. 5th ed. New York; Chemical and Pharmaceutical Publishing, 1989. 2266p. (1st ed., 1985.)	Second

Developed Countries Ranking		Third World Ranking
	Crop Tolerance to Suboptimal Land Conditions. Proceedings of a symposium, Houston, Tex., Nov.-Dec. 1976, sponsored by the American Society of Agronomy, Crop Science Society of America, and Soil Science Society of America. Madison, Wis.; American Society of Agronomy, Crop Science Society of America, and Soil Science Society of America, 1978. 343p. (ASA Special Publication No. 32)	Third
Third	Crowder, Loy V., and H. R. Chheda. Tropical Grassland Husbandry. London and New York; Longman, 1982. 562p.	First rank
Third	Cruger, Gerd. Pflanzenschutz im Gemusebau. 2d ed. Stuttgart, Germany; E. Ulmer, 1983. 422p.	
	Les Cultures de Tissues de Plantes. Colloques . . . , Strasbourg, July 1970. Paris; Editions du Centre National de la Recherche Scientifique, 1971. 511p. (Colloques Internationaux du Centre National de la Recherche Scientifique no. 193) (In French or English)	Third
First rank	Cummins, George B., and Yasuyuki Hiratsuka. Illustrated Genera of Rust Fungi. Rev. ed. St. Paul, Minn.; American Phytopathological Society, 1983. 152p. (1st ed. by G.B. Cummins. Minneapolis; Burgess, 1959. 131p.)	Second
Second	Cummins, George B. The Rust Fungi of Cereals, Grasses and Bamboos. New York; Springer-Verlag, 1971. 570p.	Second
Second	Cummins, George B. Rust Fungi on Legumes and Composites in North America. Tucson, Ariz.; University of Arizona Press, 1978. 424p.	
Second	Curtis, Charles R. Agricultural Benefits Derived from Pesticide Use: A Study of the Assessment Process. Columbus, Ohio; The Ohio State University, 1988. 148p.	
Third	Cutler, Horace G., ed. Biologically Active Natural Products: Potential Use in Agriculture; Developed from a symposium sponsored by the Division of Agrochemicals at the 194th Meeting of the American Chemical Society, New Orleans, La., Aug.–Sept. 1987. Washington, D.C.; The Society, 1988. 483p.	Second
Third	D'Arcy, William G., ed. Solanaceae: Biology and Systematics; Many of the papers were presented at the 2d International Symposium on the Biology and Systematics of the Solanaceae, Missouri Botanical Garden, Aug. 1983. New York; Columbia University Press, 1986. 603p.	First rank
Second	Dalrymple, Dana G. Development and Spread of High-Yielding Varieties of Wheat and Rice in the Less Developed Nations. 6th ed. Washington, D.C.; USDA, Office	Third

Developed Countries Ranking		Third World Ranking
	of International Cooperation and Development, 1978. 134p. (Foreign Agricultural Economic Report No. 95) (1st published in 1969 under title: Imports and Plantings of High-Yielding Varieties of Wheat and Rice in the Less Developed Nations. 30p.)	
Second	Dalrymple, Dana G. Development and Spread of Semi-Dwarf Varieties of Wheat and Rice in the United States: An International Perspective. Washington, D.C.; USDA, Office of International Cooperation and Development and U.S. Agency for International Development, 1980. 150p. (USDA Agricultural Economic Report No. 455)	
	Dalrymple, Dana G. Survey of Multiple Cropping in Less Developed Nations. Washington, D.C.; Foreign Economic Development Service, 1971. 108p.	Second
	Daly, Herman E., ed. Economics, Ecology, Ethics: Essays Toward a Steady-State Economy. San Francisco, Calif.; W.H. Freeman, 1980. 372p. (In part a revision of: Toward a Steady State Economy, 1973, edited by H.E. Daly.)	Third
Second	Daly, Howell V., John T. Doyen, and Paul R. Ehrlich. Introduction to Insect Biology and Diversity. New York; McGraw-Hill, 1978. 564p.	
Second	Daly, J. M., and B. J. Deverall, eds. Toxins and Plant Pathogenesis; Prepared in conjunction with a symposium at the 4th International Congress of Plant Pathology, Melbourne, Australia, Aug. 1983. Sydney, Australia and New York; Academic Press, 1983. 181p.	Second
Second	Daly, J. M., and Ikuzo Uritani, eds. Recognition and Specificity in Plant Host-Parasite Interactions. Tokyo, Japan; Japan Scientific Societies Press and Baltimore, Md.; University Park Press, 1979. 355p.	Third
Second	Daniels, M. J., and P. G. Markham, eds. Plant and Insect Mycoplasma Techniques. London, Eng.; Croom Helm and New York; Wiley, 1982. 369p.	
	Darch, J. P., ed. Drained Field Agriculture in Central and South America; Proceedings of the 44th International Congress of Americanists, Manchester, 1982. Oxford, Eng.; B.A.R., 1983. 263p. (BAR International Series No. 189) (English and Spanish.)	Third
Second	Darlington, Cyril D. Chromosome Botany and the Origins of Cultivated Plants. 3d rev. ed. London; Allen and Unwin, 1973. 237p. (1st ed., 1956, entitled Chromosome Botany. 186p.)	Second
Second	Darlington, Cyril D. Evolution of Genetic Systems. 2d ed.	Third

Developed Countries Ranking		Third World Ranking
	Edinburgh, Scotland; Oliver and Boyd, 1958. 265p. (1st ed., Cambridge; Cambridge University Press, 1939. 149p.)	
Third	Darlington, Cyril D., and L. F. La Cour. Handling of Chromosomes. 6th ed. New York; J. Wiley, 1975. 201p. (1st ed., New York; Macmillan Co., 1942. 165p.)	Third
Second	Darlington, Cyril D., and A. P. Wylie. Chromosome Atlas of Flowering Plants. 2d ed., rev. and exp. London; Allen and Unwin, 1955. 519p.	
	Daubenmire, Rexford F. Plants and Environment: A Textbook of Plant Autecology. 3d ed. New York; Wiley, 1974. 422p. (1st ed., 1947. 424p.)	Second
	Daussant, J., J. Mosse, and J. Vaughan, eds. Seed Proteins. London, Eng. and New York; Academic Press, 1983. 335p. (Annual Proceedings of the Phytochemical Society of Europe, no. 20)	Third
Second	Davey, Kenneth G. Reproduction in the Insects. Edinburgh and London, Eng.; Oliver and Boyd, 1965. 96p.	
Second	Davidson, Ralph H., and William F. Lyon. Insect Pests of Farm, Garden, and Orchard. 8th ed. New York; Wiley, 1987. 640p. (5th ed. by Leonard M. Peairs and Ralph H. Davidson, 1956. 661p.)	Third
Third	Davies, Jeffrey W., ed. Molecular Plant Virology. Boca Raton, Fla.; CRC Press, 1985. 2 vols.	Third
Second	Davis, D. W., et al, eds. Biological Control and Insect Pest Management. Berkeley, Calif.; Division of Agricultural Sciences, University of California, 1979. 102p.	Second
Third	Davis, Gwenda L. Systematic Embryology of the Angiosperms. New York; Wiley, 1966. 528p.	
Third	Davis, Ronald W., David Botstein, and John R. Roth. Advanced Bacterial Genetics. Cold Spring Harbor, N.Y.; Cold Spring Harbor Laboratory, 1980. 254p.	
Second	Day, Peter R. Genetics of Host-Parasite Interaction. San Francisco; W.H. Freeman, 1974. 238p.	Second
Second	Day, Peter R., and G. J. Jellis, eds. Genetics and Plant Pathogenesis. Oxford, Eng. and Boston, Mass.; Blackwell Scientific Publications, 1987. 352p.	Second
Third	Day, W., and R. K. Atkin, eds. Wheat Growth and Modelling; Proceedings of a NATO Advanced Research Workshop, . . . Bristol, U.K., Apr. 1984. New York; Plenum Press, 1985. 407p.	
Second	Deacon, J. W. Microbial Control of Plant Pests and Diseases. Washington, D.C.; Amerian Society for Microbiology, 1983. 88p.	Second
First rank	DeBach, Paul H., and David Rosen. Biological Control by	Second

Developed Countries Ranking		Third World Ranking
	Natural Enemies. 2d ed. Cambridge, Eng. and New York; Cambridge University Press, 1991. 440p. (1st ed., 1974. 323p.)	
First rank	DeBach, Paul H., and Evert I. Schlinger, ed. Biological Control of Insect Pests and Weeds. London; Chapman and Hall, 1973. 844p. (1st published New York; Reinhold, and London; Chapman and Hall, 1964. 844p.)	First rank
Third	Decker, Heinz. Phytonematologie; Biologie und Bekampfung pflanzenparasitarer Namatoden. Berlin, Germany; Deutscher Landwirtschaftsverl., 1969. 526p.	
Second	De Datta, Surajit K. Principles and Practices of Rice Production. Reprint. Malabar, Fla.; Krieger, 1987. 618p. (Originally published, New York; Wiley, 1981.)	Third
	Degras, L. L'Igname, Plante a Tubercule Tropicale. Paris, France; G.P. Maisonneuve and Larose, Agence de Cooperation Culturelle et Technique, 1986. 408p. (In French) (The Yam, A Tropical Tuber Plant)	Third
First rank	Dekker, J., and S. G. Georgopoulos, eds. Fungicide Resistance in Crop Protection. Wageningen, Netherlands; Centre for Agricultural Publishing and Documentation, 1982. 265p.	Second
Third	Delorit, Richard J. An Illustrated Taxonomy Manual of Weed Seeds. River Falls, Wisc.; Agronomy Publications, 1970. 175p.	
First rank	Delp, Charles J., ed. Fungicide Resistance in North America. St. Paul, Minn.; American Phytopathological Society, 1988. 133p.	Third
Second	Delucchi, V. L., ed. Integrated Pest Management = Protection Integree: Quo Vadis. Geneva, Switzerland; Parasitis 86, 1987. 411p.	Third
Second	Delucchi, V. L., ed. Studies in Biological Control. New York; Cambridge University Press, 1976. 304p.	Third
	Dempsey, James M. Fiber Crops. Gainesville, Fla.; University Presses of Florida, 1975. 457p.	Second
	Dendy, D. A. V., ed. Sorghum and Millets for Human Food; Proceedings of a Symposium at the 9th Congress of the International Association for Cereal Chemistry, Vienna, Austria, May 1976. London, Eng.; Tropical Products Institute, 1977. 138p.	First rank
Third	De Nettancourt, D. Incompatibility in Angiosperms. Berlin, Germany and New York; Springer-Verlag, 1977. 230p.	
Second	Dennis, Colin, ed. Post-Harvest Pathology of Fruits and Vegetables. London, Eng. and New York; Academic Press, 1983. 264p.	Second

Developed Countries Ranking		Third World Ranking
Third	Dennis, David T., and David H. Turpin, eds. Plant Physiology, Biochemistry, and Molecular Biology. Essex, Eng.; Longman Scientific and Technical; New York; Wiley, 1990. 529p.	Third
Second	Dennis, Richard W. G. British Ascomycetes. Rev. and enl. ed. Vaduz, Liechtenstein; J. Cramer, 1978. 585p. (1st ed. entitled British Cup Fungi and Their Allies: An Introduction to the Ascomycetes. London; Ray Society, 1960. 280p. Ray SocietyPublications No. 143) (Supplement to 1978 ed., 1981. 40p.)	
Third	Dent, David. Insect Pest Management. Wallingford, U.K.; CAB International, 1991. 604p.	
Second	De Pury, J. M. S. Crop Pests of East Africa. Nairobi, Kenya; Oxford University Press, 1968. 227p.	Third
	Deuse, Jacques, and E. M. Lavabre. Le Descherbage des Cultures sous les Tropiques. Paris, France; B.P. Maisonneuve et Larose, 1979. 312p. (Techniques Agricoles et Productions Tropicales No. 28)	Third
	Devaux, A., and A. J. Haverkort. Manuel de la Culture de la Pomme de Terre en Afrique Centrale. 2d ed. Ruhengeri, Rwanda; Programme Regional d'Amelioration de la Culture de la Pomme de Terre en Afrique Centrale, 1986. (1st ed., 1983.)	Third
Third	Devon, T. K., and A. I. Scott. Handbook of Naturally Occurring Compounds. New York; Academic Press, 1972–1974. 2 vols.	Second
Second	Dickinson, C. H., and J. A. Lucas. Plant Pathology and Plant Pathogens. 2d ed. Oxford, Eng., Boston, Mass. and St. Louis, Mo.; Blackwell Scientific Publications and Mosby Book Distributors, 1982. 229p. (1st ed., New York; Wiley, 1977. 161p.)	Second
Second	Dickson, James G. Diseases of Field Crops. 2d ed. New York; McGraw-Hill, 1956. 517p. (1st ed., 1947. 429p.)	Second
Second	Diener, Theodore O. Viroids and Viroid Diseases. New York; Wiley, 1979. 252p.	Second
Second	Diener, Urban L., Richard L. Asquit, and J. W. Dickens, eds. Aflatoxin and Aspergillus Flavus in Corn. Auburn, Ala.; Dept. of Research Information, Alabama Agricultural Experiment Station, Auburn University, 1983. 112p.	Third
Third	Dixon, G. R. Plant Pathogens and Their Control in Horticulture. London; Macmillan, 1984. 253p.	Second
Second	Dixon, G. R. Vegetable Crop Diseases. American ed. Westport, Conn.; AVI Pub. Co., 1981. 404p. (Also published, London; Macmillan, 1981)	Second
Second	Dixon, R. A., ed. Plant Cell Culture: A Practical Ap-	

Developed Countries Ranking		Third World Ranking
	proach. Oxford, Eng. and Washington, D.C.; IRL Press, 1985. 236p.	
Second	Doane, Charles C., and Michael L. McManus, eds. The Gypsy Moth: Research Toward Integrated Pest Management. Washington, D.C.; U.S. Dept. of Agriculture, 1981. 757p. (USDA Technical Bulletin No. 1584)	
	Dobben, W. H. van, and R. H. Lowe-McConnell, eds. Unifying Concepts in Ecology; Report of the Plenary Sessions of the 1st International Congress of Ecology, The Hague, Netherlands, Sept. 1974. The Hague, Netherlands; W. Junk and Wageningen, Netherlands; Centre for Agricultural Publishing and Documentation, 1975. 302p.	Third
	Dobereiner, Johanna, et al, eds. Limitations and Potentials for Biological Nitrogen Fixation in the Tropics; Proceedings of a conference, . . . Brasilia, 1977. New York; Plenum Press, 1978. 398p.	Second
Third	Dobereiner, Johanna, and Fabio O. Pedrosa. Nitrogen-Fixing Bacteria in Nonleguminous Crop Plants. Madison, Wis.; Science Tech Publishers and Berlin and New York; Springer-Verlag, 1987. 155p.	Secon
Second	Dobzhansky, Theodosius G. Genetics and the Origin of Species. 3d ed. New York; Columbia University Press, 1951. 364p. (1st ed., 1937)	Third
Third	Dobzhansky, Theodosius G., et al. Evolution. San Francisco; W.H. Freeman, 1977. 572p.	Second
First rank	Doggett, Hugh. Sorghum. 2d ed. Burnt Mill, Eng.; Longman Scientific and Technical; Co-published in the United States with J. Wiley, 1988. 512p. (1st ed., 1970. 403p)	First rank
	Doll, E. C., and G. O. Mott, eds. Tropical Forages in Livestock Production Systems; Proceedings of a symposium held during the annual meetings of the American Society of Agronomy, Crop Science Society of America, and Soil Science Society of America, Las Vegas, Nev., Nov. 1973. Madison, Wis.; American Society of Agronomy, 1975. 104p.	Second
Second	Dominguez Garcia-Tejero, Francisco. Plagas y Enfermedades de las Plantas Cultivadas. 8th cor. act. ed. Madrid, Spain; Mundi-Prensa, 1989. 821p. (3d ed., Madrid; Dossat, 1965. 944p.)	
Second	Dommergues, Y. R., and H. G. Diem, eds. Microbiology of Tropical Soils and Plant Productivity. Hague, Netherlands and Boston, Mass.; M. Nijhoff; Distributors for the U.S. and Canada Kluwer Boston, 1982. 328p.	First rank
Third	Domsch, Klaus H., W. Gams, and Traute-Heidi Anderson.	Third

Developed Countries Ranking		Third World Ranking
	Compendium of Soil Fungi. London and New York; Academic Press, 1980. 2 vols.	
	Doorenbos, J., and W. O. Pruitt. Guidelines for Predicting Crop Water Requirements. Rev. ed. Rome; Food and Agriculture Organization, 1977. 144p. (FAO Irrigation and Drainage Paper No. 24) (1st ed., 1975. 179p.)	Third
	Douglas, Johnson E., ed. Successful Seed Programs: A Planning and Management Guide. Boulder, Colo.; Westview Press, 1980. 302p.	Third
Third	Dover, Michael J., and Lee M. Talbot. To Feed the Earth: Agro-Ecology for Sustainable Development. Washington, D.C.; World Resources Institute, 1987. 88p.	Third
	Dowson, Valentine H. W. Date Production and Protection: With Special Reference to North Africa and the Near East. Rome; Food and Agriculture Organization, 1982. 294p. (FAO Plant Production and Protection Paper No. 35)	Second
Third	Doyle, Jack. Altered Harvest: Agriculture, Genetics, and the Fate of the World's Food Supply. New York; Penguin Books, 1986. 502p. (1st published, New York; Viking, 1985.)	
Third	Dregne, H. E., and W. O. Willis, eds. Dryland Agriculture. Madison, Wis.; American Society of Agronomy, 1983. 622p.	Third
Second	Dropkin, Victor H. Introduction to Plant Nematology. 2d ed. New York; Wiley, 1989. 304p. (1st ed., 1980. 293p.)	Third
Third	Du Montcel, Hugues T. Plantain Bananas, translation of Le Bananier Plantain by Paul Skinner. London, Eng.; Macmillan, 1987. 106p.	Second
Second	Dudley, John W., ed. Seventy Generations of Selection for Oil and Protein in Maize. Madison, Wis.; Crop Science Society of America, 1974. 212p.	Third
Third	Duke, James A. CRC Handbook of Nuts. Boca Raton, Fla.; CRC Press, 1989. 343p.	Second
Second	Duke, James A., ed. Handbook of Legumes of World Economic Importance. New York; Plenum Press, 1981. 345p.	First rank
Second	Duke, Stephen O., ed. Weed Physiology. Boca Raton, Fla.; CRC Press, 1985. 2 vols.	First rank
Second	Durbin, R. D., ed. Toxins in Plant Disease. New York; Academic Press, 1981. 515p.	Third
Third	Dutta, S. K., and Charles Sloger, eds. Biological Nitrogen Fixation Association with Rice Production. Washington, D.C.; Howard University Press, 1991.	

Developed Countries Ranking		Third World Ranking
	Earle, Elizabeth D., and Yves Demarly, eds. Variability in Plants Regenerated from Tissue Culture. New York; Praeger, 1982. 392p.	Third
Second	Ebbels, D. L., and J. E. King. Plant Health: The Scientific Basis for Administrative Control of Plant Diseases and Pests. Oxford, Eng. and New York; Blackwell Scientific; Distributed by Halsted Press, 1979. 322p.	Third
Second	Ebeling, Walter. Subtropical Fruit Pests. Berkeley, Calif.; University of California, Div. of Agricultural Sciences, 1959. 436p. (Complete reorganization and rev. of: Subtropical Entomology by W. Ebeling.)	Third
	Eckholm, Erik P. Losing Ground: Environmental Stress and World Food Prospects. New York; Norton, 1976. 223p.	Third
	Eden, Thomas. Tea. 3d ed. London, Eng.; Longman, 1976. 236p. (1st ed., 1958. 201p.)	Second
Second	Edgerton, Claude W. Sugarcane and Its Diseases. 2d rev. ed. Baton Rouge, La.; Louisiana State University Press, 1959. 301p. (1st ed., 1955. 290p.)	Second
Third	Edmond, Joseph B. Sweet Potatoes: Production, Processing, Marketing. Westport, Conn.; AVI Publishing Co., 1971. 334p.	Second
Second	Edmunds, L. K., and N. Zummo. Sorghum Diseases in the United States and Their Control. Washington, D.C.; U.S. Agricultural Research Service, 1975. 46p. (USDA Agriculture Handbook no. 468)	
Second	Edwards, Clive A., ed. Environmental Pollution by Pesticides. London and New York; Plenum Press, 1973. 542p. (More up-to-date and authoritative version of: Persistent Pesticides in the Environment, 1970.)	Second
Third	Edwards, Clive A., et al, eds. Sustainable Agricultural Systems; Proceedings of the International Conference, . . . Ohio State University, Columbus, Ohio, Sept. 1988. Ankeny, Iowa; Soil and Water Conservation Society, 1990. 696p.	First rank
Third	Edwardson, John R. and R. G. Christie. CRC Handbook of Viruses Infecting Forage Legumes. Boca Raton, Fla.; CRC Press, 1991. 504p.	Third
Second	Ehrlich, Paul R., Anne H. Ehrlich, and John P. Holdren. Ecoscience: Population, Resources, Environment. San Francisco, Calif.; W.H. Freeman, 1977. 1051p. (1st-2d ed. published under title: Population, Resources, Environment.)	Third
Third	Elkan, Gerald H., ed. Symbiotic Nitrogen Fixation Technology. New York; M. Dekker, 1987. 440p.	

Developed Countries Ranking		Third World Ranking
Second	Ellis, R. H., T. D. Hong, and E. H. Roberts. Handbook of Seed Technology for Genebanks. Rome; International Board for Plant Genetic Resources, 1985. 2 vols. 667p. (Handbooks for Genebanks No. 2-3)	Second
Third	Ellis, R. J., ed. Chloroplast Biogenesis. Cambridge and New York; Cambridge Univesity Press, 1984. 346p.	
Third	Elton, Charles S. The Ecology of Invasions by Animals and Plants. London, Eng.; Methuen, 1958. 181p.	
First rank	Engelhard, Arthur W., ed. Soilborne Plant Pathogens: Management of Diseases with Macro- and Microelements. St. Paul, Minn.; APS Press, 1989. 217p.	First rank
	Englund, Paul T., and Alan Sher, eds. The Biology of Parasitism: A Molecular and Immunological Approach. New York; Liss, 1988. 544p.	Second
Second	Ennis, W. B., ed. Introduction to Crop Protection. Madison, Wis.; American Society of Agronomy, 1979. 524p.	Second
Second	Entwistle, Philip F. Pests of Cocoa. London, Eng.; Longman, 1972. 779p.	Second
	Epstein, Emanuel. Mineral Nutrition of Plants: Principles and Perspectives. New York; Wiley, 1972. 412p.	Second
Second	Erwin, D. S., S. Bartnicki-Garcia, and P. H. Tsao, eds. Phytophthora: Its Biology, Taxonomy, Ecology, and Pathology. St. Paul, Minn.; American Phytopathological Society, 1983. 392p.	Third
Second	Esau, Katherine. Anatomy of Seed Plants. 2d ed. New York; Wiley, 1977. 550p. (1s ed., 1960. 376p.)	
Second	Esau, Katherine. Plant Anatomy. 2d ed. New York; Wiley, 1965. 767p. (1st ed., 1953. 735p.) (Also available in Spanish as Anatomia Vegetal. Barcelona; Ediciones Omega, 1959. 729p.)	Second
Third	Esquinas-Alcazar, Jose T. Genetic Resources of Tomatoes and Wild Relatives: A Global Report. Rome; IBPGR Secretariat, 1981. 65p.	
Third	Eto, Morifusa. Organophosphorus Pesticides: Organic and Biological Chemistry. Cleveland, Oh.; CRC Press, 1974. 387p.	Second
Third	EUCARPIA. Science for Plant Breeding; Proceedings of the XII Congress of EUCARPIA, Feb.-Mar. 1989, Gottingen, Germany. Berlin; P. Parey, 1989. 477p.	
Third	Evans, Dale O., and Peter P. Rotar. Sesbania in Agriculture. Boulder, Colo.; Westview Press, 1987. 192p. (Westview Tropical Agriculture Series No. 8)	Second
Third	Evans, K., D. L. Trudgill, and J. M. Webster, eds. Plant	

Developed Countries Ranking		Third World Ranking
	Parasitic Nematodes in Temperate Agriculture. Wallingford; CAB International, 1993. 648p.	
First rank	Evans, L. T., ed. Crop Physiology: Some Case Histories. 1st paperback ed. Cambridge, Eng. and New York; Cambridge University Press, 1978. 374p. (1st published, 1975. 374p.)	
Second	Evans, L. T., ed. The Induction of Flowering: Some Case Histories. South Melbourne, Australia; Macmillan, 1969. 488p.	
Third	Evans, L. T., and W. J. Peacock, eds. Wheat Science: Today and Tomorrow. Cambridge, Eng. and New York; Cambridge University Press, 1981. 290p.	Third
Third	Evered, David, and Sara Harnett, eds. Plant Resistance to Viruses; Symposium held at the Ciba Foundation, London, Eng., Mar.–Apr. 1987. Chichester, Eng. and New York; Wiley, 1987. 215p.	Second
	Fadeev, IU N., and K. V. Novozhilov, eds. Integrated Plant Protection (Translation of Integriovannaia Zashchita Rastenii). New Delhi, India; Amerind Publ. Co., 1987. 333p.	Third
	Faegri, Knut, and L. van der Rijl. The Principles of Pollination Ecology. 3d rev. ed. Oxford and New York; Pergamon, 1979. 244p. (1st ed., 1966. 248p.)	Third
Third	Fageria, N. K., V. C. Baligar, and Charles A. Jones. Growth and Mineral Nutrition of Field Crops. New York; M. Dekker, 1991. 476p.	Second
Second	Fahy, P. C., and G. J. Persley, eds. Plant Bacterial Diseases: A Diagnostic Guide. Sydney, Australia and New York; Academic Press, 1983. 393p.	Second
	Falcon, Louis A., and Ray F. Smith. Guidelines for Integrated Control of Cotton Insect Pests. Rome; Food and Agriculture Organization, 1973. 92p.	Third
First rank	Falconer, Douglas S. Introduction to Quantitative Genetics. 3d ed. Burnt Mill, Eng.; Longman, Scientific and Technical and New York; Wiley, 1989. 438p. (1st ed., Edinburgh; Oliver and Boyd, 1960. 365p.)	First rank
Second	FAO/IRRI Workshop on Judicious and Efficient Use of Insecticides on Rice, Los Banos, Philippines, 1983. Proceedings, . . . jointly sponsored by the International Rice Research Institute and Food and Agriculture Organization, Intercountry Programme for Integrated Pest Control in Rice in South and Southeast Asia. Los Banos, Philippines; International Rice Research Institute, 1984. 180p.	Third

Developed Countries Ranking		Third World Ranking
Third	Farnsworth, Marjorie W. Genetics. 2d ed. New York; Harper and Row, 1988. 806p. (1st ed., 1978. 625p.)	
Second	Farr, David F., et al. Fungi on Plants and Plant Products in the United States. St. Paul, Minn.; American Phytopathological Society, 1989. 1252p.	Third
	Fasoulas, A. Principles and Methods of Plant Breeding. Thessaloniki, Greece; Dept. of Genetics and Plant Breeding, Aristotelian University of Thessaloniki, 1981. 147p. (Dept. of Genetics and Plant Breeding, Aristotelian University of Thessaloniki Publication No. 11)	Third
	Fauquet, C., and J. C. Thouvenel. Maladies Virales des Plantes en Cote d'Ivoire = Plant Viral Diseases in the Ivory Coast. Paris; ORSTOM, 1987. 243p. (Collection Initiations Documentations Techniques no. 46)	Third
	Feakin, Susan D., ed. Pest Control in Bananas. New ed., rev. London, Eng.; Centre for Overseas Pest Research, Foreign and Commonwealth Office, Overseas Development Administration, 1972. 128p. (PANS Manual No. 1)	Second
	Feakin, Susan D., ed. Pest Control in Groundnuts. 3d ed., comp. rev. London, Eng.; Centre for Overseas Pest Research Foreign and Commonwealth Office, Overseas Development Administration, 1973. 197p. (PANS Manual No. 2)	Second
Second	Fehr, W. R., ed. Genetic Contributions to Yield Gains of Five Major Crop Plants; Proceedings of a Symposium sponsored by Division C-1, Crop Science Society of America, Atlanta, Georgia, Dec. 1981. Madison, Wis.; American Society of Agronomy, 1984. 101p. (CCSA Special Publication No. 7)	
Second	Fehr, Walter R., ed. Principles of Cultivar Development. New York and London; Macmillan, 1987. 2 vols.	Second
First rank	Fehr, Walter R., and Henry H. Hadley, eds. Hybridization of Crop Plants. Rev. ed. Madison, Wis.; American Society of Agronomy and Crop Science Society of America, 1982. 765p. (1st published 1980.)	First rank
Third	Feistritzer, Walther P., ed. Cereal Seed Technology: A Manual of Cereal Seed Production, Quality Control, and Distribution. Rome; Food and Agriculture Organization, 1975. 238p. (FAO Agricultural Development Paper No. 98)	Second
Third	Fenemore, Peter G. Plant Pests and Their Control. Wellington, N.Z.; Butterworths, 1982. 271p.	Second
Third	Fields, Bernard, Malcolm A. Martin, and Daphne Kamely, eds. Genetically Altered Viruses and the Environment.	

Developed Countries Ranking		Third World Ranking
	Cold Spring Harbor, N.Y.; Cold Spring Harbor Laboratory, 1985. 362p.	
Second	Fincham, J. R. S., P. R. Day, and A. Radford. Fungal Genetics. 4th ed. Oxford; Blackwell Scientific Publications, 1979. 636p. (1st ed., 1963. 300p.)	Third
Second	Fischer, George W., and Charles S. Holton. Biology and Control of the Smut Fungi. New York; Ronald Press, 1957. 622p.	Second
Second	Fitter, Alastair, and R. K. M. Hay. Environmental Physiology of Plants. 2d ed. London and San Diego, Calif.; Academic Press, 1987. 423p. (1st ed., 1981. 355p.)	Third
Third	Fitzsimmons, R. W., et al, eds. Australian Cereal Identification: Recent Varieties of Wheat, Triticale, Barley and Oats. E. Melbourne, Australia; Commonwealth Scientific and Industrial Research Organization, 1986. 64p.	
Second	Flaherty, Donald L., et al, eds. Grape Pest Management. Berkeley, Calif.; Division of Agriculture and Natural Resources, University of California, 1982. 312p. (Publication No. 4105)	Third
Second	Fletcher J.T. Diseases of Greenhouse Plants. London, Eng. and New York; Longman, 1984. 351p.	
Second	Fletcher, William W. Recent Advances in Weed Research. Farnham Royal, Eng.; Commonwealth Agricultural Bureaux, 1983. 266p.	
Second	Fokkema, N. J., and J. van den Heuvel, eds. Microbiology of the Phyllosphere. Cambridge and New York; Cambridge University Press, 1986. 392p.	
	Food and Agriculture Organization. Agricultural and Horticultural Seeds: Their Production, Control and Distribution. Rome; Food and Agriculture Organization, 1978. 531p. (FAO Plant Production and Protectin Series No. 12 and FAO Agricultural Studies No. 55) (1st published 1961)	Third
Third	Food and Agriculture Organization. Breeding for Durable Disease and Pest Resistance, held in cooperation with the International Institute of Tropical Agriculture, Ibadan, Nigeria, Oct. 1982. Rome; Food and Agriculture Organization, 1984. 167p. (FAO Plant Production and Protection Paper No. 55)	Third
Third	Food and Agriculture Organization. Expert Consultation on Environmental Criteria for Registration of Pesticides, 2d, Rome, Italy, 1981. Rome; Food and Agriculture Organization, 1981. 60p. (FAO Plant Production and Protection Paper No. 28)	Third
	Food and Agriculture Organization. Post-Harvest Losses in	Third

Developed Countries Ranking		Third World Ranking
	Quality of Food Grains. Rome; Food and Agriculture Organization, 1984. 103p. (FAO Food and Nutrition Paper No. 29)	
	Food and Agriculture Organization. Report on the 1st FAO/SIDA Seminar on Improvement and Production of Field Food Crops for Plant Scientists from Africa and the Near East, Cairo, Egypt, Sept. 1973. Rome, Italy; Food and Agriculture Organization, 1973. 48p.	Second
Third	Food and Agriculture Organization. A Review and Analysis of Insect Pest, Plant Disease and Weed Complexes in High Yielding Varieties and Hybrids under Intensified Agricultural Practices in Asia and the Pacific. Bangkok, Thailand; Food and Agriculture Organization, 1981. 136p. (RAPA No. 45)	Third
Third	Food and Agriculture Organization. Status and Management of Major Vegetable Pests in the Asia-Pacific Region; With Special Focus Towards Integrated Pest Management, edited by Lim Guan-Soon and Di Yuan-Bo. Bangkok, Thailand; Food and Agriculture Organization, 1990. 125p.	Third
	Food and Agriculture Organization. The Use of Viruses for the Control of Insect Pests and Disease Vectors. Rome; Food and Agriculture Organization, 1973. 48p. (FAO Agricultural Series No. 91; WHO Technical Report Series No. 531)	Second
Second	Food and Agriculture Organization. Plant Protection Service. Elements of Integrated Control of Sorghum Pests. Rome; Food and Agriculture Organization, 1980. 159p. (FAO Plant Production and Protection Paper No. 19)	Second
	Food and Agriculture Organization/IAEA. Sterile-Male Technique for Eradication or Control of Harmful Insects; Proceedings of a Panel on Application, . . . organized by the joint FADivision of Atomic Energy in Food and Agriculture, Vienna, Austria, May 1968. Vienna, Austria; International Atomic Energy Agency, 1969. 142p.	Third
	Food and Fertilizer Technology Center for the Asian and Pacific Region. The Biological Control of Plant Diseases; Proceedings of the International Seminar on Biological Control of Plant Diseases and Virus Vectors, Tsukuba, Japan, 1990, sponsored by the Food and Fertilizer Technology Center for the ASPAC Region, National Agriculture Research Center (NARC), Japan, and the National Institute of Agro- Environmental Sciences, Japan. Taipei, Taiwan; Food and Fertilizer Technology Center for the Asian and Pacific Region, 1991. 215p.	Third

Developed Countries Ranking		Third World Ranking
Second	Food and Fertilizer Technology Center for the Asian and Pacific Region. The Rice Brown Planthopper. Taipei; Food and Fertilizer Technology Center for the Asian and Pacific Region, 1977. 258p.	Third
Second	Ford, M. G., et al, eds. Combating Resistance to Xenobiotics: Biological and Chemical Approaches. Chichester, Eng.; E. Horwood; Distributor for U.S. and Canada, New York; VCH Publishers, 1987. 320p.	
Third	Forsberg, Robert A., ed. Triticale; Proceedings of a symposium sponsored by the Crop Science Society of America, Fort Collins, Colo., Aug. 1979. Madison, Wis.; Crop Science Society of America and American Society of Agronomy, 1985. 82p. (CSSA Special Publication No. 9)	Third
	Fouque, A. Especes Fruitieres d'Amerique Tropicale. Paris, France; Institut Francais de Recherches Fruitieres Outremer, 1976. 320p.	Third
Third	Fowke, L. C., and F. Constabel, eds. Plant Protoplasts. Boca Raton, Fla.; CRC Press, 1985. 245p.	
Third	Fraley, Robert T., Nicholas M. Frey, and Jeff Schell, eds. Genetic Improvement of Agriculturally Important Crops: Progress and Issues. Cold Spring Harbor, N.Y.; Cold Spring Harbor Laboratory, 1988. 116p.	Third
First rank	Francis, Charles A., ed. Multiple Cropping Systems. New York; Macmillan and London, Eng.; Collier Macmillan, 1986. 383p.	First rank
Second	Francis, Charles A., Cornelia B. Flora, and Larry D. King, eds. Sustainable Agriculture in Temperate Zones. New York; Wiley, 1990. 487p.	
Second	Francki, R. I. B., T. Hatta, and Robert G. Milne. Atlas of Plant Viruses. Boca Raton, Fla.; CRC Press, 1985. 2 vols.	Third
Second	Frankel, O. H., and J. G. Hawkes, eds. Crop Genetic Resources for Today and Tomorrow. Cambridge and New York; Cambridge University Press, 1975. 492p.	Second
Third	Frankel, O. H., and Michael E. Soul. Conservation and Evolution. Cambridge, Eng. and New York; Cambridge University Press, 1981. 327p.	Third
Second	Frankel, R., ed. Heterosis: Reappraisal of Theory and Practice. Berlin, Germany and New York; Springer-Verlag, 1983. 290p.	Second
Second	Frankel, R., and E. Galun. Pollination Mechanisms, Reproduction, and Plant Breeding. Berlin, Germany; Springer, 1977. 281p. (Monograph Theoretical and Applied Genetics No. 2)	Third

Developed Countries Ranking		Third World Ranking
Second	Franz, Jost M., ed. Biological Plant and Health Protection: Biological Control of Plant Pests and of Vectors of Human and Animal Diseases; International Symposium of the Akademie der Wissenschaften und der Literatur, Mainz, Nov. 1984. Stuttgart and New York; G. Fischer, 1986. 341p.	
Second	Franz, Jost M., and Aloysius Kreig. Biologische Schädlingsbekämpfung. 3d ed. Berlin, Germany; P. Parey, 1982. 252p. (1st ed., 1972. 208p.)	
	Fraser, R. S. S., ed. Mechanisms of Resistance to Plant Diseases. Dordrecht, Netherlands and Boston, Mass.; M. Nijhoff/W. Junk; Distributors for the U.S. and Canada, Hingham, Mass.; Kluwer Academic Publishers, 1985. 462p.	Second
First rank	Frazier, N. W., et al, eds. Virus Diseases of Small Fruits and Grapevines. Berkeley, Calif.; University of California, Division of Agricultural Sciences, 1970. 290p.	Second
First rank	Frederiksen, Richard A., ed. Compendium of Sorghum Diseases. St. Paul, Minn.; American Phytopathological Society in cooperation with Dept. of Plant Pathology and Micribiology, Texas A & M University, 1986. 82p.	First rank
Second	Free, John B. Insect Pollination of Crops. London, Eng. and New York; Academic Press, 1970. 544p.	Second
	Fremond, Yan, Robert Ziller, and M. de Nuce de Lamothe. The Coconut Palm. Berne, Switzerland; International Potash Institute, 1968. 227p. (Translation of Le Cocotier.)	Second
	Fresco, Louise O. Cassava in Shifting Cultivation: A Systems Approach to Agricultural Technology Development in Africa. Amsterdam, Netherlands; Royal Tropical Institute, 1986. 240p.	Second
Third	Frey, Kenneth J., ed. Plant Breeding II; Proceedings of the 2d Plant Breeding Symposium, Iowa State University, 1979. Ames, Ia.; Iowa State University Press, 1981. 497p.	First rank
Third	Friedman, Mendel, ed. Protein Crosslinking; Proceedings of the Symposium, . . . San Francisco, Calif., Aug.–Sept. 1976. New York; Plenum Press, 1977. 2 vols.	
Second	Frisbie, Raymond E., and Perry L. Adkinsson, eds. Integrated Pest Management on Major Agricultural Systems; from a symposium sponsored by The Consortium for Integrated Pest Management and USDA/CSRS, Oct. 1985. College Station, Tex.; Texas Agricultural Experiment Station, 1985. 743p. (Texas Agricultural Experiment Station No. 1616)	Second

Developed Countries Ranking		Third World Ranking
Second	Fröhlich, Gerd. Pflanzenschutz in den Tropen. 2d ed. Zurich, Switzerland; H. Deutsch, 1974. 526p.	
Second	Fröhlich, Gerd, W. Rodewalt, and F. Bombach, et al. Pests and Diseases of Tropical Crops and Their Control, . . . transl. by H. Liebscher and F. Koehler. Edited by H. Liebscher. 1st English ed. Oxford and New York; Pergamon Press, 1970. 371p.	Third
Third	Fry, J. M. Natural Enemy Databank, 1987: A Catalogue of Natural Enemies of Arthropods Derived from Records in the CIBC Natural Enemy Databank. Wallingford, Eng.; CAB International, 1989. 185p.	Third
First rank	Fry, William E. Principles of Plant Disease Management. New York; Academic Press, 1982. 378p.	First rank
	Fryxell, Paul A. The Natural History of the Cotton Tribe (Malvaceae, Tribe Gossypieae). College Station, Tex.; Texas A&M University Press, 1979. 245p.	Third
Second	Fuxa, James R., and Yoshinori Tanada, eds. Epizootiology of Insect Diseases. New York; Wiley, 1987. 555p.	
Second	Gair, Robert, J. E. E. Jenkins, and E. Lester. Cereal Pests and Diseases. 4th ed. Ipswich, Suffolk; Farming Press, 1987. 268p. (1st ed., 1972. 184p.)	Second
Third	Gallais, A., ed. Quantitative Genetics and Breeding Methods; Proceedings of the 4th Meeting of the Section, Biometrics in Plant Breeding, Poitiers, France, Sept. 1981. Lusignan, France; Institut National de la Recherche Agronomique, 1981. 319p.	
Third	Gallais, A. Theorie de la Selection en Amelioration des Plantes (Theory of Selection in Plant Breeding). Paris; Masson, 1990. 589p.	Third
	Gaillard, Jean-Pierre. L'Avocatier: Sa Culture, ses Produits. Paris, France; G.P. Maisonneuve and Larose, 1987. 419p. (Techniques Agricoles et Productions Tropicales No. 38)	Third
Third	Gangstad, Edward O. Weed Control Methods for River Basin Management. West Palm Beach, Fla.; CRC Press, 1978. 229p.	Third
Second	Gardner, Eldon J., Michael J. Simmons, and D. Peter Snustad. Principles of Genetics. 8th ed. New York and Chichester, Eng.; Wiley, 1991. 649p. (1st ed., 1960. 366p.)	Second
Third	Gardner, Franklin P., R. Brent Pearce, and Roger L. Mitchell. Physiology of Crop Plants. Ames, Ia.; Iowa State University Press, 1985. 327p.	Third
	Garner, Robert J., and Saeed A. Chaudhri. The Propagation of Tropical Fruit Trees. Rome; Food and Agricul-	Second

Developed Countries Ranking		Third World Ranking
	ture Organization and Slough, Eng.; Commonwealth Agricultural Bureaux, 1976. 566p.	
First rank	Garrett, Stephen D. Pathogenic Root-Infecting Fungi. Cambridge, Eng.; University Press, 1970. 294p. (Sequel to the author's Biology of Root-Infecting Fungi.)	First rank
Second	Garrett, Stephen D. Soil Fungi and Soil Fertility: An Introduction to Soil Mycology. 2d ed. Oxford, Eng. and New York; Pergamon Press, 1981. 150p. (1st ed., Oxford, Eng.; Pergamon Press and New York; Macmillian, 1963. 165p.)	First rank
Second	Gates, David M. Biophysical Ecology. New York; Springer-Verlag, 1980. 611p.	
Third	Gauch, Hugh G. Inorganic Plant Nutrition. Stroudsburg, Pa.; Dowden, Hutchinson and Ross, 1972. 488p.	
	Gaudy, M. Manuel d'Agriculture Tropicale (Afrique Tropicale et Equatoriale). 2d ed. rev. and corr. Paris, France; Maison Rustique, 1965. 412p. (1st ed., 1959. 443p.)	Third
Second	Gäumann, Ernst A. Principles of Plant Infection: A Textbook of General Plant Pathology for Biologists, Agriculturists, Foresters and Plant Breeders. Authorized English ed. by William B. Brierley. London; C. Lockwood, 1950. 543p. (Added title page in German: Pflanzliche Infektionslehre Lehrbuch der Allgemeinen Pflanzenpathologie für Biologen, Landwirte, Förster und Pflanzenzüchter.)	
Third	George, Edwin F., and Paul D. Sherrington. Plant Propagation by Tissue Culture: Handbook and Directory of Commercial Laboratories. Eversley, Hants; Exegetics Ltd., 1984. 709p.	Third
First rank	Georghiou, George P., and Tetsuo Saito, eds. Pest Resistance to Pesticides; Proceedings of a U.S.-Japan Cooperative Science Program Seminar on Pest Resistance to Pesticides: Challanges (sic) and Prospects, Palm Springs, Calif., Dec. 1979. New York; Plenum Press, 1983. 809p.	
Third	Gepts, Paul, ed. Genetic Resources of Phaseolus Beans: Their Maintenance, Domestication, Evolution, and Utilization. Dordrecht and Boston; Kluwer Academic Pub., 1988. 613p.	Second
	Geus, Jan G. de. Fertilizer Guide for the Tropics and Subtropics. 2d ed. Zurich, Switzerland; Centre de'Etude de l'Azote, 1973. 774p. (1st ed. titled: Fertilizer Guide for Tropical and Subtropical Farming.)	Second
	Ghosh, T. Handbook on Jute. Rome; Food and Agriculture Organization, 1983. 219p. (FAO Plant Production and Protection Paper No. 51)	Second

Developed Countries Ranking		Third World Ranking
	Gibbs, A. J., and Bryan D. Harrison. Plant Virology: The Principles. London, Eng.; E. Arnold, 1976. 292p.	Second
	Gibson, Alan H., and William E. Newton, eds. Current Perspectives in Nitrogen Fixation; Proceedings of the 4th International Symposium on Nitrogen Fixation, Canberra, Australia, Dec. 1980. Amsterdam and New York; Elsevier/North-Holland Biomedical Press, 1981. 534p.	First rank
	Gilbert, Lawrence E., and Peter H. Raven, eds. Coevolution of Animals and Plants; Symposium V, 1st International Congress of Systematic and Evolutionary Biology, Boulder, Colo., Aug. 1973. Rev. ed. Austin, Tex.; University of Texas, 1980. 263p. (1st ed., 1973)	Third
	Gillier, P., and P. Silvestre. L'Arachide. Paris, France; G.P. Maisonneuve et Larose, 1969. 293p. (Techniques Agricoles et Productions Tropicales no. 15)	Second
	Goeltenboth, Friedhelm, ed. Subsistence Agriculture Improvement: Manual for the Humid Tropics. Weikersheim, Germany; Margraf Scientific Pub. and Wau, Papua New Guinea; Wau Ecology Institute, 1990. 228p.	Third
	Goering, T. James. Tropical Root Crops and Rural Development. Washington, D.C.; World Bank, 1979. 85p. (World Bank Staff Working Paper No. 324)	Second
Third	Goldberg, Robert B., ed. Plant Molecular Biology; Proceedings of the ARCO Solar-UCLA Symposium, Keystone, Colo., Apr. 1983. New York; A.R. Liss, 1983. 498p. (UCLA Symposia on Molecular and Cellular Biology, New Series No. 12)	Third
Third	Goldstein, J., ed. The Least is Best Pesticide Strategy: A Guide to Putting Integrated Pest Management into Action. Ammaus, Pa.; JG Press, 1978. 205p.	
	Gomez, Kwanchai A. Techniques for Field Experiments with Rice: Layout/Sampling/Sources of Error. Los Banos, Philippines; International Rice Research Institute, 1972. 46p.	Third
Second	Gomez, Kwanchai A., and Arturo A. Gomez. Statistical Procedures for Agricultural Research. 2d ed. New York; Wiley, 1984. 680p. (Previously published as: Statistical Procedures for Agricultural Research with Emphasis on Rice. Los Banos; International Rice Research Institute, 1976. 294p.)	First rank
	Gomez, M. I., L. R. House, L. W. Rooney, and D. A. V. Dendy, eds. Utilization of Sorghum and Millets. Patancheru, India; ICRISAT, 1992. 216p.	Third
Second	Good, Ronald. The Geography of the Flowering Plants. 4th ed. London; Longman, 1974. 557p. (1st ed., 1947. 403p.)	

Developed Countries Ranking		Third World Ranking
	Goodland, Robert J. A., Catharine Watson, and George Ledec. Environmental Management in Tropical Agriculture. Boulder, Colo.; Westview Press, 1984. 237p.	Second
Second	Goodman, Robert N., Zoltan Kiraly, and K. R. Wood. The Biochemistry and Physiology of Plant Disease. Columbia, Mo.; University of Missouri Press, 1986. 433p.	Third
Second	Goodwin, T. W., ed. Plant Pigments. London, Eng. and San Diego, Calif.; Academic Press, 1988. 362p. (An entirely new book succeeding: The Chemistry and Biochemistry of Plant Pigments. 2d ed. 1975.)	Third
Second	Gordon, Donald T., John K. Knoke, and Gene E. Scott, eds. Virus and Viruslike Diseases of Maize in the United States. Wooster, Oh.; Ohio Agricultural Research and Development Center, 1981. 210p.	
Third	Goren, R., and K. Mendel, eds. Citriculture; Proceedings of the 6th International Citrus Congress, Tel Aviv, Israel, Mar. 1988. Rehovot, Israel; Balaban and Weikersheim, Germany; Margraf, 1989. 4 vols.	Second
Third	Goto, Masao. Fundamentals of Bacterial Plant Pathology. San Diego, Calif.; Academic Press, 1992. 342p.	
Third	Gotsch, Carl, ed. Improving Dryland Agriculture in the Middle East and North Africa. Stanford, Calif.; Food Research Institute, Stanford University and Cairo, Egypt; Ford Foundation Middle East Regional Office, 1980. 138p.	
Second	Gottschalk, Werner. Allgemeine Genetik. 3d ed. Stuttgart, Germany and New York; Thieme, 1989. 422p.	Third
Third	Gottschalk, Werner, and Hermann P. Muller, eds. Seed Proteins: Biochemistry, Genetics, Nutritive Value. Hague, Netherlands and Boston, Mass.; M. Nijhoff/W. Junk; Distributors for the U.S. and Canada, Kluwer Boston, 1983. 531p.	
Third	Gould, Frank W., and Robert B. Shaw. Grass Systematics. 2d ed. College Station, Tex.; Texas A & M University Press, 1983. 397p. (1st ed., by F.W. Gould. New York; McGraw-Hill, 1968. 382p.)	Second
	Gourou, Pierre. The Tropical World: Its Social and Economic Conditions and Its Future Status. 5th ed. London, Eng. and New York; Longman, 1980. 190p. (Translation of: Les Pays Tropicaux by S.H. Beaver) (1st ed., 1953. 156p. Translated by E.D. Laborde.)	Second
Second	Govindjee, Jan Amesz, ed. Photosynthesis. New York; Academic Press, 1982. 2 vols.	Third
Second	Govindjee, Jan Amesz, and David C. Fork, eds. Light Emission by Plants and Bacteria. Orlando, Fla.; Academic Press, 1986. 638p.	

Developed Countries Ranking		Third World Ranking
	Goyer, Robert A., and Myron A. Mehlman, eds. Toxicology of Trace Elements. Washington, D.C.; Hemisphere Pub. Corp.; Distributed solely, New York; Halsted Press, 1977. 303p. (Advances in Modern Toxicology No. 2)	Second
	Graham, Peter H., and Susan C. Harris, eds. Biological Nitrogen Fixation Technology for Tropical Agriculture; Based on papers presented at a Workshop, Centro Internacional de Agricultura Tropical, March 1981. Cali, Colombia; Centro Internacional de Agricultura Tropical, 1982. 726p.	Third
Third	Grainge, Michael, and Saleem Ahmed. Handbook of Plants with Pest-Control Properties. Honolulu, Hawaii; Resource Systems Institute, East-West Center and New York; Wiley, 1988. 470p.	Second
Second	Granados, Robert R., and Brian A. Federici, eds. The Biology of Baculoviruses. Boca Raton, Fla.; CRC Press, 1986. 2 vols.	
	Grant, Verne E. Organismic Evolution. San Francisco, Calif.; W.H. Freeman, 1977. 418p.	Third
Third	Grant, Verne E. Plant Speciation. 2d ed. New York; Columbia University Press, 1981. 563p. (1st ed., 1971. 435p.)	Third
Second	Grayson, B. T., M. B. Green, and L. G. Copping, eds. Pest Management in Rice; Papers presented at the Conference on Pest Management in Rice, Society of Chemical Industry, London, June–July 1990. London and New York; Elsevier, 1990. 536p.	Second
Second	Greathead, David J., and J. K. Waage. Opportunities for Biological Control of Agricultural Pests in Developing Countries. Washington, D.C.; World Bank, 1983. 44p. (World Bank Technical Paper No. 11)	Third
Second	Green, Maurice B. Chemicals for Crop Improvement and Pest Management. 3d ed. Oxford and New York; Pergamon Press, 1987. 370p. (1st ed. titled: Chemicals for Crop Protection and Pest Control.)	First rank
	Green, Maurice B., and Paul A. Hedin, eds. Natural Resistance of Plants to Pests: Roles of Allelochemicals; Developed from a symposium sponsored by the Division of Pesticide Chemistry at the 189th Meeting of the American Chemical Society, Miami Beach, Fla. Apr.–May 1985. Washington, D.C.; American Chemical Society, 1986. 243p.	Second
Third	Green, Maurice B., and Douglas A. Spilker, eds. Fungicide Chemistry: Advances and Practical Applications; Developed from a Symposium sponsored by the Divi-	

Developed Countries Ranking		Third World Ranking
	sion of Pesticide Chemistry at the 188th Meeting of the American Chemical Society, Philadelphia, Penn., Aug. 1984. Washington, D.C.; American Chemical Society, 1986. 173p.	
	Green, S. K., T. D. Griggs, and B. T. McClean, eds. Tomato and Pepper in the Tropics; Proceedings of the International Symposium on Integrated Management Practices. Taiwan; Asian Vegetable Research and Development Centre (AVRDC), 1989. 619p.	Second
Second	Gregory, P. H., ed. Phytophthora Disease of Cocoa. London, Eng.; Longman, 1974. 348p.	Second
Second	Gresshoff, Peter M., L. Evans Roth, Gary Stacey, and William E. Newton, eds. Nitrogen Fixation: Achievements and Objectives; Proceedings of the 8th International Congress on Nitrogen Fixation, Knoxville, Tenn., May 1990. New York and London; Chapman and Hall, 1990. 869p.	Second
Second	Grist, Donald H. Rice. 6th ed. London, Eng. and New York; Longman, 1986. 599p. (1st ed., 1953. 331p.)	Second
Third	Grossbard, E., and D. Atkinson, eds. The Herbicide Glyphosate. London, Eng. and Boston, Mass.; Butterworths, 1985. 490p.	Third
Third	Grover, R., ed. Environmental Chemistry of Herbicides. Boca Raton, Fla.; CRC Press, 1988+. 2 vols.	
Third	Groves, R. H., and J. J. Burdon, eds. Ecology of Biological Invasions. Cambridge, Eng. and New York; Cambridge University Press, 1986. 166p.	
Second	Grubben, G. J. H. Tropical Vegetables and Their Genetic Resources, edited by H.D. Tindall and J.T. Williams. Rome; International Board of Plant Genetic Resources, 1977. 197p. (AGPE: IBPGR 77/23)	Second
Third	Gunn, Donald L., and J. G. R. Stevens, eds. Pesticides and Human Welfare. London, Eng.; Oxford University Press, 1976. 278p.	
Third	Gunstone, Frank D., and Frank A. Norris. Lipids in Foods: Chemistry, Biochemistry, and Technology. Oxford, Eng. and New York; Pergamon Press, 1983. 170p.	
	Gupta, U. S., ed. Crop Physiology: Advancing Frontiers. New Delhi, India and Oxford, Eng.; IBH Publishing Co., 1984. 392p.	Third
	Gupta, Virendra K., ed. Advances in Parasitic Hymenoptera Research; Proceedings of the 2d conference on the Taxonomy and Biology of Parasitic Hymenoptera, University of Florida, Gainesville, Fla., Nov. 1987. Leiden and New York; E.J. Brill, 1988. 546p.	Third

Developed Countries Ranking		Third World Ranking
Third	Gustafson, J. Perry, and R. Appels, eds. Chromosome Structure and Function: Impact of New Concepts; Proceedings of the 18th Stadler Genetics Symposium, Univeristy of Missouri, Columbia, Mo., 1987. New York; Plenum Press, 1988. 326p.	Third
Third	Gustafson, J. P., ed. Gene Manipulation and Plant Improvement; Proceedings of the 16th Stadler Genetics Symposia, 1984. New York; Plenum Press, 1984. 668p.	Second
Third	Haard, Norman F., and D. K. Salunkhe, eds. Symposium: Postharvest Biology and Handling of Fruits and Vegetables. Westport, Conn.; AVI Pub. Co., 1975. 193p.	Second
	Haarer, Alec E. Modern Coffee Production. 2d rev. ed. London, Eng.; L. Hill, 1962. 495p. (1st ed., 1956. 467p.)	Third
First rank	Hagedorn, Donald J., ed. Compendium of Pea Diseases. St. Paul, Minn.; American Phytopathological Society in cooperation with the Dept. of Plant Pathology, University of Wisconsin—Madison, 1984. 57p.	Second
	Hagin, Josef, and B. Tucker. Fertilization of Dryland and Irrigated Soils. Berlin, Germany and New York; Springer-Verlag, 1982. 188p.	Second
Second	Hainsworth, Ernest. Tea Pests and Diseases, and Their Control: With Special Reference to North East India. Cambridge, Eng.; W. Heffer, 1952. 130p.	Third
Second	Halevy, Abraham H., ed. CRC Handbook of Flowering. Boca Raton, Fla.; CRC Press, 1985–87. 6 vols.	
Second	Hall, A. E., G. H. Cannell, and H. W. Lawton, eds. Agriculture in Semi-Arid Environments. Berlin, Germany and New York; Springer-Verlag, 1979. 340p.	Second
	Hall, David W. Handling and Storage of Food Grains in Tropical and Subtropical Areas. Rome; Food and Agriculture Organization, 1970. 350p. (FAO Agricultural Development Paper No. 90)	First rank
Second	Hall, Robert, ed. Compendium of Bean Diseases. St. Paul, Minn.; American Phytopathological Society, 1991. 73p.	Third
First rank	Hallauer, Arnel R., and J. B. Miranda. Quantitative Genetics in Maize Breeding. 2d ed. Ames, Ia.; Iowa State University Press, 1988. 468p. (1st ed., 1981.)	First rank
Third	Halliwell, Barry. Chloroplast Metabolism: The Structure and Function of Chloroplasts in Green Leaf Cells. Rev. ed. Oxford, Eng.; Clarendon Press and New York; Oxford University Press, 1984. 259p. (1st ed., 1982. 257p.)	Third
Second	Hallsworth, E. G., ed. Nutrition of the Legumes; Proceedings of the 5th Easter School in Agricultural Science,	Second

Developed Countries Ranking		Third World Ranking
	University of Nottingham, 1958. London, Eng.; Butterworths, 1958. 359p.	
Third	Hammerschlag, F. A., and R. E. Litz, eds. Biotechnology of Perennial Fruit Crops. Wallingford, U.K.; CAB International, 1992. 539p.	Third
	Hance, R. J., and K. Holly, eds. Weed Control Handbook: Principles. 8th ed. Oxford, Eng. and Boston, Mass.; Blackwell Scientific Publications, 1990. 582p. (1st ed. by British Weed Control Council, 1958. 245p. 6th and 7th eds. by J. D. Fryer with others, 1968–70 and 1977–78. 2 vols.)	Second
	Handbook of Plant Cell Culture. New York; Macmillan Pub. Co., 1983–86. 6 vols.	Second
Second	Hanson, Angus A., D. K. Barnes, and R. R. Hill, eds. Alfalfa and Alfalfa Improvement. Madison, Wis.; American Society of Agronomy, Crop Science Society of America, Soil Science Society of America, 1988. 1084p.	
First rank	Hanson, Clarence H., ed. Alfalfa Science and Technology. Madison, Wis.; American Society of Agronomy, 1972. 812p.	
Second	Hanson, Haldore E., Norman E. Borlaug, and R. Glenn Anderson. Wheat in the Third World. Boulder, Colo.; Westview Press, 1982. 174p. (Also in Spanish, 1985.)	First rank
Third	Hanson, J. Procedures for Handling Seeds in Genebanks. Rome, Italy; International Board for Plant Genetic Resources Secretariat, 1985. 115p. (Practical Manuals for Genebanks No. 1)	Third
Third	Harborne, Jeffrey B., ed. Biochemical Aspects of Plant and Animal Coevolution; Proceedings of the Phytochemical Society Symposium, Reading, Apr. 1977. London, Eng. and New York; Academic Press, 1978. 435p.	
Second	Harborne, Jeffrey B. Introduction to Ecological Biochemistry. 4th ed. London and San Diego, Calif.; Academic Press, 1993. 318p. (1st ed., London and New York; Academic Press, 1977. 243p.)	
Third	Harborne, Jeffrey B., ed. Plant Phenolics. London, Eng. and San Diego, Calif.; Academic Press, 1989. 552p. (Methods in Plant Biochemistry No. 1)	Third
Second	Harborne, Jeffrey B., D. Boulter, and B. L. Tuner, eds. Chemotaxonomy of the Leguminosae. London and New York; Academic Press, 1971. 612p.	Second
	Hardy, R. W. F., and W. S. Silver, eds. A Treatise on Dinitrogen Fixation. New York; Wiley, 1977–1983. 3 vols.	Second

Developed Countries Ranking		Third World Ranking
	Harlan, Jack R. Crops and Man. 2d ed. Madison, Wis.; American Society of Agronomy, 1992. 289p. (1st ed., 1975. 295p.)	First rank
	Harlan, Jack R., Jan M. J. de Wet, and Ann B. L. Stemler, eds. Origins of African Plant Domestication; Papers from a Conference, Burg Wartenstein, Austria, Aug. 1972. Hague, Netherlands; Mouton, 1976. 498p.	Third
Third	Harley, J. L., and S. E. Smith. Mycorrhizal Symbiosis. London and New York; Academic Press, 1983. 483p.	
	Harmond, Jesse E., N. Robert Brandenburg, and Leonard M. Klein. Mechanical Seed Cleaning and Handling. Washington, D.C.; U.S. Agricultural Reesarch Service, 1968. 56p. (USDA Agriculture Handbook No. 354)	Third
Second	Harper, John L., ed. The Biology of Weeds; A symposium of the British Ecological Society, Oxford, Apr. 1959. Oxford, Eng.; Blackwell Scientific Pub., 1960. 256p.	
First rank	Harper, John L. Population Biology of Plants. London and New York; Academic Press, 1977. 892p.	First rank
Third	Harris, David R., and Gordon C. Hillman, eds. Foraging and Farming: The Evolution of Plant Exploitation; A symposium, World Archaeological Congress, Southampton, Eng., Sept. 1986. London and Boston, Mass.; Unwin Hyman, 1989. 733p.	
Second	Harris, Kerry F., and Karl Maramorosch, eds. Aphids as Virus Vectors. New York; Academic Press, 1977. 559p.	Third
First rank	Harris, Kerry F., and Karl Maramorosch, eds. Pathogens, Vectors, and Plant Diseases: Approaches to Control. New York; Academic Press, 1982. 310p.	First rank
Second	Harris, Kerry F., and Karl Maramorosch, eds. Vectors of Plant Pathogens. New York; Academic Press, 1980. 467p.	Second
	Harris, M. K., et al, eds. Biology and Breeding for Resistance to Arthropods and Pathogens in Agricultural Plants; Proceedings of a short course entitled, "International Short Course in Host Plant Resistance," Texas A & M University, College Station, Tex., July–Aug. 1979. College Station, Tex.; Texas Agricultural Experiment Station, Texas A & M University System and Berkeley, Calif.; University of California- Agency for International Development, Pest Management Project, 1980. 605p.	Second
Second	Harris, P. M., ed. The Potato Crop: The Scientific Basis for Improvement. London, Eng.; Chapman and Hall and New York; Wiley, 1978. 730p.	Second
Third	Harrison, Pauline M., ed. Metalloproteins. Weinheim and Deerfield Beach, Fla.; Verlag Chemie, 1985. 2 vols.	

Developed Countries Ranking		Third World Ranking
	Hartley, Charles W. S. The Oil Palm (Elqeis Guineensis Jacq.). 3d ed. London, Eng.; Longman Scientific and Technical and New York; Wiley, 1988. 761p. (1st ed. 1967. 706p.)	First rank
Second	Hartmann, Hudson T., et al. Plant Science: Growth, Development, and Utilization of Cultivated Plants. 2d ed. Englewood Cliffs, N.J.; Prentice Hall, 1988. 674p. (1st ed., 1981. 676p.)	Second
Third	Harwood, J. L., and J. R. Bowyer, eds. Lipids: Membranes and Aspects of Photobiology. London, Eng. and San Diego, Calif.; Academic Press, 1990. 353p. (Methods in Plant Biochemistry No. 4)	
	Harwood, John L., and Nicholas J. Russell. Lipids in Plants and Microbes. London and Boston; G. Allen and Unwin, 1984. 162p.	Third
Second	Harwood, Richard R. Small Farm Development: Understanding and Improving Farming Systems in the Humid Tropics. Boulder, Colo.; Westview Press, 1979. 160p.	Second
Second	Haskell, P. T., ed. Pesticide Application: Principles and Practice. Oxford, Eng.; Clarendon Press and New York; Oxford University Press, 1985. 494p.	First rank
Second	Hassall, Kenneth A. The Chemistry of Pesticides: Their Metabolism, Mode of Action, and Uses in Crop Protection. Weinheim and Deerfield Beach, Fla.; Verlag Chemie, 1982. 372p.	Second
	Hatfield, Jerry L., and Ivan J. Thomason, eds. Biometeorology in Integrated Pest Management; Proceedings of a Conference, . . . University of California, Davis, July 1980, sponsored by the University of California Integrated Pest Management Program. New York; Academic Press, 1982. 491p.	Third
Third	Hatzios, Kriton K., and Donald Penner. Metabolism of Herbicides in Higher Plants. Minneapolis, Minn.; CEPCO Division, Burgess Pub. Co., 1982. 142p.	Third
	Hauck, Roland D., ed. Nitrogen in Crop Production; Proceedings of a Symposium, Sheffield, Ala., May 1982, sponsored by the National Fertilizer Development Center of the Tennessee Valley Authority. Madison, Wis.; American Society of Agronomy, 1984. 804p.	Third
	Havard-Duclos, Bernard. Les Plantes Fourrageres Tropicales . . . Paris, France; G.P. Maisonneuve et Larose, 1967. 399p. (Techniques Agricoles et Productions Tropicales No. 10)	Third
Second	Hawkes, John G. The Diversity of Crop Plants. Cambridge, Mass.; Harvard University Press, 1983. 184p.	Third

Developed Countries Ranking		Third World Ranking
	Hawksworth, D. L., ed. Advancing Agricultural Production in Africa; Proceedings of CAB's 1st Scientific Conference, Arusha, Tanzania, Feb. 1984. Farnham Royal, Eng.; Commonwealth Agricultural Bureaux, 1985. 454p.	Second
Second	Hawksworth, D. L. Mycologist's Handbook: An Introduction to the Principles of Taxonomy and Nomenclature in the Fungi and Lichens. Kew, Eng.; Commonwealth Mycological Institute, 1974. 231p.	Third
First rank	Hawksworth, D. L., B. C. Sutton, and G. C. Ainsworth. Ainsworth and Bisby's Dictionary of the Fungi (Including the Lichens). 7th ed. Kew, Eng.; Commonwealth Mycological Institute, 1983. 445p. (1st–4th eds. published under title: A Dictionary of the Fungi. 1st ed. by G.C. Ainsworth and G.R. Bisby. Kew, Eng.; The Imperial Mycological Institute, 1943. 359p.)	Second
Second	Hay, Robert K. M., and Andrew J. Walker. An Introduction to the Physiology of Crop Yield. Harlow, Eng.; Longman and New York; Wiley, 1989. 292p.	Third
	Hayes, A. Wallace, ed. Principles and Methods of Toxicology. 2d ed. New York; Raven Press, 1989. 929p. (1st ed., 1982. 750p.)	Third
Second	Hayes, Herbert K. A Professor's Story of Hybrid Corn. Minneapolis, Minn.; Burgess Publ. Co., 1963. 237p.	Third
Second	Hayes, Herbert K., Forrest R. Immer, and David S. Smith. Methods of Plant Breeding. 2d ed. New York; McGraw-Hill, 1955. 551p. (1st ed., 1942. 432p.)	Second
Third	Hebblethwaite, P. D. Seed Production; Proceedings of the 28th Easter School in Agricultural Science, University of Nottingham, Sept. 1978. London, Eng. and Boston, Mass.; Butterworths, 1980. 694p.	Third
	Hecht, Susanna B., ed. Amazonia: Agriculture and Land Use Research; Proceedings of the International Conference, CIAT, Cali, Colombia, Apr. 1980, sponsored by the Rockefeller Foundation. Cali, Colombia; Centro Internacional de Agricultura Tropical, 1982. 428p.	Third
Second	Hedin, Paul A., ed. Host Plant Resistance to Pests. Washington, D.C.; American Chemical Society, 1977. 286p.	Third
Third	Hedin, Paul A., ed. Naturally Occurring Pest Bioregulators; Developed from Symposia sponsored by the Division of Agrochemicals at the 197th & 198th National Meetings, American Chemical Society and the 1989 International Chemical Congress of Pacific Basin Societies. Washington, D.C.; American Chemical Society, 1991. 456p. (ACS Symposium Series No. 449)	Third

Developed Countries Ranking		Third World Ranking
Third	Hedin, Paul A., Julius J. Menn, and Robert M. Hollingworth, eds. Biotechnology for Crop Protection; Developed from a symposium at the 3d Special Conference of and sponsored by the Division of Agrochemicals of the American Chemical Society, Snowbird, Utah, June 1987. Washington, D.C.; American Chemical Society, 1988. 471p.	Third
Third	Hegnauer, Robert. Chemotaxonomie der Pflanzen. Eine Ubersicht uber die Verbreitung und die Systematische Bedeutung der Pflanzenstoffe. Basel; Birkhauser, 1962–92. 10 vols.	
Second	Heinrichs, E. A., ed. Management of Rice Insects. New York; Wiley, 1992.	Third
Second	Heinrichs, E. A., ed. Plant Stress—Insect Interactions. New York; Wiley, 1988. 492p.	Third
Second	Heinrichs, E. A., et al. Manual for Testing Insecticides on Rice. Los Banos, Philippines; International Rice Research Institute, 1981. 134p.	
Second	Heinrichs, E. A., and T. A. Miller, eds. Rice Insects: Management Strategies. New York; Springer-Verlag, 1991. 347p.	Second
Second	Heiser, Charles B. Seed to Civilization: The Story of Man's Food. New ed. Cambridge, Mass.; Harvard University Press, 1990. 228p. (1st ed., San Francisco; W. H. Freeman, 1973. 243p.)	Second
	Heiser, Charles B. The Sunflower. Norman, Okla.; University of Oklahoma Press, 1976. 198p.	Second
Second	Heitefuss, Rudolf. Crop and Plant Protection: The Practical Foundations. English ed. Chichester, Eng.; Ellis Horwood and New York; Wiley, 1989. 261p. (Translation by Jacquie Welch of: Pflanzenschultz. 2d ed., rev. and enl. Stuttgart, Germany; Georg Thieme Verlag, 1987. 326p.)	Third
Third	Heitefuss, Rudolf, et al. Pflanzenkrankheiten und Schadlinge im Ackerbau. 2d ed. Frankfurt, Germany; DLG and Munchen; BLV, 1987. 132p.	
Second	Heitefuss, Rudolph, and P. H. Williams, eds. Physiological Plant Pathology. Berlin, Germany and New York; Springer-Verlag, 1976. 890p. (Encyclopedia of Plant Physiology No. 4)	Second
Second	Helle, W., and M. W. Sabelis, eds. Spider Mites: Their Biology, Natural Enemies, and Control. Amsterdam, Netherlands and New York; Elsevier, 1985. 1 vol.	Second
Third	Herdt, Robert W., and C. Capule. Adoption, Spread, and Production Impact of Modern Rice Varieties in Asia.	Second

Developed Countries Ranking		Third World Ranking
	Los Banos, Philippines; International Rice Research Institute, 1983. 54p.	
	Herklots, Geoffrey A. C. Vegetables in South-East Asia. New York; Hafner Press, 1972. 525p.	Second
	Hershey, Clair H., ed. Cassava Breeding: A Multidisciplinary Review; Proceedings of a workshop, Philippines, Mar. 1985. Cali, Colombia; Centro Internacional de Agricultura Tropical, 1987. 312p.	Second
Third	Hesketh, J. D., and James W. Jones, eds. Predicting Phytosynthesis for Ecosystem Models. Boca Raton, Fla.; CRC Press, 1980. 2 vols.	
Third	Heslop-Harrison, John, ed. Pollen: Development and Physiology. London, Eng.; Butterworths, 1971. 338p.	Third
Third	Hewitt, Eric J., and C. V. Cutting, eds. Nitrogen Assimilation of Plants; Proceedings of the 6th Long Ashton Symposium, University of Bristol, Sept. 1977. London and New York; Academic Press, 1979. 708p.	
Third	Hewitt, Eric J., and T. A. Smith. Plant Mineral Nutrition. New York; Wiley, 1975. 298p.	Third
First rank	Heyne, E. G., ed. Wheat and Wheat Improvement. 2d ed. Madison, Wis.; American Society of Agronomy, Crop Science Society of America, and Soil Science Society of America, 1987. 765p. (1st ed., edited by Karl S. Quisenberry, et al., 1967. 560p.)	First rank
First rank	Hill, Dennis S. Agricultural Insect Pests of the Tropics and Their Control. 2d ed. Cambridge, Eng. and New York; Cambridge University Press, 1983. 746p. (1st ed., Cambridge and New York; Cambridge University Press, 1975. 516p.)	Second
Second	Hill, Dennis S. Pests of Stored Products and Their Control. London, Eng.; Belhaven Press, 1990. 274p.	Third
Second	Hill, Dennis S., and J. M. Waller. Pests and Diseases of Tropical Crops. London and New York; Longman, 1982–1988. 2 vols. (v. 1, Principles and Methods of Control; v. 2, Field Handbook)	Second
First rank	Hill, Lowell D., ed. World Soybean Research; Proceedings, . . . Champaign, Ill., 1975. Danville, Ill.; Interstate Printers and Publishers, 1976. 1073p.	
	Hill, Stuart, and Pierre Ott. Basic Technics in Ecological Farming; Papers presented at the 2d International Conference, IFOAM, Montreal, Oct. 1978 and The Maintenance of Soil Fertility: Exposes Presentes a la 3 eme Conference Internationale Organisee, IFOAM, Bruxelles, Sept. 1980. Basel, Switzerland; Birkhauser Verlag, 1982. 365p. (English and French.)	Third

Developed Countries Ranking		Third World Ranking
Second	Hill, Thomas A. The Biology of Weeds. London, Eng.; E. Arnold, 1977. 64p.	
	Hinson, K., and E. E. Hartwig. Soybean Production in the Tropics. Rev. by Harry C. Minor. Rome, Italy; Food and Agriculture Organization, 1982. 222p. (FAO Plant Production and Protection Paper No. 4) (1st published 1977. 92p.) (Also published in French by FAO.)	First rank
Third	Hock, Bertold, and Erich Elstner, ed. Pflanzentoxicologie: Der Einfluss von Schadstoffen und Schadwirkungen auf Pflanzen. Mannheim, Germany; Bibliographisches Institut, 1984. 346p.	
Second	Hodek, Ivo. Biology of Coccinellidae. Hague, Netherlands; Junk and Prague; Academia, 1973. 260p.	Third
Third	Hodek, Ivo, ed. Ecology of Aphidophagous Insects; Proceedings of a symposium, Liblice near Prague, Sept.–Oct. 1965. Prague; Academia, 1966. 360p.	
Second	Hodgson, Ernest, and Ronald J. Kuhr, eds. Safer Insecticides: Development and Use. New York; M. Dekker, 1990. 593p.	Third
Second	Hoffmann, Gunter M., et al, eds. Lehrbuch der Phytomedizin. 2d ed. Berlin, Germany; Parey, 1985. 488p. (1st ed., 1976. 490p.)	
Third	Hoffmann, Gunter M., and Heinrich Schmutterer. Parasitäre Krankheiten und Schädlinge an Landwirtschaftlichen Kulturpflanzen. Stuttgart, Germany; E. Ulmer, 1983. 488p.	
Second	Holden, John, James Peacock, and Trevor Williams. Genes, Crops and the Environment. Cambridge and New York; Cambridge University Press, 1993. 162p.	Second
Third	Holden, J. H. W., and J. T. Williams, eds. Crop Genetic Resources: Conservation and Evaluation; Papers presented in honor of the 10th Anniversary of the International Board for Plant Genetic Resources. London, Eng. and Boston, Mass.; Allen and Unwin, 1984. 296p.	Second
Third	Holden, Patrick W. Pesticides and Groundwater Quality: Issues and Problems in Four States. Washington, D.C.; National Academy Press, 1986. 124p.	
	Hollaender, Alexander, et al, eds. Genetic Engineering for Nitrogen Fixation; Proceedings, . . . Brookhaven National Laboratory, March 1977. New York; Plenum Press, 1977. 528p.	Second
First rank	Holliday, Paul. A Dictionary of Plant Pathology. Cambridge, Eng.; Cambridge University Press, 1989. 369p.	First rank
Second	Holliday, Paul. Fungus Diseases of Tropical Crops. Cambridge, Eng.; Cambridge University Press, 1980. 607p.	First rank
Second	Holm, LeRoy G., et al. A Geographical Atlas of World	Third

Developed Countries Ranking		Third World Ranking
	Weeds. New York; Wiley, 1979. 391p. (Reprint ed., Malabar, Fla.; Krieger Pub. Co., 1991.) (English, Arabic, Chinese, French, German, Hindi, Indonesian, Japanese, Russian, and Spanish.)	
First rank	Holm, LeRoy G., et al. The World's Worst Weeds: Distribution and Biology. Honolulu, Ha.; University Press of Hawaii, 1977. 609p.	Second
Second	Holzner, W., and M. Numata, eds. Biology and Ecology of Weeds. Hague, Netherlands and Boston, Mass.; W. Junk; Distributors for the U.S. and Canada, Boston; Kluwer, 1982. 461p.	
First rank	Hooker, W. J., ed. Compendium of Potato Diseases. St. Paul, Minn.; American Phytopathological Society, 1981. 125p.	First rank
Third	Hopkins, John C. F. Tobacco Diseases, With Special Reference to Africa. Kew, Eng.; Commonwealth Mycological Institute, 1956. 178p. (An up-to-date and enlarged ed. of department handbook: Diseases of Tobacco in Southern Rhodesia, pub. in 1931.)	Third
Second	Horn, David J. Ecological Approach to Pest Management. New York; Guilford Press, 1988. 285p.	Second
Third	Horn, W., et al, eds. Genetic Manipulation in Plant Breeding; Proceedings of an International Symposium organized by EUCARPIA, Berlin, Germany, Sept. 1985. Berlin and New York; W. de Gruyter, 1986. 909p.	Third
Second	Hornby, D., ed. Biological Control of Soil-Borne Plant Pathogens; Papers presented at the 5th International Congress of Plant Pathology, Kyoto, Aug. 1988. Wallingford, Eng.; CAB International, 1990. 479p.	Second
First rank	Horsfall, James G., and Ellis B. Cowling, eds. Plant Disease: An Advanced Treatise. New York; Academic Press, 1977–1980. 5 vols.	First rank
Second	Horst, R. Kenneth. Compendium of Rose Diseases. St. Paul, Minn.; American Phytopathological Society in cooperation with Dept. of Plant Pathology, Cornell Univeristy, 1983. 50p.	
Third	Hoseney, R. Carl. Principles of Cereal Science and Technology. 2d ed. St. Paul, Minn.; American Association of Cereal Chemists, 1994. 378p. (1st ed., 1986. 327p.)	Third
Third	House, Leland R. A Guide to Sorghum Breeding. 2d ed. Andhra, India; International Crops Research Institute for the Semi-arid Tropics, 1985. 206p. (1st ed., 1980's. 238p.)	Second
Third	Howard, Harold W. Genetics of the Potato: Solanum Tuberosum. London, Eng.; Logos, 1970. 126p.	Third
	Howes, Frank N. A Dictionary of Useful and Everyday	Second

Developed Countries Ranking		Third World Ranking
	Plants and Their Common Names. Cambridge, Eng.; Cambridge University Press, 1974. 290p.	
Third	Hoy, Marjorie A., Gary L. Cunningham, and Lloyd Knutson, eds. Biological Control of Pests by Mites; Proceedings of a conference, University of California, Berkeley, Calif., Apr. 1982, jointly sponsored by the University of California, Berkeley, APHIS-PPQ, USDA, and Agricultural Research Service, USDA. Berkeley, Calif.; Division of Agriculture and Natural Resources, University of California, 1983. 185p. (Agriculture Experiment Station, Div. of Agriculture and Natural Resources, University of California, Special Publication No. 3304)	
First rank	Hoy, Marjorie A., and Donald C. Herzog, eds. Biological Control in Agricultural IPM Systems; Proceedings of a conference, Citrus Research and Education Center, University of Florida at Lake Alfred, June 1984. Orlando, Fla.; Academic Press, 1985. 589p.	
Third	Hu Han, and Hongyuan Yang, eds. Haploids of Higher Plants *in Vitro*. Beijing; China Academic Pub. and Berlin, Germany and New York; Springer-Verlag, 1986. 211p.	Third
First rank	Huffaker, Carl B., ed. New Technology of Pest Control. New York; Wiley, 1980. 500p.	Second
First rank	Huffaker, Carl B., and P. S. Messenger, eds. Theory and Practice of Biological Control. New York; Academic Press, 1976. 788p.	First rank
First rank	Huffaker, Carl B., and Robert L. Rabb, eds. Ecological Entomology. New York; Wiley, 1984. 844p.	
Second	Humbert, Roger P. The Growing of Sugar Cane. Rev. ed. Amsterdam, Holland and London, Eng.; Elsevier, 1968. 780p. (Summaries in Spanish at the end of each chapter. 1st published 1963. 710p.)	Third
Second	Humburg, N. E., and S. R. Colby, et al. Herbicide Handbook of the Weed Science Society of America. 6th ed. Champaign, Ill.; Weed Science Society of America, 1989. 301p. (1st ed. titled: Herbicide Handbook of the Weed Society of America. Geneva, N.Y.; W.F. Humphrey Press, 1967. 293p.)	Third
Second	Humphreys, L. R. Tropical Pasture Utilization. Cambridge, Eng. and New York; Cambridge University Press, 1991. 206p.	Second
	Humphreys, L. R. Tropical Pastures and Fodder Crops. 2d ed. London, Eng.; Longman Scientific and Technical and New York; Wiley, 1987. 155p. (1st ed., 1978. 135p.)	First rank

Developed Countries Ranking		Third World Ranking
Second	Hunt, Roderick. Plant Growth Curves: The Functional Approach to Plant Growth Analysis. Baltimore, Md.; University Park Press, 1982. 248p.	
Third	Hurd, R. G., P. V. Biscoe, and C. Dennis, eds. Opportunities for Increasing Crop Yields. Boston, Mass.; Pitman, 1980. 410p.	Second
	Hurt, R. Douglas. Indian Agriculture in America: Prehistory to the Present. Lawrence, Kan.; University Press of Kansas, 1987. 290p.	Third
	Hussein, M. Y., and A. G. Ibrahim, eds. Biological Control in the Tropics; Proceedings of the 1st Regional Symposium on Biological Control, Universiti Pertanian Malaysia, Serdang, Sept. 1985, jointly sponsored by Universiti Pertanian Malaysia and the Malaysian Plant Protection Society. Serdang, Malaysia; Penerbit Universiti Pertanian Malaysia, 1986. 516p.	Third
Second	Hussey, N. W., and N. Scopes, eds. Biological Pest Control: The Glasshouse Experience. Ithaca, N.Y.; Cornell University Press, 1985. 240p.	
	Hutchinson, John. The Genera of Flowering Plants, Angiospermae Dicotyledones. Based principally on the genera plantarum of G. Bentham and J.D. Hooker. Oxford, Eng.; Carendon Press, 1964–67. 2 vols.	Third
Third	Hutchinson, Joseph B. Evolutionary Studies in World Crops: Diversity and Change in the Indian Subcontinent. Cambridge, Eng.; Cambridge University Press, 1974. 175p.	Third
Second	Huxley, Julian. Evolution: The Modern Synthesis. 3d ed. London, Eng.; Allen and Unwin, 1974. 705p. (1st ed., New York and London; Harper & Brothers, 1942. 645p.)	Third
Third	Huxley, Julian, ed. The New Systematics. Oxford, Eng.; Clarendon Press, 1940. 583p. (Reprint, Hampton, Eng.; Classey for the Systematics Association, 1971.)	
	Indian Council of Agricultural Research. Handbook of Agriculture: Facts and Figures for Farmers, Students and All Engaged or Interested in Farming. 5th ed. New Delhi, India; Indian Council of Agricultural Research, 1980. 1303p. (Rev. ed., 1966. 877p. 3d ed., 1967. 877p.)	Second
Third	Inglett, George E., ed. Wheat: Production and Utilization. Westport, Conn.; AVI Pub. Co., 1974. 500p.	Second
	Inglett, George E., and George Charalambous, eds. Tropical Foods: Chemistry and Nutrition; Contains papers representing the proceedings of an international confer-	Third

Developed Countries Ranking		Third World Ranking
	ence, . . . Honolulu, Hawaii, Mar. 1979. New York; Academic Press, 1979. 2 vols.	
Third	Ingold, Cecil T. Spore Liberation. Oxford, Eng.; Clarendon Press, 1965. 210p.	
Third	Ingram, David S., and J. P. Helgeson, eds. Tissue Culture Methods for Plant Pathologists. Oxford, Eng., Boston, Mass. and New York; Blackwell Scientific; Distributed in the U.S.A. and Canada by Halsted Press, 1980. 272p.	Second
Second	Inoue, Yorinao, et al, eds. The Oxygen Evolving System of Photosynthesis; Proceedings of the International Symposium on Photosynthetic Water Oxidation and Photosystem II Photochemistry, Institute of Physical and Chemical Research, Wako, Saitama, Japan, March 1983, sponsored by Science and Technology Agency of Japan. Tokyo and Orlando, Fla.; Academic Press, 1983. 459p.	
Second	International Barley Genetics Symposium, 5th, Okayama, Japan, 1986. Proceedings, . . . edited by S. Yasuda and T. Konishi. Okayama, Japan; Local Organizing Committee, 5th International Barley Genetics Symposium, 1987. 1065p.	Second
	International Board for Plant Genetic Resources. Crop Genetic Resources of Africa. Rome, Italy; International Board for Plant Genetic Resources, 1991. 2 vols. (Vol. 1. Proceedings of an International Conference on Crop Genetic Resources of Africa, Nairobi, Kenya, Sept. 1988; Vol. 2. Proceedings of an International Conference on Crop Genetic Resources of Africa, Ibadan, Nigeria, Oct. 1988)	Third
Second	International Board for Plant Genetic Resources. Directory of Germplasm Collections. 2d ed. Rome, Italy; IBPGR Secretariat, 1980–89. 7 vols. in several parts; some in revised editions.	Second
	International Conference on Plant Pathogenic Bacteria, 6th, College Park, Md., 1985. Proceedings . . . , edited by E.L. Civerolo, et al., sponsored by the International Society of Plant Pathology, Bacteria Section and U.S. Dept. of Agriculture, Agricultural Research Service. Dordrecht, Netherlands and Boston, Mass.; M. Nijhoff; Distributors for the U.S. and Canada, Kluwer, Hingham, Mass., 1987. 1050p.	Third
	International Conference on Plant Protection in the Tropics, Kuala Lumpur, Malaysia, 1982. Proceedings, . . . edited by K.L. Heong, et al. Kuala Lumpur, Malaysia; Malaysian Plant Protection Society, 1982. 743p.	First rank

Developed Countries Ranking		Third World Ranking
Third	International Congress of Entomology, 18th, Vancouver, B.C., Canada, 1988. Proceedings . . . 1988. 499p.	
Third	International Congress of Plant Pathology, 5th, Kyoto, 1988. Present Status of Plant Disease Control in Japan: 5th ICPP Exhibition by Prefectural Isntitutions. Tokyo, Japan; Phytopathological Society of Japan, 1989. 115p.	Second
Third	International Congress on Photosynthesis, 8th, Stockholm, Sweden, 1989. Current Research in Photosynthesis; Proceedings, . . . edited by M. Baltscheffsky. Dordrecht and Boston; Kluwer Academic Publishers, 1990. 4 vols.	Third
	International Crops Research Institute for the Semi-Arid Tropics. Use of Tropical Grain Legumes; Proceedings of a Consultants Meeting, ICRISAT Center, India, Mar. 1989, sponsored by ICARDA, FAO, and ICRISAT. Patancheru, India; ICRISAT, 1991. 350p.	Second
Third	International Crops Research Institute for the Semi-Arid Tropics. Adaptation of Chickpea and Pigeonpea to Abiotic Stresses; Proceedings of the Consultants' Workshop, ICRISAT Center, India, Dec. 1984, organized by the ICRISAT. Patancheru, India; ICRISAT, 1987. 178p.	Third
Third	International Grassland Congress, 15th, Kyoto, Japan, 1985. Proceedings, . . . edited by T. Okubo and M. Shiyomi. Nishinasuno, Japan; The Science Council of Japan and The Japanese Society of Grassland Science, 1985. 1401p.	Third
	International Institute of Tropical Agriculture. Cowpea Germplasm Catalog. Ibadan, Nigeria; International Institute of Tropical Agriculture, 1974. 209p.	Second
	International Maize and Wheat Improvement Center. Wheats for More Tropical Environments; Proceedings of the International Symposium, Mexico, Sept. 1984, sponsored by the U.N. Development Programme and CIMMYT. Mexico City, Mexico; International Maize and Wheat Improvement Center, 1985. 354p.	Second
	International Neem Conference, 3d, Nairobi, Kenya, 1986. Natural Pesticides from the Neem Tree (Azadirachta Indica A. Juss) and Other Tropical Plants; Proceedings, . . . edited by H. Schmutterer and K.R.S.Ascher. Eschborn, Germany; Deutsche Gesellschaft fur Technische Zusammenarbeit, 1987. 703p. (Schriftenreihe der GTZ No. 206)	Second
Second	International Organization of Citrus Virologists (IOCV). Proceedings . . . Riverside, Calif.; International Organization of Citrus Virologists (IOCV), 1957–88. 10 vols. (Vol. for 1957 issued by the Conference on Citrus Virus	Third

Developed Countries Ranking		Third World Ranking
	Diseases, Citrus Experiment Station, Riverside, Calif.)	
Second	International Organization of Citrus Virologists. Committee on Indexing Procedures, Diagnosis, and Nomenclature. Indexing Procedures for 15 Virus Diseases of Citrus Trees. Washington, D.C.; Agricultural Research Service, U.S. Dept. of Agriculture, 1968. 96p. (USDA Agriculture Handbook No. 333)	
	International Pearl Millet Workshop, CRISAT Center, India, 1986. Proceedings, . . . sponsored by USAID Title XII Collaborative Research Support Program on Sorghum and Pearl Millet (INTSORMIL) and International Crops Research Institute for the Semi-Arid Tropics (ICRISAT). Patancheru, India; ICRISAT, 1987. 354p.	First rank
Second	International Plant Nutrition Colloquium, 11th, Wageningen, Netherlands, 1989. Plant Nutrition: Physiology and Applications; Proceedings, . . . edited by M.L. van Beusichem. Dordrecht and Boston, Mass.; Kluwer Academic, 1990. 819p.	
	International Potato Center. Major Potato Diseases, Insects, and Nematodes. Lima, Peru; International Potato Center, 1983. 95p. (A rev. and expansion of: The Potato: Major Diseases and Nematodes, 1977. 66p.)	Second
	International Potato Center. Planning Conference on Bacterial Diseases of the Potato, 1987. Bacterial Diseases of the Potato; Report . . . Lima, Peru; International Potato Center, 1988. 233p.	Third
	International Potato Center. Potatoes for the Developing World: A Collaborative Experience; with a 12-page precis. Lima, Peru; International Potato Center, 1984. 148p.	Second
	International Rice Genetics Symposium, 1st, Los Banos, Philippines, 1985. Rice Genetics; Proceedings . . . Manila, Philippines; International Rice Research Institute, 1986. 932p.	Second
First rank	International Rice Research Institute. Brown Planthopper: Threat to Rice Production in Asia; Proceedings of a symposium, IRRI, May 1977. Los Banos, Philippines; International Rice Research Institute, 1979. 369p.	Third
	International Rice Research Institute. Catalog of Rice Cultivars and Breeding Lines (Oryza sativa L.) in the World Collection of the International Rice Research Institute. Los Baños, Philippines; International Rice Research Institute, 1970. 281p.	Second
	International Rice Research Institute. Climate and Rice; Proceedings of a symposium, Los Banos, Philippines,	Second

Developed Countries Ranking		Third World Ranking
	1974. Los Baños, Philippines; International Rice Research Institute, 1976. 565p.	
Third	International Rice Research Institute. Drought Resistance in Crops with Emphasis on Rice; Papers presented during the IRRI symposium, . . . May 1981. Los Baños, Philippines; International Rice Research Institute, 1982. 414p.	Third
	International Rice Research Institute. Field Problems of Tropical Rice. Rev. ed. Los Baños, Philippines; International Rice Research Institute, 1983. 172p. (co-published in several Asian Languages) (1st ed. by K.E. Mueller, 1970.)	First rank
Third	International Rice Research Institute. Innovative Approaches to Rice Breeding; Selected papers from the 1979 International Rice Research Conference. Los Baños, Philippines; International Rice Research Institute, 1980. 182p.	Third
First rank	International Rice Research Institute. The Major Insect Pests of the Rice Plant; Proceedings of a symposium, International Rice Research Institute, 1964. Baltimore, Md.; Johns Hopkins Press, 1967. 729p.	Second
Second	International Rice Research Institute. Natural Enemies of Insect Pests of Rice. Los Baños, Philippines; International Rice Research Institute, 1983.	Second
	International Rice Research Institute. Progress in Upland Rice Research; Proceedings of the 2d International Upland Rice Conference, Jakarta, Indonesia, Mar. 1985, organized by the URIRCC and the Agency for Agricultural Research and Development of Indonesia, edited by Frank J. Shidelar. Manila, Philippines; International Rice Research Institute, 1986. 567p.	Second
	International Rice Research Institute. Rainfed Lowland Rice; Selected papers from the 1978 International Rice Research Conference. Los Baños, Philippines; International Rice Research Institute, 1979. 341p.	Third
	International Rice Research Institute. Soils and Rice; Proceedings of a symposium . . . Los Baños, Philippines; International Rice Research Institute, 1978. 825p.	First rank
	International Rice Research Institute. The Virus Diseases of the Rice Plant; Proceedings of a symposium, Apr. 1967. Baltimore, Md.; Johns Hopkins Press, 1969. 354p.	Second
	International Society for Tropical Root Crops Symposium, 4th, CIAT, Cali, Columbia, 1976. Proceedings, . . . edited by James Cock, Reginald MacIntyre, and Michael	First rank

Developed Countries Ranking		Third World Ranking
	Graham. Ottawa, Canada; International Development Research Centre, 1977. 277p.	
	International Society of Sugar Cane Technologists. Pathology Section. Sugar-cane Diseases of the World; Vol. 1 edited by J.P. Martin, E.V. Abbott, and C.G. Hughes; Vol. 2 edited by C.G. Hughes, E.V. Abbott, and C.A. Wismer. Amsterdam, Netherlands and New York; Elsevier Pub. Co., 1961–. (Summaries in Spanish.)	Third
	International Sunflower Conference, 10th, Surfers Paradise, Australia, 1982. Proceedings, . . . sponsored by the Australian Sunflower Association in cooperation with I.S.A. 1982. 324p.	Third
	International Symposium on Aquatic Weeds, 5th, Amsterdam, 1978. Proceedings . . . (= 5me Symposium International sur les Mauvaises Herbes Aquatiques = 5. Internationales Symposium uber Wasserunkrauter). Wageningen, Netherlands; European Weed Research Society, 1978. 447p. (Papers in English, German or French with English, French or German summaries.)	Third
First rank	International Symposium on Biological Control of Weeds, 5th, Brisbane, 1980. Proceedings, . . . edited by E.S. Delfosse. Melbourne, Australia; Commonwealth Scientific and Industrial Research Organization, Australia, 1981. 649p.	
First rank	International Symposium on Biological Control of Weeds, 6th, University of British Columbia, 1984. Proceedings, . . . edited by Ernest S. Delfosse. Ottawa, Canada; Agriculture Canada, 1986. 885p.	Second
Third	International Symposium on Genetic Aspects of Plant Mineral Nutrition, 2d, Madison, Wis., 1985. Genetic Aspects of Plant Mineral Nutrition; Proceedings, . . . organized by the University of Wisconsin, Madison, June 1985, edited by W.H. Gabelman and B.C. Loughman. Dordrecht and Boston, Mass.; M. Nijhoff, 1987. 629p.	Third
	International Symposium on Sorghum Grain Quality, ICRISAT Center, Patancheru, India, Nov. 1981. Sorghum in the Eighties; Proceedings, . . . sponsored by USAID Title XII Collaborative Research Support Program on Sorghum and Pearl Millet (INTSORMIL), International Crops Research Institute of the Semi-Arid Tropics (ICRISAT), and Indian Council of Agricultural Research (ICAR), edited by J. V. Mertin. Patancheru, India; ICRISAT, 1982. 2 vols. 783p.	First rank
Third	International Symposium on the Molecular Genetics of the	

Developed Countries Ranking		Third World Ranking
	Bacteria-Plant Interaction, 2d, Ithaca, N.Y., 1984. Advances in Molecular Genetics of the Bacteria-Plant Interaction; Proceedings, . . . edited by Aladar A. Szalay and Roman P. Legocki. Ithaca, N.Y.; Media Services, Cornell University Publishers, 1985. 217p.	
Third	International Symposium on Virus Diseases of Rice and Leguminous Crops in the Tropics. Proceedings of a Symposium on Tropical Agricultural Research, Tsukuba, Oct. 1985. Tsukuba, Japan; Tropical Agriculture Research Center, Ministry of Agriculture, Forestry and Fisheries, 1986. 262p. (Tropical Agriculture Research Series No. 19)	Third
Third	International Wheat Genetics Symposium, 7th, Cambridge, Eng., 1988. Proceedings, . . . edited by T.E. Miller and R.M.D. Koebner. Cambridge, Eng.; Institute of Plant Science Research, 1988. 1423p.	Third
Second	International Workshop on Biotaxonomy, Classification and Biology of Leafhoppers and Planthoppers (Auchenorrhyncha) of Economic Importance, 2d, Brigham Young University, 1986. Proceedings, . . . edited by M.R. Wilson and L.R. Nault. London, Eng.; CAB International Institute of Entomology, 1987. 368p.	
Second	International Workshop on Crop Loss Assessment to Improve Pest Management in Rice and Rice-Based Cropping Systems in South and Southeast Asia, Oct. 1987. Crop Loss Assessment in Rice; Papers given at the, . . . sponsored by International Rice Research Institute, Food and Agriculture Organization, Deutsche Gesellschaft fur Technische Zusammenarbeit (German Agency for Technical Cooperation), and Consortium for International Crop Protection. Manila, Philippines; International Rice Research Institute, 1990. 334p.	
	International Workshop on Grain Legumes, 1st, Hyderabad, India, 1975. Proceedings . . . Hyderabad, India; International Crops Research Institute for the Semi-Arid Tropics, 1975. 350p.	First rank
	International Workshop on Intercropping, Hyderabad, India, 1979. Proceedings . . . Andhra, India; International Crops Research Institute for the Semi-Aird Tropics, 1981. 401p.	First rank
	International Workshop on Sorghum Diseases, Hyderabad, India, 1978. Proceedings . . . Patancheru, India; International Crops Research Institute for the Semi-Arid Tropics, 1980. 469p.	Third

Developed Countries Ranking		Third World Ranking
Second	Ivens, Giles W. East African Weeds and Their Control. 2d ed. Nairobi, Kenya; Oxford University Press, 1989. 289p. (1st ed. 1967. 244p.)	Second
	Ivens, Giles W., K. Moody, and J. K. Egunjobi. West African Weeds. Ibadan and New York; Oxford University Press, 1978. 225p.	Third
	Jackson, Ian J. Climate, Water, and Agriculture in the Tropics. 2d ed. Burnt Mill, Eng.; Longman Scientific and Technical; and New York; Wiley, 1989. 377p. (1st ed., New York; Longman Group, 1977. 248p.)	Second
Third	Jackson, Michael B., ed. New Root Formation in Plants and Cuttings. Dordrecht and Boston, Mass.; M. Nijhoff, 1986. 265p.	
Third	Jacobson, M., ed. Alternative Methods of Insect Control to Improve Environmental Quality. New York; Marcel Dekker, 1975. 88p.	Third
Second	Jacobson, Martin. Insect Sex Pheromones. 2d ed. New York; Academic Press, 1973. 382p. (Published in 1965 under title: Insect Sex Attractants.)	Third
	Jacobson, Martin, and D. G. Crosby, eds. Naturally Occurring Insecticides. New York; Marcel Dekker, 1971. 585p.	Third
	Jacobson, Martin, ed. The Neem Tree. Boca Raton, Fla.; CRC Press, 1989. 178p.	Third
Third	Jacquard P., G. Heim, and J. Antonovics, eds. Genetic Differentiation and Dispersal in Plants; Proceedings of the NATO Advanced Research Workshop on Population Biology of Plants, Montpellier, May 1984. Berlin, Germany and New York; Springer-Verlag, 1985. 452p.	
	Jameson, J. D., ed. Agriculture in Uganda. 2d ed. London, Eng.; Oxford University Press, 1970. 395p. (1st ed., 1940, edited by John D. Tothill.)	Third
Third	Janick, Jules, ed. Classic Papers in Horticultural Science. Englewood Cliffs, N.J.; Prentice Hall, 1989. 589p.	Second
Second	Janick, Jules. Horticultural Science. 4th ed. New York; W.H. Freeman, 1986. 746p. (1st ed., 1963. 472p.)	Second
Second	Janick, Jules, et al. Plant Science: An Introduction to World Crops. 3d ed. San Francisco, Calif.; W.H. Freeman, 1981. 868p. (1st ed. 1969. 629p.)	First rank
Second	Janick, Jules, and James E. Simon, eds. New Crops. Proceedings of the Second National Symposium NEW CROPS, Indianapolis, Ind., 1991. New York; Wiley, 1993. 710p.	Third
Third	Janossy, A., and F. G. H. Lupton, eds. Heterosis in Plant Breeding; Proceedings of the 7th Congress of	

Developed Countries Ranking		Third World Ranking
	EUCARPIA, Budapest, June 1974. Amsterdam and New York; Elsevier Scientific Pub. Co., 1976. 365p.	
Third	Jansson, Richard K., and Kandukuri V. Raman, eds. Sweet Potato Pest Management, A Global Perspective; Proceedings of the International Conference, . . . Miami, Fla., June 1989. Boulder, Colo.; Westview Press, 1990. 458p.	Second
Second	Jarvis, William R. Managing Diseases of Greenhouse Crops. St. Paul, Minn.; American Phytopathological Society, 1992. 310p.	Third
Third	Jauhar, Prem P. Cytogenetics and Breeding of Pearl Millet and Related Species. New York; A.R. Liss, 1981. 289p. (Progress and Topics in Cytogenetics No. 1)	Second
	Jayaraj, S., ed. Integrated Pest and Disease Management; Proceedings of the National Seminar, Tamil Nadu Agricultural University, Coimbatore, 1984. Coimbatore, India; Tamil Nadu Agricultural University, 1985. 433p.	Third
Third	Jeffrey, Charles. Biological Nomenclature. 3d ed. London, Eng.; E. Arnold, 1989. 128p. (1st ed., London, Eng.; E. Arnold and New York; Crane Russak, 1973. 69p.)	Third
Second	Jeger, Michael J., ed. Spatial Components of Plant Disease Epidemics. Englewood Cliffs, N.J.; Prentice Hall, 1989. 243p.	Second
	Jenkins, William R., and D. P. Taylor. Plant Nematology. New York; Reinhold Pub. Corp., 1967. 270p.	Second
First rank	Jenkyn, J. F., and R. T. Plumb, eds. Strategies for the Control of Cereal Disease. Oxford, Eng. and Boston, Mass.; Blackwell Scientific, 1981. 219p.	Second
	Jennings, Peter R., W. Ronnie Coffman, and H. E. Kauffman. Rice Improvement. Los Banos, Philippines; International Rice Research Institute, 1979. 186p.	Second
Third	Johns, Glenn F., ed. The Organic Way to Plant Protection. Emmaus, Pa.; Rodale Press, 1966. 355p.	
	Johnson, C. B., ed. Physiological Processes Limiting Plant Productivity; Proceedings of the 13th University of Nottingham Easter School in Agricultural Science, Sutton Bonington, April 1979. London, Eng. and Boston, Mass.; Butterworths, 1981. 395p.	Second
Second	Johnston, A., and C. Booth, eds. Plant Pathologist's Pocketbook. 2d ed. Farnham Royal, Eng.; Commonwealth Agricultural Bureaux, 1983. 439p. (1st ed. by Commonwealth Mycological Institute, 1968. 267p.)	First rank
Second	Jolivet, P., E. Petitpierre, and T. H. Hsiao, eds. Biology of Chrysomelidae. Dordrecht and Boston, Mass.; Kluwer Academic Pub., 1988. 615p.	

Developed Countries Ranking		Third World Ranking
	Jones, Allen. World Protein Resources. New York; Wiley, 1974. 381p.	Third
First rank	Jones, D. Gareth, and Brian C. Clifford. Cereal Diseases: Their Pathology and Control. 2d ed. Chichester, Eng. and New York; J. Wiley, 1983. 309p. (1st ed. Ipswich, Eng.; BASF, 1978. 279p.)	Second
	Jones, D. Price, and M. E. Solomon, eds. Biology in Pest and Disease Control; Papers presented at the 13th Symposium of the British Ecological Society, Oxford, Jan. 1972. Oxford; Blackwell Scientific; and New York; Wiley, 1974. 398p.	Third
Second	Jones, F. G. W., and Margaret G. Jones. Pests of Field Crops. 3d ed. Baltimore, Md.; E. Arnold, 1984. 392p. (1st ed. London; Arnold, 1964. 406p.)	Second
Second	Jones, J. B., John Paul Jones, R. E. Stall, and T. A. Zitter, eds. Compendium of Tomato Diseases. St. Paul, Minn.; American Phytopathological Society, 1991. 100p.	
Third	Jorna, M. L., and L. A. J. Slootmaker, compilers. Cereal Breeding Related to Integrated Cereal Production; Proceedings of the Conference of the Cereal Section of EUCARPIA (European Association for Research on Plant Breeding), Wageningen, Netherlands, Feb. 1988. Wageningen; PUDOC, 1988. 244p.	
	Judd, B. Ira. Handbook of Tropical Forage Grasses. New York; Garland STPM Press, 1979. 116p.	Third
Second	Juliano, Bievenido O., ed. Rice: Chemistry and Technology. 2d ed. rev. St. Paul, Minn.; American Association of Cereal Chemists, 1985. 774p. (1st ed. edited by Dave Houston, 1972. 517p.)	Second
Second	Julien, M. H., ed. Biological Control of Weeds: A World Catalogue of Agents and Their Target Weeds. 3d ed. Wallingford, Eng.; CAB International, 1992. 160p. (1st ed. Farnham Royal, Eng.; Commonwealth Agricultural Bureaux, 1982. 108p.)	First rank
Third	Juma, Calestous. The Gene Hunters: Biotechnology and the Scramble for Seeds. London; Zed Books, Ltd.; and Princeton, N.J.; Princeton University Press, 1989. 288p.	
Third	Juniper, Barrie E., and Richard Southwood, eds. Insects and the Plant Surfaces. London, Eng.; E. Arnold, 1986. 360p.	Second
Second	Justice, Oren L., and Louis N. Bass. Principles and Practices of Seed Storage. Washington, D.C.; U.S. Dept. of Agriculture, 1978. 289p. (Reprinted by Castle House, London, 1979.)	Second

Developed Countries Ranking		Third World Ranking
	Kadir, Abdul Aziz S. A., and Henry S. Barlow, eds. Pest Management and the Environment in 2000; International Seminar on Pest Management and the Environment in the Year 2000, Kuala Lumpur, Malaysia, 1991. Wallingford, Eng.; CAB International, 1992. 401p.	Third
First rank	Kahn, Robert P., ed. Plant Protection and Quarantine. Boca Raton, Fla.; CRC Press, 1988–1989. 3 vols. (Vol. 1. Biological Concepts; Vol. 2. Selected Pests and Pathogens of Quarantine Significance; Vol. 3. Special Topics)	First rank
	Kang, B. T., and J. van der Heide, eds. Nitrogen Management in Farming Systems in Humid and Subhumid Tropics; Proceedings of a Symposium, Ibadan, Nigeria, Oct. 1984. Harlen, Netherlands; Institute for Soil Fertility; Ibadan, Nigeria; International Institute of Tropical Agriculture, 1985. 362p.	Third
	Kasasian, L. Weed Control in the Tropics. London, Eng.; Leonard Hill, 1971. 307p.	Second
Second	Kasasian, L., et al. Pest Control in Tropical Root Crops. London, Eng.; Centre for Overseas Pest Research, Ministry of Overseas Development, 1978. 235p. (PANS Manual No. 4)	Second
Second	Kasha, Ken J., ed. Haploids in Higher Plants: Advances and Potential; Proceedings of the 1st International Symposium, Guelph, Ont., June 1974. Guelph, Ontario; Office of Continuing Education, University of Guelph, 1974. 421p.	
	Kass, Donald C. L. Polyculture Cropping Systems: Review and Analysis. Ithaca, N.Y.; NYS College of Agriculture and Life Sciences, Cornell University, 1978. 69p. (Cornell International Agriculture Bulletin No. 32)	Second
Third	Kay, Daisy E. Food Legumes. London; Tropical Products Institute, 1979. 435p. (Tropical Products Institute Crop and Product Digest No. 3)	Second
Third	Kay, Daisy E. Root Crops. 2d ed., revised by E. G. B. Gooding. London, Eng.; Tropical Products Institute, 1987. 380p. (1st ed., 1973. 245p.) (Tropical Products Institute Crop and Product Digests No. 2)	Second
Third	Kays, Stanley J. Postharvest Physiology of Perishable Plant Products. New York; Van Nostrand Reinhold, 1991. 532p.	
	Keane, P. J., and C. A. J. Putter, eds. Cocoa Pest and Disease Management in Southeast Asia and Australasia. Rome; Food and Agriculture Organization, 1992. 223p. (FAO Plant Production and Protection Paper no. 112)	Third

Developed Countries Ranking		Third World Ranking
Second	Kearney, P. C., and D. D. Kaufman, eds. Herbicides: Chemistry, Degradation and Mode of Action. 2d ed., rev. and exp. New York; Marcel Dekker, 1975–88. 3 vols. (1st ed., 1969. 394p.) (Published in 1969 under title: Degradation of Herbicides.)	Third
Third	Keen, Noel T., Tsune Kosuge, and Linda L. Walling, eds. Physiology and Biochemistry of Plant-microbial Interactions; Proceedings of the 11th Annual Symposium in Plant Physiology, University of California, Riverside, Jan. 1988. Rockville, Md.; American Society of Plant Physiologists, 1988. 187p.	
Second	Kelleher, J. S., and M. A. Hulme, eds. Biological Control Programmes Against Insects and Weeds in Canada, 1969–80. Farnham Royal, Eng.; Commonwealth Agricultural Bureaux, 1984. 410p.	
Second	Kempthorne, Oscar. An Introduction to Genetic Statistics. 2d ed. Ames, Ia.; Iowa State University Press, 1969. 545p. (1st ed., New York; Wiley, 1957.)	
Second	Kendrick, Bryce, ed. Taxonomy of Fungi Imperfecti; Proceedings of the 1st International Specialists' Workshop-Conference on Criteria and Terminoloyg in the Classification of Fungi Imperfecti, Univeristy of Calgary, 1969. Toronto, Canada; University of Toronto Press, 1971. 309p.	
Third	Kendrick, R. E., and G. H. M. Kronenberg, eds. Photomorphogenesis in Plants. Dordrecht, Netherlands and Boston, Mass.; Kluwer Academic, 1994. 828p. (1st ed., Dordrecht, Netherlands and Boston, Mass.; M. Nijhoff; Distributors for the U.S. and Canada, Norwell, Mass., 1986. 580p.)	Third
Third	Kennedy, John S., M. F. Day, and V. F. Eastop. A Conspectus of Aphids as Vectors of Plant Viruses. London, Eng.; Commonwealth Institute of Entomology, 1962. 114p.	Third
Third	Kent, Norman L. Technology of Cereals: An Introduction for Students of Food Science and Agriculture. 4th ed. Oxford, Eng. and New York; Pergamon, 1994. 334p. (2nd ed. as Technology of Cereals with Special Reference to Wheat. 1975. 1st ed. 1966. 262p.)	Third
Second	Kenya. Coffee Research Service. An Atlas of Coffee Pests and Diseases: Illustrations of the Common Insect Pests, Diseases, and Deficiency Syndromes of Coffee Arabica in Kenya. Nairobi, Kenya; Coffee Board of Kenya, 1964–68. 1 vol.	Second
	Kerkut, G. A., and L. I. Gilbert, eds. Comprehensive In-	Second

Developed Countries Ranking		Third World Ranking
	sect Physiology, Biochemistry, and Pharmacology. Oxford, Eng. and New York; Pergamon Press, 1985. 13 vols.	
Second	Keswani, C. L., and B. J. Ndunguru, eds. Intercropping; Proceedings of the 2d Symposium on Intercropping in Semi-Arid Areas, Morogoro, Tanzania, Aug. 1980. Ottawa, Canada; International Development Research Centre, 1982. 168p. (IDRC No. 186e)	Second
Third	Key, Joe L., and Tsune Kosuge, eds. Cellular and Molecular Biology of Plant Stress; Proceedings of an ARCO Plant Cell Research Institute-UCLA sympsium, Keystone, Colo., Apr. 1984. New York; A.R. Liss, 1985. 494p.	
Third	Khan, A. A., ed. The Physiology and Biochemistry of Seed Development, Dormancy, and Germination. Amsterdam, Holland and New York; Elsevier Biomedical Press, 1982. 547p. (Rev. ed. of: The Physiology and Biochemistry of Seed Dormancy and Germination. Amsterdam; North-Holland Pub. Co., 1977. 447p.)	Third
Third	Khan, M. A. Q. Pesticides in Aquatic Environments. New York; Plenum, 1977. 257p. (Environmental Science Research No. 10)	Third
Second	Khan, Shahamat U. Pesticides in the Soil Environment. Amsterdam, Holland and New York; Elsevier, 1980. 240p.	
Third	Khoo, K. C., et al, eds. Rodent Pests of Agricultural Crops in Malaysia. Kuala Lumpur, Malaysia; Malaysian Plant Protection Society, 1982. 69p.	
Third	Khush, Gurdev S. Cytogenetics of Aneuploids. New York; Academic Press, 1973. 301p.	
Second	Kiesselbach, T. A. The Structure and Reproduction of Corn. Lincoln, Neb.; University of Nebraska Press, 1980. 96p. (Reprint of 1949 ed., University of Nebraska Agricultural Experiment Station Bulletin No. 161.)	Third
Second	King, E. G., and N. C. Leppla, eds. Advances and Challenges in Insect Rearing. New Orleans, La.; Agricultural Research Service (Southern Region), U.S. Dept. of Agriculture, 1984. 306p.	Second
Second	King, Lawrence J. Weeds of the World: Biology and Control. London, Eng. and New York; L. Hill Interscience, 1966. 526p.	Second
Second	King, Robert C., and William D. Stansfield. A Dictionary of Genetics. 4th ed. New York; Oxford University Press, 1990. 406p. (1st ed., 1968. 291p.)	Second
	Kirby, E. J. M., and Margaret Appleyard. Cereal Develop-	Second

Developed Countries Ranking		Third World Ranking
	ment Guide. 2d ed. Kenilworth, Eng.; Arable Unit, National Agricultural Centre, 1986. 96p. (1st ed., 1981. 1 vol.)	
Third	Kirby, Richard H. Vegetable Fibres: Botany, Cultivation, and Utilization. London, Eng.; L. Hill and New York; Interscience Publishers, 1963. 464p.	Third
	Klatt, A. R., ed. Wheat Production Constraints in Tropical Environments; Proceedings of the International Conference, Chiang Mai, Thailand, Jan. 1987. Mexico; International Maize and Wheat Improvement Center, 1988. 410p. (Abstracts in Spanish and French.)	First rank
Third	Klinkowski, Maximilian, ed. Phytopathologie und Pflanzenschutz (Plant Pathology and Diseases). 2d ed. Berlin, Germany; Akademie-Verlag, 1974–76. 3 vols.	
Second	Kloppenburg, Jack R., ed. Seeds and Sovereignty: The Use and Control of Plant Genetic Resources. Durham, N.C.; Duke University Press, 1988. 368p.	Third
Second	Klotz, L. J. Color Handbook of Citrus Diseases. 4th ed. Berkeley, Calif.; University of California Press, 1973. 121p. (1st ed., 1941. 1 vol.)	Second
	Knight, W. J., et al, eds. 1st International Workshop on Biotaxonomy, Classification and Biology of Leafhoppers and Planthoppers (Auchenorrhyncha) of Economic Importance; Proceedings, . . . London, Oct. 1982. London, Eng.; Commonwealth Institute of Entomology, 1983. 500p.	Third
Second	Knipling, E. F. The Basic Principles of Insect Population Suppression and Management. Washington, D.C.; U.S. Dept. of Agriculture, 1979. 659p. (USDA Agriculture Handbook No. 512)	Second
Third	Knipling, E. F. Principles of Insect Parasitism Analyzed from New Perspectives: Practical Implications for Regulating Insect Populations by Biological Means. Washington, D.C.; U.S. Dept. of Agriculture, Agricultural Research Service, 1992. 337p. (USDA Agriculture Handbook no. 693)	
Third	Knorr, Dietrich, ed. Sustainable Food Systems. Westport, Conn.; AVI Pub. Co., 1983. 416p.	Third
Second	Knott, James E. Knott's Handbook for Vegetable Growers. 3d ed. New York; Wiley, 1988. 456p. (1st ed., 1957. 238p.)	
Third	Knutson, Lloyd, and Allan K. Stoner, eds. Biotic Diversity and Germplasm Perservation: Global Imperatives. Dordrecht and Boston; Kluwer Academic Publishers, 1989. 530p. (Beltsville Symposia in Agricultural Research No. 13)	Third

Developed Countries Ranking		Third World Ranking
First rank	Kogan, Marcos, ed. Ecological Theory and Integrated Pest Management Practice. New York; Wiley, 1986. 362p.	Second
Second	Kohel, R. J., and C. F. Lewis, eds. Cotton. Madison, Wis.; American Society of Agronomy, 1984. 605p.	Third
Second	Kolte, S. J. Diseases of Annual Edible Oilseed Crops. Boca Raton, Fla.; CRC Press, 1984–85. 3 vols.	Third
Third	Kommedahl, Thor, and Paul H. Williams, eds. Challenging Problems in Plant Health. St. Paul, Minn.; American Phytopathological Society, 1983. 538p.	
Second	Kosuge, Tsune, and Eugene W. Nester, eds. Plant-Microbe Interactions: Molecular and Genetic Perspectives. New York and London, Eng.; Macmillan, 1984–89. 3 vols. (Vol. 3 published by McGraw- Hill Pub. Co.)	First rank
	Kotschi, Johannes, ed. Ecofarming Practices for Tropical Smallholdings. Weikersheim, Germany; J. Margraf, 1990. 185p.	Third
Third	Kozlowski, Theodore T., ed. Seed Biology. New York; Academic Press, 1972. 3 vols.	
First rank	Kozlowski, Theodore T., ed. Water Deficits and Plant Growth. New York; Academic Press, 1968–1983. 7 vols.	Second
Second	Kramer, Paul J. Water Relations of Plants. 3d rev. ed. New York; Academic Press, 1983. 489p. (1st ed. 1949: 2d ed. 1969 titled: Plant and Soil Water Relationships. 347p.)	Third
First rank	Kranz, Jurgen, ed. Epidemics of Plant Diseases: Mathematical Analysis and Modeling. 2d completely rev. ed. Berlin, Germany and New York; Springer-Verlag, 1990. 268p. (1st ed., 1974. 170p)	Second
	Kranz, Jurgen, H. Schmutterer, and W. Koch, eds. Diseases, Pests and Weeds in Tropical Crops. Chichester, Eng. and New York; Wiley-Interscience, 1978. 704p. (Translated from 1977 German ed.) (Also available in French, Parley, 1981.)	First rank
Third	Krebs, J. R., and N. B. Davies, eds. Behavioural Ecology: An Evolutionary Approach. 3d ed. Oxford, Eng. and Boston, Mass.; Blackwell Scientific Publications, 1991. 482p. (1st ed. 1978. 494p.)	Third
Third	Krupa, S. V., and Y. R. Dommergues, eds. Ecology of Root Pathogens. Amsterdam and New York; Elsevier, 1979. 281p.	Third
Second	Kuhr, Ronald J., and H. Wyman Dorough. Carbamate Insecticides: Chemistry, Biochemistry, and Toxicology. Cleveland, Oh.; CRC Press, 1976. 301p.	
	Kumble, Vrinda, ed. Proceedings of the International Workshop on Heliothis Management, ICRISAT Center,	Second

Developed Countries Ranking		Third World Ranking
	Patancheru, India, Nov. 1981. Patancheru, India; International Crops Research Institute for the Semi-Arid Tropics, 1982. 418p. (Summaries in English and French.)	
	Kurata, Hiroshi, and Yoshio Ueno, eds. Toxigenic Fungi: Their Toxins and Health Hazard; Proceedings of the Mycotoxin Symposia held in the 3d International Mycological Congress, Tokyo, Aug.-Sept. 1983. Amsterdam, Holland and New York; Elsevier, 1984. 363p.	Third
Second	Kurstak, Edouard. Handbook of Plant Virus Infections: Comparative Diagnosis. Amsterdam and New York; Elsevier/ North-Holland Biomedical Press; Sole distributors for the U.S.A. and Canada, Elsevier North-Holland, 1981. 943p.	Second
First rank	Kurstak, Edouard. Microbial and Viral Pesticides. New York; M. Dekker, 1982. 720p.	Second
Third	Kydonieus, Agis F., and Morton Beroza, eds. Insect Suppression with Controlled Release Pheromone Systems. Boca Raton, Fla.; CRC Press, 1982. 2 vols.	Third
Third	Kyle, D. J., C. B. Osmond, and C. J. Arntzen, eds. Photoinhibition. Amsterdam and New York; Elsevier, 1987. 315p. (Topics on Photosynthesis No. 9)	
	Lagiere, Robert. Le Cotonnier. Paris, France; G.P. Maisonneuve and Larose, 1966. 306p. (Techniques Agricoles et Productions Tropicales no. 9)	Second
Second	Laird, Marshall, Lawrence A. Lacey, and Elizabeth W. Davidson, eds. Safety of Microbial Insecticides. Boca Raton, Fla.; CRC Press, 1990. 259p.	Second
Third	Lal, R., ed. Insecticide Microbiology. Berlin, Germany and New York; Springer-Verlag, 1984. 268p.	Third
Third	Lal, R., ed. Soil Tillage and Crop Production. Ibadan, Nigeria; International Institute of Tropical Agriculture, 1979. 361p. (International Institute of Tropical Agriculture Proceedings Series No. 2)	Second
Third	Lal, R. Tropical Ecology and Physical Edaphology. Chichester, Eng. and New York; Wiley, 1987. 732p.	Second
Second	Lal, R., and D. J. Greenland, eds. Soil Physical Properties and Crop Production in the Tropics. Chichester, Eng. and New York; Wiley, 1979. 551p.	First rank
	Lal, R., P. A. Sanchez, and R. W. Cummings, eds. Land Clearing and Development in the Tropics; Proceedings of the 1st International Conference on Land Clearing and Development, International Institute for Tropical Agriculture, Ibadan, Nigeria, Nov. 1982. Rotterdam and Boston, Mass.; A.A. Balkema, 1986. 450p.	Second

Developed Countries Ranking		Third World Ranking
Second	Lamb, Kenneth P. Economic Entomology in the Tropics. London, Eng. and New York; Academic Press, 1974. 195p.	Second
Third	Lambers, H., J. J. Neeteson, and I. Stulen, eds. Fundamental, Ecological and Agricultural Aspects of Nitrogen Metabolism in Higher Plants; Proceedings of a symposium, Haren, Apr. 1985, organized by the Dept. of Plant Physiology, University of Groningen and the Institute for Soil Fertility. Dordrecht and Boston, Mass.; M. Nijhoff, 1986. 508p.	Third
Third	Lamberti, F., and C. E. Taylor, eds. Root-Knot Nematodes (Meloidogyne Species): Systematics, Biology and Control. London and New York; Academic Press, 1979. 477p.	
	Lamberti, F., C. E. Taylor, and J. W. Seinhorst, eds. Nematode Vectors of Plant Viruses; NATO Advanced Study Institute, Riva dei Tessali, Italy, 1974. London, Eng. and New York; Plenum Press, 1975. 460p.	Third
Third	Lamberti, F., C. De Giorgi, and David McK. Bird, eds. Advances in Molecular Plant Nematology. Proceedings of a NATO Advanced Research Workshop on Molecular Plant Nematology, Martina Franca, Italy, November 20–27, 1993. New York; Plenum Press, 1994. 309p. (NATO Advanced Study Institutes Series, Series A, Life Sciences No. 268)	
Third	Landsberg, J. J., and C. V. Cutting, eds. Environmental Effects on Crop Physiology; Proceedings of the 5th Long Ashton Symposium, Long Aston Research Station, University of Bristol, Apr. 1975. London and New York; Academic Press, 1977. 388p.	
	Lange, W., A. C. Zeven, and W. G. Hogenboom, eds. Efficiency in Plant Breeding; Proceedings of the 10th Congress of the European Association for Research on Plant Breeding, EUCARPIA, Wageningen, Netherlands, June 1983. Wageningen, Netherlands; PUDOC, 1984. 383p.	Third
Third	Larcher, Walter. Physiological Plant Ecology. Translated by M. A. Biederman-Thorson. 2d ed., corrected. Berlin, Germany and New York; Springer-Verlag, 1983. 303p. (1st ed., 1975. 252p.)	Third
Second	Large, Ernest C. The Advance of the Fungi. New York; Dover Publications, 1962. 488p. (1st published, New York; H. Holt and Co., 1940.)	Third
Second	Lashomb, James H., and Richard Casagrande, eds. Advances in Potato Pest Management; Proceedings of a	

Developed Countries Ranking		Third World Ranking
	conference, . . . Rutgers University, Jan. 1980. Stroudsburg, Pa. and New York; Hutchinson Ross; Distributed world wide by Academic Press, 1981. 288p.	
Third	Lazenby, Alec, and E. M. Matheson, eds. Australian Field Crops. Sydney, Australia; Anus and Robertson, 1975–79. 2 vols.	Third
Second	Le Pelley, R. H. Pests of Coffee. London, Eng.; Longmans, 1968. 590p.	Second
Second	Lea, P. J., ed. Enzymes of Primary Metabolism. London, Eng. and San Diego, Calif.; Academic Press, 1990. 414p. (Methods in Plant Biochemistry No. 3)	Third
Third	Lea, P. J., and G. R. Stewart, eds. The Genetic Manipulation of Plants and its Application to Agriculture. Oxford, Eng.; Clarendon Press, 1984. 313p.	Second
Third	Leahey, John P., ed. The Pyrethroid Insecticides. London, Eng.; Taylor and Francis, 1985. 440p.	Second
Second	Leakey, C. L. A., and J. B. Wils, eds. Food Crops of the Lowland Tropics. Oxford, Eng.; Oxford University Press, 1977. 345p.	Second
Third	Leaver, C. J., ed. Genome Organization and Expression in Plants; Proceedings of the NATO Advanced Study Institute, Edinburgh, Scotland, July 1979. New York; Plenum Press, 1980. 607p.	
	Leaver, Christopher, and Heven Sze, eds. Plant Membranes: Structure, Function, Biogenesis; Proceedings of the ARCO Plant Cell Research Institute-UCLA Symposium, Park City, Utah, Feb. 1987. New York; A. Liss, 1987. 461p.	Third
Third	LeBaron, Homer M., et al, eds. Biotechnology in Agricultural Chemistry; Developed from a symposium sponsored by the Divisions of Agrochemicals, Agricultural and Food Chemistry, and Fertilizer and Soil Chemistry at the 190th Meeting of the American Chemical Society, Chicago, Ill., Sept. 1985. Washington, D.C.; American Chemical Society, 1987. 367p.	Second
First rank	LeBaron, Homer M., and Jonathan Gressel, eds. Herbicide Resistance in Plants. New York; Wiley, 1982. 401p.	Second
	Lee, B. S., W. H. Loke, and K. L. Heong, eds. Proceedings of the Seminar on Integrated Pest Management in Malaysia, Kuala Lumpur, Malaysia, Jan. 1984, jointly organized by Malaysian Plant Protection Society (MAPPS), Malaysian Agricultural Research and Development Institute (MARDI). Kuala Lumpur, Malaysia; Malaysian Plant Protection Society, 1985. 335p.	Third

Developed Countries Ranking		Third World Ranking
Third	Lee, R. E., ed. Air Pollution from Pesticides and Agricultural Processes. Cleveland, Oh.; CRC Press, 1976. 264p.	
	Leihner, Dietrich. Management and Evaluation of Intercropping Systems with Cassava. Cali, Colombia; Centro Internacional de Agricultura Tropical, 1983. 70p. (Spanish version also available.)	First rank
	Lemon, Edgar R., ed. Carbon Dioxide and Plants: The Response of Plants to Rising Levels of Atmospheric Carbon Dioxide; Proceedings of a meeting, Athens, Ga., May 1982. Boulder, Colo.; Westview Press for the American Association for the Advancement of Science, 1983. 280p.	Third
Second	Lenne, Jillian M. A World List of Fungal Diseases of Tropical Pasture Species. Wallingford, Eng.; Centro Internacional de Agricultura Tropical and International Mycologial Institute, CABI, 1990. 162p. (CABI Phytopathological Papers No. 31)	Second
Second	Leon, Carlos de. Maize Diseases: A Guide for Field Identification. 3d ed. Mexico; CIMMYT, 1984. 114p. (Previous eds. published as International Maize and Wheat Improvement Center Information Bulletin No. 11) (1st ed., 1974. 77p.) (Also in Spanish.)	First rank
	Leon, Jorge. Botanica de los Cultivos Tropicales. San Jose, Costa Rica; Instituto Interamericano de Cooperacion para la Agricultura, 1987. 445p. (Coleccion Libros y Materiales Educativos No. 84)	Second
	Leon, Jorge, and Eugenio Sgaravatti. Tropical Pastures: Grasses and Legumes = Paturages Tropicaux: Graminees et Legumineuses = Pastos tro Picales: Gramineas y Leguminosas. Rome; Food and Agriculture Organization, 1971. 74p. (FAO AGP/PFC/17	Third
	Leon, Jorge, and Lyndsey A. Withers, eds. Guidelines for Seed Exchange and Plant Introduction in Tropical Crops. Rome; Food and Agriculture Organization, 1986. 207p. (FAO Plant Production and Protection Paper No. 76)	Third
Second	Leonard, Kurt J., and William E. Fry, eds. Plant Disease Epidemiology. New York and London, Eng.; Macmillan, 1986–88. 2 vols.	Second
First rank	Leopold, A. Carl, and Paul E. Kriedemann. Plant Growth and Development. 2d ed. New York; McGraw-Hill, 1975. 545p. (1st ed. A. Carl Leopold, 1964. 466p.)	First rank
Third	Lerner, Isadore M. The Genetic Basis of Selection. New York; Wiley, 1958. 208p.	

Developed Countries Ranking		Third World Ranking
	Leroy, Jean-Francois. Les Fruits Tropicaux. Paris; Presses Universitaires de France, 1968. 128p.	Third
	Lever, Robert J. A.W. Pests of the Coconut Palm. Rome; Food and Agriculture Organization, 1969. 180p. (FAO Agricultural Studies No. 77)	Third
First rank	Levitt, Jacob. Responses of Plants to Environmental Stresses. New York; Academic Press, 1980. 2 vols. (1st ed., 1972. 697p.)	Second
Third	Levy, Stuart B., and Robert V. Miller, eds. Gene Transfer in the Environment. New York; McGraw-Hill, 1989. 434p.	
Third	Lewin, Benjamin. Genes IV. 4th ed. Oxford, Eng. and New York; Oxford University Press, 1990. 857p. (1st ed. titled: Genes. New York; Wiley, 1983. 715p.)	Third
Second	Lewis, Trevor. Thrips: Their Biology, Ecology and Economic Importance. London, Eng. and New York; Academic Press, 1973. 349p.	Second
Third	Lewontin, Richard C. The Genetic Basis of Evolutionary Change. New York; Columbia University Press, 1974. 346p.	
Third	Li, Paul H., and A. Sakai, eds. Plant Cold Hardiness and Freezing Stress: Mechanisms and Crop Implications; Proceedings of an International Plant Cold Hardiness Seminar, St. Paul, Minn., Nov. 1977, sponsored by U.S. National Science Foundation, Japan Society for the Promotion of Science, and College of Agriculture, University of Minnesota. Vol. 2 based on the proceedings of the 2d International Seminar on Plant Cold Hardiness, Sapporo Educational and Cultural Hall, Sapporo, Japan, Aug. 1981. New York; Academic Press, 1978–. 2 vols.	
Third	Li, Paul H., ed. Potato Physiology. Orlando, Fla.; Academic Press, 1985. 586p.	Third
	Libby, John L. Insect Pests of Nigerian Crops. Madison, Wis.; University of Wisconsin, 1968. 68p. (Univ. of Wis. Research Bulletin No. 269)	Third
	Lieberman, Morris, ed. Post-Harvest Physiology and Crop Preservation; Proceedings of a NATO Advanced Study Institute, Sounion, Greece, Apr.–May 1981. New York; Plenum Press, 1983. 572p. (NATO Advanced Study Institute Series, Series A, Life Sciences No. 46)	Third
	Liener, Irvin E., ed. Toxic Constituents of Plant Food Stuffs. 2d ed. New York; Academic Press, 1980. 502p. (1st ed., 1969. 500p.)	Second
Third	Lindsey, K., and M. G. K. Jones. Plant Biotechnology in Agriculture. Englewood Cliffs, N.J.; Prentice Hall,	Third

Developed Countries Ranking		Third World Ranking
	1990. 241p. (1st published, Milton Keynes, Eng.; Open University Press, 1989. 241p.)	
	Little, E. C. S. Handbook of Utilization of Aquatic Plants: A Review of World Literature. Rome; Food and Agriculture Organization, 1979. 176p. (FAO Fisheries Technical Paper No. 187)	Third
	Little, Thomas M., and F. Jackson Hills. Statistical Methods in Agricultural Research. Davis, Calif.; University of California, 1972. 242p.	Second
Second	Littlefield, Larry J., and Michele C. Heath. Ultrastructure of Rust Fungi. New York; Academic Press, 1979. 277p.	
	Litzenberger, Samuel C., ed. Guide for Field Crops in the Tropics and the Subtropics. Washington, D.C.; Office of Agriculture, Technical Assistance Bureau, Agency for International Development, 1974. 321p.	Second
Third	Lockeretz, William, ed. Environmentally Sound Agriculture; Selected papers from the 4th International Conference of the International Federation of Organic Agriculture Movements, Cambridge, Mass., Aug. 1982. New York; Praeger, 1983. 426p.	
	Logan, W. J. C., and J. Biscoe, eds. Coffee Handbook. Harare, Zimbabwe; Cannon Press, 1987. 182p.	Third
	Lomer, C. J., and C. Prior, eds. Biological Control of Locusts and Grasshoppers; Proceedings of a Workshop, International Institute of Tropical Agriculture, Cotonou, Republic of Benin, Apr.-May, 1991. Wallingford, Eng.; CAB International, 1992. 394p.	Third
	Lopez-Real, J. M., and R. D. Hodges, eds. The Role of Microorganisms in a Sustainable Agriculture; Selected papers from the 2d International Conference of Biological Agriculture, University of London, Wye College, Wye, UK, 1984. Berkhamsted, Eng.; A.B. Academic, 1986. 246p.	Third
Third	Lorenz, Klaus J., and Karel Kulp, eds. Handbook of Cereal Science and Technology. New York; M. Dekker, 1991. 882p.	Second
	Lorenzi, Harri. Manual de Identificacao e Controle de Plantas Daninhas: Plantio dereto e Convencional. 2d ed. (Weed Identification and Control Manual). Nova Odessa, Sao Paulo; Harri Lorenzi, 1986. 220p. (1st ed., Londrina, Brazil; Fundacao Instiuto Agronomico do Parana, 1984. 237p.) (In Portuguese)	Third
Second	Lorenzi, Harri J., and Larry S. Jeffery. Weeds of the United States and Their Control. New York; Van Nostrand Reinhold, 1987. 355p.	

Developed Countries Ranking		Third World Ranking
	Loussert, R., and G. Brousse. L'Olivier (The Olive Tree). Paris, France; Maisonneuve et Larose, 1978. 464p. (In French.)	Third
Second	Luc, Michael, Richard A. Sikora, and John Bridge, eds. Plant Parasitic Nematodes in Subtropical and Tropical Agriculture. Wallingford, Eng.; CABI, 1990. 629p.	First rank
First rank	Lucas, George B. Diseases of Tobacco. 3d ed. Raleigh, N.C.; Biological Consulting Associates, 1975. 621p. (1st ed., New York; Scarecrow Press, 1958. 498p.)	Second
Third	Ludden, Paul W., and John E. Burris, eds. Nitrogen Fixation and CO/Metabolism; Proceedings of the 14th Steenbock Symposium, University of Wisconsin-Madison, Madison, Wis., June 1984. New York; Elsevier, 1985. 445p.	
Third	Lugtenberg, Ben J. J., ed. Signal Molecules in Plants and Plant-Microbe Interactions; Proceedings of the NATO Advanced Research Workshop on Molecular Signals in Microbe-Plant Symbiotic and Pathogenic Systems, Biddinghuizen, Netherlands, May 1989. Berlin, Germany and New York; Springer-Verlag, 1989. 425p.	Third
Second	Luh, Bor S., ed. Rice. 2d. New York; Van Nostrand Reinhold, 1991. 2 vols. (1st ed., Rice—Production and Utilization. Westport, Conn.; AVI Pub. Co., 1980. 925p.)	First rank
Second	Lumpkin, Thomas A., and Donald L. Plucknett. Azolla as a Green Manure: Use and Management in Crop Production. Boulder, Colo.; Westview Press, 1982. 230p.	Second
First rank	Lyr, H., ed. Modern Selective Fungicides: Properties, Applications, Mechanisms of Action. 2d rev. ed., Jena, Germany, and New York; Gustav Fischer, 1994. 544p. (1st ed., Harlow, Eng.; Longman Scientific and Technical and New York; Wiley, 1987. 383p.)	Second
First rank	Maas, J. L., ed. Compendium of Strawberry Diseases. St. Paul, Minn.; American Phytopathological Society in cooperation with Agricultural Research Service, U.S. Dept. of Agriculture, 1984. 138p.	Second
	MacDicken, Kenneth G., and Napoleon T. Vergara, eds. Agroforestry, Classification and Management. New York; Wiley, 1990. 382p.	Third
First rank	Mace, Marshall E., Alois A. Bell, and Carl H. Beckman. Fungal Wilt Diseases of Plants. New York; Academic Press, 1981. 640p.	Second
Second	Mackauer, Manfred, Lester E. Ehler, and Jens Roland, eds. Critical Issues in Biological Control; Based on papers presented during two complementary symposia at the 18th International Congress of Entomology, Van-	Second

Developed Countries Ranking		Third World Ranking
	couver, Canada, July 1988. Andover, Eng.; Intercept and New York; VCH Publishers, 1990. 330p.	
	Macmillan, Hugh F. Tropical Planting and Gardening, rev. by H. S. Barlow, I. Enoch, and R. A. Russell. 6th ed. Kuala Lumpur, Malaysia; Malayan Nature Society, 1991.767p. (1st ed. titled: A Handbook of Tropical Gardening and Planting with Special Reference to Ceylon. 5th ed., London; Macmillan, 1949. 560p.)	Second
Third	MacMillan, J., ed. Molecular Aspects of Plant Hormones. Berlin, Germany and New York; Springer-Verlag, 1980. 681p. (Encyclopedia of Plant Physiology, New Series, vol. 9)	Third
Second	Mai, William F. Pictorial Key to Genera of Plant-Parasitic Nematodes. 4th ed., rev. Ithaca, N.Y.; Constock Pub. Associates, 1975. 219p. (1st ed., 1960. Ithaca, New York; Dept. of Plant Pathology, Cornell University. 150p.)	First rank
	Maistre, Jacques. Les Plantes a Epices. Paris, France; G.P. Maisonneuve and Larose, 1964. 289p.	Third
Second	Mandava, N. Bhushan, ed. CRC Handbook of Natural Pesticides: Methods. Boca Raton, Fla.; CRC Press, 1985. 8 vols. (Vol. 3–4, and 6 titled: Handbook of Natural Pesticides edited by E. David Morgan and N. Bhushan Mandava. Vol. 5, pt. A edited by Carlo M. Ignoffo.)	Second
Second	Mangelsdorf, Paul C. Corn: Its Origin, Evolution, and Improvement. Cambridge, Mass.; Belknap Press, 1974. 262p.	Second
	Manica, I. Fruiticultura Tropical. Vol. 1. Maracuja; Vol. 2. Manga; and Vol. 3. Pawpaw. Sao Paulo, Brazil; Editora Agronomica Ceres Ltda., 1981–82. 3 vols. (In Portuguese)	Third
	Manners, J. G. Principles of Plant Pathology. 2d ed. Cambridge, Eng. and New York; Cambridge University Press, 1993. 343p. (1st ed., 1982. 264p.)	Third
Third	Mantell, S. H., J. A. Matthews, and R. A. McKee. Principles of Plant Biotechnology: An Introduction to Genetic Engineering in Plants. Oxford, Eng. and Boston, Mass.; Blackwell Scientific Publications, 1985. 269p.	Second
Second	Maramorosch, Karl, ed. Biotechnology in Invertebrate Pathology and Cell Culture. San Diego, Calif.; Academic Press, 1987. 511p.	Third
Third	Maramorosch, Karl, ed. Mycoplasma and Mycoplasma-like Agents of Human, Animal, and Plant Diseases; Papers presented at a conference sponsored by the Acad-	

Developed Countries Ranking		Third World Ranking
	emy's Section of Biological and Medical Sciences, Jan. 1973. New York; 1973. 532p.	
Second	Maramorosch, Karl, and Kerry F. Harris, eds. Plant Diseases and Vectors: Ecology and Epidemiology. New York; Academic Press, 1981. 368p.	Second
Third	Maramorosch, Karl, and Hiroyuki Hirumi, eds. Practical Tissue Culture Applications. New York; Academic Press, 1979. 426p.	
Second	Maramorosch, Karl, and Hilary Koprowski, eds. Methods in Virology. New York; Academic Press, 1967–1984. 8 vols.	Second
Second	Maramorosch, Karl, and John J. McKelvey, eds. Subviral Pathogens of Plants and Animals: Viroids and Prions; Proceedings of a Symposium sponsored by the Rockefeller Foundation, Bellagio, Italy, June-July 1983. Orlando, Fla.; Academic Press, 1985. 549p.	Third
Second	Maramorosch, Karl, and S. P. Raychaudhuri, ed. Mycoplasma Diseases of Trees and Shrubs. New York; Academic Press, 1981. 362p.	Third
Second	Maramorosch, Karl, and K. E. Sherman, eds. Viral Insecticides for Biological Control. Orlando, Fla.; Academic Press, 1985. 809p.	Second
Third	Marcus, Abraham, ed. Proteins and Nucleic Acids. New York; Academic Press, 1981. 658p.	
Third	Martin, Franklin W., ed. CRC Handbook of Tropical Food Crops. Boca Raton, Fla.; CRC Press, 1984. 296p.	Second
Third	Martin, Franklin W., et al. Vegetables for the Hot, Humid Tropics; pt. 1. The Winged Bean, Psophocarpus Tetragonolobus; pt. 2. Okra, Abelmoschus Esculentus; pt. 3. Chaya, Cnidoscolus Chayamansa; pt. 4. Sponge and Bottle Gourds, Luffa and Lagenera; p. 5. Eggplant, Solanum Melongena; pt. 6. Amaranth and Celosia, Amaranthus an Celosia; pt. 7. the Peppers, Capsicum Species. New Orleans, La.; Southern Region, Agricultural Research Service, U.S. Dept. of Agriculture, 1978–79. 128p. (7 parts)	First rank
	Martin, Franklin W., and Ruth M. Ruberte. Edible Leaves of the Tropics. 2d ed. Agency for International Development, Department of State, and the Agricultural Research Service, USDA, 1979. (1st ed., 1975. 235p.)	Second
Second	Martin, Hubert, and David Woodcock. Scientific Principles of Crop Protection. 7th ed. London, Eng.; E. Arnold, 1983. 486p. (1st ed., 1928. 316p. 1st-3d eds. titled: The Scientific Principles of Plant Protection. 1st-6th eds. by H. Martin.)	Second

Developed Countries Ranking		Third World Ranking
Second	Martin, J. P., E. V. Abbott, and C. G. Hughes, eds. Sugar-Cane Diseases of the World. Amsterdam, Holland and New York; Elsevier, 1961–64. 2 vols.	First rank
Second	Marz, Ulrich. The Economics of Neem Production and Its Use in Pest Control. Kiel; Wissenschaftsverlag Vauk Kiel, 1989. 153p.	Second
Second	Mather, Kenneth, and John L. Jinks. Biometrical Genetics: The Study of Continuous Variation. 3d ed. London, Eng. and New York; Chapman and Hall, 1982. 396p. (1st ed., London; Methuen & Co., 1949. 162p.)	Third
Second	Mather, Kenneth, and John L. Jinks. Introduction to Biometrical Genetics. Ithaca, N.Y.; Cornell University Press, 1977. 231p.	Third
First rank	Mathre, Don E., ed. Compendium of Barley Diseases. St. Paul, Minn. and Bozeman, Mont.; American Phytopathological Society; Dept. of Plant Pathology, Montana State University, 1982. 78p.	Second
	Matsubayashi, Minaru, et al, eds. Theory and Practice of Growing Rice. 2d ed. Tokyo; Fuji Publishing Co., 1968. 527p. (1st ed., 1963.) (Translated into English by Suzuki, Yoshio with the aid of N. Yamada.)	Second
First rank	Matsumura, Fumio. Toxicology of Insecticides. 2d ed. New York; Plenum Press, 1985. 598p. (1st ed., 1975. 503p.)	Second
Third	Matsumura, Fumio, and C. R. Krishna Murti, eds. Biodegradation of Pesticides. New York; Plenum Press, 1982. 312p.	Second
Third	Matsuo, T., ed. Gene Conservation: Exploration, Collection, Preservation, and Utilization of Genetic Resources. Tokyo, Japan; University of Tokyo Press, 1975. 229p.	Third
Second	Matthews, G. A. Cotton Insect Pests and Their Management. Harlow, Eng.; Longman Scientific and Technical and New York; copublished in the U.S.A. with Wiley, 1989. 199p.	Second
Second	Matthews, G. A. Pest Management. London and New York; Longman, 1984. 231p.	Third
	Matthews, G. A. Pesticide Application Methods. Harlow, Essex; Longman Scientific and Technical, and New York; Wiley, 1992. 405p. (1st ed., London, Eng. and New York; Longman, 1979. 334p.)	Second
First rank	Matthews, Richard E. F. Plant Virology. 3d ed. San Diego, Calif.; Academic Press, 1991. (1st ed., New York; Academic Press, 1970. 778p.)	Second
Third	Matthews, Richard E. F. Diagnosis of Plant Virus Diseases. Boca Raton, Fla.; CRC Press, 1993. 374p.	Third

Developed Countries Ranking		Third World Ranking
Second	Maxwell, Fowden G., and Peter R. Jennings, eds. Breeding Plants Resistant to Insects. New York; Wiley, 1980. 683p.	First rank
Third	May, Robert M., ed. Theoretical Ecology: Principles and Applications. 2d ed. Sunderland, Mass.; Sinauer Associates, 1981. 489p. (1st ed., 1976. 317p.)	
Third	Mayer, A. M., and A. Poljakoff-Mayber. The Germination of Seeds. 4th ed. Oxford, Eng. and New York; Pergamon Press, 1989. 270p. (1st ed., New York; Macmillan, 1963. 236p.)	Third
Second	Mayo, Oliver. The Theory of Plant Breeding. 2d ed. Oxford, Eng.; Clarendon Press and New York; Oxford University Press, 1987. 334p. (1st ed., 1980. 293p.)	Second
	Mazliak, Paul. Physiologie Vegetale: Nutrition et Metabolisme. Paris; Hermann, 1974. 349p.	Third
Second	McEwen, F. L., and G. R. Stephenson. The Use and Significance of Pesticides in the Environment. New York; J. Wiley, 1979. 538p.	Third
Third	McFarlane, N. R., ed. Herbicides and Fungicides—Factors Affecting Their Activity. London, Eng.; The Chemical Society; Bangor, Wales, 1976. 141p. (Chemical Society Special Publication No. 29)	
Second	McGee, Denis C. Soybean Diseases: A Reference Source for Seed Technologists. St. Paul, Minn.; American Phytopathological Society, 1992. 160p.	Third
	McGregor, Samuel E. Insect Pollination of Cultivated Crop Plants. Washington, D.C.; Agricultural Research Service, 1976. 411p. (USDA Agriculture Handbook No. 496)	First rank
Third	McKelvey, John J. Man Against Tsetse: Struggle for Africa. Ithaca, N.Y.; Cornell University Press, 1973. 306p.	Third
Third	McLaren, J. S., ed. Chemical Manipulation of Crop Growth and Development; Proceedings of the 33d Easter School of Agricultural Science, Sutton Bonington, Mar. 1981. London, Eng.; Butterworth Scientific, 1982. 564p.	
Second	McLean, George D., Ronald G. Garrett, and William G. Ruesink, eds. Plant Virus Epidemics: Monitoring, Modelling and Predicting Outbreaks. Sydney, Australia and Orlando, Fla.; Academic Press, 1986. 550p.	
Third	McWhorter, C. G., and M. R. Gebhardt, eds. Methods of Applying Herbicides. Champaign, Ill.; Weed Science Society of America, 1987. 358p.	Third
Second	Mellanby, Kenneth. Pesticides and Pollution. 2d rev. ed.	

Developed Countries Ranking		Third World Ranking
	London, Eng.; Collins, 1970. (New Naturalist. A Survey of British Natural History No. 50) (1st ed., 1967. 221p.)	
Second	Mendel, Gregor. Experiments in Plant Hybridisation (Versuche uber Pflanzenhybriden); Mendel's original paper in English translation, with commentary and assessment by R.A. Fisher, together with a reprint of W. Bateson's biographical notice of Mendel, edited by J.H. Bennett. Edinburgh, Eng.; Oliver and Boyd, 1965. 95p.	Third
Second	Mengel, Konrad, and Ernest A. Kirkby. Principles of Plant Nutrition. 4th ed. Bern, Switzerland; International Potash Institute, 1987. 687p. (1st ed. 1978. 593p.)	Third
	Menninger, Edwin A. Edible Nuts of the World. Stuart, Fla.; Horticultural Books, 1977. 175p.	Second
Third	Menzinger, Walther, and Herbert Sanftleben. Parasitare Krankheiten und Schaden an Geholzen. Berlin, Germany; P. Parey, 1980. 216p.	
Third	Meredith, Donald S. Banana Leaf Spot Disease (Sigatoka) Caused by Mycosphaerelle Musicola Leach. Kew, Eng.; Commonwealth Mycological Institute, 1970. 147p. (Phytopathological Papers No. 11)	Third
	Merlier, H., and J. Montegut. Adventices Tropicales: Flore aux Stades Plantule et Adulte de 123 e Speces Africaines ou Pantropicales. Paris; Ministere des Relations Exterieures, Cooperation et Developpement, 1982. 490p.	Third
Second	Messiaen, Charles-M., et al. Les Maladies des Plantes Maraicheres. 3d ed. Paris, France; Institut National de la Recherche Agronomique, 1991. 552p. (1st ed., 1963. 2 vols.)	Second
	Messiaen, Charles-M. Le Potager Tropical. 2d ed. Paris; Presses Universitaires de France, 1974, 1978 printing. 2 vols.	Third
First rank	Metcalf, Clell L., and W. P. Flint. Destructive and Useful Insects: Their Habits and Control. 4th ed. New York; McGraw-Hill, 1962. 1087p. (1st ed., 1928. 918p.)	Second
First rank	Metcalf, Robert L., and William H. Luckmann, eds. Introduction to Insect Pest Management. 3d ed. New York; J. Wiley, 1994. 650p. (1st ed., 1975. 587p.)	First rank
Third	Metcalfe, Charles R. Anatomy of the Monocotyledons. Oxford, Eng.; Clarendon Press, 1960–72. 7 vols. (Vols. 2, 3, and 7 by P.B. Tomlinson, vols. 4 by D. F. Cutler, and vol. 6 by E. S. Ayensu.)	
Third	Mettler, Lawrence E., Thomas G. Gregg. and Henry E. Schaffer. Population Genetics and Evolution. En-	

Developed Countries Ranking		Third World Ranking
	glewood Cliffs, N.J.; Prentice-Hall, 1988. 325p. (1st ed. edited by Lawrence E. Mettler and Thomas G. Gregg, 1969. 212p.)	
Third	Metzner, H., ed. Photosynthetic Oxygen Evolution; Symposium, . . . Tubingen, Germany, 1977. London, Eng. and New York; Academic Press, 1978. 532p.	
Third	Miflin, B. J., ed. Amino Acids and Derivatives. New York; Academic Press, 1980. 670p.	
Second	Miflin, B. J., and Peter J. Lea, eds. Intermediary Nitrogen Metabolism. San Diego, Calif.; Academic Press, 1990. 402p.	Third
Third	Miller, Darrell A. Forage Crops. New York; McGraw-Hill, 1984. 530p.	Second
First rank	Miller, James R., and Thomas A. Miller, eds. Insect-Plant Interactions. New York; Springer-Verlag, 1986. 342p.	Third
Second	Miller, Paul R., and Hazel L. Pollard. Multilingual Compendium of Plant Diseases. St. Paul, Minn.; American Phytopathological Society for the U.S. Agency for International Development in cooperation with the Agricultural Research Service, U.S. Dept. of Agriculture, 1976–77. 2 vols. 457p., 434p. (Vol. 2: Viruses and Nematodes.)	Third
	Miller, Ross. Insect Pests of Wheat and Barley in West Asia and North Africa. Aleppo, Syria; International Center for Agricultural Research in the Dry Areas, 1987. 209p. (ICARDA Technical Manual No. 9, Rev. 1)	Third
Second	Milner, Max, ed. Nutritional Improvement of Food Legumes by Breeding; based on proceedings of a symposium sponsored by PAG, Food and Agriculture Organization, Rome, July 1972. New York; J. Wiley, 1975. 399p.	Second
Third	Milthorpe, Frederick L., and J. D. Ivins, eds. The Growth of Cereals and Grasses; Proceedings of the 12th Easter School in Agricultural Science, University of Nottingham, 1965. London, Eng.; Butterworths, 1966. 358p.	Third
	Milthorpe, Frederick L., and J. Morby. An Introduction to Crop Physiology. 2d ed. Cambridge, Eng. and New York; Cambridge University Press, 1979. 244p. (1st ed., 1974. 202p.)	Third
Second	Minks, A. K., and P. Gruys, eds. Integrated Control of Insect Pests in the Netherlands. Wageningen, Netherlands; Centre for Agricultural Publishing and Documentation, 1980. 304p.	
First rank	Minks, A. K., and P. Harrewijn, eds. Aphids: Their Biol-	First rank

Developed Countries Ranking		Third World Ranking
	ogy, Natural Enemies, and Control. Amsterdam, Netherlands and New York; Elsevier, 1987–89. 3 vols.	
	Miracle, Marvin P. Maize in Tropical Africa. Madison, Wis.; University of Wisconsin Press, 1966. 327p.	Second
	Mitchell, D. S., ed. Aquatic Vegetation and its Use and Control. Paris; UNESCO, 1974. 135p.	Third
	Mitchell, Rodger D. The Analysis of Indian Agro-Ecosystems. New Delhi, India; Interprint, 1979. 180p.	Third
Third	Miyamoto, J., and P. C. Kearney, eds. Pesticide Chemistry, Human Welfare and the Environment; Proceedings of the 5th International Congress of Pesticide Chemistry, Kyoto, Japan, Aug.–Sept. 1982. Oxford, Eng. and New York; Pergamon Press, 1983. 4 vols.	Third
	Moldenhauer, W. C., and N. W. Hudson, eds. Conservation Farming on Steep Lands. Ankeny, Ia.; Soil and Water Conservation Society and World Association of Soil and Water Conservation, 1988. 296p.	Third
	Moll, H. A. J. The Economics of Oil Palm. Wageningen, Netherlands; PUDOC, 1987. 288p.	Third
Third	Moncur, Michael W. Floral Initiation in Field Crops: An Atlas of Scanning Electron Micrographs. Canberra, Australia; Division of Land Use Research, CSIRO, 1981. 135p.	
Third	Monselise, Shaul P., ed. CRC Handbook of Fruit Set and Development. Boca Raton, Fla.; CRC Press, 1986. 568p.	Third
	Montaldo, Alvaro. La Yuca o Mandioca: Cultivo, Industrializacion, Aspectos Economicos, Empleo en la Alimentacion Animal, Mejoramiento. San Jose, Costa Rica; Instituto Inter-americano de Ciencias Agricolas, 1979. 386p. (Serie Libros y Materiales Educativos No. 38)	Third
Second	Monteith, John L., ed. Vegetation and the Atmosphere. London, Eng. and New York; Academic Press, 1975–76. 2 vols.	
Second	Monteith, John L., and M. H. Unsworth. Principles of Environmental Physics. 2d ed. London, Eng. and New York; E. Arnold, 1990. 291p. (1st ed., John L. Monteith. New York; American Elsevier, 1973. 241p.)	Second
	Monyo, J. H., A. D. R. Ker, and Marilyn Campbell, eds. Intercropping in Semi-Arid Areas; Report of a Symposium, Faculty of Agriculture, Forestry and Veterinary Science, University of Dar es Salaam, Morogoro, Tanzania, May 1976. Ottawa, Canada; International Development Research Centre, 1976. 72p.	Third
Third	Moo-Young, Murray, ed. Comprehensive Biotechnology:	Second

Developed Countries Ranking		Third World Ranking
	The Principles, Applications, and Regulations of Biotechnology in Industry, Agriculture, and Medicine. Oxford, Eng. and New York; Pergamon, 1985. 4 vols.	
	Moody, Keith, C. E. Monroe, R. T. Lubigan, and E. C. Paller. Major Weeds of the Philippines. Los Banos, Philippines; Weed Science Society of the Philippines, University of the Philippines, 1984. 328p.	Third
	Moody, Keith, ed. Weed Control in Tropical Crops: Symposium; Papers presented at the 9th Pest Control Council of the Philippines Conference commemorating the 10th Anniversary of the Weed Science Society of the Philippines, Inc., Philippine International Convention Center, Manila, May 1978. Manila; Weed Science Society of the Philippines and Philippine Council for Agriculture and Resources Research, 1979. 203p.	Third
Third	Mooney, Harold A., and James A. Drake, eds. Ecology of Biological Invasions of North America and Hawaii; Based on a symposium, Asilomar, Calif. Oct. 1984. New York; Springer-Verlag, 1986. 321p.	
Third	Mooney, Patrick R. The Law of the Seed: Another Development and Plant Genetic Resources. Uppsala; Dag Hammarskjold Foundation, 1983. 172p.	
	Mooney, Patrick R. Seeds of the Earth: A Private or Public Resource? Rev. ed. San Francisco, Calif.; Institute for Food and Development Policy and Ottawa, Canada; Inter Pares, published by Inter Pares for the Canadian Council for International Coalition for Development Action, 1983, 1980. 126p. (1st published 1979.)	Third
Second	Moore, P. D., and S. B. Chapman, eds. Methods in Plant Ecology. 2d ed. Oxford, Eng. and Boston, Mass.; Blackwell Scientific Pub., 1986. 589p. (1st ed. S. B. Chapman, ed., 1976. 536p.)	Second
Second	Moore, Raymond M., ed. Australian Grasslands. Canberra, Australia; Australian National University Press, 1970. 455p.	
Third	Moore, Thomas C. Biochemistry and Physiology of Plant Hormones. 2d ed. New York; Springer-Verlag, 1989. 330p. (1st ed., 1979. 274p.)	Third
	Moreau, Fernand. Les Champignons: Physiologie, Morphologie, Developpement et Systematique. Paris, France; P. Lechevalier, 1952. 2 vols.	Third
Third	Moreland, Donald E., Judith B. St. John, and F. Dana Hess, eds. Biochemical Responses Induced by Herbicides; Based on a symposium, sponsored by the Division of Pesticide Chemistry at the 181st ACS National	

Developed Countries Ranking		Third World Ranking
	Meeting Atlanta, Ga., Mar.–Apr. 1981. Washington, D.C.; American Chemical Society, 1982. 274p.	
Second	Morgan, W. M., M. S. Ledieu, and G. Stell. Pest and Disease Control in Glasshouse Crops. Croydon, Eng.; BCPC Publications, 1979. 108p. (Reprinted from Pest and Disease Control Handbook, ed. by N. Scopes.)	
Second	Moriarty, F. Ecotoxicology: The Study of Pollutants in Ecosystems. 2d ed. London, Eng. and San Diego, Calif.; Academic Press, 1988. 289p. (1st ed., 1983. 233p.)	Third
First rank	Morse, Roger A., and Richard Nowogrodzki, eds. Honey Bee Pests, Predators, and Diseases. 2d ed. Ithaca, N.Y.; Comstock, 1990. 474p. (1st ed., 1978. 430p.)	
Second	Mortensen, Ernest, and Ervin T. Bullard. Handbook of Tropical and Subtropical Horticulture. Rev. Washington, D.C.; Agency for International Development, 1970. 186p. (1st published, 1964.)	Second
	Morton, Julia F. Major Medicinal Plants: Botany, Culture, and Uses. Springfield, Ill.; Thomas, 1977. 431p.	Second
Second	Mortvedt, J. J. et al., eds. Micronutrients in Agriculture. 2d ed. Madison, Wis.; Soil Science Society of Agronomy, 1991. 670p. (1st ed., 1972. Proceedings of a Symposium, Muscle Shoals, Ala., Apr. 1971. 666p.)	First rank
	Moss, J. P., ed. Biotechnology and Crop Improvement in Asia. Patancheru, Andhra Pradesh, India; International Crops Research Institute for the Semi-Arid Tropics, 1992. 385p.	Third
Second	Mount, Mark S., and George H. Lacy, eds. Phytopathogenic Prokaryotes. New York; Academic Press, 1982. 2 vols.	First rank
Third	Muchow, Russell C., and Jennifer A. Bellamy, eds. Climatic Risk in Crop Production: Models and Management for the Semiarid Tropics and Subtropics; Proceedings of the International Symposium, . . . Brisbane, Australia, July 1990. Wallingford, Eng.; CAB International, 1991. 548p.	Second
Second	Muhammed, Amir, Rustem Aksel, and R. C. von Borstel, eds. Genetic Diversity in Plants; Proceedings of an International Symposium on Genetic Control of Diversity in Plants, Lahore, 1976. New York; Plenum Press, 1977. 506p.	
	Mukerji, K. G., V. P. Agnihotri, and R. P. Singh, eds. Progress in Microbial Ecology: J.N. Rai Festschriften. Lucknow; Print House (India), 1984. 653p.	Third
Second	Mukerji, K. G., and K. L. Garg, eds. Biocontrol of Plant	First rank

Developed Countries Ranking		Third World Ranking
	Diseases. Boca Raton, Fla.; CRC Press, Inc., 1988. 3 vols.	
Third	Mulcahy, D. L., ed. Symposium on Gamete Competition in Plants and Animals; Proceedings of the Symposium, . . . Lake Como, Italy, Aug. 1975. Amsterdam, Netherlands; North Holland Publishing Co., 1975. 288p.	
	Mulder, D., ed. Soil Disinfestation. Amsterdam, Netherlands and New York; Elsevier, 1978. 368p.	Third
Second	Munro, James W. Pests of Stored Products. London, Eng.; Hutchinson, 1966. 234p.	Second
Second	Munro, John M. Cotton. 2d ed. Harlow, Eng.; Longman Scientific and Technical and New York; Wiley, 1987. 436p. (1st ed., 1972 by A.N. Prentice.	Third
Third	Murray, David R., ed. Seed Physiology. Sydney, Australia and Orlando, Fla.; Academic Press, 1984. 2 vols.	Third
Third	Murton, R. K., and E. N. Wright, eds. The Problems of Birds as Pests; Proceedings of a Symposium, Royal Geographical Society, London, Sept. 1967. London and New York; Academic Press, 1968. 254p.	Third
Second	Mussell, Harry, and Richard C. Staples, eds. Stress Physiology in Crop Plants, co-sponsored by the Boyce Thompson Institute for Plant Research and the Rockefeller Foundation. New York; Wiley, 1979. 510p.	Second
Third	Musselman, Lytton J., ed. Parasitic Weeds in Agriculture. Boca Raton, Fla.; CRC Press, 1987. 1 vol.	Second
Third	Nagy, Steven, Philip E. Shaw, and Matthew K. Veldhuis, eds. Citrus Science and Technology. Westport, Conn.; AVI Pub. Co., 1977. 2 vols.	Second
Third	Nakas, James P., and Charles Hagedorn, eds. Biotechnology of Plant-Microbe Interactions. New York; McGraw-Hill, 1990. 348p.	Third
Third	Narahashi, T., and Janice E. Chambers, eds. Insecticide Action: From Molecule to Organism. Proceedings of a symposium held as part of the 196th National Meeting of the Agrochemicals Division of the American Chemical Society, September 25–30, 1988, in Los Angeles, California. New York; Plenum, 1989. 275p.	Third
Third	Nash, Thomas A. M. Africa's Bane: The Tsetse Fly. London, Eng.; Collins, 1969. 224p.	Third
	National Academy of Sciences (U.S.). Leucaena: Promising Forage and Tree Crop for the Tropics; Report of an Ad Hoc Panel of the Advisory Committee on Technology Innovation, Board on Science and Technology for International Development, Office of Internationao Affairs in cooperation with the Nitrogen Fixing Tree Asso-	Second

Developed Countries Ranking		Third World Ranking
	ciation. 2d ed. Washington, D.C.; National Academy of Sciences, 1984. 100p. (1st ed., 1977. 115p.)	
Second	National Academy of Sciences (U.S.). Pest Control: An Assessment of Present and Altenative Technologies. Washington, D.C.; National Academy of Sciences, 1975–76. 5 vols.	Third
	National Research Council (U.S.). Board of Agriculture. Genetic Engineering of Plants: Agricultural Research Opportunities and Policy Concerns; Based on a convocation, National Academy of Sciences, May 1983, sponsored by the Council for Research Planning and Biological Sciences, Inc. and the Board on Agriculture, National Research Council. Washington, D.C.; National Academy Press, 1984. 83p.	Second
	National Research Council (U.S.). Board on Science and Technology for International Development. Amaranth: Modern Prospects for an Ancient Crop; Report of an ad hoc panel of the Advisory Committee on Technology Innovation. Washington, D.C.; National Academy Press, 1984. 81p.	Second
	National Research Council (U.S.). Board on Science and Technology for International Development. Plant Biotechnology for Developing Countries. Washington, D.C.; National Academy of Sciences, 1990.	First rank
Second	National Research Council (U.S.). Board on Science and Technology for International Development. Postharvest Food Losses in Developing Countries. Washington, D.C.; National Academy of Sciences, 1978. 206p. (Text in English with summaries in French and Spanish.)	Third
	National Research Council (U.S.). Board on Science and Technology for International Development. Tropical Legumes: Resources for the Future; Report of an ad hoc panel of the Advisory Committee on Technology Innovation. Washington, D.C.; National Academy of Sciences, 1979. 331p.	First rank
	National Research Council (U.S.). Board on Science and Technology for International Development. Commission on International Relations. Underexploited Tropical Plants with Promising Economic Value; Report of an Ad Hoc Panel of the Advisory Committee on Technology Innovation. Washington, D.C.; National Academy of Sciences, 1975. 188p. (Resumes in French and Spanish.)	First rank
	National Research Council (U.S.). Committee on Managing Global Genetic Resources (Agricultural Impera-	Third

Developed Countries Ranking		Third World Ranking
	tives). The U.S. National Plant Germplasm System. Washington, D.C.; National Academy Press, 1991. 171p.	
Second	National Research Council (U.S.). Committee on Plant and Animal Pests. Principles of Plant and Animal Pest Control. Washington, D.C.; National Academy of Sciences, 1968–70. 6 vols.	Third
Second	National Research Council (U.S.). Committee on Scientific and Regulatory Issues Underlying Pesticide Use Patterns and Agricultural Innovation. Regulating Pesticides in Food: the Delaney Paradox. Washington, D.C.; National Academy Press, 1987. 272p.	Third
	National Research Council (U.S.). Committee on Selected Biological Problems in the Humid Tropics. Ecological Aspects of Development in the Humid Tropics. Washington, D.C.; National Academy Press, 1982. 297p.	Third
Third	National Research Council (U.S.). Committee on Strategies for the Management of Pesticide Resistant Pest Populations. Pesticide Resistance: Strategies and Tactics for Management. Washington, D.C.; National Academy Press, 1986. 471p.	First rank
	National Research Council (U.S.). Committee on Sustainable Agriculture and the Environment in the Humid Tropics. Sustainable Agriculture and the Environment in the Humid Tropics. Washington, D.C.; National Academy Press, 1993. 702p.	Third
Second	National Research Council (U.S.). Committee on the Role of Alternative Farming Methods in Modern Production Agriculture. Alternative Agriculture. Washington, D.C.; National Academy Press, 1989. 448p.	Second
	National Research Council (U.S.). Food Protection Committee. The Use of Chemicals in Food Production, Processing, Storage, and Distribution. Washington, D.C.; National Academy of Sciences, 1973. 34p.	Third
	National Research Council (U.S.). Office of International Affairs. Priorities in Biotechnology Research for International Development; Proceedings of a Workshop, Washington, D.C. and Berkeley Springs, W.V., July 1982. Washington, D.C.; National Academy Press, 1982. 261p.	Third
	National Research Council (U.S.). Panel on Saline Agriculture in Developing Countries. Saline Agriculture: Salt-Tolerant Plants for Developing Countries; Report of a Panel of the Board on Science and Technology for International Development, National Research Council.	Second

Developed Countries Ranking		Third World Ranking
	Washington, D.C.; National Academy Press, 1990. 143p.	
Third	National Research Council (U.S.). Subcommittee on Insect Pests. Insect-Pest Management and Control. Washington, D.C., 1969. 508p. (National Academy of Sciences Publication No. 1695)	Second
Second	Nault, L. R., and J. G. Rodriguez, eds. The Leafhoppers and Planthoppers. New York; Wiley, 1985. 500p.	First rank
Second	Naumann, Klaus. Synthetic Pyrethroid Insecticides: Chemistry and Patents. Berlin, Germany and New York; Springer-Verlag, 1990. 390p. (Chemistry of Plant Protection No. 5)	Second
Second	Naumann, Klaus. Synthetic Pyrethroid Insecticides: Structures and Properties. Berlin, Germany and New York; Springer-Verlag, 1990. 241p. (Chemistry of Plant Protection No. 4)	
	Nayar, N. M., ed. Coconut Research and Development; Proceedings of the International Symposium, . . . Central Plantation Crops Research Institute, Kasaragod 670 124, Kerala, Dec. 1976. New Delhi, India; Wiley, 1983. 518p.	Third
Third	Neal, John W., ed. Guidelines for Control of Insect and Mite Pests of Food, Fibers, Feeds, Ornamentals, Livestock, Forests, and Forest Products. Washington, D.C.; U.S. Dept. of Agriculture, 1979. 822p. (USDA Agriculture Handbook No. 554)	Third
First rank	Neergaard, Paul. Seed Pathology. New York; J. Wiley, 1977. 2 vols. 1187p.	Second
First rank	Nelson, P. E., T. A. Toussoun, and R. J. Cook, eds. Fusarium Diseases, Biology, and Taxonomy. University Park, Pa.; Pennsylvania State University, 1981. 457p.	Second
	Nene, Y. L., Susan D. Hall, and V. K. Sheila, eds. The Pigeonpea. Wallingford, U.K.; CAB International, 1990. 490p.	Second
	Nestel, Barry, and Michael Graham, eds. Cassava as Animal Feed; Proceedings of a workshop, University of Guelph, Apr. 1977, co-sponsored by the International Development Research Centre and the University of Guelph. Ottawa, Canada; International Development Research Centre, 1977. 147p.	Third
Third	Neuffer, M. G., E. H. Coe, and S. R. Wessler. Mutants of Maize. Plainview, N.Y.; Cold Spring Harbor Laboratory Press, 1995. 450p. (Replaces 1968 ed. published by Crop Science Society of America)	
Third	Newman, David W., and Kenneth G. Wilson, eds. Models	Third

Developed Countries Ranking		Third World Ranking
	in Plant Physiology and Biochemistry. Boca Raton, Fla.; CRC Press, 1987. 3 vols.	
Second	Newton, William E., and William H. Orme-Johnson, eds. Nitrogen Fixation; Proceedings of the 3d Kettering International Symposium, Madison, Wis., 1978. Baltimore, Md.; University Park Press, 1980. 2 vols.	
	Ng, Siew K. The Oil Palm: Its Culture, Manuring and Utilization. Rev. ed. Berne; International Potash Institute, 1972. 142p.	Second
Third	Nickell, Louis G. Plant Growth Regulators: Agricultural Uses. Berlin, Germany and New York; Springer-Verlag, 1982. 173p.	Third
Third	Nickle, William R., ed. Manual of Agricultural Nematology. New York; M. Dekker, 1991. 1035p. (Companion volume to Plant and Insect Nematodes, 1984.)	Second
	Nickle, William R., ed. Plant and Insect Nematodes. New York; M. Dekker, 1984. 925p.	First rank
Third	Nienhaus, Franz. Virus and Similar Diseases in Tropical and Subtropical Areas. Eschborn, Germany; German Agency for Technical Cooperation, 1981. 216p.	Second
Third	Nieuwhof, M. Cole Crops: Botany, Cultivation, and Utilization. Translated from the Dutch by G. Berends. London; L. Hill, and Cleveland; CRC Press, 1969. 353p.	Third
Second	Nishida, T., and T. Torri. A Handbook of Field Methods for Research on Rice Stem-Borers and Their Natural Enemies. Oxford; Blackwell Scientific and Philadelphia, Pa.; Davis, 1970. 132p. (IBP Handbook No. 14)	Third
First rank	Noble, Mary, and M. J. Richardson. An Annotated List of Seed-Borne Diseases. 2d ed. Kew, Eng.; Commonwealth Mycological Institute and Wageningen, Netherlands; International Seed Testing Association, 1968. 191p. (1st ed. by M. Noble, J. de Tempe and P. Neergaard, 1958. 159p.)	Third
First rank	Norman, A. Geoffrey, ed. Soybean Physiology, Agronomy, and Utilization. New York; Academic Press, 1978. 249p. (1st published under title: The Soybean: Genetics, Breeding, Physiology, Nutrition, Management, 1963. 239p.)	Third
	Norman, Michael J. T. Annual Cropping Systems in the Tropics: An Introduction. Gainesville, Fla.; University Presses of Florida, 1979. 276p.	Third
Second	Norman, Michael J. T., C. J. Pearson, and P. G. E. Searle. The Ecology of Tropical Food Crops. Cambridge, Eng. and New York; Cambridge University Press, 1984. 369p.	Second

Developed Countries Ranking		Third World Ranking
Third	Norton, Don C. Ecology of Plant-Parasitic Nematodes. New York; Wiley, 1978. 268p.	Second
Third	Norton, G., et al. Plant Proteins; Proceedings of the 24th Easter School in Agricultural Science, University of Nottingham, 1976. London, Eng. and Boston, Mass.; Butterworths, 1978. 352p.	
Second	Nutman, P. S., ed. Symbiotic Nitrogen Fixation in Plants. Cambridge, Eng. and New York; Cambridge University Press, 1976. 584p.	Third
Second	Nyvall, Robert F. Field Crop Diseases Handbook. 2d ed. New York; Van Nostrand Reinhold, 1989. 817p. (1st ed., Westport, Conn.; AVI Pub. Col, 1979. 436p.)	Third
Second	O'Brien, Richard D. Insecticides, Action and Metabolism. New York; Academic Press, 1967. 332p.	
Third	O'Connor, B. A. Exotic Plant Pests and Diseases; A Handbook of Plant Pests and Diseases to be Excluded From or Prevented From Spreading within the Area of the South Pacific Commission. Noumea, New Caledonia; South Pacific Commission, 1969. 460p.	
	Ochse, J. J., et al. Tropical and Subtopical Agriculture. New York; Macmillan, 1961. 2 vols. 1446p.	Second
Third	Ogawa, Joseph M., and Harley English. Diseases of Temperate Zone Tree Fruit and Nut Crops. Berkley, Calif.; Div. of Agriculture and Natural Resources, University of California, 1991. 461p.	
Third	Ohnesorge, Bernhart. Tiere als Pflanzenschadlinge. Stuttgart, Germany; Thieme, 1976. 288p.	
	Oka, Hikoichi. Origin of Cultivated Rice. Tokyo, Japan; Japan Scientific Societies Press; and Amsterdam and New York; Elsevier, 1988. 254p.	Third
Third	Olby, Robert C. Origins of Mendelism. 2d ed. Chicago, Ill.; University of Chicago Press, 1985. 310p. (1st ed., London; Constable, 1966. 204p.)	Third
Second	Old, R. W., and S. B. Primrose. Principles of Gene Manipulation: An Introduction to Genetic Engineering. 5th ed. Oxford, Eng. and Boston, Mass.; Blackwell Scientific, 1994. 474p. (1st ed., Berkeley; University of California Press, 1980. 138p.)	Second
Third	Olien, Charles R., and Myrtle N. Smith, eds. Analysis and Improvement of Plant Cold Hardiness. Boca Raton, Fla.; CRC Press, 1981. 215p.	
First rank	Olson, R. A., and K. J. Frey, eds. Nutritional Quality of Cereal Grains: Genetic and Agronomic Improvement. Madison, Wis.; American Society of Agronomy, 1987. 511p.	First rank

Developed Countries Ranking		Third World Ranking
	Onwueme, I. C. The Tropical Tuber Crops: Yam, Cassava, Sweet Potato, and Cocoyam. Chichester, Eng. and New York; Wiley, 1978. 234p.	Second
Third	Opeke, Lawrence K. Tropical Tree Crops. Chichester, Eng. and New York; Wiley, 1982. 312p.	Second
	Opena, Romeo T., C. G. Duo, and J. Y. Yoon. Breeding and Seed Production of Chinese Cabbage in the Tropics and Subtropics. Shanhua, Taiwan; Asian Vegetable Research and Development Center, 1988. 92p. (Technical Bulletin No. 17)	Third
Second	Ordish, George. The Constant Pest: A Short History of Pests and Their Control. New York; Scribner, 1976. 240p.	
First rank	Ou, Shu H. Rice Diseases. 2d ed. Kew, Eng.; CAB Commonwealth Mycological Institute, 1985. 380p. (1st ed., 1972. 368p.)	First rank
Third	Packer, Lester, George C. Papageorgiou, and Achim Trebst, eds. Bioenergetics of Membranes; Proceedings of the International Symposium on Membrane Bioenergetics, Island of Spetsai, Greece, July 1977. Amsterdam, Netherlands and New York; Elsevier, 1977. 538p.	
Second	Painter, Reginald H. Insect Resistance in Crop Plants. Lawrence, Kan.; University Press of Kansas, 1968. 520p. (1st printing, New York; Macmillan, 1951. 520p.)	
Third	Pal, R., and M. J. Whitten. The Use of Genetics in Insect Control. Amsterdam, Netherlands; Elsevier/North-Holland, 1974. 241p.	Third
Second	Palacios, Rafael, and D. P. S. Verma, eds. Molecular Genetics of Plant-Microbe Interactions 1988; Proceedings of the 4th International Symposium . . . Acapulco, Mexico, 1988. St. Paul, Minn.; APS Press, 1988. 401p.	Third
Second	Paleg, L. G., and D. Aspinall, eds. The Physiology and Biochemistry of Drought Resistance in Plants. Sydney, Australia and New York; Academic Press, 1981. 492p.	
Third	Palti, J., and J. Kranz, eds. Comparative Epidemiology: A Tool for Better Disease Management; Proceedings of the Session on Comparative Epidemiology, 3d International Congress of Plant Pathology, Munich, Aug. 1978. Wageningen, Netherlands; Centre for Agricultural Publishing and Documentation, 1980. 122p.	Third
First rank	Palti, Josef. Cultural Practices and Infectious Crop Diseases. Berlin, and New York; Springer-Verlag, 1981. 243p. (Advanced Series in Agricultural Sciences No. 9)	

Developed Countries Ranking		Third World Ranking
First rank	Panopoulos, Nickolas J., ed. Genetic Engineering in the Plant Sciences. New York; Praeger, 1981. 271p.	Third
	Pantastico, B., ed. Postharvest Physiology, Handling and Utilization of Tropical and Subtropical Fruits and Vegetables. Westport, Conn.; AVI Pub. Co., 1975. 560p.	Third
Third	Papa, S., B. Chance, and L. Ernster, eds. Cytochrome Systems: Molecular Biology and Bioenergetics; Proceedings of the UNESCO International Workshop, Bari, Italy, Apr. 1987. New York; Plenum Press, 1987. 805p.	
First rank	Papavizas, George C., ed. Biological Control in Crop Production; Papers presented at an Annual Symposium organized by BARC Symposium V Committee and sponsored by Beltsville Agricultural Research Center, May 1980. Totowa, N.J.; Allanheld, Osmun and London; Granada, 1981. 461p.	Third
First rank	Papendick, R. I., P. A. Sanchez, and G. B. Triplett, eds. Multiple Cropping; Proceedings of a Symposium, sponsored by the American Society of Agronomy, Crop Science Society of America, and Soil Science Society of America. Madison, Wis.; American Society of Agronomy, 1976. 378p. (ASA Special Publication No. 27)	First rank
Third	Parry, David W. Plant Pathology in Agriculture. Cambridge, Eng. and New York; Cambridge University Press, 1990. 385p.	Third
	Parry, Georges. Le Cotonnier et ses Produits. Paris, France; G.P. Maisonneuve and Larose, 1982. 502p. (Techniques Agricoles et Productions Tropicales no. 30)	Second
	Parry, John W. Spices. New York; Chemical Pub. Co., 1969. 2 vols. Vol. 1, 228p.; Vol. 2, 229p. (Vol. 2 previously issued separately under title: Spices, Their Morphology, Histology and Chemistry.)	Third
First rank	Pattee, Harold E., and Clyde T. Young, eds. Peanut Science and Technology. Yoakum, Tex.; American Peanut Research and Education Society, 1982. 825p. (Rev. ed. of: Peanuts—Culture and Uses.)	Third
	Paulino, Leonardo A. Food in the Third World: Past Trends and Projections to 2000. Washington, D.C.; International Food Policy Research Institute, 1986. 76p. (IFPRI Research Report No. 52)	Second
	Peachey, J. E., ed. Nematodes of Tropical Crops; Papers presented at the Caribbean Symposium on Nematodes of Tropical Crops, Trinidad, Mar.–Apr. 1968. Farnham Royal, Eng.; Commonwealth Agricultural Bureaux, 1969. 355p. (Commonwealth Bureau of Helminthology. Technical Communication No. 40)	First rank

Developed Countries Ranking		Third World Ranking
	Peachey, J. E., and M. R. Chapman. Chemical Control of Plant Nematodes. Farnham Royal, Eng.; Commonwealth Agricultural Bureaux, 1966. 119p. (Technical Communication No. 36)	Second
Second	Pearson, C. J., ed. Control of Crop Productivity. Sydney, Australia and Orlando, Fla.; Academic Press, 1984. 315p.	Third
Second	Pearson, C. J., and R. L. Ison. Agronomy of Grassland Systems. Cambridge, Eng. and New York; Cambridge University Press, 1987. 169p.	
Third	Peberdy, John F., and Lajos Ferenczy, eds. Fungal Protoplasts: Applications in Biochemistry and Genetics. New York; M. Dekker, 1985. 354p.	Third
Second	Pedgley, David E. Windborne Pests and Diseases: Meteorology of Airborne Organisms. Chichester, Eng.; E. Horwood and New York; Halsted Press, 1982. 250p.	
	Peng, Sheng-yang. The Biology and Control of Weeds in Sugarcane. Amsterdam, Holland and New York; Elsevier, 1984. 336p.	Third
Third	Perring, F. H., and K. Mellanby, eds. Ecological Effects of Pesticides. New York and London; Academic Press, 1977. 193p.	
	Persley, G. J., and E. A. De Langhe, eds. Banana and Plantain Breeding Strategies; Proceedings of an international workshop, Cairns, Australia, Oct. 1986, co-sponsored by the Autralian Centre for International Agricultural Research (ACIAR), International Network for Improvement of Banana and Platain (INIBAP), Queensland Department of Primary Industries (QDPI). Canberra, Australia; Australian Centre for International Agricultural Research, 1987. 187p.	Second
	Peters, W. J., and Leon F. Neuenschwander. Slash and Burn: Farming in the Third World Forest. Moscow, Idaho; University of Idaho Press, 1988. 113p.	Third
Third	Peterson, Rudolph F. Wheat: Botany, Cultivation and Utilization. London; L. Hill and New York; Interscience Publishers, 1965. 422p.	Third
First rank	Pfadt, Robert E., ed. Fundamentals of Applied Entomology. 4th ed. New York; Macmillan and London, Eng.; Collier Macmillan, 1985. 742p. (1st ed., 1962. 668p.)	Second
Third	Phillips, D. H., and D. A. Burdekin. Diseases of Forest and Ornamental Trees. 2nd ed. London; Macmillan, 1992. 581p. (1st ed., 1982. 435p.)	
	Phillips, Ronald E., and Shirley H. Phillips, eds. No-Tillage Agriculture: Principles and Practices. New York; Van Nostrand Reinhold, 1984. 306p.	Third

Developed Countries Ranking		Third World Ranking
Third	Pierik, R. L. M. In Vitro Culture of Higher Plants, translated from the German: Plantenteelt in Kweekbuizen. Dordrecht and Boston; M. Nijhoff, 1987. 344p.	
Third	Pierre, William H., and A. G. Norman, eds. Soil and Fertilizer Phosphorus in Crop Nutrition. New York; Academic Press, 1953. 492p.	
Second	Pijl, L. van der. Principles of Dispersal in Higher Plants. 3d rev. and exp. ed. Berlin, Germany and New York; Springer, 1982. 214p. (1st ed., 1969. 153p.)	Third
First rank	Pimentel, David, ed. CRC Handbook of Pest Management in Agriculture. 2d ed. Boca Raton, Fla.; CRC Press, 1991. 3 vols. (1st ed., 1981)	First rank
Second	Pimentel, David. Ecological Effects of Pesticides on Non-Target Species. Washington, D.C.; USDA Office of Science and Technology, 1971. 220p.	
Second	Pimentel, David, ed. Some Aspects of Integrated Pest Management. Ithaca, N.Y.; Dept. of Entomology, Cornell University, 1986. 368p.	First rank
Third	Pimentel, David. World Food, Pest Losses, and the Environment. Boulder, Colo.; Westview Press, 1978. 206p. (AAAS Selected Symposium No. 13)	First rank
Second	Pimentel, David, et al. Alternatives for Reducing Insecticides on Cotton and Corn: Economic and Environmental Impact. Athens, Ga.; Environmental Research Laboratory, Office of Research and Development, U.S. Environmental Protection Agency, 1979. 145p.	
Second	Pimentel, David, and Carl W. Hall, eds. Food and Natural Resources. San Diego, Calif.; Academic Press, 1989. 512p.	Third
Third	Pinowski, Jan, and S. Charles Kendeigh, eds. Granivorous Birds in Ecosystems: Their Evolution, Populations, Energetics, Adaptations, Impact, and Control. Cambride, U.K. and New York; Cambridge University Press, 1977. 431p.	
Second	Pirone, Pascal P. Diseases and Pests of Ornamental Plants. 5th ed. New York; Wiley, 1978. 566p. (1st-3d eds. by B.O. Dodge and H.W. Rickett. 1st ed., 1943. 638p.)	
Second	Plakidas, Antonios G. Strawberry Diseases. Baton Rouge, La.; Louisiana State University Press, 1964. 195p. (Louisiana State University Studies, Biological Science Series No. 6)	
Second	Plant Breeding Reviews, ed. by Jules Janick. Westport, Conn.; AVI Pub. Co., 1983–90. 8 vols.	Second
	Plant Resources of South-East Asia: A Handbook. Vol. 1. Pulses, edited by L. J. G. van der Maesen and Sadikin Somaatmadja. Vol. 2. Edible Fruits and Nuts, edited by	Third

Developed Countries Ranking		Third World Ranking
	E. W. M. Verheij and R. E. Coronel. Vol. 3. Dye and Tannin-Producing Plants, by R. H. M. J. Lemmensand N. Wulijarni. Vol. 4. Forages, by L. Mannetje and R. M. Jones. Wageningen, Netherlands; PUDOC, 1989–92. 3 vols.	
Second	The Plant Viruses. Vol. 1, Polyhedral Virions with Tripartite Genomes, edited by R.I.B. Francki; Vol. 2, The Rod-shaped Plant Viruses, edited by M.H.V. Van Regenmortel and H. Fraenkel-Courat; Vol. 3, Polyhedral Virions with Monopartite RNA Genomes, edited by R. Koenig; and Vol. 4, The Filamentous Plant Viruses, edited by R.G. Milne. New York; Plenum Press, 1985–. 4 vols.	Second
	Platen, Henning H. von. Appropriate Land Use Systems of Smallholder Farms on Steep Slopes in Costa Rica: A Study on Situation and Development Possibilities. Kiel; Wissenschaftsverlag Vauk Kiel, 1985. 187p.	Third
	Ploetz, R. C., et al., eds. Compendium of Tropical Fruit Diseases. St. Paul, Minn.; APS Press, The American Phytopathological Society, 1994. 88p.	Third
First Third	Plucknett, Donald L., et al. Gene Banks and the World's Foods. Princeton, N.J.; Princeton University Press, 1987. 247p.	Third
	Plucknett, Donald L., and Halsey L. Beemer, eds. Vegetable Farming Systems in China; Report of the visit of the Vegetable Farming Systems Delegation to China. Boulder, Colo.; Westview Press, 1981. 386p.	Third
Second	Plumb, R. T., and J. M. Thresh, eds. Plant Virus Epidemiology: The Spread and Control of Insect-Borne Viruses. Oxford, Eng. and Boston, Mass.; Blackwell Scientific Pub., 1983. 377p.	Second
First rank	Poehlman, John M. Breeding Field Crops. 3d ed. Ames, Iowa; Iowa State University Press, 1994. 724p. Reprint of 1987 ed. (1st ed. New York; Holt, 1959. 427p.)	Second
Third	Poehlman, John M., and Dhirendranath Borthakur. Breeding Asian Field Crops: With Special Reference to Crops of India. Calcutta, India; Oxford and IBH Pub. Co., 1969. 385p.	Third
Second	Poinar, George O. Nematodes for Biological Control of Insects. Boca Raton, Fla.; CRC Press, 1979. 277p.	Second
Second	Poinar, George O., and Hans-Borje Jansson, eds. Diseases of Nematodes. Boca Raton, Fla.; CRC Press, 1988. 2 vols.	Third
Second	Poljakoff-Mayber, Alexandra, and J. Gale, eds. Plants in Saline Environments. New York; Springer-Verlag, 1975. 213p.	

Developed Countries Ranking		Third World Ranking
First rank	Pomeranz, Yeshajahu, ed. Wheat: Chemistry and Technology. 3d ed. St. Paul, Minn.; American Association of Cereal Chemists, 1988. 2 vols. (1st ed. ed. by Isydore Hlynka, 1964. 603p.)	First rank
First rank	Porter, Morris D., Donald H. Smith, and R. Rodriguez-Kabana, eds. Compendium of Peanut Diseases. St. Paul, Minn.; Amerian Phytopathological Society, 1984. 73p.	First rank
Second	Postgate, John R. The Fundamentals of Nitrogen Fixation. Cambridge, Eng. and New York; Cambridge University Press, 1982. 252p.	
Second	Postgate, John R. Nitrogen Fixation. 2d ed. London, Eng.; E. Arnold, 1987. 73p.	
Third	Power, J. F., ed. The Role of Legumes in Conservation Tillage Systems; Proceedings of a National Conference, University of Georgia, Athens, Apr. 1987. Ankeny, Iowa; Soil Conservation Society of America, 1987. 153p.	Second
Third	Pradhan, Shyamsunderlal. Agricultural Entomology and Pest Control. New Delhi, India; Indian Council of Agricultural Research, 1983. 267p.	Third
	Praloran, J. C. Les Agrumes. Paris, France; G.P. Maisonneuve and Larose, 1971. 565p. (Techniques Agricoles et Productions Tropicales no. 21/22)	Third
	Prentice, A. N. Cotton. Essex, Eng.; Longman, 1987. 436p. (1st ed. with title: Cotton: With Special Reference to Africa. London; Longman, 1972. 282p.)	Second
Second	Prescott, J. M., et al. Wheat Diseases and Pests: A Guide for Field Identification. Mexico City, Mexico; International Maize and Wheat Improvement Center, 1986. 135p.	First rank
Second	Price, Peter W. Insect Ecology. 2d ed. New York; Wiley, 1984. 607p. (1st ed., 1975. 514p.)	Second
Third	Pühler, A., ed. Molecular Genetics of the Bacteria-Plant Interaction. Berlin, Germany and New York; Springer-Verlag, 1983. 393p.	
Third	Purchase, I. F. H., ed. Mycotoxins. Amsterdam, Netherlands and New York; Elsevier, 1974. 443p.	
First rank	Purseglove, John W. Tropical Crops: Dicotyledons. London, Eng.; Longmans, 1974. 719p. (1st published 1968. 2 vols.)	First rank
Second	Purseglove, John W. Tropical Crops: Monocotyledons. London; Longmans and New York; Wiley, 1972. 2 vols. 607p. (Reprinted with corrections, 1988.)	
First rank	Purseglove, John W., et al. Spices. London, Eng. and New York; Longman, 1981. 2 vols. 813p.	Third
	Pushparajah, E., and Chew P. Soon, eds. The Oil Palm in	Second

Developed Countries Ranking		Third World Ranking
	Agriculture in the Eighties: A Report of the Proceedings of the International Conference, . . . Kuala Lumpur, June 1981, sponsored by Palm Oil Research Institute of Malaysia and the Incorporated Society of Planters. Kuala Lumpur; Incorporated Society of Planters, 1982. 2 vols.	
Second	Putnam, Alan R., and Chung-Shih Tang, eds. The Science of Allelopathy. New York; Wiley, 1986. 317p.	Third
	Py, Claude, Jean J. Lacoeuilhe, and Claude Teisson. L'Ananas: Sa Culture, ses Produits. Paris, France; G.P. Maisonneuve and Larose and Agence de Cooperation Culturelle et Technique, 1984. 562p. (Techniques Agricoles et Productions Tropicales No. 33)	Third
	Quimio, T. H., S. T. Chang, and D. J. Royse. Technical Guidelines for Mushroom Growing in the Tropics. Rome; Food and Agriculture Organization, 1990. 156p. (FAO Plant Production and Protection Paper No. 106)	Third
Third	Quraishi, Mohammed S. Biochemical Insect Control: Its Impact on Economy, Environment, and Natural Selection. New York; Wiley, 1977. 280p.	Third
Third	Rabbinge, R., et al. Theoretical Production Ecology: Reflections and Prospects. Wageningen, Netherlands; PUDOC, 1990. 350p.	
Third	Rabbinge, R., S. A. Ward, and H. H. van Laar, eds. Simulation and Systems Management in Crop Protection. Wageningen; PUDOC, 1989. 420p.	
	Rachie, Kenneth O. The Millets: Importance, Utilization and Outlook. Hyderabad, India; International Crops Research Institute for the Semi-arid Tropics, 1975. 63p.	Second
Third	Rachie, Kenneth O., and Judith M. Lyman, eds. Genetic Engineering for Crop Improvement; A Rockefeller Foundation Conference, May 1980. New York; Rockefeller Foundation, 1981. 254p.	Second
Second	Rachie, Kenneth O., and J. V. Majmudar. Pearl Millet. University Park, Pa.; Pennsylvania State University Press, 1980. 307p.	First rank
Second	Radosevich, Steven R., and Jodie S. Holt. Weed Ecology: Implications for Vegetation Management. New York; Wiley, 1984. 265p.	Second
	Ramakhrishnan, Taracad S. Diseases of Millets. New Delhi, India; Indian Council of Agricultural Research, 1963. 152p.	Second
	Rangaswami, Govindachetty. Diseases of Crop Plants in India. 2d ed. New Delhi, India; Prentice-Hall of India, 1979. 520p.	Second

Developed Countries Ranking		Third World Ranking
Third	Raper, C. David, and Paul J. Kramer, eds. Crop Reactions to Water and Temperature Stresses in Humid, Temperate Climates; Proceedings of a Workshop, Duke University, Durham, N.C., Oct. 1980, co-sponsored by NSF and Duke University. Boulder, Colo.; Westview Press, 1983. 373p.	Second
First rank	Rasmusson, Donald C., ed. Barley. Madison, Wis.; American Society of Agronomy, 1985. 522p.	First rank
Second	Raychaudhuri, Syama P., and J. P. Verma, eds. Review of Tropical Plant Pathology. New Delhi, India; Today and Tomorrow's Printers and Publishers, 1984–88. 6 vols.	Third
	Raychaudhuri, Syama P., and T. K. Nariani. Virus and Mycoplasma Diseases of Plants in India. New Delhi, India; Oxford and IBH Pub. Co., 1977. 102p.	Third
Third	Real, Leslie, ed. Pollination Biology. Orlando, Fla.; Academic Press, 1983. 338p.	Third
	Reddy, D. Bap. Plant Protection in India. Bombay and New York; Allied Publishers, 1968. 454p.	Third
Third	Reddy, D. Bap, ed. Reviews on Pest, Disease, and Weed Problems in Rainfed Crops in Asia and the Far East. Bangkok, Thailand; Food and Agriculture Organization, 1975. 254p.	
Third	Redei, George P. Genetics. New York; Macmillan, 1982. 736p.	Third
Third	Rehm, Sigmund, and Gustav Espig. The Cultivated Plants of the Tropics and Subtropics: Cultivation, Economic Value, Utilization = Die Kulturpflanzen der Tropen und Subtropen. Weikersheim, Germany; Margraf, 1991. 552p.	Second
Third	Reinert, J., and Y. P. S. Bajaj, eds. Applied and Fundamental Aspects of Plant Cell, Tissue and Organ Culture. Berlin and New York; Springer-Verlag, 1977. 803p.	
	Reissig, W. H., et al. Illustrated Guide to Integrated Pest Management in Rice in Tropical Asia. Los Banos, Philippines; International Rice Research Institute, 1985. 411p. (Translated into several Asian languages.)	Second
	Reuter, D. J., and J. B. Robinson, eds. Plant Analysis: An Interpretation Manual. Melbourne, Australia; Inkata Press, 1986. 218p.	Third
Second	Reuther, Walter, Herbert J. Webber, and Leon D. Batchelor, eds. The Citrus Industry. Rev. ed. Berkeley, Calif.; University of California, Division of Agricultural Sciences, 1967–89. 5 vols. (A substantial rev. and exp. of 1st ed., 1943, edited by Webber and Batchelor.)	First rank
Second	Ricaud, C., et al, eds. Diseases of Sugarcane: Major Dis-	Second

Developed Countries Ranking		Third World Ranking
	eases. Amsterdam, Netherlands and New York; Elsevier, 1989. 389p. (Published under auspices of the International Society of Sugarcane Technologists)	
	Rice Information Cooperative Effort. Rice Production Manual. Rev. ed. Laguna, Philippines; University of the Philippines, 1970. 382p. (1st ed., Quezon, Philippines; University of Philippines, College of Agriculture, in co-operation with the International Rice Research Institute, 1967. 345p.)	Second
Second	Rice, Elroy L. Allelopathy. 2d. Orlando, Fla.; Academic Press, 1984. 422p. (1st ed., New York; Academic Press, 1974. 353p.)	Second
	Rice, Robert P. Fruit and Vegetable Production in Warm Climates. London; Macmillan Publishers, 1990. 486p. (Superseded as: Fruit and Vegetable Production in Africa, edited by R. P. Rice, L. W. Rice, and H. D. Tindall. London and Basingstoke; Macmillan Publishers, 1987. 371p.)	Second
Second	Rich, Avery E. Potato Diseases. New York; Academic Press, 1983. 238p.	Second
	Richards, Paul. Indigenous Agricultural Revolution: Ecology and Food Production in West Africa. London; Hutchinson, 1985. 192p.	Third
Second	Richardson, Michael J. An Annotated List of Seed-borne Diseases. 3d ed. Kew, Eng.; Commonwealth Mycological Institute and Zurich, Switzerland; International Seed Testing Association, 1981+. 320p. (Phytopathological Papers No. 23) (Supplement, 3d ed. Zurich, Switzerland; International Seed Testing Association, 1981+) (2d ed., 1968 by Mary Noble and J.J. Richardson. 191p.)	Second
Third	Ridgway, Richard L., E. P. Lloyd, and W. H. Cross, eds. Cotton Insect Management with Special Reference to the Boll Weevil. Washington, D.C.; U.S. Govt. Print. Off., 1983. 591p. (USDA Agriculture Handbood No. 589)	
Second	Ridgway, Richard L., and S. Bradleigh Vinson, eds. Biological Control by Augmentation of Natural Enemies; Insect and Mite Control with Parasites and Predators; Proceedings of a Symposium held at the 15th International Congress of Entomology, Washington, D.C., Aug. 1976. New York; Plenum Press, 1977. 480p. (Environmental Science Research Series No. 11)	Third
Second	Rieger, Rigomar, A. Michaelis and M. M. Green. Glossary of Genetics: Classical and Molecular. 5th ed. Berlin and New York; Springer-Verlag, 1991. 533p. (1st ed.,	

Developed Countries Ranking		Third World Ranking
	Genetisches und Cytogenetisches Worterbuch, 1954. 140p.)	
Third	Ritchie, Gary A., ed. New Agricultural Crops. Boulder, Colo.; Westview Press, 1979. 259p. (AAAS Special Symposium 38)	Third
Second	Ritchie, J. T. CERES-Maize (Machine-Readable Data File): A Simulation Model of Maize Growth and Development, edited by C.A. Jones and J.R. Kiniry. College Station, Tex.; Texas A & M University Press, 1986. 194p.	
Second	Rivnay, Ezekiel. Field Crop Pests in the Near East. Den Haag; W. Junk, 1962. 450p. (Monographiae Biologicae No. 10)	
Third	Roane, Martha K., Gary J. Griffin, and John R. Elkins. Chestnut Blight, Other Endothia Diseases, and the Genus Endothia. St. Paul, Minn.; APS Press, American Phytopathological Society, 1986. 53p.	
Third	Robb, D. A., and W. S. Pierpoint, eds. Metals and Micronutrients: Uptake and Utilization by Plants. London, Eng. and New York; Academic Press, 1983. 341p. (Annual Proceedings of the Phytochemical Society of Europe No. 21)	Second
Second	Röbbelen, Gerhard, R. Keith Downey, and Amram Ashri, eds. Oil Crops of the World: Their Breeding and Utilization. New York; McGraw-Hill, 1989. 553p.	Third
Second	Roberts, Daniel A. Fundamentals of Plant-Pest Control. San Francisco, Calif.; Freeman, 1978. 242p.	
	Roberts, Daniel A., and Carl W. Boothroyd. Fundamentals of Plant Pathology. 2d ed. New York; W.H. Freeman and Co., 1984. 432p. (1st ed., San Francisco; W.H. Freeman, 1972. 402p.)	Second
Second	Roberts, Donald W., and James R. Aist, eds. Infection Processes of Fungi. New York; Rockefeller Foundation, 1984. 201p.	Third
First rank	Roberts, Donald W., and R. R. Granados, eds. Biotechnology, Biological Pesticides, and Novel Plant-Pest Resistance for Insect Pest Management; Proceedings of a conference, July 1988, sponsored by Insect Pathology Resource Center and Boyce Thompson Institute for Plant Research. Ithaca, N.Y.; Insect Pathology Resource Center and Boyce Thompson Institute for Plant Research, Cornell University, 1989. 175p.	Second
First rank	Roberts, Eric H. Viability of Seeds. London, Eng.; Chapman and Hall, 1972. 448p.	
Third	Robertson, Hugh D., et al, eds. Plant Infectious Agents:	Third

Developed Countries Ranking		Third World Ranking
	Viruses, Viroids, Virusoids, and Satellites. Cold Spring Harbor, NY; Cold Spring Harbor Laboratory, 1983. 230p.	
Second	Robinson, J. B. D., ed. Diagnosis of Mineral Disorders in Plants. 1st Amer. ed. New York; Chemical Pub. Co., 1984–87. 3 vols. (Vol. 1 by C. Bould, E.J. Hewitt and P. Needham and Vol. 2 by Alan Scaife and Mary Turner. Exp. and updating of T. Wallace's the Diagnosis of Mineral Deficiences in Plants, 1st pub. in 1943 by H.M.S.O.)	Third
First rank	Robinson, Raoul A. Plant Pathosystems. Berlin, Germany and New York; Springer-Verlag, 1976. 184p. (Advanced Series in Agricultural Science No. 3)	Second
	Rockstein, Morris, ed. The Physiology of Insecta. 2d ed. New York; Academic Press, 1973–74. 6 vols. (1st ed., 1964–65. 3 vols.)	Second
Second	Rodriguez, J. G., ed. Insect and Mite Nutrition: Significance and Implications in Ecology and Pest Management; Proceedings of the International Conference, . . . University of Kentucky, 1972. Amsterdam, Netherlands; North-Holland Pub. Co.; Distributors for the U.S.A. and Canada: New York; Elsevier, 1973. 702p.	
Second	Rogers, L. J., ed. Amino Acids, Proteins and Nucleic Acids. London, Eng.; Academic Press, 1991. 341p. (Methods in Plant Biochemistry No. 5)	
Third	Rom, Roy C., and Robert F. Carlson, eds. Rootstocks for Fruit Crops. New York; Wiley, 1987. 494p.	
	Roman, Jesse. Fitonematologia Tropical. Rio Piedras; Universidad de Puerto Rico, Recento Universitario de Mayaguez, Colegio de Cinecias Agricolas, Estacion Experimental Agricola, 1978. 256p.	Third
Third	Rosenberg, Norman J., Blaine L. Blad, and Shashi B. Verma. Microclimate: The Biological Environment. 2d ed. New York; Wiley, 1983. 495p. (1st ed. by N.J. Rosenberg, 1974. 315p.)	
Third	Rosengarten, Frederic. The Book of Edible Nuts. New York; Walker, 1984. 384p.	Second
Third	Ross, Merrill A., and Carole A. Lembi. Applied Weed Science. Minneapolis, Minn.; Burgess Pub. Co., 1985. 340p.	Third
Third	Rothfos, Bernhard. Coffee Production = Kaffee, die Produktion. Translated by Sabine Buken. Hamburg, Germany; Gordian-Max-Rieck, 1980. 366p.	Second
	Rothwell, Norman V. Understanding Genetics; A Molecu-	Second

Developed Countries Ranking		Third World Ranking
	lar Approach. New York; Wiley-Liss, 1993. 656p. (1st ed., Baltimore, Md.; Williams and Wilkins Co., 1976. 486p.)	
First rank	Roush, Richard T., and Bruce Tabashnik, eds. Pesticide Resistance in Arthropods. New York; Chapman and Hall, 1990. 303p.	Second
Third	Rubenstein, Irwin, et al, eds. The Plant Seed: Development, Preservation, and Germination; Proceedings of a Symposium on the Development, Preservation, and Germination of the Plant Seed, University of Minnesota, St. Paul, Minn., Mar. 1978. New York; Academic Press, 1979. 266p.	
	Ruck, Harold C. Deciduous Fruit Tree Cultivars for Tropical and Sub-tropical Regions. Oxford, Eng.; Commonwealth Agricultural Bureaux, 1975. 91p.	Third
Second	Rudd, R. L. Pesticides and the Living Landscape. Madison, Wis.; University of Wisconsin Press, 1964. 320p.	Third
	Ruddle, Kenneth, and Gongufu Zhong. Integrated Agriculture-Aquaculture in South China: The Dike-Pond System of the Zhujiang Delta. Cambridge, Eng. and New York; Cambridge University Press, 1988. 173p.	Third
First rank	Russell, Edward J. Soil Conditions and Plant Growth, edited by Alan Wild. 11th ed. Burnt Mill, Eng.; Longman Scientific and Technical; New York; Wiley, 1988. 991p. (Rev. ed. of: Soil Conditions and Plant Growth, 10th ed., 1973. 1st ed., London and New York; Longmans, Green, 1912. 168p.)	First rank
Third	Russell, G., B. Marshall, and P. G. Jarvis, eds. Plant Canopies: Their Growth, Form, and Function. Cambridge, Eng. and New York; Cambridge University Press, 1989. 178p.	
Second	Russell, G. E. Plant Breeding for Pest and Disease Resistance. London, Eng. and Boston, Mass.; Butterworth, 1978. 485p.	Third
Second	Russell, G. E., ed. Progress in Plant Breeding—1. London, Eng. and Boston, Mass.; Butterworths, 1985. 325p.	Second
Second	Russell, J. S., and E. L. Greacen, eds. Soil Factors in Crop Production in a Semi-Arid Environment. St. Lucia, Queensland; University of Queensland Press, 1977. 327p.	Second
Second	Russell, Robert S. Plant Root Systems: Their Functions and Interaction with the Soil. 1st ed. London, Eng. and New York; McGraw-Hill, 1977. 298p.	First rank

Developed Countries Ranking		Third World Ranking
Second	Ruthenberg, Hans. Farming Systems in the Tropics. 3d ed. Oxford, Eng.; Clarendon Press and New York; Oxford University Press, 1980. 424p. (1st ed., 1971. 313p.)	First rank
Third	Safeeulla, K. M. Biology and Control of the Downy Mildews of Pearl Millet, Sorghum and Finger Millet. Mysore, India; Downy Mildew Research Laboratory, Manasagangothri, Mysore University, 1976. 304p.	Third
Third	Salaman, Redcliffe N. The History and Social Influence of the Potato, edited by J. G. Hawkes. Rev. impression. Cambridge, Eng. and New York; Cambridge University Press, 1985. 685p. (1st ed., 1949. 685p.)	Third
First rank	Salisbury, Frank B., and Cleon W. Ross. Plant Physiology. 4th ed. Belmont, Calif.; Wadsworth Pub. Co., 1992. 682p. (1st ed., 1969. 747p.)	
Third	Sambrook, J., E. F. Fritsch, and Tom Maniatis. Molecular Cloning: A Laboratory Manual. 2d ed. Cold Spring Harbor, N.Y.; Cold Spring Harbor Laboratory, 1989. 3 vols. (1st ed., 1982. 545p.)	
	Samson, Jules A. Tropical Fruits. 2d ed. Essex, Eng. and New York; Longman, 1986. 335p. (1st ed., 1980. 250p.)	First rank
Third	Samson, Robert A., Just M. Vlak, and Dick Peters, eds. Fundamental and Applied Aspects of Invertebrate Pathology; Proceedings of the 4th International Colloquium of Invertebrate Pathology, in conjunction with the 19th Annual Meeting of the Society of Invertebrate Pathology, Congress Centre Koningshof, Veldhoven, Netherlands, Aug. 1986. Wageningen, Netherlands; Foundation of the 4th International Colloquium of Invertebrate Pathology, 1986. 711p.	
Third	San Pietro, Anthony G., ed. Biochemical and Photosynthetic Aspects of Energy Production. New York; Academic Press, 1980. 231p.	
Second	San Pietro, Anthony G., Francis A. Greer, and Thomas J. Army, eds. Harvesting the Sun: Photosynthesis in Plant Life; Proceedings of a Symposium sponsored by the International Minerals and Chemical Corporation. New York; Academic Press, 1967. 342p.	
	Sanchez, Pedro A. Properties and Management of Soils in the Tropics. New York; Wiley, 1976. 618p.	First rank
	Sanchez, Pedro A., and Luis E. Tergas, eds. Pasture Production in Acid Soils of the Tropics; Proceedings of a seminar, CIAT, Cali, Colombia, Apr. 1978. Cali, Colombia; Beef Program, Centro Internacional de Agricultura Tropical, 1979. 488p.	Second

Developed Countries Ranking		Third World Ranking
Second	Sauer, D. B., ed. Storage of Cereal Grains and Their Products. 4th ed. St. Paul, Minn.; American Association of Cereal Chemists, 1992. 615p. (American Association of Cereal Chemists Monograph Series No. 2) (1st ed. edited by J. A. Anderson and A. W. Alcock, 1954. 515p.; 3d ed. edited by Clyde M. Christensen, 1982. 544p.)	Second
Third	Saxena, Mohan C., and K. B. Singh, eds. The Chickpea. Wallingford, Eng.; CAB International, 1987. 409p.	Second
Second	Scandalios, John G., ed. Molecular Genetics of Plants. Orlando, Fla.; Academic Press, 1984. 288p.	Third
Third	Scannerini, Silvano, et al, eds. Cell to Cell Signals in Plant, Animal, and Microbial Symbiosis; Proceedings of the NATO Advanced Research Workshop, Villa Gualino, Trin, Italy, May 1987. Berlin, Germany and New York; Springer-Verlag, 1988. 414p. (NATO ASI Series, Series H, Cell Biology No. 17)	Second
	Schaad, N. W., ed. Laboratory Guide for Identification of Plant Pathogenic Bacteria. 2d ed. St. Paul, Minn.; American Phytopathological Society Press, 1988. 164p. (1st. ed., 1980. 72p.)	Third
Third	Scheer, Hugo, and Siegfried Schneider, eds. Photosynthetic Light-Harvesting Systems: Organization and Function; Proceedings of an international workshop, Freising, Germany, Oct. 1987. Berlin, Germany and New York; W. de Gruyter, 1988. 631p.	Third
	Schenck, N. C., ed. Methods and Principles of Mycorrhizal Research. St. Paul, Minn.; American Phytopathological Society, 1982. 244p.	Second
	Schmutterer, H. Pests of Crops in Northeast and Central Africa with Particular Reference to the Sudan. Stuttgart and Portland, Or.; G. Fischer, 1969. 296p.	Third
Second	Schneider, R. W., ed. Suppressive Soils and Plant Disease; Based on presentations from a Symposium on the Nature of Soils Suppressive to Soilborne Plant Diseases, New Orleans, La., Aug. 1981. St. Paul, Minn.; American Phytopathological Society, 1982. 88p.	Second
	Schnell, R. Plantes Alimentaires et vie Agricole de l'-Afrique Noire. Paris; Ed. Larose, 1957.	Second
	Schwartz, Howard F., and Marcial A. Pastor-Corrales, eds. Bean Production Problems in the Tropics. 2d ed. Cali, Colombia; Centro Internacional de Agricultura Tropical (CIAT), 1989. 424p. (1st ed. by H.F. Schwartz, and G.E. Galvez titled: Bean Production Problems: Disease, Insect, Soil and Climatic Constraints of Phaseolus Vulgaris, 1980. 424p.)	First rank

Developed Countries Ranking		Third World Ranking
Second	Schwartz, P. H., ed. Guidelines for the Control of Insect and Mite Pests of Foods, Fibers, Feeds, Ornamentals, Livestock, and Households. Washington, D.C.; U.S. Agricultural Research Service, 1982. 734p. (Agricultural Handbook No. 584)	Second
Third	Schwintzer, Christa R., and John D. Tjepkema, eds. The Biology of Frankia and Actinorhizal Plants. San Diego, Calif.; Academic Press, 1990. 408p.	
Second	Scopes, Nigel E. A., and Lorna Stables, eds. Pest and Disease Control Handbook. 3d ed. Thornton Heath, Surrey, Eng.; BCPC Publications, 1989. 732p. (1st ed., 1979. 361p.)	Second
Second	Scott, K. J., and A. K. Chakravorty, eds. The Rust Fungi. London and New York; Academic Press, 1982. 288p.	Third
Second	Scott, P. R., and Bainbridge A., eds. Plant Disease Epidemiology; Based on Selected Papers presented at a Symposium, London School of Economics, organized by the Federation of British Plant Pathologists in conjunction with the Association of Applied Biologists and the British Mycological Society, Dec. 1977. Oxford, Eng.; Blackwell Scientific Publications and New York; Halsted Press, 1978. 329p.	Third
Second	Scott, Walter O., and Samuel R. Aldrich. Modern Soybean Production. 2d ed. Champaign, Ill.; S and A Publications, 1983. 230p. (1st ed., Cincinnati, Ohio; Farm Quarterly, 1970. 192p.)	
Third	Sears, E. R. The Aneuploids of Common Wheat. Columbus, Mo.; 1954. 58p. (Missouri Agricultural Experiment Station Research Bulletin No. 572)	
Second	Setlow, Jane K., and Alexander Hollaender, eds. Genetic Engineering: Principles and Methods. New York; Plenum Press, 1979–1993. 15 vols.	Second
Third	Shanmugasundaram, S., ed. Mungbean; Proceedings of the 2d International Symposium, Bangkok, Thailand, Nov. 1987. Shanhua, Taiwan; Asian Vegetable Research and Development Center, 1988. 730p.	Second
	Shanmugasundaram, S., ed. Soybean in Tropical and Subtropical Cropping Systems; Proceedings of the International Symposium on Soybeans in the Tropics and Subtropics, Tsukuba, Japan, Sept.—Oct. 1983, sponsored jointly by the Tropical Agriculture Research Center and the Asian Vegetable Reseach and Development Center. Rev. ed. Shanhua, Taiwan; Asian Vegetable Research and Development Center, 1986. 471p. (AVRDC Publication No. 86–253)	First rank

Developed Countries Ranking		Third World Ranking
Third	Sharp, W. R., et al, eds. Plant Cell and Tissue Culture: Principles and Applications; 4th Biosciences Colloquium, Ohio State University, 1977. Columbus, Oh.; Ohio State University Press, 1979. 892p.	
	Shaw, N. H., and W. W. Bryan, eds. Tropical Pasture Research: Principles and Methods. Farnham Royal, Eng.; Commonwealth Agricultural Bureaux, 1976. 454p. (Commonwealth Bureaux of Pastures and Field Crops Bulletin No. 51) (Rev. and expanded version of the work by Cunningham Laboratory published in 1964 under title: Some Concepts and Methods in Sub-Tropical Pasture Research)	Second
Third	Sheail, John. Pesticides and Nature Conservation: The British Experience, 1950–75. Oxford, Eng.; Clarendon Press and New York; Oxford University Press, 1985. 276p.	Third
Third	Sheets, T. J., and David Pimentel, eds. Pesticides: Contemporary Roles in Agriculture, Health, and Environment; Outgrowth of a symposium at the meeting of the American Association for the Advancement of Science, Feb. 1978. Clifton, N.J.; Humana Press, 1979. 186p.	Third
	Shepard, Barclay M., A. T. Barrion, and J. A. Litsinger. Helpful Insects, Spiders, and Pathogens: Friends of the Rice Farmer. Los Banos, Philippines; International Rice Research Institute, 1987. 136p.	Third
First rank	Sherf, Arden F., and Alan A. MacNab. Vegetable Diseases and Their Control. 2d ed. New York; J. Wiley, 1986. 728p. (1st ed. by Charles Chupp and Arden F. Sherf. New York; Ronald Press Co., 1960. 693p.)	First rank
Third	Sheridan, William F., ed. Maize for Biological Research. Charlottesville, Va.; Plant Molecular Biology Association, 1982. 434p.	Third
Third	Shew, H. D., and G. B. Lucas, eds. Compendium of Tobacco Diseases. St. Paul, Minn.; APS Press, 1991. 68p.	Third
	Shideler, F. S., and H. Ricon, eds. Proceedings of the 6th Symposium of the International Society for Tropical Root Crops, Lima, Peru, Feb. 1983. Lima, Peru; International Potato Center, 1984. 671p.	Third
Second	Shorey, Harry H. Animal Communication by Pheromones. New York; Academic Press, 1976. 167p.	
First rank	Shorey, Harry H., ed. Chemical Control of Insect Behavior: Theory and Application. New York; J. Wiley, 1977. 414p.	Third
First rank	Shurtleff, Malcolm C., ed. Compendium of Corn Diseases. 2d ed. St. Paul, Minn.; American Phytopathologi-	Second

Developed Countries Ranking		Third World Ranking
	cal Society, 1980. 105p. (1st ed. prepared by Cooperative Extension Service, University of Illinois, 1973. 64p.)	
Third	Sill, W. H. The Plant Protection Discipline: Problems and Possible Developmental Strategies. New York; Allanheld, Osmun, 1978. 190p.	
	Silvestre, P., and M. Arraudeau. Le Manioc. Paris, France; Maisonneuve et Larose, Agence de Cooperation Culturelle et Technique, 1983. 262p. (Techniques Agricoles et Productions Tropicales no. 32)	
First rank	Simmonds, Norman W. The Evolution of the Bananas. London, Eng.; Longmans, 1962. 170p.	Second
Second	Simmonds, Norman W. Principles of Crop Improvement. London, Eng. and New York; Longman, 1979. 399p.	Second
Second	Simmonds, Norman W., and S. Rajaram, eds. Breeding Strategies for Resistance to the Rusts of Wheat. Mexico; Centro Internacional de Mejoramiento de Maiz y Trigo, 1988. 151p. (Summaries in Spanish and French.)	
Third	Simpson, Beryl B., and Molly Conner-Ogorzaly. Economic Botany: Plants in Our World. New York; McGraw-Hill, 1986. 640p.	Second
First rank	Sinclair, James B., and P. A. Backman, eds. Compendium of Soybean Diseases. 3d ed. St. Paul, Minn.; American Phytopathological Society, 1989. 106p. (1st ed., 1975. 69p.)	First rank
Third	Sinclair, Walton B. The Biochemistry and Physiology of the Lemon and Other Citrus Fruits. Oakland, Calif.; University of California, Division of Agriculture and Natural Resources, 1984. 946p.	
Second	Singer, Rolf, and Bob Harris. Mushrooms and Truffles: Botany, Cultivation, and Utilization. 2d rev. and enl. ed. Koenigstein, Germany; Koeltz Scientific Books, 1987. 389p. (1st ed. by R. Singer, London; L. Hill and New York; Interscience Publishers, 1961. 272p.)	Second
First rank	Singh, Pritam, and R. F. Moore, eds. Handbook of Insect Rearing. Amsterdam, Netherlands and New York; Elsevier, 1985. 2 vols.	Third
	Singh, S. R., ed. Insect Pests of Tropical Food Legumes. Chichester, Eng. and New York; J. Wiley, 1990. 451p.	Third
Second	Singh, S. R., and K. O. Rachie, eds. Cowpea Research: Production and Utilization. Chichester, Eng. and New York; Wiley, 1985. 460p.	Second
	Singh, S. R., K. O. Rachie, and K. E. Dashiell, es. Soybeans for the Tropics: Research, Production, and Utilization; Based on papers presented at the Tropical Soy-	Second

Developed Countries Ranking		Third World Ranking
	bean Workshop, IITA , Sept.—Oct. 1985. Chichester, Eng. and New York; Wiley, 1987. 230p.	
First rank	Singh, S. R., H. F. van Emden, and T. Ajibola Taylor, eds. Pests of Grain Legumes: Ecology and Control. London, Eng. and New York; Academic Press, 1978. 454p.	Third
Third	Singh, Uma S., et al, eds. Plant Diseases of International Importance. Englewood Cliffs, N.J.; Prentice Hall, 1992–93. 4 vols. (Vol. 3 edited by Jatinder Kumar, et al.)	Third
Third	Singhal, G. S., et al, eds. Photosynthesis: Molecular Biology and Bioenergetics; Proceedings of the International Workshop on Application of Molecular Biology and Bioenergetics of Photosynthesis, Jawaharlal Nehru University, 1988. New Delhi, India; Springer-Verlag and Narosa Publishing House, 1989. 441p.	Third
	Sinha, Suresh K. Food Legumes: Distribution, Adaptability and Biology of Yield. Rome, Italy; Food and Agriculture Organization, 1977. 124p. (FAO Plant Production and Protection Paper No. 3)	Second
Second	Sivanesan, A., and J. M. Waller. Sugarcane Diseases. Slough, Eng.; CAB International, 1986. 88p. (Phytopatholgical Paper No. 29)	Second
Second	Skerman, P. J., D. G. Cameron, and F. Riveros. Tropical Forage Legumes. 2d ed., rev. and exp. Rome; Food and Agriculture Organization, 1988. 692p. (FAO Plant Production and Protection Series No. 2) (1st ed., 1977. 609p.)	Second
Second	Skerman, P. J., and F. Riveros. Tropical Grasses. Rome; Food and Agriculture Organization, 1990. 832p. (FAO Plant Production and Protection Series No. 23)	First rank
	Skinner, F. A., and D. W. Lovelock, eds. Identification Methods for Microbiologists. 2d ed. London, Eng. and New York; Academic Press, 1979. 315p. (Society for Applied Bacteriology Technical Series No. 14)	Third
	Smart, Grover C., and V. G. Perry, eds. Tropical Nematology; Papers presented at the 5th Annual Meeting of the Society of Nematologists, Daytona Beach, Fla., Aug. 1966. Gainesville, Fla.; University of Florida Press, 1968. 153p. (Summaries in Spanish)	Second
Second	Smartt, J. Grain Legumes: Evolution and Genetic Resources. Cambridge, Eng. and New York; Cambridge University Press, 1990. 379p.	
Second	Smartt, J. Tropical Pulses. London, Eng.; Longman, 1976. 348p.	Second
First rank	Smartt, J. and Norman W. Simmonds, eds. Evolution of	First rank

Developed Countries Ranking		Third World Ranking
	Crop Plants. 2d ed., Harlow, Essex; Longman Scientific and Technical, and New York; Wiley, 1995. (1st ed. edited by Norman W. Simmonds, London and New York; Longman, 1976. 339p.)	
First rank	Smiley, Richard W., Peter H. Doernolden, and Bruce B. Clarke. Compendium of Turfgrass Diseases. 2d ed. St. Paul, Minn.; American Phytopathological, 1992. 128p. (1st ed., 1983. 102p.)	
Second	Smith, Allan K., and Sidney J. Circle, eds. Soybeans: Chemistry and Technology. Revised 2d printing. Westport, Conn.; AVI Pub. Co., 1978. 1 vol. 470p. (1st ed., 1972. 1 vol.)	Second
First rank	Smith, Edward H., and D. Pimentel, eds. Pest Control Strategies. New York; Academic Press, 1978. 342p.	Second
Second	Smith, I. M., et al., eds. European Handbook of Plant Diseases; Sponsored by the British Society for Plant Pathology and the European and Mediterranean Plant Protection Organization. Oxford, Eng. and Boston, Mass.; Blackwell Scientific Publications, 1988. 583p.	Third
Third	Smith, I. M., D. G. McNamara, P. R. Scott, and K. M. Harris, eds. Quarantine Pests for Europe; Data Sheets on Quarantine Pests for the European Communities and for the European and Mediterranean Plant Protection Organization, prepared by CABI and EPPO for the European Communities. Wallingford, U.K.; CAB International, 1992. 1032p.	
Second	Smith, Jeffrey D., N. Jackson, and A. R. Woolhouse. Fungal Diseases of Amenity Turf Grasses. 3d ed. London, Eng. and New York; E. and F.N. Spon, 1989. 401p. (Previous ed. titled: Fungal Diseases of Turf Grasses. Bingley, Eng.; Sports Turf Research Institute, 1965. 90p.)	
	Smith, Kenneth M. Plant Viruses. 6th ed. New York; J. Wiley, 1977. 241p.	Second
	Smith, Nigel J. H., et al. Tropical Forests and Their Crops. Ithaca, N.Y.; Comstock Pub. Associates, 1992. 568p.	Third
Second	Smith, Ora. Potatoes: Production, Storing, Processing. 2d ed. Westport, Conn.; AVI Pub. Co., 1977. 776p. (1st ed., 1968. 642p.)	Second
	Smith, R. L., and E. A. Bababunmi, eds. Toxicology in the Tropics; Papers presented at the 1st International Symposium, . . . Ibadan, Nigeria, Feb. 1979, sponsored by the World Health Organization, Toxicology Forum of U.S.A. and Wellcome Trust of U.K. London, Eng.; Taylor and Francis, 1980. 280p.	Third

Developed Countries Ranking		Third World Ranking
Third	Smith, Ray F., Thomas E. Mittler, and Carroll N. Smith, eds. History of Entomology. Palo Alto, Calif.; Annual Reviews, Inc., 1973. 517p.	Second
	Smith, W. H., and Stephen J. Banta, eds. Symposium on Potential Productivity of Field Crops Under Different Environments, Los Banos, Philippines, 1980. Los Banos, Philippines; International Rice Research Institute, 1983. 526p.	Third
First rank	Snedecor, George W., and William G. Cochran. Statistical Methods. 8th ed. Ames, Ia.; Iowa State University Press, 1989. 503p. (1st ed., Ames; Collegiate Press Inc., 1937. 341p.)	Second
Third	Sneep, J., and A. J. T. Hendriksen, eds. Plant Breeding Perspectives; A Centennial Publication . . . Wageningen, Netherlands; Centre for Agricultural Publishing and Documentation, 1979. 435p.	
Third	Snelson, J. T. Grain Protectants. Canberra, Australia; Australian Center for International Agricultural Research, 1987. 448p.	Third
Second	Snodgrass, Robert E. A Textbook of Arthropod Anatomy. Ithaca, NY; Comstock Pub. Co., 1952. 363p.	
Third	Somasegaran, P., and H. J. Hoben. Methods in Legume-Rhizobium Technology. Paia, Ha.; NIFTAL Project and MIRCEN, Department of Agronomy and Soil Science, College of Tropical Agriculture and Human Resources, University of Hawaii, 1985. 367p.	Second
First rank	Southwood, Richard. Ecological Methods: With Particular Reference to the Study of Insect Populations. 2d ed. London, Eng.; Chapman and Hall and New York; Wiley, 1978. 524p. (1st ed., London; Methuen, 1966. 391p.)	
Third	Spaar, Dieter, et al. Bakteriosen der Kulturpflanzen. Berlin, Germany; Akademie-Verlag, 1977. 276p.	
Second	Spedding, Colin R. W. The Biology of Agricultural Systems. London and New York; Academic Press, 1975. 261p.	Third
	Spedding, Colin R. W., ed. Vegetable Productivity: The Role of Vegetables in Feeding People and Livestock; Proceedings of a Symposium, Royal Geographical Society, London, Sept. 1979. London, Eng.; Macmillan, 1981. 268p.	Second
Second	Spencer, D. M., ed. The Downy Mildews. London, Eng. and New York; Academic Press, 1981. 636p.	Second
First rank	Spencer, D. M., ed. The Powdery Mildews. London, Eng. and New York; Academic Press, 1978. 565p.	First rank
First rank	Sprague, George F., and J. W. Dudley, eds. Corn and	First rank

Developed Countries Ranking		Third World Ranking
	Corn Improvement. 3d ed. Madison, Wis.; American Society of Agronomy, 1988. 986p. (Agronomy No. 18) (1st ed., New York, Academic Press, 1955. 699p.)	
	Sprague, Howard B., ed. Hunger Signs in Crops; a Symposium, prepared by S.A. Barber, et al. 3d ed. New York; David McKay, 1964. 461p. (1st ed. prepared by George M. Bahrt, et al., edited by Gove Hambidge. Washington, D. C.; American Society of Agronomy and National Fertilizer Association, 1941. 327p.)	Second
Second	Sprague, Milton A., and Glover B. Triplett, eds. No-Tillage and Surface-Tillage Agriculture: The Tillage Revolution. New York; Wiley, 1986. 467p.	
Third	Sprague, Victor. Systematics of the Microsporidia, edited by Lee A. Bulla and Thomas C. Cheng. New York; Plenum Press, 1977. 510p.	
Third	Sprent, Janet I., and Peter Sprent. Nitrogen Fixing Organisms: Pure and Applied Aspects. London, Eng. and New York; Chapman and Hall, 1990. 256p.	Third
	Squire, Geoffrey R. The Physiology of Tropical Crop Production. Wallingford, U.K.; CAB International, 1990. 236p.	Third
	Srivastava, P. D., et al, eds. Agricultural Entomology. New Delhi, India; All India Scientific Writers Society, 1982–83. 2 vols.	Second
	Ssali, H., and S. O. Keya, eds. Biological Nitrogen Fixation in Africa; Proceedings of the 1st Conference of the African Association for Biological Nitrogen Fixation (AABNF), Nairobi, Kenya, July 1984. Nairobi, Kenya; Nairobi Rhizobium Microbiological Resources Centre, Department of Soil Science, University of Nairobi, 1985. 540p.	Second
Third	Stace, Helen M., and L. A. Edye, eds. The Biology and Agronomy of Stylosanthes. Sydney, Australia and Orlando, Fla.; Academic Press, 1984. 636p.	
Second	Stacey, Gary, Robert H. Burris, and Harold J. Evans, eds. Biological Nitrogen Fixation. New York; Chapman and Hall, 1992. 943p.	Third
Third	Stadler Genetics Symposium. Columbia, Mo.; University of Missouri, Agricultural Experiment Station, 1969/70–90. 19 vols.	
Third	Stafleu, F. A., et al, eds. International Code of Botanical Nomenclature; Adopted by the 11th International Botanical Congress, Seattle, Aug. 1969. Utrecht, Netherlands; A. Oostoek's Uitgeversmaatsch appij N.V., 1972. 426p. (Regnum Vegetabile No. 82)	Second

Developed Countries Ranking		Third World Ranking
Third	Stanhill, G., ed. Energy and Agriculture. Berlin, Germany and New York; Springer, 1984. 192p.	
	Staples, Richard C., and Ronald J. Kuhr, eds. Linking Research to Crop Production; Boyce Thompson Institute for Plant Research Conference on Linking Basic Research to Crop Improvement Programs for Less Developed Countries, Cornell University, 1979. New York; Plenum Press, 1980. 235p.	Second
Second	Staples, Richard C., and Gary H. Toenniessen, eds. Plant Disease Control: Resistance and Susceptibility; Papers drawn from a Conference, Rockefeller Foundation's Bellagio Study and Conference Center, Aug. 1979. New York; Wiley, 1981. 339p.	
Third	Stapley, J. H., F. C. H. Gayner, and K. A. Hassal, eds. World Crop Protection. Cleveland, Oh.; CRC Press, 1969. 2 vols.	Third
Second	Starr, Mortimer P., ed. Phytopathogenic Bacteria. New York; Springer-Verlag, 1983. 1 vol.	Third
Second	Starr, Mortimer P., et al, eds. The Prokaryotes: A Handbook on the Biology of Bacteria: Ecophysiology, Isolation, Identification, Applications. 2d ed. Berlin, Germany and New York; Springer-Verlag, 1991. 4 vols. (1st ed., 1981 entitled: The Prokaryotes: A Handbook on Habitats, Isolation, and Identification of Bacteria. 2284p.)	
Third	Staskawicz, Brian, Paul Ahlquist, and Olen Yoder, eds. Molecular Biology of Plant-Pathogen Interactions; Proceedings of a UCLA Colloquium, Steamboat Springs, Colo., Mar.—Apr. 1988. New York; A.R. Liss, 1989. 307p.	Second
Third	Stebbins, G. Ledyard. Chromosomal Evolution in Higher Plants. London, Eng.; E. Arnold, 1971. 216p.	Third
	Stebbins, G. Ledyard. Flowering Plants: Evolution Above the Species Level. Cambridge, Mass.; Belkap Press of Harvard University Press, 1977. 397p.	Third
	Stebbins, G. Ledyard. Variation and Evolution in Plants. New York; Columbia University Press, 1950. 643p.	Third
First rank	Steel, Robert G. D., and James H. Torrie. Principles and Procedures of Statistics: A Biometrical Approach. 2d ed. New York; McGraw-Hill, 1980. 633p.	
	Steiner, Kurt G. Intercropping in Tropical Smallholder Agriculture with Special Reference to West Africa. 2d ed. Eschborn, Germany; Deutsche Gesellschaft fur Technische Zusammenarbeit, 1984. 304p. (Schriftenreihe der GTZ No. 137) (1st ed., 1982. 303p.)	Second

Developed Countries Ranking		Third World Ranking
First rank	Steinhaus, Edward A. Insect Pathology: An Advanced Treatise. New York; Academic Press, 1963. 2 vols.	
Second	Steinhaus, Edward A. Principles of Insect Pathology. New York; McGraw-Hill, 1949. 757p. (Reprinted New York; Hafner Pub. Co., 1967.)	
Second	Steward, Frederick C., ed. Plant Physiology: A Treatise. New York; Academic Press, 1959–91. 10 vols. (Vols. 7–8 edited by F.C. Steward & R.G.S. Bidwell.)	
Second	Stewart, B. A., and D. R. Nielsen, eds. Irrigation of Agricultural Crops. Madison, Wisc.; American Society of Agronomy, 1990. 1,246p. (Agronomy Monograph Number 30)	Second
	Stewart, W. D. P., and J. R. Gallon, eds. Nitrogen Fixation; Proceedings of the Phytochemical Society of Europe Symposium, Sussex, Eng. Sept. 1979. London, Eng. and New York; Academic Press, 1980. 451p.	Second
Third	Steyn, Pieter S., ed. The Biosynthesis of Mycotoxins: A Study in Secondary Metabolism. New York; Academic Press, 1980. 432p.	
Third	Stirling, G. R., Biological Control of Plant Parasitic Nematodes. Wallingford; CAB International, 1991. 282p.	Third
Third	Stirton, C. H., and J. L. Zarucchi, eds. Advances in Legume Biology; Proceedings of the Second International Legume Conference, St. Louis, Mo., June 1986. St. Louis, Mo.; Missouri Botonical Garden, 1989. 842p. (Missouri Botanical Garden Monographs in Systematic Botany No. 29)	Second
	Stoll, Gary. Natural Crop Protection: Based on Local Farm Resources in the Tropics and Subtropics. 3d improved ed. Weikersheim, Germany; Margraf, 1988. 183p. (Tropical Agroecology No. 1) (Available in German, French, Spanish, and Thai.)	Second
	Stone, John F., and Wayne O. Willis, eds. Plant Production and Management Under Drought Conditions; Papers presented at the symposium, Tulsa, Okla., Oct. 1982, sponsored jointly by the Oklahoma State University Center for Water Research and others. Amsterdam, Holland and New York; Elsevier, 1983. 389p.	Second
Second	Stoskopf, Neal C. Cereal Grain Crops. Reston, Va.; Reston Pub. Co., 1985. 516p.	Third
First rank	Stover, Robert H. Banana, Plantain, and Abaca Diseases. Kew, Eng.; Commonwealth Mycological Institute, 1972. 316p.	First rank
Second	Stover, Robert H. Fusarial Wilt (Panama Disease) of Ba-	Second

Developed Countries Ranking		Third World Ranking
	nanas and Other Musa Species. Kew, Eng.; Commonwealth Mycological Institute, 1962. 117p. (Phytopathological Paper No. 4)	
First rank	Stover, Robert H., and Norman W. Simmonds. Bananas. 3d ed. Harlow, Eng.; Longman Scientific & Technical, 1987. 468p. (1st ed., by N.W. Simmonds. London; Longman, 1959. 466p. 1966 and 1982 eds., N.W. Simmonds.)	First rank
Second	Strickberger, Monroe W. Genetics. 3d ed. New York and London, Eng.; Macmillan, 1985. 842p. (1st ed., 1968. 868p.)	Third
Second	Stubbs, R. W., et al. Cereal Disease Methodology Manual. Mexico City; International Maize and Wheat Improvement Center, 1986. 46p.	Second
First rank	Stumpf, P. K., and E. E. Conn, eds. The Biochemistry of Plants: A Comprehensive Treatise. New York; Academic Press, 1980–1990. 16 vols.	First rank
Second	Stumpf, Paul K., J. Brian Mudd, and W. David Nes, eds. The Metabolism, Structure, and Function of Plant Lipids; Proceedings of the 7th International Symposium on Plant Lipids, University of California, Davis, 1986. New York; Plenum Press, 1987. 724p.	
First rank	Stuteville, D. L., and D. C. Erwin, eds. A Compendium of Alfalfa Diseases. 2d ed. St. Paul, Minn.; American Phytopathological Society, 1990. 100p. (1st ed., 1979 by Graham, Joseph H. et al. 65p.)	Second
	Stutte, C. A., ed. Plant Growth Regulators: Chemical Activity, Plant Responses, and Economic Potential. Washington, D.C.; American Chemical Society, 1977. 94p. (Advances in Chemistry Series No. 159)	Second
Third	Subba Rao, N-S, ed. Biological Nitrogen Fixation: Recent Developments. New York; Gordon and Breach, 1988. 337p.	Second
	Subramanian, Chirayathumadom V. Hyphomycetes: Taxonomy and Biology. U.S. ed. London and New York; Academic Press, 1983. 502p.	Third
First rank	Summerfield, R. J., and E. H. Roberts, eds. Grain Legume Crops. London, Eng.; Collins; Distributed in USA, Dobbs Ferry, N.Y.; Sheridan House, 1985. 859p.	Second
Third	Sussman, M., ed. The Release of Genetically-Engineered Micro-organisms; Proceedings of the 1st International Conference, . . . Cardiff, Wales, 1988, sponsored by Society for Applied Bacteriology. London, Eng. and San Diego, Calif.; Academic Press, 1988. 306p.	Second

Developed Countries Ranking		Third World Ranking
Third	Sutcliffe, James F., and J. S. Pate, ed. The Physiology of the Garden Pea. London, Eng. and New York; Academic Press, 1977. 500p.	
Second	Suzuki, David T. et al. An Introduction to Genetic Analysis. 4th ed. New York; W. H. Freeman, 1989. 768p. (1st ed., 1976. 468p.)	
Second	Swinburne, T. R., ed. Iron, Siderophores, and Plant Diseases; Proceedings of a NATO Advanced Research Workshop, . . . Wye, Eng., July 1985. New York; Plenum Press, 1986. 351p.	Third
Second	Sybesma, C., ed. Advances in Photosynthesis Research; Proceedings of the 6th International Congress on Photosynthesis, Brussels, Belgium, Aug. 1983. Hague, Netherlands and Boston, Mass.; Nijhoff/Junk; Distributors for the U.S. and Canada, Kluwer; Boston, 1984. 4 vols.	
	Symposium on Test Methods of Vertebrate Pest Control and Management Materials, 3d, Fresno, Calif., 1980. Proceedings, . . . sponsored by ASTM Committee E-35 on Pesticides, American Society for Testing and Materials, edited by E.W. Schafer and C.R. Walker. Philadelphia, Pa.; American Society for Testing and Materials, 1981. 197p. (ASTM Special Technical Publication No. 752)	Third
Second	Symposium on the Major Insect Pests of the Rice Plant, International Rice Research Institute, 1964. Proceedings . . . Baltimore, Md.; Johns Hopkins Press, 1967. 729p.	Third
Second	Tait, Joyce, and Banpot Napompeth, eds. Management of Pests and Pesticides: Farmers' Perceptions and Practices. Boulder, Colo.; Westview Press, 1987. 244p.	Second
	Talbott, I. D. Agricultural Innovation in Colonial Africa; Kenya and the Great Depression. Lewiston, N.Y.; E. Mellen Press, 1990. 188p.	Third
Third	Talekar, N. S., ed. Diamondback Moth Management; Proceedings of the 1st International Workshop, Tainan, Taiwan, Mar. 1985. Shanhau, Taiwan; Asian Vegetable Research and Development Center, 1986. 471p.	Third
Second	Talhouk, Abdul M. S. Insects and Mites Injurious to Crops in Middle Eastern Countries. Hamburg and Berlin, Germany; Verlag P. Parey, 1969. 239p.	
Third	Tanksley, Steven D., and Thomas J. Orton, eds. Isozymes in Plant Genetics and Breeding. Amsterdam, Netherlands and New York; Elsevier, 1983. 2 vols.	Third
First rank	Tarr, Sydney A. J. Diseases of Sorghum, Sudan Grass and	Second

Developed Countries Ranking		Third World Ranking
	Broom Corn. Kew, Eng.; Commonwealth Mycological Institute, 1962. 380p.	
Third	Tashiro, Haruo. Turfgrass Insects of the United States and Canada. Ithaca, N.Y.; Comstock Pub. Associates, 1987. 391p.	
	Taylor, Albert L., and J. N. Sasser. Biology: Identification and Control of Root-Knot Nematodes (Meloidogyne Species). Raleigh, N.C.; Dept. of Plant Pathology, North Carolina State University, U.S. Agency for International Development, 1978. 111p.	Second
	Taylor, Howard M., Wayne R. Jordan, and Thomas R. Sinclair, eds. Limitations to Efficient Water Use in Crop Production. Madison, Wis.; American Society of Agronomy, 1983. 538p.	Third
Second	Taylor, N. L., ed. Clover Science and Technology. Madison, Wis.; American Society of Agronomy and Crop Science Society of America, 1985. 616p.	
Second	Teng, P. S., ed. Crop Loss Assessment and Pest Management. St. Paul, Minn.; American Phytopathological Society, 1987. 270p.	First rank
	Teng, P. S., and K. L. Heong, eds. Pesticide Management and Integrated Pest Management in Southeast Asia; Proceedings of the Southeast Asia Pesticide Management and Integrated Pest Management Workshop, Pattaya, Thailand, Feb. 1987. College Park, Md.; Consortium for International Crop Protection, 1988. 473p.	
Second	Terra, G. J. A. Tropical Vegetables. Vegetable Growing in the Tropics and Subtropics Especially of Indigenous Vegetables. Amsterdam, Holland; Koninklijk Instituut voor de Tropen, 1967. 108p.	Second
Third	Terry, E. R., et al, eds. Tropical Root Crops: Production and Uses in Africa; Proceedings of the 2nd Triennial Symposium of the International Society for Tropical Root Crops, Africa Branch, Douala, Cameroon, Aug. 1983. Ottawa, Canada; International Development Research Centre, 1984. 231p. (Abstract in English, French, and Spanish.)	Second
	Terry, E. R., K. A. Oduro, and F. Caveness, eds. Tropical Root Crops: Research Strategies for the 1980s; Proceedings of the 1st Triennial Root Crops Symposium of the International Society for Tropical Root Crops—Africa Branch, Ibadan, Nigeria, Sept. 1980. Ottawa, Canada; International Development Research Centre, 1981. 279p.	Third

Developed Countries Ranking		Third World Ranking
	Terry, P. J., and R. W. Michieka. Common Weeds of East Africa = Magugu ya Afrika Mashariki. Rome; Food and Agriculture Organization, 1987. 184p. (English and Kiswahili.)	Third
	Thampan, Palakasseril K. Handbook on Coconut Palm. New Delhi, India; Oxford and IBH, 1981. 311p. (Revised ed. of The Coconut Palm and Its Products, 1975)	Third
Second	Theberge, Robert L., ed. Common African Pests and Diseases of Cassava, Yam, Sweet Potato and Cocoyam. Ibadan, Nigeria; International Institute of Tropical Agriculture, 1985. 108p.	Second
Third	Thimann, Kenneth V. Hormone Action in the Whole Life of Plants. Amherst, Mass.; University of Massachusetts Press, 1977. 448p.	
	Thomas, Paul R., and C. E. Taylor. Plant Nematology in Africa South of the Sahara. Farnham Royal, Eng.; Commonwealth Agricultural Bureaux, 1968. 83p.	Second
Second	Thompson, Alonzo C., ed. The Chemistry of Allelopathy: Biochemical Interactions Among Plants. Washington, D.C.; American Chemical Society, 1985. 470p. (Based on a symposium sponsored by the Division of Pesticide Chemistry at the 187th Meeting of the American Chemical Society, St. Louis, Mo., April 1984.)	
Third	Thompson, James N., and J. M. Thoday, eds. Quantitative Genetic Variation. San Francisco, Calif.; Academic Press, 1979. 305p.	Third
Second	Thorne, Gerald. Principles of Nematology. New York; McGraw-Hill, 1961. 553p.	Second
Second	Thornley, J. H. M. Mathematical Models in Plant Physiology: A Quantitative Approach to Problems in Plant and Crop Physiology. London, Eng. and New York; Academic Press, 1976. 318p.	
Second	Thorold, C. Diseases of Cocoa. Oxford, Eng.; Clarendon Press, 1975. 423p.	Second
	Thorpe, Trevor A., ed. Plant Tissue Culture: Methods and Application in Agriculture; Proceedings of a Symposium based on the UNESCO Training Course on Plant Tissue Culture, . . . Sao Paulo, Brazil, Nov. 1978. New York; Academic Press, 1981. 379p.	Third
Second	Thurston, H. David. Sustainable Practices for Plant Disease Management in Traditional Farming Systems. Boulder, Colo.; Westview Press, 1991. 300p.	First rank
Second	Thurston, H. David. Tropical Plant Diseases. St Paul, Minn.; American Phytopathological Society, 1984. 208p.	First rank

Developed Countries Ranking		Third World Ranking
Third	Timbrell, John A. Introduction to Toxicology. London and New York; Taylor and Francis, 1989. 155p.	Third
Third	Timm, Robert M., ed. Prevention and Control of Wildlife Damage. Lincoln, Neb.; Great Plains Agricultural Council, Wildlife Resources Committee and Nebraska Cooperative Extension Service, Institute of Agriculture and Natural Resources, University of Nebraska—Lincoln, 1983. 1 vol.	
Third	Tindall, H. D. Vegetables in the Tropics. Westport, Conn.; AVI Pub. Co., 1983. 533p.	Second
Second	Ting, Irwin P., and Martin Gibbs, eds. Crassulacean Acid Metabolism; Proceedings of the 5th Annual Symposium in Botany, University of California, Riverside, Jan. 1982. Rockville, Md.; American Society of Plant Physiologists, 1982. 316p.	
Third	Tisselli, O., J. B. Sinclair, and T. Hymowitz. Sources of Resistance to Selected Fungal, Bacterial, Viral and Nematode Diseases of Soybeans. Champaign-Urbana, Ill.; College of Agiculture, University Illinois at Urbana-Champaign, 1980. 134p.	Second
	Todaro, Michael P. Economic Development in the Third World. 4th ed. New York; Longman, 1989. 698p. (1st ed. titled: Economic Development in the Third World: An Introduction to Problems and Policies in a Global Perspective, 1977. 440p.)	Second
Third	Tomes, D. T. Application of Plant Cell and Tissue Culture to Agriculture and Industry. Guelph, Ontario; University of Guelph, 1982. 231p.	Third
	Tomiyama, K., et al, eds. Biochemistry and Cytology of Plant-Parasite Interaction. Tokyo, Japan; Kodansha; Amsterdam, Holland and New York; Elsevier, 1976. 256p.	Third
Second	Toussoun, T. A., Robert V. Bega, and Paul E. Nelson, eds. Root Diseases and Soil-Borne Pathogens; Papers of a Symposium in conjunction with the 1st International Congress of Plant Pathology. Berkeley, Calif.; University of California Press, 1970. 252p.	Third
Second	Toussoun, T. A., and Paul E. Nelson. Fusarium: A Pictorial Guide to the Identification of Fusarium Species According to the Taxonomic System of Snyder and Hansen. 2d ed. University Park, Pa.; Pennsylvania State University Press, 1976. 43p.	Second
First rank	Trebst, A., et al, eds. Photosynthesis; 1. Photosynthetic Electron Transport and Photophosphorylation, edited by A. Trebst and M. Avron; 2. Photosynthetic Carbon Metabolism and Related Processes, edited by M. Gibbs and	First rank

Developed Countries Ranking		Third World Ranking
	E. Latko; 3. Photosynthetic Membranes and Light Harvesting Systems, edited by L.A. Staehelin and C.J. Arntzen. Berlin, Germany and New York; Springer-Verlag, 1977–. (Encyclopedia of Plant Physiology; New Series No. 5–6, 19)	
Second	Trinci, A. P. J., and J. F. Ryley, eds. Mode of Action of Antifungal Agents; Symposium of the British Mycological Society, University of Manchester, Manchester, Sept. 1983. Cambridge, Eng. and New York; Cambridge University Press, 1984. 405p.	
	Tropical Development and Research Institute. Pest Control in Tropical Onions. London; Tropical Development and Research Institute, 1986. 109p.	Third
Second	Tsunoda, S., K. Hinata, and C. Gomez-Campo, eds. Brassica Crops and Wild Allies: Biology and Breeding. Tokyo, Japan; Japan Scientific Societies Press, 1980. 354p.	Third
Third	Tsunoda, S., and Norindo Takahashi, eds. Biology of Rice. Tokyo, Japan; Japan Scientific Societies Press and Amsterdam, Holland and New York; Elsevier, 1984. 380p.	Third
Third	Tucker, Richard K., and D. Glen Crabtree. Handbook of Toxicity of Pesticides to Wildlife. Washington, D.C.; U.S. Bureau of Sport Fisheries and Wildlife, 1970. 131p. (U.S. Bureau of Sport Fisheries and Wildlife Resource Publication' No. 84)	
Third	Tummala, Ramamohan L., Dean L. Haynes, and Brian A. Croft, eds. Modeling for Pest Management: Concepts, Techniques, and Applications; Papers presented at the 2d US/USSR Symposium, Michigan State University, E. Lansing, Mich., Oct. 1974. East Lansing, Mich.; Michigan State University, 1976. 247p.	
Second	Turner, B. L., and Stephen D. Brush, eds. Comparative Farming Systems. New York; Guilford Press, 1987. 428p.	Second
Second	Turner, Neil C., and Paul J. Kramer, eds. Adaptation of Plants to Water and High Temperature Stress; Proceedings of a seminar, Carnegie Institution of Washington and Department of Plant Biology, Stanford, Nov. 1978. New York; Wiley, 1980. 482p.	
Second	Turner, Peter D. Oil Palm Diseases and Disorders. Kuala Lumpur, Malaysia; Oxford University Press, 1981. 280p.	Second
	Turner, Peter D., and R. A. Gillbanks. Oil Palm Cultiva-	Second

Developed Countries Ranking		Third World Ranking
	tion and Management. Kuala Lumpur, Malaysia; Incorporated Society of Planters, 1974. 672p.	
	Uhlir, P. F. and G. C. Carter, eds. Crop Modeling and Related Environmental Data; A Focus on Applications for Arid and Semiarid Regions in Developing Countries. Work of ICSU Committee on Data for Science and Technology. Paris; CODATA Commission on Global Change Data, 1994. 239p.	Third
Third	Ullrich, W. R., et al, eds. Inorganic Nitrogen Metabolism; Papers presented at the Advanced Course, . . . Universidad de Extremadura, 1986, organized by the Federation of European Societies of Plant Physiology. Berlin, Germany and New York; Springer-Verlag, 1987. 295p.	
	Ullrich, W. R., C. Rigano, A. Fuggi, and P. J. Aparicio, eds. Inorganic Nitrogen in Plants and Microorganisms: Uptake and Metabolism. Berlin, Germany; Springer-Verlag, 1990. 359p.	Third
Second	Ullstrup, A. J. Corn Diseases in the United States and Their Control. Washington, D.C.; Dept. of Agriculture, Agricultural Research Service, 1978. 55p. (USDA Agricultural Handbook No. 199)	
Third	Ungar, E. D. Management of Agropastoral Systems in a Semiarid Region. Wageningen; PUDOC, 1990. 221p.	Third
	United Nations Industrial Development Organization, Vienna. Formulation of Pesticides in Developing Countries. New York; United Nations Industrial Development Organization, Vienna, 1983. 217p.	Third
Second	United States. Agricultural Research Service. Index of Plant Diseases in the United States. Washington, D.C.; 1960. 531p. (USDA Agriculture Handbook No. 165)	
First rank	United States. Agricultural Research Service. Virus Diseases and Noninfectious Disorders of Stone Fruits in North America. Washington, D.C.; U.S. Agricultural Research Service, 1976. 433p. (USDA Agriculture Handbook No. 437)	
	United States. Department of Agriculture. Biological Control of Pests in China. Washinton, D.C.; U.S. Dept. of Agriculture, Office of International Cooperation and Development, Scientific and Technical Exchange Div., China Program, 1982. 266p.	Third
Second	United States. Department of Agriculture. Guidelines for the Control of Plant Diseases and Nematodes. Washington, D.C.; U.S. Dept. of Agriculture, Agricultural Re-	Third

Developed Countries Ranking		Third World Ranking
	search, 1986. 274p. (USDA Agriculture Handbook No. 656)	
Third	United States. Department of Agriculture. Plant Diseases. Washington, D.C., 1953. 940p. (USDA Yearbook of Agriculture 1953)	
Third	United States. Environmental Protection Agency. Agricultural Chemicals in Ground Water: Proposed Pesticide Strategy. Washington, D.C.; U.S. Environmental Protection Agency, 1987. 150p.	
Second	United States. Department of Agriculture, World Agricultural Outlook Board. Major World Crop Areas and Climatic Profiles. Washington, D.C.; U.S. Dept. of Agriculture, Joint Agricultural Weather Facility, 1987. 159p. (USDA Agriculture Handbook No. 664)	Second
	United States. Federal Grain Inspection Service. Official United States Standards for Grain. Rev. Jan. 1984. Washington, D.C.; The Service, 1984. 1 vol. (various pagings) (Earlier ed. by Agricultural Marketing Service. 1974. 61p.)	Third
Second	University of California. The Agromedical Approach to Pesticide Management. This compilation was prepared for the United States Agency for International Development by the UC/AID Pest Management and Related Environmental Protection Project at the University of California, Berkeley. Berkeley, Calif.; University of California, 1976. 113p.	Third
Second	University of California. Integrated Pest Management for Rice. Statewide Integrated Pest Management Project, Division of Agricultural Sciences, 1983. 94p.	Third
Second	University of California. Integrated Pest Management: The Principles, Strategies and Tactics of Pest Population Regulation and Control in Major Crop Ecosystems; Progress Report. Berkeley, Calif.; University of California, International Center or Biological Control, 1975. 2 vols.	Second
Second	University of California. Study Guide for Agricultural Pest Control Advisers on Vertebrate Pests. Berkeley, Calif.; Div. of Agricultural Sciences, University of California, 1976. 124p.	
Second	Unsworth, M. H., and D. P. Ormrod, eds. Effects of Gaseous Air Pollution in Agriculture and Horticulture; Proceedings of the 32d Easter School in Agricultural Science, University of Nottingham. London, Eng. and Boston, Mass.; Butterworth Scientific, 1982. 532p.	Third

Developed Countries Ranking		Third World Ranking
	Uphof, Johannes C. T. Dictionary of Economic Plants. 2d ed., rev. and enl. Lehre; J. Cramer and New York; Stechert-Hafner, 1968. 591p. (1st ed., 1959. 400p.)	Second
	Usher, George. A Dictionary of Plants Used by Man. London, Eng.; Constable, 1974. 619p.	Second
Second	Uvarov, Boris P. Grasshoppers and Locusts: A Handbook of General Acridology. Cambridge, Eng.; Cambridge University Press, 1966–77. 2 vols.	
	Valencia, Luis, ed. Memorias del Curso Sobre: Control Integrado de Plagas de Papa. Bogota, Colombia; Centro Internacional de la Papa, 1986. 200p.	Third
Second	Van den Bosch, Robert. The Pesticide Conspiracy. Garden City, N.Y.; Doubleday, 1978. 226p.	Third
	van der Heide, J., ed. Nutrient Management for Food Crop Production in Tropical Farming Systems; Proceedings of a Symposium, Universitas Brawijaya, Malang, Indonesia, Oct. 1987. Haren, Netherlands; Institute for Soil Fertility and Malang, Indonesia; Universitas Brawijaya, 1989. 394p.	Third
First rank	Van der Plank, J. E. Disease Resistance in Plants. 2d ed. Orlando, Fla.; Academic Press, 1984. 194p. (1st ed. New York; Academic Press, 1968. 206p.)	First rank
Second	Van der Plank, J. E. Genetic and Molecular Basis of Plant Pathogenesis. Berlin, Germany and New York; Springer-Verlag, 1978. 167p.	Second
First rank	Van der Plank, J. E. Host-Pathogen Interactions in Plant Disease. New York; Academic Press, 1982. 207p.	First rank
First rank	Van der Plank, J. E. Plant Diseases, Epidemics and Control. New York; Academic Press, 1963. 349p.	First rank
First rank	Van der Plank, J. E. Principles of Plant Infection. New York; Academic Press, 1975. 216p.	First rank
Third	Van Der Zwet, Tom, and Harry L. Keil. Fire Blight: A Bacterial Disease of Rosaceous Plants. Washington, D.C.; U.S. Dept. of Agriculture, 1979. 200p. (USDA Agriculture Handbook No. 510)	Third
Second	Van Emden, Helmut F., ed. Insect/Plant Relationships; A Symposium, Imperial College, London, Sept. 1971. Oxford; Blackwell Scientific, 1973. 215p.	
Second	Van Emden, Helmut F. Pest Control. 2d ed. London, Eng. and New York; E. Arnold, 1989. 117p. (1st ed., 1974 titled: Pest Control and Its Ecology. 59p.)	Second
Third	Vandermeer, John H. The Ecology of Intercropping. Cambridge, Eng. and New York; Cambridge University Press, 1989. 237p.	Second

Developed Countries Ranking		Third World Ranking
Third	Vasil, Indra K., ed. Cell Culture and Somatic Cell Genetics of Plants. Orlando, Fla.; Academic Press, 1984–91. 9 vols.	Third
Second	Vasil, Indra K., ed. Perspectives in Plant Cell and Tissue Culture. New York; Academic Press, 1980. 2 vols.	
Second	Vasil, Indra K., William R. Scowcroft, and Kenneth J. Frey, eds. Plant Improvement and Somatic Cell Genetics. New York; Academic Press, 1982. 300p.	Third
First rank	Vavilov, Nikolavi I. The Origin, Variation, Immunity and Breeding of Cultivated Plants. Translated by K. Starr Chester. Waltham, Mass.; Chronica Botanica Co., 1951. 364p. (Chronica Botanica No. 13)	First rank
	Veech, Joseph A., and Donald W. Dickson, eds. Vistas on Nematology: A Commemoration of the 25th Anniversary of the Society of Nematologists; Includes papers presented in symposia during the 1986 Annual Meeting of the Society of Nematologists, Orlando, Fla. Hyattsville, MD; Society of Nematologists, 1987. 509p.	Third
Second	Veeger, C., and W. E. Newton, eds. Advances in Nitrogen Fixation Research; Proceedings of the 5th International Symposium on Nitrogen Fixation, Noordwijkerhout, The Netherlands, Aug.—Sept. 1983. Hague, Netherlands and Boston, Mass.; M. Nijhoff and W. Junk, 1984. 760p.	
	Veltkamp, H. J. Physiological Causes of Yield Variation in Cassava (Manihot Esculenta Crantz). Wageningen, Netherlands; Agricultural University Wageningen, 1986. 103p.	Second
Third	Verma, D. P. S., and Normand Brisson, ed. Molecular Genetics of Plant-Microbe Interactions; Proceedings of the 3d International Symposium on the Molecular Genetics of Plant-Microbe Associations, Montreal, Quebec, Canada, July 1986. Dordrecht and Boston, Mass.; M. Nijhoff, 1987. 338p.	
Second	Verma, D. P. S., and T. Hohn, eds. Genes Involved in Microbe-Plant Interactions. Vienna, Austria and New York; Springer-Verlag, 1984. 393p.	Third
Third	Vettorazze, G. International Regulatory Aspects for Pesticide Chemicals. Vol. 1. Toxicity Profiles. Boca Raton, Fla.; CRC Press, 1979. 216p.	Third
	Vickery, Margaret, and Brian Vickery. Plant Products of Tropical Africa. London and Basingstoke; The Macmillan Press Ltd., 1979. 116p.	Third
	Vijaysegaran, S., and A. G. Ibrahim, eds. Fruit Flies in the Tropics; Proceedings of the 1st International Sympo-	Third

Developed Countries Ranking		Third World Ranking
	sium, Kuala Lumpur, Malaysia, Mar. 1988. Kuala Lumpur, Malaysia; Malaysian Agricultural Research and Development Institute (MARDI) and Malaysian Plant Protection Society, 1991. 430p.	
Third	Viktorov, Georgii A. Ekologiia Parazitov-Entomofagov (Ecology of Parasitic Entomophages). Moscow; Nauka Publ., 1976. 150p. (In Russian.)	
	Villareal, Ruben L. Tomatoes in the Tropics. Boulder, Colo.; Westview Press, 1980. 174p.	Second
Third	Vince-Prue, Daphne. Photoperiodism in Plants. London and New York; McGraw-Hill, 1975. 444p.	
	Vincent, J. M., A. S. Whitney, and J. Bose, eds. Exploiting the Legume—Rhizobium Symbiosis in Tropical Agriculture; Proceedings of a Workshop, Kahului, Hawaii, Aug. 1976. Honolulu, Ha.; College of Tropical Agriculture, Department of Agronomy and Soil Science, University of Hawaii, 1977. 469p. (College of Tropical Agriculture. Miscellaneous Publication No. 145)	Second
Third	Vloten-Doting, Lous van, Gert S. P. Groot, and Timothy C. Hall, eds. Molecular Form and Function of the Plant Genome; Proceedings of a NATO Advanced Study Institute-FEBS Advanced Course, Renesse, Netherlands, July 1984. New York; Plenum Press, 1985. 693p.	
Third	Vose, P. B., and S. G. Blixt, eds. Crop Breeding: A Contemporary Basis. Oxford, Eng. and New York; Pergamon Press, 1984. 443p.	
Second	Vreden, Gerrit van, and Abdul L. Ahmadzabidi. Pests of Rice and Their Natural Enemies in Peninsular Malaysia. Wageningen, Netherlands; Centre for Agricultural Publishing and Docuemntation (PUDOC), 1986. 230p.	Third
First rank	Waage, Jeff, and David Greathead, eds. Insect Parasitoids; 13th Symposium of the Royal Entomological Society of London, Dept. of Physics Lecture Theatre, Imperial College, London, Sept. 1985. London, Eng. and Orlando, Fla.; Academic Press, 1986. 389p.	
Second	Wagner, Sheldon L. Clinical Toxicology of Agricultual Chemicals. Park Ridge, N.J.; Noyes Data Crop., 1983. 306p.	
Third	Waibel, Hermann. The Economics of Integrated Pest Control in Irrigated Rice: A Case Study from the Philippines. Berlin, Germany and New York; Springer-Verlag, 1986. 196p.	Third
Second	Walden, David B., ed. Maize Breeding and Genetics. New York; Wiley, 1978. 794p.	First rank
Second	Walkey, D. G. A. Applied Plant Virology. 2d ed. London,	Second

Developed Countries Ranking		Third World Ranking
	Eng. and New York; Chapman and Hall, 1991. 338p. (1st ed., New York; Wiley, 1985. 329p.)	
	Wall, Duncan, ed. Potentials for Field Beans and Other Food Legumes in Latin America; Seminar, Cali, Colombia, Feb.—Mar. 1973. Cali, Colombia; Centro Internacional de Agricultura Tropical, 1973. 388p.	Second
Second	Wall, Joseph S., and William M. Ross, eds. Sorghum Production and Utilization: Major Feed and Food Crops in Agriculture and Food Series. Westport, Conn.; AVI Pub. Co, 1970. 702p.	First rank
Second	Walter, Heinrich. Vegetation of the Earth and Ecological Systems of the Geo-Biosphere (Vegetationszonen und Klima), Translated from the 5th rev. German ed. by Owen Muise. 3d rev. and enl. ed. Berlin, Germany and New York; Springer-Verlag, 1985. 318p.	Second
Second	Wang, Jaw-kai, and Sally Higa, eds. Taro: A Review of Colocasia Esculenta and its Potentials. Honolulu, Ha.; University of Hawaii Press, 1983. 400p.	Third
	Wardlaw, Claude W. Banana Diseases: Including Plantains and Abaca. 2d ed. London, Eng.; Longman, 1972. 878p. (1st ed., 1961. 648p. Original version published in 1935 under title: Diseases of the Banana.)	Second
Third	Wardlaw, I. F., and J. B. Passioura, eds. Transport and Transfer Processes in Plants; Proceedings of a symposium, Canberra, Australia, Dec. 1975, under auspices of the U.S.-Australia Agreement for Scientific and Technical Cooperation. New York; Academic Press, 1976. 484p.	
Second	Ware, George W. Fundamentals of Pesticides: A Self-Instruction Guide. 3d ed. Fresno, CA; Thomson Publications, 1991. 307p. (1st ed., 1982. 257p.)	Third
Third	Ware, George W. The Pesticide Book. San Francisco, Calif.; W.H. Freeman, 1978. 197p.	
Second	Wareing, P. F., and I. D. J. Phillips. Growth and Differentiation in Plants. 3d ed. Oxford, Eng. and New York; Pergamon Press, 1981. 343p. (Previously published as: The Control of Growth and Differentiation in Plants, 1970. 303p.)	
	Watanabe, I., and Walter G. Rockwood, eds. Nitrogen and Rice; Papers presented at a symposium sponsored by the International Rice Research Institute. Los Banos, Philippines; International Rice Research Institute, 1979. 499p.	Second
Third	Waterson, A. P., and Lise Wilkinson. An Introduction to the History of Virology. Cambridge, Eng. and New York; Cambridge University Press, 1978. 237p.	Third

Developed Countries Ranking		Third World Ranking
First rank	Watkins, G. M., ed. Compendium of Cotton Diseases. St. Paul, Minn.; American Phytopathological Society, 1981. 87p.	First rank
Second	Watson, David L., and A. W. A. Brown, eds. Pesticide Management and Insecticide Resistance. New York; Academic Press, 1977. 638p.	Third
Second	Watson, Stanley A., and Paul E. Ramstad, eds. Corn: Chemistry and Technology. St. Paul, Minn.; American Association of Cereal Chemists, 1987. 605p.	Third
Third	Webb, C., and G. Hawtin, eds. Lentils. Farnham Royal, Eng.; Commonwealth Agricultural Bureaux and International Center for Agricultural Research in the Dry Areas, 1981. 216p.	Third
	Weber, Edward, Barry Nestel, and Marilyn Campbell, eds. Intercropping with Cassava; Proceedings of an International Workshop, Trivandrum, India, Nov.—Dec. 1978, co-sponsored by the Central Tuber Crops Research Institute (Indian Council for Agricultural Reseach) and the International Development Research Centre. Ottawa, Canada; International Development Research Centre, 1979. 142p.	Second
Second	Weber, George F. Bacterial and Fungal Diseases of Plants in the Tropics. Gainesville, Fla.; University of Florida Press, 1973. 673p.	First rank
	Weber, Hans C., and W. Forstreuter, eds. Parasitic Flowering Plants; Proceedings of the 4th International Symposium, . . . Philipps-Universitat, Marburg, Germany, Aug. 1987, sponsored by Gesellschaft fur Technische Zusammenarbeit. Marburg, Germany; Philipps-Universitat, 1987. 848p.	Third
Third	Webster, C. C., and W. J. Baulkwill, eds. Rubber. Burnt Mill, Eng.; Longman Scientific and Technical and New York; Wiley, 1989. 614p.	Third
Second	Webster, Cyril C., and P. N. Wilson. Agriculture in the Tropics. 2d ed. London, Eng. and New York; Longman, 1980. 640p. (1st ed., 1966. 488p.)	Second
Second	Webster, Robert K., and Pamela S. Gunnell, eds. Compendium of Rice Diseases. St. Paul, Minn.; American Phytopathological Society, 1992. 110p.	First rank
Second	Weed Science Society of America. Herbicide Handbook of the Weed Science Society of America. 6th ed. Champaign, Ill.; Weed Science Society of America, 1989. 301p. (1st ed., Geneva, N.Y.; W.F. Humphrey Press, 1967. 293p.) (Different authors with every edition.)	Third
Third	Wegler, Richard. Chemie der Pflanzenschutz- und Schad-	

Developed Countries Ranking		Third World Ranking
	lingsbekampfungsmittel. Berlin, Germany; Heidelberg and New York; Springer, 1970–82. 8 vols.	
Third	Wehmeyer, Lewis E. The Pyrenomycetous Fungi, edited by Richard T. Hanlin. Lehre, Germany; J. Cramer, 1975. 250p. (Mycologia Memoir No. 6)	Third
Third	Weir, David, and Mark Schapiro. Circle of Poison: Pesticides and People in a Hungry World. San Francisco, Cailf.; Institute for Food and Development Policy, 1981. 99p.	Third
Second	Weiss, E. A. Oilseed Crops. London, Eng. and New York; Longman, 1983. 660p.	Second
	Wellman, Frederick L. Coffee: Botany, Cultivation and Utilization. London, Eng. and New York; L. Hill Interscience Publishers, 1961. 488p.	Second
Second	Wellman, Frederick L. Dictionary of Tropical American Crops and Their Diseases. Metuchen, N.J.; Scarecrow, 1977. 495p.	Second
	Wellman, Frederick L. Tropical American Plant Disease. Metuchen, N.J.; Scarecrow Press, 1972. 989p.	Second
Third	West, S. H., ed. Physiological-Pathological Interactions Affecting Seed Deterioration; Proceedings of a symposium, Chicago, Ill., Dec. 1985, sponsored by the Crop Science Society of America. Madison, Wis.; Crop Science Society of America, 1986. 95p. (CSSA Special Publication No. 12)	
First rank	Westcott, Cynthia. Westcott's Plant Disease Handbook, rrevised by R. Kenneth Horst. 5th ed. New York; Van Nostrand Reinhold, 1990. 953p. (4th ed., 1979 by R. Kenneth Horst. 803p.)	Second
	Wharton, Clifton R., ed. Subsistence Agriculture and Economic Development. Chicago, Ill.; Aldine Pub. Co., 1969. 481p.	Second
Second	Wheeler, Bryan E. J. An Introduction to Plant Diseases. London, Eng. and New York; J. Wiley, 1969. 374p.	
	Wheeler, Harry. Plant Pathogenesis. New York; Springer-Verlag, 1975. 106p.	Second
Second	Whetzel, Herbert H. An Outline of the History of Phytopathology. New York; Arno Press, 1977. 130p.	Second
Third	Whipps, J. M., and R. D. Lumsden, eds. Biotechnology of Fungi for Improving Plant Growth; Symposium of the British Mycological Society, University of Sussex, Sept. 1988. Cambridge, Eng. and New York; Cambridge University Press, 1989. 303p.	
Third	Whitaker, Thomas W., and Glen N. Davis. Cucurbits: Botany, Cultivation, and Utilization. London, Eng.; L.	Second

Developed Countries Ranking		Third World Ranking
	Hill and New York; Interscience Publishers, 1962. 250p.	
Third	Whitcombe, John R., and William Erskine, eds. Genetic Resources and Their Exploitation: Chickpeas, Faba Beans, and Lentils. Hague, Netherlands and Boston, Mass.; M. Nijhoff/Junk Publishers, 1984. 256p.	Third
Third	White, Ian M., and Marlene M. Elson-Harrison. Fruit Flies of Economic Significance: Their Identification and Bionomics. Wallingford, Eng.; CAB International, 1992. 601p.	Third
	White-Stevens, R. Pesticides in the Environment. New York; Marcel Dekker, 1971–77. 3 vols.	Second
Third	Whitehead, David, and William S. Bowers, eds. Natural Products for Innovative Pest Management. Oxford Pergamon Press, 1983. 586p.	Second
Second	Whiteman, Peter C. Tropical Pasture Science. Oxford, Eng. and New York; Oxford University Press, 1980. 392p.	Third
Second	Whitney, E. D., and James E. Duffus, eds. Compendium of Beet Diseases and Insects. St. Paul, Minn.; American Phytopathological Society Press, 1986. 76p.	Third
	Whyte, Robert O., G. Nilsson-Leissner, and H. C. Trumble. Legumes in Agriculture. Rome; Food and Agriculture Organization, 1953. 367p. (FAO Agricultural Studies No. 21)	Second
	Wickens, G. E., J. R. Goodin, and D. V. Field, eds. Plants for Arid Lands; Proceedings of the Kew International Conference on Economic Plants for Arid Lands, Kew, Eng., July 1984. London, Eng. and Boston, Mass.; Allen and Unwin, 1985. 542p.	Third
Third	Wickens, G. E., N. Haq, and P. Day, eds. New Crops for Food and Industry. New York; Chapman and Hall, 1989. 444p.	Third
Second	Wicklow, Donald T., and George C. Carroll. The Fungal Community: Its Organization and Role in the Ecosystem. New York; M. Dekker, 1981. 855p.	Third
	Wiese, A. F., ed. Weed Control in Limited-Tillage Systems. Champaign, Ill.; Weed Science Society of America, 1985. 297p. (Monograph Series of the Weed Science Society of America No. 2)	Third
First rank	Wiese, Maurice V. Compendium of Wheat Diseases. 2d ed. St. Paul, Minn.; American Phytopathological Society Press, 1987. 112p. (1st ed., 1977. 106p.)	First rank
First rank	Wigglesworth, Vincent B. Insect Physiology. 8th ed. London and New York; Chapman and Hall, 1984. 191p. (1st ed., London; Methuen and Co., Ltd., 1934. 134p.)	Third

Developed Countries Ranking		Third World Ranking
Second	Wigglesworth, Vincent B. The Principles of Insect Physiology. 7th ed. London; Chapman and Hall, 1972. 827p. (1st ed., New York; E.P. Dutton and Co., Inc., 1939. 343p.)	
First rank	Wilcox, J. R., ed. Soybeans: Improvement, Production, and Uses. 2d ed. Madison, Wis.; American Society of Agronomy, 1987. 888p. (1st ed., B. E. Caldwell, et al., eds., 1973. 681p.)	First rank
Third	Wilding, James L., A. G. Barnett, and R. L. Amor. Crop Weeds. Melbourne, Australia; Inkata Press, 1986. 153p.	
	Wilken, Gene C. Good Farmers: Traditional Agricultural Resource Management in Mexico and Central America. Berkeley, Calif.; University of California Press, 1987. 302p.	Second
Third	Wilkins, Malcolm B., ed. Advanced Plant Physiology. London, Eng. and Marshfield, Mass.; Pitman, 1984. 514p.	Third
Second	Wilkins, Malcolm B. The Physiology of Plant Growth and Development. New York; McGraw-Hill, 1969. 695p.	
Second	Wilkins, Richard M., ed. Controlled Delivery of Crop Protection Agents. London and New York; Taylor and Francis, 1990. 322p.	
Second	Wilkinson, C. F., ed. Insecticide Biochemistry and Physiology. New York; Plenum Press, 1976. 768p.	Third
Third	Williams, C. N. The Agronomy of the Major Tropical Crops. Kuala Lumpur, Malaysia and New York; Oxford University Press, 1975. 228p.	Second
	Williams, C. N., W. Y. Chew, and J. H. Rajaratnam. Tree and Field Crops of the Wetter Regions of the Tropics. London, Eng.; Longman, 1980. 262p.	Second
	Williams, C. N., and K. T. Joseph. Climate, Soil and Crop Production in the Humid Tropics. Rev. ed. Kuala Lumpur, Malaysia and New York; Oxvford University Press, 1970. 177p.	Second
Second	Williams, J. R., et al., eds. Pests of Sugar Cane. Amsterdam, Holland and New York; Elsevier, 1969. 568p.	Third
Second	Williams, R. J., R. A. Frederiksen, and J. C. Girard. Sorghum and Pearl Millet Disease Identification Handbook. Hyderabad; International Crops Research Institute for the Semi-arid Tropics, 1978. 88p. (Information Bulletin No. 2) (Also available in French from ICRISAT.)	Third
	Willis, John C. A Dictionary of the Flowering Plants and Ferns, rev. by H.K. Airy Shaw. 8th ed. Cambridge, Eng.; Cambridge University Press, 1973. 1245p. (4th ed., 1919. 712p. 7th ed., 1966. 1214p.)	Second

Developed Countries Ranking		Third World Ranking
Second	Wilson, Charles L., and Charles L. Graham, eds. Exotic Plant Pests and North American Agriculture. New York; Academic Press, 1983. 522p.	
	Wilson, E. O., and Frances M. Peter, eds. Biodiversity; Papers from the National Forum on Biodiversity, Washington, D.C., Sept. 1986, under the auspices of the National Academy of Sciences and the Smithsonian Institution, and developed by the Board on Basic Biology of the National Research Council's Commission on Life Sciences and by the Smithsonian Institutions' Directorate of International Activities. Washington, D.C.; National Academy Press, 1988. 521p.	Second
Third	Wilson, John R., ed. Advances in Nitrogen Cycling in Agricultural Ecosystems; Proceedings of the Symposium, . . . Brisbane, Australia, May 1987. Wallingford, Eng.; CAB International, 1988. 451p.	Second
First rank	Wilson, John R., ed. Plant Relations in Pastures. East Melbourne, Australia; Commonwealth Scientific and Industrial Research Organization, 1978. 425p.	
Third	Wilson, Malcolm, and Douglas M. Henderson. British Rust Fungi. Cambridge, Eng.; Cambridge University Press, 1966. 384p.	
	Winkler, Albert J., et al. General Viticulture. Rev. and enl. ed. Berkeley, Calif.; University of California Press, 1974. 710p. (1st published 1962. 633p.)	Second
Third	Winteringham, F. P. W., ed. Environment and Chemicals in Agriculture; Proceedings of a symposium, Dublin, Oct. 1984. London, Eng. and New York; Elsevie; Distributor in the USA and Canada, Elsevier, 1985. 407p.	
Second	Wit, Cornelis T. de. Simulation of Assimilation, Respiration, and Transpiration of Crops. New York; Wiley, 1978. 140p.	
Second	Wittwer, Sylvan, et al. Feeding a Billion: Frontiers of Chinese Agriculture. E. Lansing, Mich.; Michigan State University Press, 1987. 462p.	Second
Third	Wohrmann, K., and S. K. Jain, eds. Population Biology: Ecological and Evolutionary Viewpoints. Berlin, Germany and New York; Springer-Verlag, 1990. 456p.	Second
First rank	Wolfe, M. S., and C. E. Caten, eds. Populations of Plant Pathogens: Their Dynamics and Genetics. Oxford, Eng. and Boston, Mass.; Blackwell Scientific Publication, 1987. 280p.	Second
Second	Wood, D. L., R. M. Silverstein, and M. Nakagima, eds. Control of Insect Behavior by Natural Products. New York; Academic Press, 1970. 345p.	Third

Developed Countries Ranking		Third World Ranking
Second	Wood, D. R., ed. Crop Breeding. Madison, Wis.; American Society of Agronomy and Crop Science Society of America, 1983. 294p.	First rank
	Wood, George A. R., and R. A. Lass. Cocoa. 4th ed. London, Eng. and New York; Longman, 1985. 620p. (1st and 2d eds. by Duncan H. Urquhart. 1st ed., 1955. 230p.)	Second
	Wood, R. K. S., ed. Active Defense Mechanisms in Plants; Proceedings of a NATO Advanced Study Institute, Cape Sounion, Greece, Apr.—May 1980. New York; Plenum Press in cooperation with NATO Scientific Affairs Division, 1982. 381p. (NATO Advanced Study Institutes Series, Series A, Life Sciences No. 37)	Second
Second	Wood, R. K. S., A. Ballio, and A. Graniti, eds. Phytotoxins in Plant Diseases; Proceedings of the NATO Advanced Study Institute, Pugnochiuso, Italy, June 1970. London, Eng. and New York; Academic Press, 1972. 530p.	
Second	Wood, R. K. S., and A. Graniti, eds. Specificity in Plant Diseases; NATO Advanced Study Institute, . . . Alghero, Sardinia, 1975. New York; Plenum Press, 1976. 354p. (NATO Advanced Study Institutes Series, Life Sciences No. 10)	Third
Second	Wood, R. K. S., and G. J. Jellis. Plant Diseases: Infection, Damage and Loss. Oxford, Eng. and Boston, Mass.; Blackwell Scientific, 1984. 327p.	Second
Second	Wood, R. K. S., and M. J. Way, eds. Biological Control of Pests, Pathogens and Weeds: Developments and Prospects; Proceedings of a Royal Society discussion meeting, Feb. 1987. London, Eng.; Royal Society, 1988. 266p.	Second
	Woodroof, Jasper G. Coconuts: Production, Processing, Products. 2d ed. Westport, Conn.; AVI Pub. Co., 1979. 307p. (1st ed., 1970. 241p.)	Second
Second	Woodroof, Jasper G., ed. Peanuts: Production, Processing, Products. 3d ed. Westport, Conn.; AVI Pub. Co., 1983. 414p. (1st ed., 1966. 291p.)	First rank
Third	Woodroof, Jasper G. Tree Nuts: Production, Processing, Products. 2d ed. Westport, Conn.; AVI Pub. Co., 1979. 731p. (1st ed., 1967. 2 vols.)	Second
Third	Woolhouse, H. W., ed. Dormancy and Survival. New York; Academic Press, 1969. 598p. (Its Symposia No. 23)	
	Workshop on Chemical Aspects of Rice Grain Quality, Los Banos, Philippines, 1978. Proceedings . . . Los Banos,	Second

Developed Countries Ranking		Third World Ranking
	Philippines; International Rice Research Institute, 1979. 390p.	
	Workshop on the Global Status of and Prospects for Integrated Pest Management of Root and Tuber Crops in the Tropics, Ibadan, Nigeria, 1987. Integrated Pest Management for Tropical Root and Tuber Crops; Proceedings . . ., edited by S. K. Hahn and F. E. Caveness, organized jointly by IITA, the International Center for Tropical Agriculture (CIAT) and the International Potato Center (CIP). Ibadan, Nigeria; International Institute of Tropical Agriculture, 1990. 235p.	Third
	Workshop on the Role of Legumes in the Farming Systems of the Mediterranean Areas, Tunis, Tunisia, 1988. The Role of Legumes in the Farming Systems of the Mediterranean Areas: Proceedings, . . . sponsored by UNDP/ICARDA, Tunis, June 1988, edited by A.E. Osman, M.H. Ibrahim, and M.A. Jones. Dordrecht and Boston, Mass.; Kluwer Academic Pub., 1990. 310p.	Third
	World Bank. Vetiver Grass: The Hedge Against Erosion. 3d ed. Washington, D.C.; World Bank, 1990. 78p.	Second
Third	World Soybean Research Conference III, Iowa State University, 1984. Proceedings, . . . edited by Richard Shibles. Boulder, Colo.; Westview Press, 1985. 1262p.	Third
Third	World Soybean Research Conference, 4th, Buenos Aires, Argentina, 1989. Proceedings, . . . edited by A.J. Pascale. IV Conferencia Mundial de Investigacion en Soja: Actas, Marzo 1989, Buenos Aires, Argentina, organizada por Asociacion Argentina de la Soja. Buenos Aires; Edotora S. R. L., 1990. 5 vols. 2388p.	Third
Second	Worthing, Charles R., and Raymond J. Hance, eds. The Pesticide Manual: A World Compendium. 9th ed. Thornton Heath, Eng.; British Crop Protection Council, 1991. 1141p. (1st ed.: Hubert Martin, ed. Pesticide Manual: Basic Information on the Chemicals Used as Active Components of Pesticides. London: British Crop Control Council, 1968. 464p.)	First rank
	Wren, Richard C., ed. Potter's New Cyclopaedia of Botanical Drugs and Preparations, rewritten by Elizabeth M. Williamson and Fred J. Evans. Compl. rev. ed. Saffron Walden; C.W. Daniel, 1988. 362p. (Earlier eds. titled: Potter's Cyclopaedia of Botanical Drugs and Preparations. 2d ed., 339p.)	Third
Third	Wright, Sewall. Evolution and the Genetics of Populations: A Treatise. Chicago, Ill.; University of Chicago Press, 1968–78. 4 vols.	Third

Developed Countries Ranking		Third World Ranking
Third	Wrigley, Gordon. Coffee. Harlow, Eng.; Longman Scientific and Technical; and New York; Wiley, 1988. 639p.	Second
	Wrigley, Gordon. Tropical Agriculture: The Development of Production. 4th ed. London, Eng. and New York; Longman, 1982. 496p. (1st ed., London; B.T. Batsford, 1961. 291p.)	Second
	Wuest, P. J., D. J. Royse, and R. B. Beelman, eds. Cultivating Edible Fungi; Proceedings of the International Symposium on Scientific and Technical Aspects of Cultivating Edible Fungi, Pennsylvania State University, July 1986. Amsterdam, Holland and New York; Elsevier, 1987. 677p.	Second
Second	Wyllie, T. D., and D. H. Scott, eds. Soybean Diseases of the North Central Region; Proceedings of the North Central Region Soybean Disease Workshop, Indianapolis, Ind., Mar. 1987, sponsored by the North Central Regional Committee on Soybean Diseases. St. Paul, Minn.; American Phytopathological Society Press, 1988. 149p.	Third
	Wyniger, Rene. Pests of Crops in Warm Climates and Their Control. 2d enl. and rev. ed. Basel, Germany; Verlag fur Recht und Gesellschaft, 1968. 161p. (1st ed., 1962. 555p.) (Acta Tropica; Zeitschrift fur Tropenwissenschaften und Tropenmedizin. Supplementum No. 7)	Third
Third	Yamaguchi, Mas. World Vegetables: Principles, Production, and Nutritive Values. Westport, Conn.; AVI Pub. Co.; and Chichester, Eng.; E. Harwood, 1983. 415p.	Second
Second	Yepsen, Roger B., ed. The Encylopedia of Natural Insect and Disease Control: The Most Comprehensive Guide to Protecting Plants—Vegetables, Fruit, Flowers, Trees, and Lawns—without Toxic Chemicals. Emmaus, Pa.; Rodale Press, 1984. 490p. (Rev. ed. of Organic Plant Protection, . . . 1976. 688p.)	Third
Second	Yepsen, Roger B., ed. Organic Plant Protection. Emmaus, Pa.; Rodale Press, 1976. 688p.	
Second	Yoshida, Shouichi. Fundamentals of Rice Crop Science. Los Banos, Philippines; International Rice Research Institute, 1981. 269p.	Third
	Young, Anthony. Tropical Soils and Soil Survey. Cambridge, Eng. and New York; Cambridge University Press, 1976. 468p.	Third
Second	Young, Harry M., and William Hayes. No-tillage Farming/ Minimum Tillage Farming. Brookfield, Wis.; No-till	Second

Developed Countries Ranking		Third World Ranking
	Farmer, 1982. 167p. (Original ed. titled: No-Tillage Farming by Shirley H. Phillips and H.M. Young. Milwaukee, Wis.; Reiman Associates, 1973. 224p.)	
First rank	Zadoks, Jan C. Epidemiology and Plant Disease Management. New York; Oxford University Press, 1979. 427p.	Third
Third	Zadoks, Jan C., and F. Rijsdijk. Atlas of Cereal Diseases and Pests of Europe. Wageningen, Netherlands; PUDOC, 1985. 169p.	Second
Second	Zaumeyer, W. J., and H. R. Thomas. A Monographic Study of Bean Diseases and Methods for Their Control. Washington, D.C.; U.S. Dept. of Agriculture, 1957. 255p. (USDA Technical Bulletin No. 868)	
Second	Zentmyer, George A. Phytophthora Cinnamomi and the Diseases it Causes. St. Paul, Minn.; American Phytopathological Society, 1980. 96p. (American Phytopathological Society Monograph No. 10)	Third
	Zethner, Ole, ed. Pest Management and the African Farmer; A Conference on Integrated Pest Management (IPM) in Africa, Nairobi, May 1989. Nairobi, Kenya; ICIPE Science Press and Academy Science Pub., 1989. 153p.	Third
Second	Zeven, A. C., and J. M. J. de Wet. Dictionary of Cultivated Plants and Their Regions of Diversity: Excluding Most Ornamentals, Forest Trees and Lower Plants. 2d ed., rev. Wageningen, Netherlands; PUDOC, 1982. 263p. (Rev. ed. of: Dictionary of Cultivated Plants and Their Centres of Diversity by A.C. Zeven and P.M. ZXhukovsky, 1975.)	Second
Second	Zillinsky, F. J. Common Diseases of Small Grain Cereals: A Guide to Identification. Mexico; Centro Internacional de Mejoramiento de Maiz y Trigo, 1983. 141p.	Second
	Zimdahl, R. L. Weed-Crop Competition: A Review. Corvallis, Or.; International Plant Protection Center, 1980. 196p.	Second
Third	Zimmermann, Ulrich, and Jack Dainty, eds. Membrane Transport in Plants; Proceedings of the International Workshop, . . . Julich, 1974. New York; Springer-Verlag, 1974. 473p.	
Second	Zohary, Daniel, and Maria Hopf. Domestication of Plants in the Old World: The Origin and Spread of Cultivated Plants in West Asia, Europe, and the Nile Valley. Oxford, Eng. and New York; Clarendon Press, 1988. 249p.	Third
	Zuber, M. S., E. B. Lillehoj, and B. L. Renfro, eds. Aflatoxin in Maize; Proceedings of the Workshop, El	Third

Developed Countries Ranking		Third World Ranking
	Batan, Mexico, Apr. 1986, sponsored by CIMMYT, UNDP and USAID. Mexico City, Mexico; International Maize and Wheat Improvement Center, 1987. 389p.	
Second	Zuckerman, B. M., W. F. Mai, and M. B. Harrison, eds. Plant Nematology Laboratory Manual. Rev. Amherst, Mass.; University of Massachusetts Agricultual Experiment Station, 1990. 252p. (1st ed., 1985. 212p.)	Third
First rank	Zuckerman, B. M., W. F. Mai, and R. A. Rohde, eds. Plant Parasitic Nematodes. New York; Academic Press, 1971–1981. 3 vols.	Second
Third	Zycha, Herbert, and R. Siepmann. Mucorales: Eine Beschreibung aller Gattungen und Arten Dieser Pilzgruppe. Lehre, Germany; J. Cramer, 1969. 355p.	

D. The Top Twenty Monographs

The top ranking monographs for both the developed countries and the Third World countries are displayed here in their ranking order. The first nine titles in the developed countries list are also highly ranked by the Third World, which demonstrates enormous agreement on which books are most valuable. Five of these top nine are concerned with pests and their control. This subject's importance is emphasized by the *CRC Handbook of Pest Management in Agriculture*, edited by David Pimentel, which is ranked first in both the Third World and the developed countries.

Twenty Top-Ranked Monographs for Developed Countries and Third World

Developed Countries Ranking		Third World Ranking
1	Pimentel, David, ed. *CRC Handbook of Pest Management in Agriculture*. 2d ed. Boca Raton, Fla.; CRC Press, 1991. 3 vols. (1st ed., 1981)	1
2	Horsfall, James G., and Ellis B. Cowling, eds. *Plant Disease: An Advanced Treatise*. New York; Academic Press, 1977–1980. 5 vols.	2–3
3	Fehr, Walter R., and Henry H. Hadley, eds. *Hybridization of Crop Plants*. Rev. ed. Madison, Wis.; American Society of	4

Developed Countries Ranking		Third World Ranking
	Agronomy and Crop Science Society of America, 1982. 765p. (1st published 1980.)	
4	Huffaker, Carl B., and P. S. Messenger, eds. *Theory and Practice of Biological Control*. New York; Academic Press, 1976. 788p.	7–8
5	Burges, Horace D. *Microbial Control of Pests and Plant Diseases 1970–1980*. London, and New York; Academic Press, 1981. 949p.	12–15
6	DeBach, Paul H., and Evert I. Schlinger, ed. *Biological Control of Insect Pests and Weeds*. London; Chapman and Hall, 1973. 844p. (1st published New York; Reinhold, and London; Chapman and Hall, 1964. 844p.)	12–15
7	Sprague, George F., and J. W. Dudley, eds. *Corn and Corn Improvement*. 3d ed. Madison, Wis.; American Society of Agronomy, 1988. 774p. (Agronomy No. 18) (1st ed., New York, Academic Press, 1955. 699p.)	2–3
8	Smartt, J. and Norman W. Simmonds, eds. *Evolution of Crop Plants*. 2d ed., Harlow, Essex; Longman Scientific and Technical, and New York; Wiley, 1995. (1st ed. edited by Norman W. Simmonds. London and New York; Longman, 1976. 339p.)	12–15
9	Baker, Kenneth F., and R. J. Cook. *Biological Control of Plant Pathogens*. St. Paul, Minn.; American Phytopathological Society, 1982. 433p. (Reprint. Originally published San Francisco; W.H. Freeman, 1974.)	5
10–11	Hoy, Marjorie A., and Donald C. Herzog, eds. *Biological Control in Agricultural IPM Systems; Proceedings of a Conference, Citrus Research and Education Center, University of Florida at Lake Alfred, June 1984*. Orlando, Fla.; Academic Press, 1985. 589p.	
10–11	Lyr, H., ed. *Modern Selective Fungicides: Properties, Applications, Mechanisms of Action*. 2d ed. Jena, Germany and New York; Gustav Fischer, 1994. 544p. (1st ed., Harlow, Eng.; Longman Scientific and Technical, and New York; Wiley, 1987. 383p.)	
12–13	Heyne, E. G., ed. *Wheat and Wheat Improvement*. 2d ed. Madison, Wis.; American Society of Agronomy, Crop Science Society of America, and Soil Science Society of America, 1987. 765p. (1st ed., edited by Karl S. Quisenberry, et al., 1967. 560p.)	9–10
12–13	Metcalf, Robert L., and William H. Luckmann, eds. *Introduction to Insect Pest Management*. 3d ed. New York; J. Wiley, 1994. 650p. (1st ed., 1975. 587p.)	
14–15	Baker, Ralph R., and Peter E. Dunn, eds. *New Directions in Biological Control: Alternatives for Suppressing Agricultural*	6

Developed Countries Ranking		Third World Ranking
	Pests and Diseases; Proceedings of a UCLA Colloquium, Frisco, Colo., Jan. 1989. New York; Alan R. Liss, Inc., 1990. 837p. (UCLA Symposia on Molecular and Cellular Biology, New Series No. 112)	
14–15	Papendick, R. I., P. A. Sanchez, and G. B. Triplett, eds. *Multiple Cropping; Proceedings of a Symposium, sponsored by the American Society of Agronomy, Crop Science Society of America, and Soil Science Society of America*. Madison, Wis.; American Society of Agronomy, 1976. 378p. (ASA Special Publication No. 27)	9–10
16	*Bergey's Manual of Determinative Bacteriology*. Edited by John G. Holt, et al. 9th ed. Baltimore, Md.; Williams & Wilkins, 1994. 787p.. (1923 ed. by D. H. Bergey. 442p. Also published with title: *Bergey's Manual of Systematic Biology*, edited by John G. Holt and Noel R. Krieg. 1984–1989. 4 vols.)	
17–18	Delp, Charles J., ed. *Fungicide Resistance in North America*. St. Paul, Minn.; American Phytopathological Society, 1988. 133p.	
17–18	Rasmusson, Donald C., ed. *Barley*. Madison, Wis.; American Society of Agronomy, 1985. 522p.	16–19
19–23	Charudattan, R. *Biological Control of Weeds with Plant Pathogens*. New York; J. Wiley, 1982. 293p.	
19–23	International Rice Research Institute. *Brown Planthopper: Threat to Rice Production in Asia; Proceedings of a symposium, IRRI, May 1977*. Los Banos, Philippines; International Rice Research Institute, 1979. 369p.	
19–23	Sinclair, James B., and P. A. Backman, eds. *Compendium of Soybean Diseases*. 3d ed. St. Paul, Minn.; American Phytopathological Society, 1989. 106p. (1st ed., 1975. 69p.)	
19–23	United States. Agricultural Research Service. *Virus Diseases and Noninfectious Disorders of Stone Fruits in North America*. Washington, D.C.; U.S. Agricultural Research Service, 1976. 433p. (USDA Agriculture Handbook No. 437)	
19–23	Vanderplank, J. E. *Disease Resistance in Plants*. 2d ed. Orlando, Fla.; Academic Press, 1984. 194p. (1st ed. New York; Academic Press, 1968. 206p.)	7–8
	Vanderplank, J. E. *Plant Diseases, Epidemics and Control*. New York; Academic Press, 1963. 349p.	11
	Burges, Horace D. *Microbial Control of Insects and Mites*. London and New York; Academic Press, 1971. 861p.	12–15
	Sauer, D. B., ed. *Storage of Cereal Grains and Their Products*. 4th ed. St. Paul, Minn.; American Association of Cereal Chemists, 1992. 615p. (American Association of Cereal Chemists Monograph Series No. 2) (1st ed. edited by J.A. Anderson and A.W. Alcock, 1954. 515p.; 3d ed. edited by Clyde M. Christensen, 1982. 544p.)	16–19

Developed Countries Ranking		Third World Ranking
	International Workshop on Intercropping, Hyderabad, India, 1979. Proceedings . . . Andhra Pradesh, India; International Crops Research Institute for the Semi-Arid Tropics, 1981. 401p.	16–19
	Ou, Shu H. *Rice Diseases*. 2d ed. Kew, Eng.; CAB Commonwealth Mycological Institute, 1987. 380p. (1st ed., 1972. 368p.)	16–19
	Hardy, R. W. F., and W. S. Silver, eds. *A Treatise on Dinitrogen Fixation*. New York; Wiley, 1977–1983. 3 vols.	20–21
	Zuckerman, B. M., W. F. Mai, and R. A. Rohde, eds. *Plant Parasitic Nematodes*. New York; Academic Press, 1971–1981. 3 vols.	20–21

Of the thirty monographs in this composite list, twelve were previously published in variant titles or editions. Five in the list are used extensively as textbooks, while an additional ten are used widely for supplemental classroom readings or as reference tools.

Of these top monographs, seventeen, or 57%, are currently published by commercial presses; nine, or 30%, are published by societies. The remaining four titles are published by international and governmental units. Surprisingly, there is not a single university press title, which points out the commercial application of this subject. Four publishers are very strong in this top twenty group, Academic Press with seven titles, the American Society of Agronomy and its sister crop institutions in Madison, Wis., with five titles, Wiley with four, and the American Phytopathological Society with three. All of the other publishers have one title each. The influence of U.S. societies is great in these top twenty as well as throughout the core monograph list. Academic Press is the leader in the crop science area, however, with 10.1% of all the monograph titles in the total list of 1,663.

Six of the thirty titles were published or originated in England; the remainder bear U.S. imprints except for one each from the Philippines and India. The median age of the merged top twenty titles is 1982.8, or ten years before this study was done. However, the range of years is great, with two titles from the 1960s and six from the 1970s. Of particular interest are the DeBach and Schlinger work from 1973 and Van der Plank's 1963 study. The DeBach and Schlinger is exceptional in having ranked in the top twenty in both the developed countries and the Third World. Both of these titles should be called modern classics along with the other titles which have gone through more than one edition. J. E. Van der Plank and Horace D.

Burges are the only authors with two titles each; neither has an edited work. Over half of the titles are edited works, several with more than one editor.

Among the top twenty in both communities were books on three individual crops: barley, corn, and wheat. All of these were published by the American Society of Agronomy or in conjunction with the Crop Science Society of America and the Soil Science Society of America. No comparable book on rice, the other major cereal, made the top twenty.

E. Nature of the Core Monograph Titles

The impact of Rachel Carson's 1962 book, *Silent Spring*, continues to be great. Evaluators of the core monographs in this chapter placed the book in the first and second ranks. The American public feels similarly. In 1993, twenty-two famous Americans from all walks of life judged it as "the most influential book" published in fifty years.[8] They chose Carson's book from a list of fifty titles including *Cancer Ward* and *A Room of One's Own*. *Silent Spring* also ranked in the core list of monographs on the forestry and agroforestry evaluation of the Core Agricultural Literature Project.

Commercial presses issued over half (53.2%) of the 1,663 titles in the entire list, or a greater number than all other types of publishers combined (see Table 8.2). Of the commercials, Academic Press provided 10.1% of all titles in the crop science core, a remarkable showing. The next highest publisher, with 5.1% of all titles, is also commercial, Wiley. The remaining publishers, each with over 2.0% of the titles, are primarily government organizations and societies. The Food and Agricultural Organization has 32 titles (1.9%) in the list and is not as strong a contender as it is in other agricultural subject areas. In addition to the American Phytopathological Society (2.7%), two other societies or non-profit organizations report well, the International Rice Research Institute in the Philippines and the American Society of Agronomy. Following Cambridge and Oxford University Presses, two other universities have a substantial number of titles: the University of California and Cornell University.

Determination of the country of publication was made by using the first city or country mentioned in the imprint of the publication. Although many large publishers have offices on nearly every continent, publishers tend to list the place of origination first; the data therefore reflects the origin of the

8. The Associated Press release appeared in many newspapers, among them the *Los Angeles Times*, 5 July 1993, Sec. B, under the title "New Echoes of the Silent Spring," and the *Christian Science Monitor*, July 7, 1993, p. 18, as "Silent Spring Today."

Table 8.2. Summary characteristics of core monographs

Type of publishers	
Commercial press	53.2%
Independent organization or society	19.1
University (press, department, institute)	15.1
Government (including FAO, UN)	12.6
Place of publication	
United States	47.6%
United Kingdom	23.5
Netherlands	5.7
Germany	5.2
India	2.5
Australia	2.4
Italy	2.2
Primary publishers	
Academic Press	10.1%
Wiley	5.1
U.K. Government (including CABI)	3.7
Springer	3.5
Cambridge University Press	3.3
Longman	3.2
U.S. Government	3.2
American Phytopathological Society	2.7
CRC Press	2.6
Elsevier	2.5
Oxford University Press	2.4
Plenum Press	2.2
Median age of publications	1982.5

vast majority of publications. The serious and long-standing work on plants and crops in India is demonstrated by the 2.5% of all monographs in the list coming from India.

F. Relationship of Core Holdings in Five Libraries

One of the purposes of these efforts is to determine the improvements that can be expected when the compact disks with full text of the core literature are issued. To assist in these determinations, the Core Agricultural Literature Project asked select libraries to check the core monograph lists against their library collections. Three libraries were chosen to represent library collections in the developed world and several to represent collections in the Third World, although only one from the latter group was returned fully completed. The data are these:

N = 1,050 (both lists)	Owned	Earlier eds. only
Nanking University	30.4%	3.4%
Kansas State University	75.7	4.3
Agricultural University, Wageningen, Neth.	86.9	3.0
Michigan State University	87.6	1.0

A high percentage of "earlier editions only" indicates a lack of currency. It can be argued that a second edition is nearly as valuable as a fourth edition by the same author and with the same title. It must be noted that of the 1,050 titles searched, all were searched as core whether for the developed countries or the Third World. Most of the titles not owned by the libraries of the Agricultural University at Wageningen, Kansas State University, and Michigan State University were those appropriate only for the Third World.

The composite monograph list was sent to the Library of the University of Zimbabwe, Harare, with a request that their holdings be checked. The Library instead sent a printout of all of the appropriate books in their library which covered the subjects represented in the crops list. This was matched in the Core Agricultural Literature Project against only the Third World monograph list. Counting any edition of a title, 128 of the 1282 core titles were owned by the University of Zimbabwe Library, or exactly 10% of the core titles for Third World institutions. The University of Zimbabwe will gain substantially by having the additional 90% of the titles in full text on compact disk one of the aims of the Project.

9. Primary Journals and the Core List

WALLACE C. OLSEN
Cornell University

Studies during the past twenty years concerned with the journal literature of crop improvement and protection have been fairly numerous. Many are either listings of monograph and journals or examinations of very specific subjects within the field. These generally do not give much assistance today because the titles and their importance change and many listings are not evaluative.

A. Source Documents and Methodology

The twenty-nine monographs and four journal articles used as source documents for the citation analysis for monographs (see Chapter 8) were also used to obtain data on journals, periodicals, serials and report series. The same methods, definitions, and caveats apply to the journal and serial data as outlined for monographs in Chapter 8. All journals, annuals, and select serials cited in the source documents were recorded and tabulated by title and date of publication. In these cases, each time an article within a journal was cited, a count was made for the journal or serial title. This provided a count of each time a journal or serial was cited in the entire 37,963 citations analyzed.

Proceedings volumes listed as journals or serials require clarification. Of the numerous proceedings identified, approximately 85% were tallied and evaluated as monographs because they had distinctive titles, short term or non-continuous editors, or concentrated on a specialized aspect of some area of crop science. Proceedings and annuals were counted as journals or serials when they represented the continuing deliberations of an organization or society, and when their subject foci and titles did not vary. Examples are the *Proceedings of the American Society of Horticultural Science*, which runs consistently under the same title each year, has no special subject issue, and includes technical papers on a variety of specialized subjects. Another example is *Advances in Agronomy*.

B. Journals Cited in the Source Documents

Table 8.1 (Chapter 8) indicates that 67.7% of the references in the analyzed documents were to journal articles. This is contrasted with the 86.4% of journal items indexed in the CABI database for crops. In contemporary literature of a science field, any figure of journal research literature under 75% is unusual. The concentration of this study, on both instructional and research literature, accounts for the lower journal percentage. Also influential is the analysis of titles concerned with Third World crop improvement and protection.

The sources of citations (Chapter 8) yielded 25,701 citations to journals or serials. This immense set of citations to journals identified 887 distinct titles. Of these, 100 to 150 garnered nearly all the citations, a common pattern, particularly in an application science such as agriculture. The top 10% (89) of all journals cited accounted for 59% of all the citations to journals and serials. The significance of the top three journals is demonstrated in Figure 9.1.

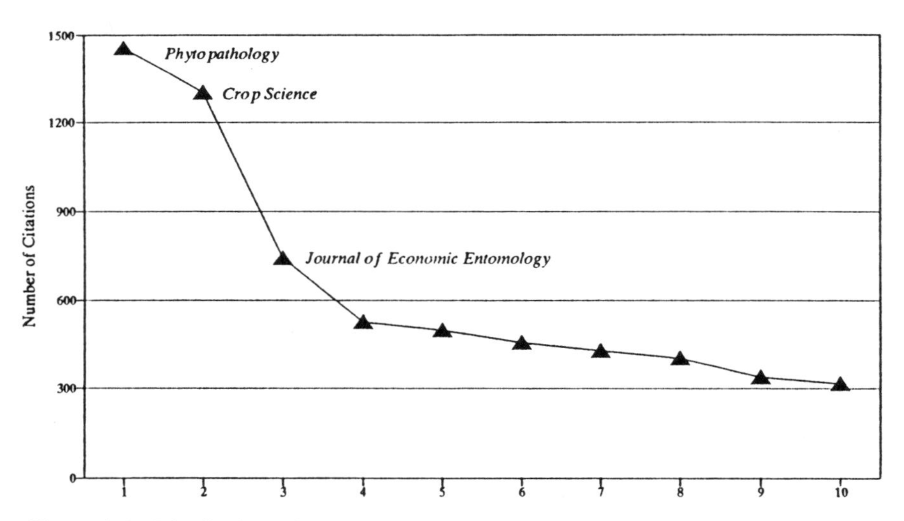

Figure 9.1. Distribution of top ten journals and serials.

Within the top 10% of journals there is considerable subject dispersion among the areas of crop science as well as general science. Two related areas show a concentration: science and general agricultural journals number thirteen, and environment/ecology four. Within the closely allied crop science subjects the specialities are:

General biology, biochemistry, and genetics	17
Plant diseases and physiology	15
Entomology, pests, weeds and controls	14
Agronomy, crop breeding and soils	12
General botany	7
Specific crops (fruits, oils, potatoes, rice, seeds)	6

One aim of this study was to determine the crop science journals most valuable for Third World teaching and research. To accomplish this, a subset of monographs written about or aimed at Third World crop science was analyzed. Eleven of these source documents were designated as relevant to the Third World only, and thirteen documents were designated as being appropriate for both the developed countries and the Third World. Twenty-four source titles were analyzed for the Third World and twenty for the developed countries. Table 9.1 ranks the top ninety-six titles showing the overall ranking at the left of the page and, to the immediate right, the rankings within the two subsets. Approximately 50% of the titles in each subset were common to both groups; this is largely the influence of the thirteen source documents deemed appropriate for both groups.

From these juxtaposed rankings, possible skews and influences within the two different geographic groups can be seen. For example, it is clear that the *Journal of Invertebrate Pathology*, which ranks twenty-sixth overall, has little or no influence in the Third World. Similarly, the *Journal of the American Oil Chemists' Society*, thirty-ninth, has heavy play in the Third World but not in the developed countries, although it is used extensively by food scientists in the developed countries and is a core journal for both groups in food science and human nutrition.[1]

The two right-hand columns of Table 9.1 need additional explanation. The *Science Citation Index* provides an analysis yearly of the citations it has recorded and counted from articles in 5,500 journals indexed.[2] These data are used extensively since there is no other large database which provides continuing counts on the numbers of times a journal is cited in the current literature. The complications with using *SCI* data are similar to those with

1. Jennie Brogdon and Wallace C. Olsen, eds. *Food Science and Human Nutrition: Contemporary and Historical Literature* (Ithaca, N.Y.: Cornell University Press, 1995), chapter 5.

2. *Science Citation Index* is issued by the Institute for Scientific Information, Philadelphia.

Table 9.1. Top journals cited by rank

Analysis Composite Rank	Journals	Core Citation Analysis		SCI[a]	
		Developed Countries	Third World	Average count[b]	Impact factor[c]
1	Phytopathology	1	1	15	27
2	Crop Science	2	2	22	59
3	Journal of Economic Entomology	3	5	21	43
4	Annual Review of Phytopathology	4	3	54	16
5	Agronomy Journal	5	6	23	47–48
6	Weed Science	12	7	44	44–45
7	Plant Disease Reports	6	4	NA	NA
8	Plant Diseases	7	14	41	37
9	Science	9	10	4	4
10–11	Environmental Entomology	10	17	37	40
10–11	Euphytica	8	13	56	61
12–13	Annals of Applied Biology	14	11	33	52
12–13	Nature	11	9	2	3
14	Plant Physiology	13	8	9	12–13
15	Economic Botany	45	16	66	64
16	Plant and Soil	15	12	32	56
17	Australian Journal of Agricultural Research	17	25	38	51
18	Proceedings of the National Academy of Sciences (USA)	16	15	3	5
19	Annual Review of Entomology	19	39–40	47	8
20	Canadian Journal of Plant Science	22	21	50	63
21	Advances in Agronomy	23	22	58	28
22	Journal of Bacteriology	18	18–19	6	11
23	Canadian Entomology	21	58	34	53–55
24–25	Applied Environmental Microbiology	24	28–29	16	18
24–25	Journal of Applied Ecology	44	18–19	48	38–39
26	Journal of Invertebrate Pathology	20	—[d]	45	36
27	Theoretical and Applied Genetics	25	28–29	43	25

28	Weed Research	89	23–24	61	57–58
29	Biochemica et Biophysica Acta	26	26–27	5	15
30–31	Canadian Journal of Botany	27–28	33–34	14	33–34
30–31	Oleagineaux	74	23–24	70	69
32	Annals of the Entomology Society of America	31	70–72	31	47–48
33	Journal of Biological Chemistry	27–28	20	1	6
34	International Rice Research Newsletter	75–76	33–34	NA	NA
35	Journal of Agricultural & Food Chemistry	29–30	—	17	24
36	Bulletin of Entomological Research	69	30	57	42
37	Entomophaga	42	42–44	67	65
38	Physiological and Molecular Plant Pathology	29–30	53–55	53	20
39	Journal of the American Oil Chemists' Society	—	26–27	27	33–34
40–41	Annual Review of Plant Physiology	34	35	25	1
40–41	PANS	—	31–32	NA	NA
42	Journal of Heredity	32–33	—	40	41
43–44	American Journal of Botany	56–58	49	20	29
43–44	Transactions of British Mycological Society	50	36–38	36	53–55
45	Journal of Nematology	39–40	—	60	38–39
46	Canadian Journal of Microbiology	32–33	42–44	24	30
47	Soil Biology and Biochemistry	35	39–40	39	31
48–49	American Potato Journal	39–40	—	62	53–55
48–49	Journal of Agricultural Science (Cambridge)	37	62–63	30	50
50	Indian Phytopathology	90	36–38	NA	NA
51–52	Journal of General Microbiology	36	65–67	13	19
51–52	Plant Pathology	43	50–51	65	66
53	Journal of Experimental Botany	38	59–60	28	23
54–56	Indian Farming	—	31–32	NA	NA
54–56	Journal of the American Society of Horticultural Science	46–47	—	35	46
54–56	Molecular and General Genetics (MGG)	41	45	12	10
57–58	Botanical Review	51–52	56–57	55	14
57–58	Seed Science Technology	—	41	69	70
59	Entomologia Experimentalis et Applicata	56–58	—	51	44–45
60	Physiologia Plantarum	48–49	46	19	22
61–63	Current Science	—	47	49	68

Table 9.1. Continued

Analysis Composite Rank	Journals	Core Citation Analysis		SCI[a]	
		Developed Countries	Third World	Average count[b]	Impact factor[c]
61–63	EMBO Journal	46–47	50–51	10	2
61–63	Pesticide Science	53–54	—	57	35
64–65	Annals of Botany (London)	51–52	—	29	32
64–65	HortScience	59–60	62–63	46	62
66–67	American Naturalist	61	—	18	12–13
66–67	FEBS Letter	48–49	42–44	7	9
68	Fruits	—	36–38	72	71
69–71	Indian Journal of Agricultural Science		65–67	68	72
69–71	Journal of the American Society of Agronomy	53–54	—	NA	NA
69–71	Phytopathologische Zeitschrift = Journal of Phytopathology	77–78	48	59	67
72	Madras Agricultural Journal	—	52	NA	NA
73–74	Ecology	70–71	68–69	11	17
73–74	Hilgardia	67–68	—	64	49
75–76	Journal of Environmental Quality	59–60	—	42	21
75–76	New Phytology	72–73	—	26	26
77–78	Field Crop Research	—	61	71	60
77–78	Zeitschrift für Pflanzenuchtung = Journal of Plant Breeding	62–63	—	63	57–58

79	Nucleic Acid Research	55	53–55	8	7
80–81	Oecologia	79–81	56–57		
80–81	Pest Biochemistry and Physiology	56–58	—		
82–85	Heredity	66	—		
82–85	Journal of Agricultural Research	70–71	—		
82–85	Planta	64–65	70–72		
82–85	Tropical Agriculture	—	64		
86	Biochemistry Journal	62–63	65–67		
87–88	European Journal of Biochemistry	67–68	59–60		
87–88	Plant Breeding Abstracts	64–65	53–55		
89–90	Australian Journal of Experimental Agricultural & Animal Husbandry	72–73	—		
89–90	Crop Protection	82			
91	Genetics	77–78	—		
92	Experimental Agriculture		70–72		
93–96	California Agriculture				
93–96	Proceedings of the American Society for Horticultural Science	75–76	—		
93–96	Proceedings of the Southern Weed Science Society	—	—		
93–96	Soil Science	85			

[a]*Science Citation Index*; Data in both columns were gathered for each title for the years 1979, 1982, and 1988 from *SCI Journal Citation Reports* and then ranked.
[b]Average of citations to the journals in *Journal Citation Reports (SCI)* for the years 1979, 1982, and 1988.
[c]Impact factors averaged from those supplied in *Journal Citation Reports* for the same three years.
[d]—in Developed Countries column = above rank of 90; —in Third World column = above rank of 72.

most statistical and indexing operations. The Core Agricultural Literature Project did the same type of analysis of source document literature but did not focus on journal research literature as does *SCI*. Instead, the Project looked at the literature of greatest value to both instruction and research today. The *SCI* citation analysis must be perceived as a reflection of research literature, while the Core Agricultural Literature Project has a broader focus. Therefore, the two columns must be read carefully with these parameters kept well in mind. A second precaution is that the Core Agricultural Literature Project set out to identify the most valuable literature for the Third World as well. No separate evaluation for the Third World can be derived easily from the *SCI* summary data. The *SCI* basically represents the literature cited by developed country literature with but a small percentage of the total coming from or applicable to the Third World. Stated another way, while the *SCI* data give quality returns on the international and developed countries research literature, they represent the Third World applications with less accuracy.

Another word of explanation. The influence of sheer numbers provides a complication in citation analysis because citation counts are often the direct result of the numbers of articles published or of pages published. The greater the number of articles or pages published per year, the greater the potential for being cited. Quality of the literature does influence citings heavily also, as well as growth and changes in the scope or specificity of the literature. *SCI* provides a measurement to somewhat equalize this problem by calculating an "impact factor." This is the division of the citations *to* the journal *by* the number of articles it published in the previous two years. An impact of 1.31 means that the number of citations in the literature to the journal exceeded the number of articles published in the cited journal in the previous two years; a factor of 1.0 indicates an equal number of both.

We are fortunate to have a recent summary of crops and soils data using *SCI* information by the creator of the database.[3] The work was done at the behest of the American Society of Agronomy and presented at an annual meeting symposium concerned with research ethics and journal quality. E. Garfield provides several comparisons and tables which are of interest to crop researchers including a list of the twenty most cited articles from four journals of the Society or its affiliates: *Agronomy Journal, Crop Science,*

3. Eugene Garfield, "The Effectiveness of American Society of Agronomy Journals: A Citationist's Perspective," in H. F. Mayland and R. E. Sojka, eds., *Proceedings of a Symposium on the Peer Review-Editing Process: Research Ethics, Manuscript Review, and Journal Quality* (Madison, Wis.: American Society of Agronomy, Crop Science Society of America, Soil Science Society of America, 1992). Garfield published the same data with additions in *Current Contents* no. 51 (Dec. 17, 1990). This was reprinted in his *Journalology, KeyWords Plus, and Other Essays* (Philadelphia: ISI Press, 1990 [*Essays of an Information Scientist: 1990*]), pp. 455– 467.

Journal of Environmental Quality, and *Soil Science Society of America Journal* for the period 1945–1988. Garfield identified sixty-six core journals in agronomy, including crops and soils. From the *SCI* data on scores from its core journals, the citations to the core journals, along with SCI impact factors, these eight core journals are ranked among the top fifteen in all three categories:

Agricultural and Biological Chemistry
Agronomy Journal
Australian Journal of Agricultural Research
Journal of Agricultural and Food Chemistry
Journal of Economic Entomology
Journal of Soil Science
Soil Biology and Biochemistry
Soil Science Society of America Journal.[4]

Fifteen years earlier, Garfield published a list of fifty highly cited botany journals which are mostly agricultural botany.[5]

CAB International abstracts about 20,000 citation on crop protection annually. These citations are drawn from about 2,300 journal or serial titles according to C. J. Hamilton, who also lists primary titles. Hamilton provides the fifty titles most often cited by *CAB Abstracts* for a two-year period under each of five crop protection subjects: biological control, entomology, plant nematology, plant pathology, and weed science.[6]

Among the many serial lists covering agriculture and crop improvement and protection, one of the most valuable is that of the libraries of the Biologische Bundesanstalt für Land- und Forstwirtschaft.[7] Now ten years old, the compilation covers 4,638 serials or journals. Its strongest point is its extensive and thorough coverage of biological agriculture and forestry titles from Eastern Europe, particularly the former Soviet Block countries.

E. O. Akinboro studied recent articles published by the scientists of the National Cereals Research Institute in Nigeria. One finding indicated that the NCRI "research scientists publish more in international than in local Nigerian journals. Out of the 206 articles analyzed, only 51 (24.8% of the

4. Garfield, "The Effectiveness," pp. 10– 11.

5. "Journal Citation Studies. 18: Highly Cited Botany Journals" in Eugene Garfield, *Essays of an Information Scientist*, vol. 3, 1974–1976. Philadelphia, Penn.; ISI Press, 1977. pp. 305–309. Originally published in *Current Contents* (Jan. 13, 1975, no. 2)

6. C. J. Hamilton, compiler, *Pest Management: A Directory of Information Sources, Volume 1: Crop Protection* (Wallingford, U.K.; CAB International, 1991), pp. 159– 165.

7. *Verzeichnis der Zeitschriften und Serien in den Bibliotheken der Biologischen Bunesanstalt fur Land- und Forstwirtschaft = List of Serials in the Libraries of the Federal Biological Research Centre for Agriculture and Forestry*. (Berlin; Paul Parey, 1984). (*Mitteilungen aus der Biologischen Bundesanstalt fur Land- und Forstwritschaft*. Heft 224)

total number) were published in Nigerian journals."[8] The top five international journals with nine or more contributions each from the NCRI staff were:

Tropical Agriculture
Plant Disease Reporter (currently *Plant Disease*)
Plant and Soil
Experimental Agriculture
Journal of Economic Entomology

C. Current Core Journal List

This recommended core list is not a match of the top ninety-six titles listed earlier. The primary reason for dropping a title is that it is no longer published. Table 9.1 ranks journals by the number of citations to the journal in the source literature over a twenty-five-year period, ending in 1992. The following list identifies and ranks those journals or serials currently published which constitute the core today. Each list has been divided into top-ranked titles (A) and second-ranked ones (B). Forty-one titles (53%) of all core titles are common to both developed and developing groups.

Core Journals for Developed and Third World Countries (N = 79)

	Developed Countries	Third *World*
Advances in Agronomy. Vol. 1 (1949)+. New York; Academic Press. Annual.	A	B
Agronomy Journal. Vol. 1 (1907/09)+. Madison, Wis.; American Society of Agronomy. Bimonthly.	A	A
American Journal of Botany. Vol. 1 (1914)+. Lawrence, Kansas; Botanical Society of America. Monthly.	B	B
American Naturalist. Vol. 1 (1867)+. Chicago, Ill.: Published by the University of Chicago Press for the American Society of Naturalists. Bimonthly.	B	—
American Potato Journal. Vol. 3 (1926)+. Orono, Maine; Potato Association of America. Monthly.	B	—
Annals of Applied Biology. Vol. 1 (1914/15)+. London; Association of Applied Biologists. Monthly.	A	A
Annals of Botany. New series, Vol. 1 (1937)+. London and New York; Academic Press. 5 no. a year.	B	—

8. E. O. Akinboro, "Analysis of Publications of Research Scientists of the National Cereals Research Institute, Nigeria," *IAALD Quarterly Bulletin* 35 (2) (1990): 78.

	Developed Countries	Third World
Annals of the Entomological Society of America. Vol. 1 (1908)+. Lanham, Md.; Entomological Society of America. Bimonthly.	B	—
Annual Review of Entomology. Vol. 1 (1956)+. Palo Alto, Calif.; Annual Reviews.	—	B
Annual Review of Phytopathology. Vol. 1 (1963)+. Palo Alto, Calif.; Annual Reviews.	A	A
Annual Review of Plant Physiology and Plant Molecular Biology. Vol. 39 (1988)+. Palo Alto, Calif.; Annual Reviews Inc. Annual. Continues: *Annual Review of Plant Physiology*.	B	B
Applied and Environmental Microbiology. Vol. 31, no. 1 (1976)+. Washington, D.C.; American Society for Microbiology. Monthly. Continues: *Applied Microbiology*.	A	B
Australian Journal of Agricultural Research. Vol. 1 (1950)+. Melbourne; Commonwealth Scientific and Industrial Research Organization. Bimonthly, 1974+.	A	B
The Biochemical Journal. Vol. 1 (1906)+. London; The Biochemical Society. Semimonthly.	B	—
Biochimica et Biophysica Acta. Vol. 1 (1947)+. Amsterdam; Elsevier. Monthly.	A	B
The Botanical Review; Interpreting Botanical Progress. Vol. 1 (1935)+. Bronx, New York; New York Botanical Garden. Quarterly.	B	—
Brighton Crop Protection Conference—Weeds. Proceedings. Vol. 1 (1989)+. Farnham, U.K.; British Crop Protection Council. Bimonthly. Continues: *British Crop Protection Conference— Weeds*.	B	—
Bulletin of Entomological Research. Vol. 1 (1910)+. Farnham Royal; Published by the Commonwealth Institute of Entomology for the Commonwealth Agricultural Bureaux. Quarterly.	B	B
The Canadian Entomologist. Vol. 1 (1868)+. Ottawa; Entomological Society of Canada. Monthly.	A	—
Canadian Journal of Botany. Vol. 29 (1951)+. Ottawa; National Research Council of Canada. Monthly 1981+, Semimonthly 1975–80, Monthly 1951–75. Continues: *Canadian Journal of Research*. Bimonthly 1929–50.	A	B
Canadian Journal of Microbiology. Vol. 1 (1954)+. Ottawa; National Research Council of Canada. Monthly.	B	B
Canadian Journal of Plant Science. Vol. 37 (1957)+. Ottawa; Agricultural Institute of Canada. Quarterly.	A	B
Crop Science. Vol. 1 (1961)+. Madison, Wis.; Crop Science Society of America. Bimonthly.	A	A
Current Science. Vol. 1 (1932)+. Bangalore; Published by the Indian Institute of Science for the Current Science Association. Semimonthly.	—	B

	Developed Countries	Third World
Economic Botany. Vol. 1 (1947)+. New York; New York Botanical Garden. Quarterly.	B	A
Entomologia, Experimentalis et Applicata. Vol. 1 (1958)+. Dordrecht, The Netherlands; Published by Kluwer Academic for Nederlandse Entomologische Vereniging. In English. Quarterly.	B	—
Entomophaga. Vol. 1 (1956)+. Paris; Le Francoise. Quarterly.	B	B
Environmental Entomology. Vol. 1 (1972)+. Lanham, Md.; Entomological Society of America. Bimonthly.	A	A
Euphytica; International Journal on Plant Breeding. Vol. 1 (1952)+. Dordrecht, Netherlands; Kluwer Academic Publishers. 18 times/year.	A	A
European Journal of Biochemistry. Vol. 1 (1967)+. Berlin, New York; Federation of European Biochemical Societies. Semimonthly.	B	—
The European Molecular Biology Organization Journal (EMBO J). Vol. 1 (1982)+. Oxford; IRL Press. Monthly.	B	B
Experimental Agriculture. Vol. 1 (1965)+ London and New York; Cambridge University Press. Quarterly. Supersedes: *Empire Journal of Experimental Agriculture*. (1933–1964)	—	B
FEBS Letters (Federation of European Biochemical Societies). Vol. 1 (1968)+. Amsterdam; North-Holland Pub. Co. Semimonthly.	B	B
Fruits; Fruits d'Outre-Mer. Vol. 1 (1945)+. Paris; Institut de Recherches sur les Fruits et Agrumes. Monthly (Supersedes: Fruits d'Outre Mer.)	—	B
Heredity; An International Journal of Genetics. Vol. 1 (1947)+. Edinburgh; Longman Group. Bimonthly.	B	—
Hilgardia. Vol. 1 (1925)+. Berkeley; California Agricultural Experiment Station. Irregular frequency.	B	—
HortScience. Vol. 1 (1966)+. Mt. Vernon, Va.; American Society for Horticultural Science. Bimonthly.	B	—
Indian Farming. New Series; Vol. 1 (1951)+. New Delhi; Indian Council of Agricultural Research. Monthly.	—	B
Indian Phytopathology. Vol. 1 (1948)+. New Delhi; Indian Phytopathological Society. Quarterly.	—	B
International Rice Research Notes. Vol. 18 (1993)+. Manila, Phillipines; International Rice Research Institute. Continues: *IRR Newsletter*. Quarterly.	—	B
Journal of Agricultural and Food Chemistry. Vol. 1, no. 1 (1953)+. Easton, Pa.; American Chemical Society. Bimonthly, 1960–.	A	—
The Journal of Agricultural Science. Vol. 1 (1905/06)+. Cambridge; Cambridge University Press. Bimonthly.	B	—
Journal of Applied Ecology. Vol. 1 (1964)+. Oxford; Blackwell Scientific Pub. Three times a year.	B	A

	Developed Countries	Third World
Journal of Bacteriology. Vol. 1. (Jan. 1985)+. Baltimore, Md.; American Society for Microbiology. Monthly.	A	A
Journal of Biological Chemistry. Vol. 1. (Oct. 1905)+. Baltimore; American Society of Biological Chemists. 3 times/month.	A	A
Journal of Economic Entomology. Vol. 1 (1908)+. Lanham, Md.; Entomological Society of America. Bimonthly.	A	A
Journal of Environmental Quality. Vol. 1 (1972)+. Madison, Wis.; Published cooperatively by American Society of Agronomy, Crop Science Society of America, and Soil Science Society of America. Quarterly.	B	—
Journal of Experimental Botany. Vol. 1 (1950)+. Oxford; Clarendon Press. Three times a year.	B	—
Journal of Genetics and Breeding; A Journal Devoted to Agricultural Genetics. Vol. 43 (1989)+. Rome; Istituto Sperimentale per la Cerealicoltura. Quarterly. Continues: *Genetica Agraria*.	B	B
Journal of Heredity. Vol. 1. (1910)+. New York; Oxford University Press for the American Genetic Association. Bimonthly. Continues: *American Breeder's Magazine*.	B	—
Journal of Invertebrate Pathology. Vol. 7 (1965)+. New York; Academic Press. Bimonthly. Continues: *Journal of Insect Pathology*.	A	—
Journal of Nematology. Vol. 1 (1969)+. Orlando, Fla.; Society of Nematologists. Quarterly.	B	—
Journal of Phytopathology = Phytopathologische Zeitschrift. Vol. 115 (1986)+. Berlin and Hamburg; Paul Parey. Quarterly. (Names were inverted prior to 1986, or in German only.)	—	B
Journal of the American Oil Chemists' Society. Vol. 57 (1980)+. Champaign, IL; American Oil Chemists' Society. Continues: *Journal of the American Oil Chemists Society*. Monthly.	—	B
Journal of the American Society for Horticultural Science. Vol. 94 (1969)+. Alexandria, Va.; American Society for Horticultural Science. Bimonthly. Continues: *Proceedings of the American Society for Horticultural Science*.	B	—
Microbiology. Vol. 140 (1994)+. London and New York; Cambridge University Press. Monthly. Continues: *The Journal of General Microbiology*.	B	—
Molecular and General Genetics. Vol. 99 (1967)+. Berlin and New York; Springer-Verlag. Irregular. Continues: *Zeitschrift fur Vererbungslehre*.	B	B
Mycological Research; An International Journal of Fungal Biology. Vol. 92 (1989)+. Cambridge; British Mycological Society and Cambridge University Press. Monthly. Continues: *Transactions of the British Mycological Society*.	B	B

	Developed Countries	Third World
Nature. Vol. 1 (1869)+. London; Macmillan. Weekly.	A	A
Nucleic Acids Research. Vol. 1 (Jan. 1974)+. Oxford; Information Retrieval Limited. Semimonthly, May 1979+.	B	B
Oleagineaux. Vol. 1 (Sept. 1946)+. Paris; Institut de Recherches pour les Huiles et Oleagineux. Monthly.	—	B
Pesticide Biochemistry and Physiology; An International Journal. Vol. 1 (Mar. 1971)+. Orlando, Fla.; Academic Press. 9 no. a year.	B	—
Pesticide Science. Vol. 1 (1970)+. Oxford; Blackwell Scientific Publications for the Society of Chemical Industry. Bimonthly.	B	—
Physiologia Plantarum. Vol. 1 (1948)+. Copenhagen; Munksgaard. Monthly.	B	B
Physiological and Molecular Plant Pathology. Vol. 28 (1986)+. London and Orlando, Fla.; Academic Press. Bimonthly. Continues: *Physiological Plant Pathology*.	A	B
Phytopathology. Vol. 1 (1911)+. St. Paul, Minn.; American Phytopathological Society. Bimonthly, 1911–17, Monthly 1918+.	A	A
Plant and Soil. Vol. 1 (1948)+. The Hague; Kluwer Academic. Bimonthly, 1976+.	A	A
Plant Breeding. Vol. 97 (1986)+. Berlin; Paul Parey. Bimonthly. Continues: *Zeitschrift für Pflanzenzüchtung*.	B	—
Plant Disease. Vol. 64 (1980)+. St. Paul, Minn.; American Phytopathological Society. Monthly. Continues: *Plant Disease Reporter;* Continues: *Plant Disease Bulletin*.	A	A
Plant Pathology. Vol. 1 (1952)+. Oxford; Blackwell Scientific Pub. Quarterly.	B	B
Plant Physiology. Vol. 1 (1926)+. Lancaster, Pa., American Society of Plant Physiologists. Monthly; 3 vols. a year.	A	A
Planta; An International Journal of Plant Biology. Bd. 1 (1925)+. Berlin; Springer Int. Monthly. Until 1988 had the subtitle: *Archiv für Wissenschaftliche Botanik*.	B	—
Proceedings of the National Academy of Sciences (U.S.). Vol. 1. (Jan. 1915)+. Washington, D.C.; National Academy of Sciences. Biweekly.	A	A
Science. New Series, Vol. 1, no. 1 (1895)+ Washington, D.C.; American Association for Advancement in Science. Weekly. Supersedes: Science; A Weekly Record of Scientific Progress; Absorbed: Scientific Monthly, Jan. 1958.	A	A
Seed Science and Technology. Vol. 1 (1973)+. Wageningen; International Seed Testing Association. Quarterly.	—	B
Soil Biology and Biochemistry. Vol. 1 (1969)+. Oxford; Pergamon Press. Bimonthly.	B	B
Theoretical and Applied Genetics. Vol. 1 (1929)+. Berlin; Springer. Bimonthly.	A	B

	Developed Countries	Third World
Weed Research. Vol. 1 (Mar. 1961)+. Oxford; Published by Blackwell Scientific Publications for the European Weed Research Society. Bimonthly 1974–.	—	B
Weed Science. (1968)+. Weed Science Society of America. Bimonthly since 1970, Quarterly 1968–69. Continues: *Weeds*.	A	A

These points must be noted about the core list of journals:

(1) The core list is not a match with the rankings in Table 9.1, left side of the table, or any other columns. Additional influences were incorporated, such as the inordinate weight of research journals, lack of substantive data about correlations between the developed countries and the Third World, shifts in subject coverage in the last decade, and reactions of collaborating educators in the field. Influential titles that have not been published for five years are not included as core.

(2) A total of seventy-nine journals are identified as core for the two combined groups. Forty-one of these constitute the most valuable journals, since they are in both communities; this is a 53% overlap of titles. Twenty-six additional journals are core for the developed countries, but only eleven more are core for the Third World. It must be emphasized that the literature of crop science in developing countries is relatively site-specific, small in quantity (except for Brazil and India), and dispersed in several formats, especially reports.

(3) The subject areas of the forty-one common journals demonstrate the broad scope of crop science and its close relationship to biology and biochemistry. Using the same subject divisions as for the titles in Table 9.1, these forty-one titles have the following subject concentrations:

General biology, biochemistry, and genetics	15
Plant diseases and physiology	10
Entomology, pests, weeds, and control	5
Agronomy, crop breeding, and soils	8
General botany	3

(4) There are only seventeen journals top rated A by both communities of scholars. These journals must be considered as the elite and most valuable for all of scholarly crop science throughout the world:

Agronomy Journal
Annals of Applied Biology
Annual Review of Phytopathology
Crop Science
Environmental Entomology
Euphytica
Journal of Bacteriology
Journal of Biological Chemistry
Journal of Economic Entomology

Nature
Phytopathology
Plant Disease
Plant Physiology
Plant and Soil
Proceedings of the National Academy of Sciences (U.S.)
Science
Weed Science

D. Observations on Trends

Within agricultural literature several trends in journal publishing are evident, particularly with crop improvement and protection journals. The roots of the changes are easily identified, and most result from economics.

The Dutch have been energetic publishers in agriculture for over fifty years. Many of the publications in the last twenty-five have come from commercial publishers in the Netherlands. Journal titles began to change thirty years ago from the use of Dutch and sometimes Latin to English. At first these had both English and another language (e.g. German) and more recently English only. This trend has accelerated within the past five years and spread to other countries, the *lingua franca* of science today. The trend can be seen in the Current Core Journal List, immediately preceding. The *Journal of Genetics and Breeding*, an Italian journal, was known before 1989 as *Genetica Agraria*. Similarly, *Phytopathologische Zeitschrift* went through 114 volumes before putting the English title first, *Journal of Phytopathology*, in 1986. Other German examples are *Molecular and General Genetics* and the well respected *Zeitschrift fur Pflanzenzuchtung*, which since 1986 has become *Plant Breeding*. Most of these changes also include greater use of English in the text. A recent French illustration is *Revue de Nematologie*, which in 1992 became *Fundamental and Applied Nematology* still published by ORSTOM, the French government agency.

Additional influences, both political and scientific, are changing the titles of journals as well. One is the effort to broaden the value and application of the journal by making known that it is "international" in coverage. The word is usually used at the beginning of a new or changed title or in subtitles. Three words are changing agricultural and biological titles today: "biotechnology," "molecular," and "microbiology." Biotechnology has not been around long enough in the crops area to be readily acceptable in changing journal titles, but the word is commonly used in the chemical and biological areas of agriculture. One illustration is *Physiological Plant Pathology*, which changed in 1986 to *Physiological and Molecular Plant Pathology*. The word "European" is beginning to appear more often in journal titles as journals seek to represent an assortment of these countries or their economic and professional interests.

A third major influence is publication in electronic format. This is not yet done in the crops field although some biology-related titles such as the *American Journal of Physiology* now have their article titles and abstracts online on the Internet as soon as the journal issue is published.[9] One of the first agricultural journals published only in electronic format is *Livestock Research for Rural Development*, distributed on floppy disks.[10] New or converted electronic publications will increase as scientists and publishers become more familiar with this new environment. Another action in electronic publishing is to substitute electronic media for print, dropping the latter completely. This was recently done by Agriculture Canada with its *The Dairy Market Report*, which was moved to an in-house electronic bulletin board system that necessitates special access arrangements with Agriculture Canada.[11] This latter move will not be an advancement for scholarship unless coupled with more ready access by scholars and libraries.

A more useful arrangement is the placement of a wealth of standard agricultural and economic statistics from the U.S.Dept. of Agriculture online on the Internet (gopher//:usda.mannlib.cornell.edu:3242) for free searching of 274 data sets covering as many subjects. This is a joint project of the USDA and the Albert R. Mann Library, Cornell University, and is an effort to make agricultural data readily available to scholars worldwide. Most of the data are published in continuing USDA publications, but the online version allows for data sets to be downloaded and manipulated at a scholar's workstation. This probably will alter the publishing in paper of some of the data in the future. In 1995 the files became available on the World Wide Web (http://www.mannlib.cornell.edu).

Arlene E. Luchsinger, writing from a background in plant pathology and several years of librarianship, has recently provided a summary of where the scientific journal may be headed and what the new technology's implications and options are for crop journals.[12]

9. Personal communication with Martin Frank, Executive Director of the American Physiological Society, July 1993.

10. Wallace C. Olsen, ed. *The Literature of Animal Science and Health* (Ithaca, N.Y.: Cornell University Press, 1993), p. 218.

11. Letter to the Albert R. Mann Library from Jaclyn Jerome, Agri-Food Development Branch, Agriculture Canada, dated September 21, 1993.

12. Arlene E. Luchsinger, "The Future of Scientific Communications: More Questions Than Answers," *Annals of the Missouri Botanical Garden* 80 (2) (1993): 297–303.

E. New Journal Titles

Growth of new and more specialized journals in crop science is demonstrated by the number of titles which have begun publication since 1980. Without exception, they are not represented in the top-ranked titles since these titles are relatively new and were not cited frequently in the source documents analyzed.

These titles are provided as a reference list which, although not exhaustive, covers most new scholarly journals or serials. Popular literature, such as newsletters, is excluded; titles which began before 1980 but changed names thereafter are not included.

Crop Improvement and Protection Journals, 1980+

Acta Agriculturae Boreali-Sinica. Vol. 8 (1993)+. Shijiazhuang, China; Acta Agriculturae Boreali Sincia, sponsored by Academies of Agricultural Sciences. Quarterly. Begun in an English ed. with a Latin title in 1993 of *Hua Pei Nung Hsueh Pao* (begun in 1986) to which it is a Suppl.

Advances in Plant Cell Biochemistry and Biotechnology. Vol. 1 (1992)+. London and Greenwich, Conn.; JAI Press. Annual.

Advances in Plant Pathology. Vol. 1 (1982)+. London and New York; Academic Press. Annual.

AG Chem New Compound Review. (1990)+. Indianapolis; Ag Chem Information Services. Annual. Supersedes *Agricultural Chemical New Products Development Review* begun in 1983.
"An annual review covering the latest new product developments in the world's crop protection industry." Provides nomenclature, chemistry, mode of action, usage and stage of development information. Gives manufacturers' addresses and lists of suspended and newly marketed products.

Agricultural Chemical Usage: Field Crops Summary. (1991)+. Washington, D.C.; USDA, National Agricultural Statistics Service, Economic Research Service. Annual. Also available on diskette.

Agricultural Chemical Usage: Fruits and Nuts Summary. (1991)+. Washington, D.C.; USDA, National Agricultural Statistics Service, Economic Research Service. Biennial.

Agricultural Chemical Usage. Vegetables Summary. (1990)+. USDA, Agricultural Statistics Board. Biennial.

Agronomie. Vol. 1 (1981)+. Paris; Institut National de la Recherche Agronomique. Formed by the union of *Annales de Phytopathologie; Anales de l'Amelioration des Plantes;* and *Annales Agronomiques*.

Agrotropica; Revista de Agricultura dos Tropicos Umidos = Journal of Agriculture for the Humid Tropics. Vol. 1 (1989)+. Iltheus, Brazil; Centro de Pesquisas do Cacau. Quarterly.

Applied Plant Science = Toegepaste Plantwetenskap. Vol. 1 (1987)+. Sunnyside; Southern African Weed Science Society. Two issues yearly.

Basic and Applied Mycology: BAM. Vol. 1 (1991)+. Padova, Italy; Unipress. Quarterly.

Biocontrol Science and Technology. Vol. 1 (1991)+. Abingdon, Eng.; Carfax Publishing. Quarterly.

Covers biological control of pests, weeds and plant diseases.

Biological and Cultural Test for Control of Plant Diseases: B & C Tests. (1986)+. St. Paul, Minn.; APS Press. Annual.

"Designed to improve biological and cultural testing and reporting. Each volume contains reports based on replicated disease control tests involving nonchemical means. Each volume includes biological and cultural tests for the following plant groups: Fruits and Nuts, Vegetables, Corn and Sorghum, Soybeans, Small Grains, Field Crops, Turfgrass, ornamentals and Trees, Citrus and Tropical."

Biological Agriculture and Horticulture. Vol. 1 (1982)+. Berkhamsted, Eng.; A. B. Academic Pub. Quarterly. The official journal of the International Institute of Biological Husbandry.

Biological Control: Theory and Applications in Pest Management. Vol. 1 (1991)+. San Diego, Calif.; Academic Press.

A muldidisciplinary journal devoted to the science and technology of biological control of pests. It includes the theory and application of the control of viruses, nematodes, insects, mites and other arthropods, weeds, and vertebrates in agricultural, aquatic, forest, natural resources, stored products and urban environments.

Biology and Fertility of Soils. Vol. 1 (1985)+. Berlin and New York; Springer International. Four issues yearly.; 2 vols. a year.

Biology of Metals. Vol. 1 (1988–91) New York; Springer-Verlag. Four nos. a year.

Conservation Biology; Journal of the Society of Conservation Biology. Vol. 1 (1987)+. Boston, Mass.; Blackwell Scientific Pub. Quarterly.

Critical Reviews in Plant Sciences. Vol. 1 (1983)+. Boca Raton, Fla.; CRC Press. Four issues yearly.

Covers, among others, the areas of molecular biology, molecular biochemistry, cell biology, plant physiology, genetics, classical botany, ecology, and agricultural applications. List of articles from recent earlier volumes in each issue.

Crop Protection. Vol. 1 (1982)+. Guilford, Eng.; Butterworth. Quarterly.

A quarterly journal whose major focus is the practical aspects of disease, weed and pest control in the field. Topics covered include: control of diseases caused by microorganisms; control of animal pests and weeds; assessment of pest and disease damage; epidemiology of pests and diseases in relation to control; pesticide development and application; biorational pesticides; effects of plant growth regulators; genetic engineering applications; biological control; integrated control; agronomic control; environmental effects of pesticides; and abiotic crop damage.

Current Topics in Plant Physiology. Vol. 1 (1989)+. Rockville, Md.; American Society of Plant Physiologists. Annual.

Diversity. Vol. 1 (1982)+. Washington, D.C.; Genetic Resources Communications Systems. Quarterly.

European Journal of Agronomy; The Journal of the European Society for Agronomy. Vol. 1 (1993)+. Montrouge Cedex, France; Gauthier-Villars.

FloraCulture International. (1990)+. Mt. Morris, Ill.; International Horticulture Publications. Bimonthly.

Fundamental and Applied Nematology. Vol. 15, no. 1 (1992)+. Montrouge Cedex, France; Gauthier-Villars and ORSTOM. Bimonthly. Continues: *Revue de Nematologie*. Biannual 1978–91.

Herbs, Spices, and Medicinal Plants. Vol. 1 (1986)+. Phoenix, Ariz.; Oryx Press. Annual.

The Horticulturist. Vol 1, no. 1 (1992)+. London; Institute of Horticulture. Quarterly. Formed by union of *Scientific Horticulture*, *Professional Horticulture*, and *IoH News*.

Indonesian Journal of Crop Science. Vol. 1., no. 1 (1985)+. Jakarta, Indonesia; Agency for Agricultural Research and Development, Ministry of Agriculture. Semiannual.

Indonesian Journal of Tropical Agriculture. Vol. 1 (1989)+. Bogor, Indonesia; Institut Pertanian Bogor. Semiannual.

Industrial Crops and Products; An International Journal. Vol. 1 (1992)+. Amsterdam; Elsevier. Quarterly.

International Journal of Genome Research. Vol. 1 (1991)+. Singapore; River Edge, NJ; World Scientific. Bimonthly.

International Journal of Tropical Plant Diseases. Vol. 1, no. 1 (1983)+. New Delhi, India; Today and Tomorrow's Printers and Pub.

Journal of Agricultural Entomology. Vol. 1 (1984)+. Clemson, S.C.; South Carolina Entomological Society. Quarterly.

Publishes research relating to insects and other arthropods of agricultural significance, including those affecting livestock and poultry as well as crops.

Journal of Applied Seed Production. Vol. 1 (1983)+. Corvallis, Oreg.; Oregon Seed Growers League, International Herbage Seed Production Research Group. Annual.

Journal of Plant Protection in the Tropics. Vol. 1 (1984)+. Kuala Lumpur; Malaysian Plant Protection Society. Semiannual.

Journal of Small Fruit and Viticulture. Vol. 1 (1992)+. Binghamton, N.Y.; Food Products Press. Quarterly.

Journal of Vegetable Crop Production. (1993)+. Binghamton, N.Y.; Haworth Press Inc. Quarterly.

Journal of Vegetation Science (JVS): Official Organ of the International Association for Vegetation Science. Vol. 1 (1990)+. Knivsta, Sweden; OPULUS Press.

Journal of Wine Research. Vol. 1 (1990)+. Abingdon, U.K.; Carfax Pub. Co. for The Institute of Masters of Wine. Three issues yearly.

The *Journal of Wine Research* is a new international and interdisciplinary journal publishing the results of recent research on all aspects of viticulture, oenology and the wine trade. Founded by the Institute of Masters of Wine to enhance and encourage scholarly and scientific interdisciplinary research in these fields. The main areas covered include archaeology, biochemistry, botany, classics, economics, geography, geology, history, medicine, oenology, psychology, sociology and viticulture.

Molecular Plant-Microbe Interactions: MPMI. Vol. 1 (1988)+. St. Paul, Minn.; APS Press. Monthly.

Oxford Surveys of Plant Molecular and Cell Biology. Vol. 1 (1984)+. Oxford; Clarendon Press. Annual.

Pakistan Journal of Weed Science Research; A Journal of Pakistan Weed Science Society. Vol. 1 (1989)+. Islamabad; Pakistan Weed Science Society, National Agricultural Research Centre. One or two issues a year.

The Peanut Grower. (1989)+. Tifton, Ga.; Agri-Publications. Monthly.

Plant Breeding Reviews. Vol. 1 (1983)+. Westport, Conn.; AVI. Twelve vols. by 1995.

The Plant Cell. Vol. 1 (1989)+. Rockville, Md.; American Society of Plant Physiologists. Monthly.

Plant Cell Reports. Vol. 1 (1981)+. Berlin and New York; Springer International. Six issues yearly.

Plant Cell, Tissue and Organ Culture. Vol. 1 (1981)+. The Hague and Boston; Nijhoff/Junk. Quarterly. An international journal on *in vitro* culture of higher plants.

Plant Genetic Resources Abstracts. Vol. 1 (1992)+. Wallingford, Eng.; CAB International in association with the International Board for Plant Genetic Resources. Quarterly.

Plant Growth Regulation. Vol. 1 (1982)+. The Hague and Boston; M. Nijhoff/W. Junk. Quarterly.
Research articles and some reviews from international scientists, with an agricultural focus. Beginning with volume 11, "papers on all aspects of the chemical, molecular, and environmental regulation of plant growth and development."

Plant Molecular Biology. Vol. 1 (1981)+. The Hague and Boston; Nijhoff/Junk. Quarterly.

Plant Varieties and Seeds. Vol. 1 (1988)+. Oxford; Blackwell Scientific for the National Institute of Agricultural Botany (U.K). Three nos. yearly.

Plant Varieties Journal. Vol. 1 (1988)+. Canberra, A.C.T.; Registrar of Plant Variety Rights, Bureau of Rural Science. Quarterly. Official journal of the Australian Plant Variety Rights Office.

Reviews of Pesticide Toxicity. Vol. 1 +. Raleigh, N.C.; Toxicology Communications, Inc. Irregular (v. 1, 1991; v. 2, 1993)

Revista Brasileira de Fisiologia Vegetal = Brazilian Journal of Plant Physiology. (1989)+. Londrina; Sociedade Brasileira de Fisiologia Vegetal. Twice a year.

Seed Pathology and Microbiology. Vol. 1 (1990)+. Wallingford, U.K.; CAB International; Hellerup, Denmark; Danish Govt. Institute of Seed Pathology for Developing Countries. Annual.
Compilation from the *CAB Abstracts* database prepared by the CAB International Bureau of Crop Protection and published in association with the Danish Government Institute of Seed Pathology for Developing Countries.

Seed Science Research. Vol. 1, no. 1 (1991)+. Wallingford, U.K.; CAB International. Quarterly.

Soilless Culture. Vol. 1 (1985)+. Wageningen, Netherlands; International Society for Soilless Culture. Semiannual.

Transactions of the Mycological Society of Republic of China. Vol. 1 (1985)+. Taipei, Taiwan. Irregular.

Trends in Biotechnology. Vol. 1 (1983)+. Amsterdam; Elsevier Science Pub. Bimonthly.

Trends in Cell Biology. Vol. 1 (1991)+. Cambridge; Elsevier Science Pub. Monthly.

Tropical Plant Science Research. Vol. 1, no. 1 (1983)+. New Delhi; Allied Publishers. Quarterly.

Vector. no. 1 (1986)+. Chapingo, Mexico; Sociedad Mexicana de Fitopatologia.

Weed Technology; A Journal of the Weed Science Society of America. Vol. 1 (1987)+. Champaign, Ill.; Weed Science Society of America. Quarterly.

World Journal of Microbiology and Biotechnology. Vol. 6, no. 1 (1990)+. Oxford, U.K.; Rapid Communications of Oxford Ltd. in association with UNESCO and in collaboration with the International Union of Microbiological Societies. Quarterly. Vol. 1–5 (1985–89) as *MIRCEN Journal of Applied Microbiology and Biotechnology*.

10. Reference Update

Barbara A. DiSalvo
Cornell University

Phyllis Reich*
University of Minnesota
*deceased 1994

Keeping alert to the numerous reference or working tools that continuously appear is a difficult task. Handbooks, cumulative indexes of long-term value, data collections and lengthy analyses of data, authoritative histories, and valuable bibliographies are not read chapter by chapter or as a reading assignment for a class. Rather, they provide reference points of authoritative data, literature citations, sources of detailed information, or general background.

Guide to Sources for Agricultural and Biological Research is the most recent major compilation of reference tools for the crop sciences. It was published in 1981 and edited by J. Richard Blanchard and Lois Farrell (Berkeley, University of California Press). This immense work serves as the basic guide to the literature of the agricultural sciences. Two very large chapters in the *Guide* covered the crop improvement and protection subjects: Chapter B (Plant Sciences) and Chapter C (Crop Protection—Pesticides and Pest Control). Chapter B listed 1,529 reference tools and Chapter C had 507. This chapter is a supplement and complement to those two chapters. The subject matter in this reference update follows the subject coverage of Blanchard and Farrell's chapters, but emphasis shifts to more current and emerging topics of critical interest to researchers and practitioners. The reference materials cited in this chapter are directed toward researchers, academicians, advanced students, and crop professionals.

This update chapter lists major reference work published since 1980. Older materials in the *Guide* are not replicated in this update. However, those titles which have been updated are noted in this chapter (the letters B and C refer to the item in Blanchard and Farrell). When a title is a direct successor, the reader may wish to consult the original volume for a complete annotation and description. Some important titles such as textbooks and handbooks are in the core monograph list of this book and are not included here as extensively as in Blanchard and Farrell. Titles in the Core Monograph List (Chapter 8) are identified in this chapter by the word CORE in the citation.

Quotes in descriptions of entries are from the literature provided by the literature creators, or in the publication being described.

Databases

Since the publication of Blanchard and Farrell's volume, both the number of resources once available only in printed format as well as entirely new compilations of information have proliferated at an incredible rate in a variety of electronic formats. The following indexes, abstracts, and statistical data sets listed in Blanchard and Farrell (reference in brackets) are now available electronically on CD-ROM:

Abstracts on Tropical Agriculture	[A136]	as TropAg & Rural
Agrindex	[A037]	as *AGRIS* [I020]
Bibliography of Agriculture (not all of *AGRICOLA*)	[A041]	as *AGRICOLA* [I019]
Biological & Agricultural Index	[A045]	[same title]
Census of Agriculture	[H192]	[same title]
Field Crop Abstracts	[B716]	as CABCD or HORTCD
Genetics Abstracts	[B872]	as CSA LifeSciences
Herbage Abstracts	[B787]	as CABCD of HORTCD
Horticultural Abstracts	[B448]	as CABCD or HORTCD
Plant Breeding Abstracts	[B945]	as CABCD
Potato Abstracts	[B567]	as CABCD or HORTCD
Production Yearbook (FAO)	[H168]	as *AGRIS* [I020]
Review of Plant Pathology	[B971]	as CABCD or CABPESTCD
Weed Abstracts	[C009]	as CABCD or CABPESTCD

Agricultural Statistics. Washington, D.C.: Slater Hall Information Products, 1989. 1 compact disk.
Data from the 1982 and 1987 *Censuses of Agriculture* and the complete *U.S. Bureau of the Census County* File, as well as farm income time series from 1969 to the present. Gives United States agricultural statistics, including acreage, operating expenses, production, sales, crops, and livestock. Available on CD-ROM; a users' guide provides instruction, examples, a list of data items. Searching software and menus are poorly done; cumbersome and expensive.

Agrochemicals Handbook. Cambridge: Royal Society of Chemistry, 1983+. 1 disk; also online, and available as a diskette.
Contains data on approximately 750 substances that are active components of

pesticides: herbicides, fungicides, insecticides, nematicides, acaricides, rodenticides, molluscides and bactericides; plant growth regulators, pest repellents and synergists. Provides chemical, physical, analytical and toxicity information. Print equivalent: *Agrochemicals Handbook*. Produced by the Royal Society of Chemistry and offered by DIALOG and Data Star.

Biotechnology Abstracts. London: Derwent Publications, 1982+. Online.
Currently available only online through commercial search services, this electronic version of the semimonthly publication, *Biotechnology Abstracts,* provides citations and abstracts of articles from about 1,100 international journals, conference proceedings, and patent literature. Subjects covered include genetic engineering, biochemical engineering, fermentation, tissue culture, plant cell culture, and waste disposal. Updated monthly.

CABPESTCD. London: SilverPlatter. 1 disk, first issued 1993. 1973+.
A database on CD-ROM compiled from the *CAB Abstracts* database relating to crop protection. It covers all aspects of crop pests and diseases. Includes all relevant information added to the *CAB Abstracts* database since 1973.

CD-GENE. Brisbane, Calif.: Hitachi America. 1 disk with semi-annual updates.
Numeric database of genetic sequencing data on DNA, nucleic acids, amino acids, and proteins. Combines information from GenBank, EMBL Data Library, NBRF-PIR Protein Sequence Database, and SWISS-PROT.

Current Biotechnology Abstracts. Cambridge: Royal Society of Chemistry, 1983+. 1 disk; also online.
Abstracts of over 400 journals, newsletters, patents, reports, proceedings, and monographs are available for the period 1983 to the present. Scientific, technical and commercial aspects of biotechnology are covered. Over 40,00 records are updated with an average 450 new citations monthly.

Entrez. Bethesda: National Center for Biotechnology Information, 1992+. On diskette. (GenInfo Compact Library Series)
". . . a retrieval system developed at the National Center for Biotechnology Information (NCBI) that provides an integrated approach for gaining access to nucleotide and protein sequence databases (GenBank, PIR, and SWISS-PROT), to the MEDLINE citations in which the sequences were published, and to a sequence-associated subset of MEDLINE." Must have Windows software to operate.

European Directory of Agrichemical Products. London: Royal Society of Chemistry, 1984+. 1 disk, also online
Data on agrochemicals registered for use and marketed in twenty-three European countries. Provides chemical, physical, analytical toxicity and manufacturer information. Covers herbicides, fungicides, nematicides, acaricides, rodenticides, molluscicides, bactericides, plant growth regulators, pest repellents and synergists. File size of 23,000 products. Print equivalent: *The European Directory of Agrochemical Products*.

HORTCD. Wallingford, U.K.: CAB International, 1992 with annual updates planned. 1 disk.
An historic and current subset of *CAB Abstracts*, this CD-ROM covers the world's horticultural literature from 1973 to 1991. Subjects include temperate tree fruit, vegetables, ornamental plants, minor industrial crops, small fruits, vi-

ticulture, plantation crops, and tropical fruits. The current abstracts in *HORTCD* also appear in the composite master file on CD of *CAB Abstracts*.

Maize Germplasm Bank Enquiry System. Mexico: International Maize and Wheat Improvement Center (CIMMYT), 1988. Diskette.
Available free from CIMMYT, the database provides passport data, in English, on 11,000 varieties of maize. Written in dBase III.

National Pesticide Information Retrieval System (NPIRS) Online by membership fee. Subset on 1 disk as *Pest-Bank* available from SilverPlatter.
Information on about 57,000 products registered with the EPA, as well as similar information from thirty-six states. Consists of the following eight files: EPA fact sheets; Material Safety Data Sheets (OSHA); Pesticide Document Management System (165,000 studies supporting EPA registration); Federal Product Database (about 40,000 products currently registered with the EPA); State Product Database; Pesticide Tolerance Index; Registration Standards/Data. Developed by the National Pesticide Information System at Purdue University. A subset which includes the registration numbers, composition, typical application sites, vulnerable pests, and maximum permanent residue levels of all pesticides registered in the United States is available as *Pest-Bank*.

PEST (Pesticide Education, Safety, and Training). Beltsville: National Agricultural Library, 1991. 1 disk.
A hypertext database designed for trainers of pesticide applicators. Contains information developed for the certification of pesticide applicators and links trainers to many resources such as contact names, bibliographies of citations on toxicity, water quality, and pest management from *Pest Management Principles for the Wisconsin Farmer* and *Safe and Effective Use of Pesticides*, and databases such as *AGRICOLA*, PEST-BANK, and TOXLINE. In addition, the EXTOXNET (Extension Toxicological Network) manual, which gives Pesticide Information Profiles (PIPs) and Toxicology Information Briefs (TIBs), is included. Available on diskette, menu-driven and easy-to-use despite lack of documentation other than that on the diskette. No updates planned.

Pest-Bank.
See *National Pesticide Information Retrieval System*.

Pest Management Research Information System (PRIS). Ottawa: Agriculture Canada, 1981+. Available as 1 disk or online from CCINFO.
A full-text, bibliographic directory and numeric database of information on pest and organism control products from their introduction into Canada to the time they are registered for use in Canada. Print equivalent: *Pesticide Information*.

Pesticide Use Data. Washington, D.C.: U.S. Department of Agriculture, Economic Research Service. Diskette in ASCII.
Data for major field crops, reporting active ingredients applied by crop and region, acres treated and total quantities applied.

The Pesticides Disc. London: Royal Chemical Society, 1988+. Available as 1 disk from Pergamon Press, Ltd. Updated semiannually.
Provides technical commercial data. Consists of the following files: Agrochemicals Handbook; European Directory of Agrochemical Products; Pesticides Index; and World Directory of Pesticide Control Organizations.

SATCRIS Database. Patancheru, India: International Crops Research Institute for the Semi-Arid Tropics, 1990. Batch access only from ICRISAT.

English language bibliographic database of citations and abstracts to the international literature on sorghum, millets, chickpeas, pigeonpeas, groundnuts, and farming systems in the semi-arid tropics. Updated continuously from the *CAB Abstracts* and *AGRIS* databases as well as from books, theses, reports, and nonconventional literature acquired at the ICRISAT library, which is the only place the database is available.

SOYA. Lafayette, Calif.: The Soyfoods Center, 1985+. Batch access only from Center.

Online version of *Bibliographies of Soya Series*. Contains over 20,000 records (80% in English) dating from B.C. to the present on all aspects of soybean utilization, history, nutrition, processing, marketing, and production.

Walker, M. M. and L. H. Keith. *EPA's Pesticide Fact Sheet Database*. Boca Raton, Fla.: Lewis Publishers, 1992. Diskette.

"A comprehensive source of information on several hundred pesticides and pesticide formulation. The database provides: chemical descriptors, which include chemical name, CAS number, synonyms and trade names, year of registration; United States and foreign producers; use patterns and formulation; chemical characteristics; acute toxicological data; chronic toxicological data; physiological and biochemical characteristics; environmental characteristics; ecological characteristics; tolerance assessments for agricultural crops; required labeling regulations; and major data gap summaries." Available on 3 1/2 and 5 1/4 inch diskettes for use on IBM or IBM- compatible computers.

World Agriculture Supply and Disposition (WASD). Bala Cynwyd, Pa.: The WEFA Group, data from 1960 to the present. Diskette. Updated quarterly.

Data covering stocks, areas harvested, yield, production, domestic use, imports, and exports of agricultural commodities including coffee, cotton, grains, oilseeds, and sugar for more than 195 countries. Also includes world totals for rice, wheat, coarse grains, and soybeans. Data are gathered from the USDA Foreign Agricultural Service. Available on diskette as *WADB*, special software enables the user to print balance sheet format reports for any country and commodity.

World WeatherDisc: Climate Data for the Planet Earth. Seattle: WeatherDisc Associates, 1988. 1 disk.

Seven data sets and software providing: world monthly surface station climatology, including temperature, precipitation, and pressure for 3,265 stations around the world, with some information dating back to the 1700s; worldwide airfield summaries, including temperature, precipitation, thunderstorms, humidity, and wind for 5,717 airports worldwide; United States monthly normals for temperature and precipitation from 1951 to 1980; climatography of the United States for 1862 weather stations in the conterminous United States; detailed local climatological data for the 288 primary stations in the United States; data on monthly-averaged temeperature, precipitation; and Palmer drought indices for several hundred United States climatic divisions from 1895 to 1988; and daily weather observations of maximum and minimum temperatures and total daily precipitation at 205 stations from the mid 1940s through the mid 1980s. Many data sets provide both graphical and numerical displays. Menu-driven and very easy to use.

Guides:

Allkin, R., and F. A. Bisby., eds. *Databases in Systematics*. London and Orlando, Fla.: Published for the Systematics Association by Academic Press, 1984. 329p. More than a directory to existing and future databases, this volume describes the impact of computer technology in the field of systematics. Articles cover general principles to consider in designing and creating a database, ways to take advantage of technology to store and manipulate statistical and bibliographic data, and ways to customize programs for specific uses. Software, such as TAXIR, and databases, including PRECIS, a databank of herbarium specimen information in the National Herbarium in Pretoria, South Africa, and BRASS BAND, the Brassicaceae data bank at the University of Notre Dame in Indiana, are described in detail.

Shurtleff, William, and Akiko Aoyagi. *Thesaurus for SOYA: Computerized Database on Soybean Utilization, Processing, Marketing, Nutrition, Production, and History, 1100 B.C. to the 1980s*. 2d ed. Lafayette, Calif.: Soyfoods Center, 1986. 28 leaves.
Contains a description of the SOYA data base, services offered by the Soyfoods Center, directions on obtaining information from the database, as well as a listing of descriptive terminology used. See also the description of the database.

Major Subject Groupings:

A. Crop Improvement
- Abstracts
- Reviews
- Union Lists
- Dictionaries and Terminology
- Crop Anatomy and Taxonomy
- Crop Physiology and Nutrition
- Crop Germplasm Resources
- Distribution and Ecology
- Economic Botany
- Crop Culture, General
- Fruit and Nut Culture
- Vegetable and Mushroom Culture
- Field and Plantation Crops
- Pasture Crops
- Hydroponics
- Organic and Alternative Methods
- Poisonous Plants
- Seed Science
- Genetics, Cytology and Breeding
- Biotechnology and *in vitro* Propagation
- Plant Patents

B. Crop Protection
- Literature
- Guides
- Directories
- Dictionaries and Terminology
- Manuals and Handbooks, General
- Pesticides
- Crop Diseases, General
- Bacterial Diseases
- Fungi and Fungal Diseases
- Mycoplasmas and Prokaryotes
- Nematodes
- Insect and Other Arthropod Diseases
- Virus Diseases
- Weeds
- Nutritional Diseases
- Biological Control, Integrated Pest Management, and Allelopathy
- Tests and Analytical Methods
- Quarantine
- Safety

A. Crop Improvement

Barbara A. DiSalvo

Abstracts

Plant Genetic Resources Abstracts. June 1992– . Wallingford, Eng. and Tucson, Ariz.: CAB International. Quarterly.

Produced in collaboration with the International Plant Genetic Resources Institute (IPGRI). Over 2,500 summaries yearly of the research literature on: policy and planning; integrated conservation strategies; taxonomy, evolution, and origin of crop plants; genetics and genetic diversity; biodiversity; seed science and conservation; *in vitro* conservation and propagation; biotechnology; gene banks; germplasm documentation, research, enhancement, and utilization. Covers journals, books, reports, and conference proceedings. Available as part of *CAB Abstracts* in CD-ROM or from an online search service.

Reviews

Annual Review of Plant Physiology and Plant Molecular Biology. Vol. 39– . Palo Alto, Calif.: Annual Reviews, 1988– . Annual.

Continuation of *Annual Review of Plant Physiology.* Expanded to reflect the influence of the field of molecular biology, especially in the area of technique. [B119]

Critical Reviews in Plant Sciences. Vol. 1– . Four issues yearly. Boca Raton, Fla.: CRC Press, 1983–.

Covers, among others, the areas of molecular biology, molecular biochemistry, cell biology, plant physiology, genetics, classical botany, ecology, and agricultural applications. List of articles from recent earlier volumes in each issue.

Current Plant Science and Biotechnology in Agriculture. Vol. 1– . Dordrecht, The Netherlands and Boston: Kluwer Academic, 1986– .

Series covering research on various aspects of agricultural plant science, with some volumes focused on specific crops: *Genetic Resources of Phaseolus Beans: Their Maintenance, Domestication, Evolution, and Utilization; World Crops: Cool Season Legumes.*

Developments in Crop Science. Vol. 1–. Amsterdam and New York: Elsevier Scientific, 1976– .

To date, a twenty-two volume series treating topics in general crop plant science, with most volumes covering specific crops such as sugarcane, rice, wheat, oil palm and hops.

Union Lists

Three useful and unique lists of agricultural serials have been published since 1990. While there is some overlap among them, each provides unique information with regard to journals in the agricultural fields.

Basavaraj, Nalini and Sandhya Gir, comp. *Union Catalog of Serials in International Agricultural Research Centers (IARCs)*. 2 vols. Patancheru, India: ICRISAT, 1991.
A database created for the use and benefit of National Agricultural Research Systems (NARSs), this alphabetical listing, by title, of 5,401 serials held by the AVRDC, CIAT, CIMMYT, CIP, ICARDA, ICIPE, ICRISAT, IFPRI, IIMI, ILCA, ILRAD, IRRI, ISC, ISNAR, and CGIAR Secretariat covers a number of titles not indexed by the major online database producers, CAB, the National Agricultural Library of the U.S., and the FAO. Citations include title, publisher, place of publication, ISSN, frequency, starting date, previous titles, library codes, and the holdings of each library. Given the importance of journal publications as information sources for agricultural research and development and the costliness of maintaining a complete library of current periodical publications, this cooperative project is intended as an aid to NARSs in locating needed information. The database is also available to IARCs and NARSs in microcomputer-readable form with UNESCO's Micro CDS/ISIS software.

International Union List of Agricultural Serials. Wallingford, U.K.: CAB International, 1990. 767 p.
IULAS is a compilation of 11,567 serials from 129 countries in 53 languages indexed in *AGRICOLA*, AGRIS, and/or *CAB Abstracts* as of 1988. Titles which have changed, merged, or split prior to 1988 are not indexed, but may be included in the citation for the current form of the title. Information contained in the record for each title includes publisher, place of publication, beginning numbers and dates (when known), language of the text, ISSN, frequency, NAL call number, *AGRICOLA* subject code, and whether indexed in *AGRICOLA*, *AGRIS* or *CAB Abstracts*.

World List of Agricultural Serials. Boston: SilverPlatter Information Services, 1991. CD-ROM.
Purported to be the electronic equivalent of IULAS, this database actually contains more information than its print counterpart. Preceding and succeeding titles with beginning/ending dates, translated titles, and the OCLC number are included in all applicable records. In addition, all titles from NAL's serials list not indexed in any of the three major online databases, *AGRICOLA*, AGRIS, and *CAB Abstracts*, are included even if they have ceased. WLAS provides subject access beyond the *AGRICOLA* code in a descriptor field.

Dictionaries and Terminology

Barnes, Robert F., and James B. Beard. *Glossary of Crop Science Terms*. Madison, Wis.: Crop Science Society of America, 1992. 88p.
Divided into three sections: general terminology, cell biology and molecular genetics terminology, and crop terminology. Crop terms are listed by common and scientific names. Cross references and usage notes are given. Compound terms are usually cross-referenced when given in inverse order. There is some overlap of terms from the *Terminology for Grazing Lands and Grazing Animals*. Bibliographic references for additional sources are included at the end of the volume.

Clason, W. E. *Elsevier's Dictionary of Wild and Cultivated Plants in Latin,*

English, French, Spanish, Italian, Dutch, and German. Amsterdam and New York: Elsevier, 1989. 1016p.
Scientific and vernacular names of wild and cultivated plants found in temperate and alpine zones of Europe. Indexes provided for each language. Rare plants are not listed. Intended as a tool for facilitating translation.

Elsevier's Dictionary of Horticultural and Agricultural Plant Production: In Ten Languages, English, Dutch, French, German, Danish, Swedish, Italian, Spanish, Portuguese, and Latin. Rev. ed. Amsterdam and New York: Elsevier, 1990. 817p.
Covers newly introduced terms, particularly the technological, as well as new topics: tropical and subtropical horticulture, meteorology, apiculture, arable farming, grasses, and others. Stronger emphasis on economics and education. Pests and diseases are not covered. [B462]

Facciola, Stephen. *Cornucopia: A Source Book of Edible Plants*. Vista, Calif.: Kampong, 1990. 677p.
Guide to fifty categories of edible plants including fruits, seeds, leaves, roots, oils. Detailed cultivar listings for 110 major crops. Taxonomic nomenclature of families, genera, and species follows that of Tanaka and Kunkel. Classification of cultivars is modified after Lewis and *Hortus Third*. In three sections: botanical listings, cultivar listings, and sources.

Hackett, Clive, and Julie Carolane. *Edible Horticultural Crops: A Compendium of Information on Fruit, Vegetable, Spice, and Nut Species*. 4 vols. Sydney and New York: Academic Press, 1982. 661p.

Based on Terrell's *A Checklist of 3000 Vascular Plants of Economic Importance*. Designed to provide information enabling the fitting of appropriate crops to land type. The four volumes must be used together to extract the characteristics and attribute data of 150 of the most important crops. [B358]

Harborne, Jeffrey B. and Herbert Baxter, eds. *Phytochemical Dictionary: A Handbook of Bioactive Compounds from Plants*. London and Washington, D.C.: Taylor & Francis, 1993. 791 p.
"Information on the most widely encountered plant constituents, with emphasis on those which are biologically active and/or of economic value." Covers substances ranging from phenolics and alkaloids through carbohydrates and glycosides to essential oils and triterpenoids. Compounds are listed by common name and synonyms and the class and subclass to which they belong are indicated. Major source plants, frequency of occurrence, chemical structure, biological activity, and use by humans is included. Entries are arranged biogenetically into five parts with compounds grouped in alphabetical order according to class and subclass. References to phytochemical literature for location of more detailed information are provided.

Koster, H., and F. Schneider, comps. *A Multilingual Glossary of Common Plant-Names, Vol. 1*. 2d ed. Zurich, Switzerland: International Seed Testing Association, 1982.
Revisions for this edition include addition of the list of agricultural and horticultural species from the International Rules for Seed Testing. Important species have been subdivided into infra-specific groups. Number of countries covered has been extended to sixty-one. Transliterations have been omitted. [B061]

Kunkel, Gunther. *Plants for Human Consumption: An Annotated Checklist of the Edible Phanerogams and Ferns.* Koenigstein: Koeltz Scientific Books, 1984. 393p.
In part, an update of Willis' *A Dictionary of Flowering Plants and Ferns*, but with a focus on edibles. Does not include cryptogams or those plants used for beverages or smoking. Not an exhaustive list, but covers 12,650 species, 3,100 genera, and 400 families. Includes taxonomic name, phytogeography, and utilization.

Mabberley, D. J. *The Plant-Book: A Portable Dictionary of the Higher Plants Utilizing Cronquist's An Integrated System of Classification of Flowering Plants (1981) and Current Botanical Literature Arranged Largely on the Principles of Editions 1–6 (1896/97–1931) of Willis's A Dictionary of the Flowering Plants and Ferns.* Cambridge: Cambridge University Press, 1990. 706p.
Intended as an update of Willis' work. "A handy text covering the higher plants, their botany and relationships, their uses and their common names." Appendices of abbreviations of terms and names. Bibliographical references and Cronquist's system of taxonomy included. [B065]

Rehm, Sigmund, and Gustav Espig. *The Cultivated Plants of the Tropics and Subtropics: Cultivation, Economic Value, Utilization.* Weikersheim, West Germany: Margraf, 1991. 552p. CORE
Translation of *Kulturpflanzen der Tropen und Subtropen.* Covers 1,000 plants, excluding forest and ornamental. Major crops are treated in detail, covering breeding, ecophysiology, cultivation practices, diseases, etc. Properties such as economics, production trends, nutritional aspects, chemical and technological features, etc. are provided for each group of plants. Extensive bibliography.

Soule, James. *Glossary for Horticultural Crops.* New York: Wiley, 1985. 898p.
Compiled for the purpose of bridging the communications gap among the horticultural disciplines by providing access to technical terms in the plant-related sciences. "Related terms [are] grouped alphabetically in sections, cross-referenced within and between sections, and indexed by terms and crops." Includes bibliographical references.

System Steinmetz. *Pflanzenproduktion: Mehrsprachen-Bildworterbuch = Plant Production: Multilingual Illustrated Dictionary.* Weikersheim, Germany: Margraf, 1992. 682p.
Contains 5,000 terms in German, English, French, Dutch, Italian, and Spanish. Intended as a standard terminology source to facilitate communication between scientists and extension workers. Separate indexes for each language. Includes 1,000 illustrations.

U.S. Soil Conservation Service. *National List of Scientific Plant Names.* Rev. ed. 2 vols. Washington, D.C.: USDA Soil Conservation Service, 1982.
Now divided into three lists of accepted names for each of three regions: the Caribbean, Hawaii, and the contiguous United States and Canada. The second volume is a synonymy volume for names and symbols used incorrectly, with pointer to accepted names. [B071]

Crop Anatomy and Taxonomy

Fahn, A. *Plant Anatomy*. 4th ed. Oxford: Pergamon, 1990. 588p.
Solid, basic text with an emphasis on functional interpretation of anatomical features. This edition covers structure and development of tissues and cells, ecological anatomy, cambial activity, and new technologies and instrumentation for measuring. Five hundred added references, an extended glossary, and excellent illustrations make this a worthy successor to Esau's *Plant Anatomy*.

Soltis, Pamela S., Douglas E. Soltis, and Jeff J. Doyle, eds. *Molecular Systematics of Plants*. New York: Chapman and Hall, 1992. 434p.
Although not crop specific, covers the application of new molecular techniques in plant systematics and phylogeny and provides a contrast with more traditional morphological techniques. An important introduction to an expanding field. Includes bibliographic references.

Crop Physiology and Nutrition

Baligar, V. C., and R. R. Duncan, eds. *Crops as Enhancers of Nutrient Use*. San Diego: Academic, 1990. 574p.
Describes how plants respond genetically and physiologically to nutrients at the cellular and whole-plant levels and under stress. Essential information for the implementation of low-impact sustainable agriculture globally.

Bothe, H., F. J. de Bruijn, and W. E. Newton, eds. *Nitrogen Fixation: Hundred Years After; Proceedings of the 7th International Congress on (Triple-bond) Nitrogen Fixation, Cologne, Germany, Mar. 1988*. Stuttgart and New York: G. Fischer, 1988. 878p. CORE
Papers and poster summaries from the 1988 conference on nitrogen fixation. Emphasis is on microbiological aspects of nitrogen fixation, properties of alternative nitrogenases, and the application of nitrogen fixation research in improving agricultural productivity. Includes bibliographies.

Fageria, N. K., V. C. Baligar, and Charles Allan Jones. *Growth and Mineral Nutrition of Field Crops*. New York: M. Dekker, 1991. 476p. CORE
Designed to aid improvement of yield, the volume contains both basic and applied research information on the importance of cultivars and environment on economic crops such as wheat, barley, rice, corn, sorghum, soybean, field beans, peanuts, sugarcane, cassava, potato, cotton, and forages.

Goldsworthy, Peter B., and N. M. Fisher, eds. *The Physiology of Tropical Field Crops*. Chichester and New York: Wiley, 1984. 664p.
Provides an overview, in early chapters, on environment, water relations, growth and development of roots, and vegetative and reproductive parts. Individual crops of tropical origin (the chickpea excepted) ranging from cereals, legumes, and root crops to oil, fiber, and other economic crops are covered in later chapters.

International Congress on Photosynthesis. (Various places and publishers, 1968– .)
Held every three years, these meetings are devoted entirely to worldwide re-

search in photosynthesis and the published proceedings have grown from 200 papers for the first to over 1,000 papers and poster summaries for the eighth, which was held in 1989. Only plenary lectures were published for the fifth Congress.

Mengel, Konrad, and Ernest A. Kirkby. *Principles of Plant Nutrition*. 4th ed. Bern: International Potash Institute, 1987. 687p. CORE

Retains the same contents' arrangement as the earlier edition, but now with a greater emphasis on ecological factors. Seventy pages of references. [B180]

Mortvedt, J. J., ed. *Micronutrients in Agriculture*. 2d ed. Madison, Wis.: Soil Science Society of America, 1991. 760p. (*The Soils Science Society of America Book Series* no. 4) CORE

Covers and updates these areas of the first edition: chemistry of micronutrients in soils; micronutrient uptake, translocation, functions, and interactions in plants; diagnosis and correction of micronutrient deficiencies; micronutrient fertilizer technology; and trace elements in animal nutrition. New areas covered are micronutrients and disease resistance or tolerance in plants; trace elements in human nutrition; and beneficial elements, functional elements, and possible new essential elements. [B132]

Plant, Cell and Environment. Vol. 1– . Oxford: Blackwell Scientific, 1978– . Quarterly.

Original contributions from any field dealing with the physiology of green plants including plant biochemistry, molecular biology, biophysics, cell physiology, whole plant physiology, crop physiology, and physiological ecology. Also covers the structural, genetic, pathological, and micrometeorological aspects of plant function. Abstracted and indexed in *AGRICOLA, Current Awareness in Biological Sciences, CAB Abstracts, Current Contents, ASCA, Environmental Periodicals Bibliography,* and *Science Citation Index.*

Plant Growth Regulation. Vol. 1– . The Hague and Boston: M. Nijhoff/W. Junk, 1982– . Quarterly.

Research articles and some reviews from international scientists, with an agricultural focus. Beginning with volume 11, "papers on all aspects of the chemical, molecular, and environmental regulation of plant growth and development."

Rendig, Victor V., and Howard M. Taylor. *Principles of Soil-Plant Interrelationships*. New York: McGraw-Hill, 1989. 275p.

Covers "the nature and dynamic features of the soil-plant root interface and the feasibility of using breeding techniques, tissue cultures, and genetic manipulation to develop new cultivars that are more efficient in obtaining water and nutrients and in resisting toxins in different soil environments." Gives crop-specific examples. Nearly thirty pages of references.

Stacey, Gary, Robert H. Burris, and Harold J. Evans, eds. *Biological Nitrogen Fixation*. New York: Chapman & Hall, 1992. 943p. CORE

Current summaries and literature reviews on all areas of biological nitrogen fixation research, with an emphasis on recent advances. Includes a chapter on the history of BNF research. Extensive bibliographies follow each chapter.

Stumpf, P. K., and E. E. Conn, eds. *The Biochemistry of Plants: A Comprehensive Treatise*. 16 vols. New York: Academic Press, 1980–90. CORE

Review series which replaces the three editions of Bonner's *Plant Biochemistry*. Original series consists of eight volumes: the plant cell; metabolism and respiration; carbohydrates, structure and function; lipids, structure and function; amino acids and derivatives; proteins and nucleic acids; plant products; and, photosynthesis. These are updated in additional volumes by the same authors, with new information added. [B173]

Teare, I. D., and M. M. Peet, eds. *Crop-Water Relations*. New York: Wiley, 1983. 547p.
Review of research on the physical and physiological aspects of water transfer, including evaporation, soil factors, distribution, and movement and function of water in plant cells, tissues, and organs. Twelve food and fiber crops are used as models for describing root growth, canopy structure, interaction with environment, and irrigation.

Crop Germplasm Resources

Diversity. Vol. 1– . Washington, D.C.: Genetic Resources Communications Systems, 1982– . Quarterly.
"News journal for the international plant genetic resources community."

Bernard, Richard L., Gail A. Juvik, and Randall L. Nelson. *USDA Soybean Germplasm Collection Inventory, 1987–1989*. 2 vols. Urbana; Illinois Agricultural Experiment Station, 1990. (*INTSOY* Series, nos. 30 and 31.) CORE
"Information on the origins of soybean and wild soybean germplasm including introduced and old United States and Canadian domestic varieties and foreign and domestics strains identified by FC and PI numbers up to PI150.000 acquired through 1944 and maintained by the United States Department of Agriculture."—Vol. 1. "Information on the origins of soybean and wild soybean germplasm acquired from 1945 to 1985, identified by PI numbers between PI 150.000 and PI 500.000 . . ."—Vol. 2.

Bettencourt, E., and P. M. Perret. *Directory of European Institutions Holding Crop Genetic Resources Collections*. 3d ed. Rome: International Board for Plant Genetic Resources, 1986. 367p.
Information from 312 institutions in twenty-eight European countries. Countries listed alphabetically and institutions listed by location. Private organizations are listed at the end of the directory. Name of institution given in an official language of the country followed by an English translation. Some staff are listed. Some listings are cross-referenced to IBPGR crop directories, where more detail is provided.

Erskine, W., and J. R. Witcombe. *Lentil Germplasm Catalog*. Aleppo, Syria: International Center for Agricultural Research in the Dry Areas, 1984. 363p.
Characteristics and import information on germplasm resources of lentil.

Hawkes, J. G., J. T. Williams, and R. P. Croston. *A Bibliography of Crop Genetic Resources*. Rome: IBPGR, 1983. 492p.
Nearly three times as large as its predecessor due to an increase in the number of crops (particularly fruits and vegetables) covered, this newest version of the *Bib-*

liography . . . comprises nearly all the references from the 1976 version and supplement, as well as references from 1975 to mid-1983 given in *Plant Breeding Abstracts, Herbage Abstracts, Field Crop Abstracts,* and *Horticultural Abstracts*. Emphasis is on crop evolution, agricultural origins, crop taxonomy, the genetic basis of diversity, crop ecology, exploration, conservation, evaluation, documentation, data preparation and management, and genebanks. Plant breeding work and variety trials have been excluded, but pre-breeding and germplasm enhancement are included. Crop sections have been rearranged and there is greater emphasis on tropical crops. [B929]

International Board for Plant Genetic Resources. *Crop Genetic Resources of Africa*. 3 vols. Rome: IBPGR, 1991.

Proceeedings of two workshops, the first held in Nairobi, Kenya in September 1988 and the second in Ibadan, Nigeria in October 1988; organized jointly by IBPGR, United Nations Environment Programme (UNEP), International Institute of Tropical Agriculture (IITA), and Consiglio Nazionale delle Ricerche (CNR). The first covered developments in the field of plant genetic resources and addressed ways of collecting, conserving, evaluating, and using the wide range of plant diversity of Africa. The second focused on crops of special interest to IITA, i.e., cassava, cowpea, maize, plantain, rice, soybean, yams, and bambara groundnut, as well as issues of quarantine regulation and safe movement of germplasm. Respresentatives from the thirty-eight participating African countries presented status reports which comprise volume 3.

International Board for Plant Genetic Resources. *Directory of Germplasm Collections*. 7 vols. Rome: IBPGR Secretariat, 1982–89. CORE

To date, a series of fourteen parts covering food legumes, root crops, cereals, vegetables, industrial crops, tropical and subtropical fruits and tree nuts, temperate fruits and tree nuts, and forages. Provides information on where to obtain resources. International in scope.

International Board for Plant Genetic Resources. *Elsevier's Dictionary of Plant Genetic Resources*. Amsterdam and New York: Elsevier, 1991. 187p.

Dictionary of English terms specific to plant genetic resources; a multilingual version is expected.

International Board for Plant Genetic Resources. *Phaseolus Actifolius Descriptors*. Rome: IBPGR Secretariat, 1985. 26p.

A list using standard international format for the collection of characteristic and evaluation data for documentation of new germplasm resources of tepary bean. Other titles in a series of lists published by IBPGR covering the world's most important economic crop plants include: *Barley Descriptors, Chickpea Descriptors, Cotton Descriptors, Cowpea Descriptors, Descriptor List for Forage Grasses, Descriptor List for Forage Legumes, Descriptors for Groundnut, Descriptors for Oilpalm, Descriptors for Rice, Descriptors for Rice and Triticale, Faba Bean Descriptors, Lentil Descriptors,* and *Sorghum Descriptors*.

Oram, R. N. *Register of Australian Herbage Plant Cultivars*. 3d ed. Melbourne: Commonwealth Scientific and Industrial Research Organization, 1990. 304p.

New edition of Barnard's *Register of Australian Herbage Plant Cultivars*. Includes cultivars listed in the 1982 supplement as well as fifty-five cultivars (except lupins) registered since 1982 through June 1989. Some descriptions have been changed to include additional references and updated botanical names. [B920]

Plant Resources of South-East Asia. Wageningen, The Netherlands: Pudoc, 1989–92. CORE

One outgrowth of an international program (PROSEA) to make available information on the plant resources of Southeast Asia. The first four volumes of this multi-volume handbook cover pulses, edible fruits and nuts, dye and tannin-producing plants, and forages. Good references follow sections on individual plants. Expected to be made available in an online format.

Pundir, R. P. S. *ICRISAT Chickpea Germplasm Catalog*. 2 vols. Patancheru, India: ICRISAT, 1988.

A reference work to aid in the development of chickpea as an international economic crop. The first volume covers evaluation and analysis; the second, passport information.

Remanandan, P., D. V. S. S. R. Sastry, and Melak H. Mengesha. *ICRISAT Pigeonpea Germplasm Catalog*. Patancheru, India: ICRISAT, 1988. 2 vols.

Evaluation and analysis and passport information for resources of pigeonpea held by ICRISAT.

Robertson, Larry D. *Faba Bean Germplasm Catalog: Pure Line Collection*. Aleppo, Syria: International Center for Agricultural Research in the Dry Areas, 1988. 140p.

Evaluation data of 840 pure lines of faba bean germplasm through use of forty-three standard descriptors.

Singh, K. B. *Kabuli Chickpea Germplasm Catalog*. Aleppo, Syria: International Center for Agricultural Research in the Dry Areas, 1983. 284p.

Covers over 4,400 accessions from thirty-two countries with listing of passport information, descriptors, evaluation and analysis, donor, origin, and synonyms. Includes directions on querying the online version.

Smartt, J. *Grain Legumes: Evolution and Genetic Resources*. Cambridge and New York: Cambridge University Press, 1990. 379p. CORE

"A comprehensive survey of all the major grain legumes, their evolution and their potential for further development and improvement as economically important food crops." Considers known genetic resources and the role of herbaceous, shrub, and tree legume species in restoring biological productivity to areas not presently under cultivation or in minimal use. Includes bibliographical references.

Somaroo, B. H., Y. J. Adham, and M. S. Mekni. *Barley Germplasm Catalog I*. Aleppo: International Center for Agricultural Research in the Dry Areas, 1986. 413p.

Intended to "provide breeders with germplasm material characterized for important traits." Eight thousand entries of spring and winter barley from forty-five countries evaluated for quality, morphology, and physiology.

Srivastava, J. P., and A. B. Damania, eds. *Wheat Genetic Resources: Meeting Diverse Needs*. Chichester and New York: Wiley, 1990. 391p.

Papers presented by worldwide experts at a symposium held in May 1989, evaluating the biodiversity of resources, explaining constraints to evaluation and germplasm networks, describing the research at national genebanks, and discussing utilization of resources. Forty pages of references.

Sweet Potato Planning Conference. *Exploration, Maintenance, and Utilization of Sweet Potato Genetic Resources*. Lima, Peru: International Potato Center (CIP), 1988. 369p.

With a focus on improvement and expansion of sweet potato as a world crop, covers all aspects of sweet potato production, including germplasm resources. Contributions are from a global perspective. Includes extensive bibliographical references.

Distribution and Ecology

Major World Crop Areas and Climate Profiles. Rev. ed. Washington, D.C.: U.S. Dept. of Agriculture, World Agricultural Outlook Board, 1987. 159p. (*USDA Agriculture Handbook* no. 664)

"This reference provides a framework for assessing the weather's impact on world crop production by providing benchmark climate and crop data for key regions. For each major crop, maps define the key production regions and pinpoint production concentrations within the region. Histograms show normal temperature and precipitation by month at representative locations. Tables report historical averages and trends of crop area, yield and production. Coverage includes twelve major agricultural regions and crops of coarse grains, winter and spring wheat, rice, cotton, and major oilseeds. World maps show the normal stage of regional crops for each month."

Norman, M. J. T., C. J. Pearson, and P. G. E. Searle. *The Ecology of Tropical Food Crops*. Cambridge and New York: Cambridge University Press, 1984. 369p. CORE

Treats the effects of cropping systems, climate, and soil on cereal, legume, and non-cereal food energy crops by focusing on the four most important crops in each group. Sixty-five page bibliography.

Zeven, A. C., and J. M. J. de Wet. *Dictionary of Cultivated Plants and Their Regions of Diversity: Excluding Most Ornamentals, Forest Trees and Lower Plants*. 2d ed., rev. Wageningen, The Netherlands: Pudoc, 1982. 263p. CORE

Revised edition of *Dictionary of Cultivated Plants and Their Centres of Diversity* (1975). In addition to new data collected, chromosome numbers (from Bolkhovskikh, 1969) and genome formulae are given when known. Taxonomy is based on Willis and *Hortus Third*. [B395]

Economic Botany

Janick, Jules, and James E. Simon, eds. *New Crops: Proceedings of the Second National Symposium NEW CROPS; Exploration, Research and Commercialization, Indianapolis, 1991*. New York: John Wiley, 1993. 710p. CORE

New crops as a potential solution to malnutrition, deforestation, and erosion. Part

I: Policy and Programs. Part II: Research & Development; Part III: Paths Toward Commercialization. Extensiveness of references varies by article. Companion volume to first NEW CROPS proceedings held in Indianapolis, 1988 (Portland: Timber Press, 1990. 560p.)

Roecklein, John C., and PingSun Leung. *A Profile of Economic Plants*. New Brunswick, N.J.: Transaction Books, 1987. 623p.

Excellent reference source originally developed for use in agricultural planning in Hawaii. Builds on the list of 1,000 crops described by Duke and Terrell in 1974 with the addition of 163 crops, not including tropical and subtropical timber or nursery and cut flowers. Crop profile contents include common name, scientific name, geographical regions (from a list of twenty-six areas), end-use categories (such as fiber, cereal, medicinal, etc.), cultivation, and citation numbers (all citations are listed in Index C). Each crop is summarized. Indexed by common name, scientific name, and reference code. [B405]

Crop Culture, General

Rice, Robert P. *Fruit and Vegetable Production in Warm Climates*. London: Macmillan, 1990. 486p. CORE

Crop production, including indigenous crops, discussed within the framework of systems specific to warm climates. Appendices of data distilled from an extensive literature review cover chilling and pollination requirements, nutritional values, chemical weed control, recommended techniques for propagation, symptoms of nutrient deficiencies, diseases and pests and their control, and seed sowing and growth rates.

Fruit and Nut Culture

Bose, T. K., and S. K. Mitra, eds. *Fruits: Tropical and Subtropical*. Calcutta: Naya Prokash, 1990. 838p.

Describes in detail the composition and uses, origin and distribution, species and cultivars, soil and climate, propagation, cultivation, flowering, pollination, and fruit set and drop, growth and development, pests and diseases, harvesting and yield, and breeding and improvement of the major economically important fruits.

Childers, Norman Franklin. *Modern Fruit Science: Orchard and Small Fruit Culture*. 9th ed., completely revised. Gainesville, Fla.: Horticultural Publications, 1983. 583p.

A complete revision including new bibliographic references. Still with strong emphasis on fruits grown in the United States, particularly the apple. [B533]

Faust, Miklos. *Physiology of Temperate Zone Fruit Trees*. New York: Wiley, 1989. 338p.

Up-to-date information on most aspects of fruit tree physiology including photosynthesis, nutrition and water use, fruiting, effects of pruning, and hardiness. Includes bibliographies.

Hammerschlag, F. A., and R. E. Litz, eds. *Biotechnology of Perennial Fruit Crops*. Wallingford, U.K.: CAB International, 1991. 550p.
A systematic review of the latest techniques of plant biotechnology. Emphasizes techniques facilitating genetic manipulation, not plant propagation. Written by international authors from the United States, Europe, South America, India, and Bangladesh, who cover general techniques in the part one and techniques for specific fruit crops in part two. Crops covered include the major temperate, subtropical, and tropical fruits. Extensive references with each chapter.

Morton, Julia F. *Fruits of Warm Climates*. Miami: Julia F. Morton; and Winterville, N.C.: Creative Resources Systems, 1987. 505p.
"A condensation, largely of the information in journal articles which have accumulated in the subject files of the Morton Collectanea." An extensive bibliography provides references to Collectanea monographs which detail specific fruits and deal in depth with their chemistry, pathology, entomology, and other technical aspects of fruit growing. Covers not only the major economic fruit crops, but many of the minor ones as well, emphasizing breeding and culture.

Rom, Roy C., and Robert F. Carlson, eds. *Rootstocks for Fruit Crops*. New York: Wiley, 1987. 494p. CORE
Intended for use by teachers, researchers, and practitioners, this volume provides comprehensive coverage of characteristics, usefulness, and availability of current rootstocks of tree and vine crops. Individual fruits are covered in separate chapters, with bibliographies at the end of each.

Whealy, Kent, ed. *Fruit, Berry and Nut Inventory: An Inventory of Nursery Catalogs Listing All Fruit, Berry and Nut Varieties Available by Mail Order in the United States*. Decorah, Iowa: Seed Saver Publications, 1989. 366p.
Compiled by the staff of the Seed Savers Exchange, a listing of germplasm resources of old-time varieties of fruits, berries, and nuts with information on how to order them.

Wright, C. J., ed. *Manipulation of Fruiting*. London and Boston: Butterworths, 1989. 414p.
Papers from the 47th Easter School in Agricultural Science. Reviews the physiological processes in fruit production from flower initiation to fruit ripening. Discusses methods of manipulating these processes to improve yield and quality. Deals with high economic value fruit crops and the promotion of fruiting for early yield or suppression of fruiting for fodder purposes. Covers intraplant competition, flower initiation, dormancy, and anthesis, pollination, fruit set, light interception and canopy manipulation, genetic regulation, and chemical manipulation.

Vegetable and Mushroom Culture

Bajaj, Y. P. S., ed. *Potato*. Berlin and New York: Springer-Verlag, 1987. 509p. (*Biotechnology in Agriculture and Forestry* no. 3) CORE
A collection of state-of-the-art information on in-vitro technology for improvement of the potato. Thirty-four articles by experts from twenty countries on successful biotechnology approaches to mass production of new varieties. Five sec-

tions cover: micropropagation, virus-free plants, haploid production and field trials; physiology, biochemistry and nutritional studies and molecular genetics; protoplast isolation, culture and somatic hybridization; somaclonal variation, selection of mutants and resistant plants; conservation and exchange of germplasm.

Chang, S. T., and T. H. Quimio, eds. *Tropical Mushrooms: Biological Nature and Cultivation Methods*. Hong Kong: Chinese University Press, 1982. 493p.

An expansion specific to tropical species of part of Chang's *The Biology and Cultivation of Edible Mushrooms*. Covers research into increasing yield of tropical *Agaricus* varieties. Includes bibliographic references. [B627]

Delmas, Jacques. *Les Champignons et Leur aCulture*. Paris: La Maison Rustique, Flammarion, 1989. 969p.

Describes current knowledge of those species of mushrooms important for human consumption globally. Indexed by species and geographic location. Includes bibliographies.

Hawksworth, D. L., and J. C. David, comps. *Family Names: Index of Fungi Supplement*. Oxon, England: CAB, 1989. 75p.

Fourth Supplement to the *Index of Fungi*. Endeavors "to compile all names published since the starting point of fungal nomenclature (1 May 1753) in one particular taxonomic rank—that of the family (familia). This list catalogs 1,917 names . . ." published prior to Jan. 1, 1988. [B1131]

Jellis, G. J., and D. E. Richardson, eds. *The Production of New Potato Varieties: Technological Advances*. Cambridge, England and New York: Cambridge University Press, 1987. 358p.

Papers from the EAPR/EUCARPIA Breeding Variety and Assessment Meeting of Dec. 16–20, 1985. Outlines current strategies used in breeding and testing using traditional and new techniques, including genetic manipulation, tissue culture and protoplast fusion. "Opportunities for breeding varieties suitable for propagation from true seed, of particular significance to the agricultural systems of developing countries . . . together with an extensive and up-to-date bibliography."

Kalloo, G. *Vegetable Breeding*. Boca Raton, Fla.: CRC Press, 1988. 3 vols.

The three volumes comprise distant hybridization, disease and pest resistance and stress, and genetic resources and breeding. Reproduction pollination, control mechanisms, natural breeding systems, floral biology, hybridization techniques, and biometrical approaches for all major crops are covered. Substantial number of bibliographic references.

Kalloo, G., and B. O. Bergh, eds. *Genetic Improvement of Vegetable Crops*. Oxford and New York: Pergamon Press, 1993. 833p.

Details the results of genetic, breeding, and biotechnology research on allium, cole, cucurbit, leaf, legume, root, salad, solanaceous, tropical underground, and other miscellaneous vegetable crops. Each chapter concentrates on one crop, covering cytology, genetics, breeding objectives, germplasm resources, reproductive biology, selection/breeding methods, heterosis and hybrid seed production, quality and processing attributes, and biotechnology. Biotic and abiotic stress resistance in also included. Bibliographical references follow each chapter.

Li, Paul H., ed. *Potato Physiology*. Orlando, Fla.: Academic Press, 1985. 586p. CORE

A comprehensive reference book on the physiology of the potato. Covers photosynthesis, photoassimilate partitioning, respiration, tuberization, carbohydrate and protein metabolism, rest disorders, environmental responses, frost hardiness, and tissue culture. Postharvest physiology is not included. Written by active researchers. References at the end of each chapter.

Loon, C. D. van, and D. G. van der Heij, eds. *Potato Terms: Trilingual Dictionary of the Potato = Dreisprachiges Worterverzeichnis der Kartoffel = Dictionnaire Trilingue de la Pomme de Terre*. Wageningen, The Netherlands: Pudoc, 1989. 402p.
Actually three dictionaries; one each in English, German, and French, and a quadralingual list of Latin and vernacular names as appropriate. Terms are based on a 1969 work and the contents of ten volumes of the EAPR journal *Potato Research*, expanded to include related areas within biology, chemistry, physiology, and technology.

Messiaen, Charles-Marie. *The Tropical Vegetable Garden: Principles for Improvement and Increased Production, with Applications to the Main Vegetable Types*. London: Macmillan Academic and Professional, 1991. 488p.
Describes the effects of climate and soils, principles of plant physiology, plant breeding, and plant protection, and cultivation techniques of wide range of tropical vegetables.

Singer, Rolf, and Bob Harris. *Mushrooms and Truffles: Botany, Cultivation, and Utilization*. 2d ed., rev. and enl. Koenigstein, Germany: Koeltz Scientific Books, 1987. 389p. CORE
Expansion of the earlier edition with more of a focus on the economic importance of and changes in species being cultivated. References have been extensively updated. [B638]

Wang, Jaw-Kai, ed. *Taro, A Review of Colocasia Esculenta and Its Potentials*. Honolulu, Hawaii: University of Hawaii Press, 1983. 400p. CORE
Compilation of knowledge about production and use. Excellent figures and tables of statistical information on composition and analysis, pests, and more. Twenty-seven pages of references.

Field and Plantation Crops

General

Advances in Cereal Science and Technology. 10 vols. St. Paul, Minn.: American Association of Cereal Chemists, 1976–1990. Annual
Definitive review series covering production as well as products.

Bajaj, Y. P. S., ed. *Legumes and Oilseeds Crops I*. Berlin and New York: Springer-Verlag, 1990. 682p. (*Biotechnology in Agriculture and Forestry* no. 10)
Contributions from experts from twenty countries on the biotechnology, in vitro and conventional propagation of soybean, pea, chickpea, alfalfa, lupines, brassica, sunflower, and other important economic crops. Includes chapters on germplasm, hybridization, transformation, and cryopreservation.

Olson, R. A., and K. J. Frey, eds. *Nutritional Quality of Cereal Grains: Genetic and Agronomic Improvement*. Madison, Wis.: American Society of Agronomy, 1987. 511p. (*Agronomy* no. 28) CORE

Evaluations of the value of the major cereals. Assesses the opportunities and limitations for genetic and agronomic manipulation of these values. Intended for use by plant breeders, agronomists, nutritionists, and biochemists. Includes references.

Barley

Rasmusson, Donald C., ed. *Barley*. Madison, Wis.: American Society of Agronomy, 1985. 522p. CORE

Covers distribution, taxonomy, anatomy, morphology, physiology, genetics, breeding, diseases and pests, and cultivation of barley. Includes subject index and substantial bibliography.

Shewry, Peter R., ed. *Barley—Genetics, Biochemistry, Molecular Biology and Biotechnology*. London: C.A.B. International, 1992. 610p. (*Biotechnology in Agriculture Series* no. 5)

A combination of current review articles and special topic papers on the state of genetic engineering on barley.

Corn

Hallauer, Arnel R., and J. B. Miranda. *Quantitative Genetics in Maize Breeding*. 2d ed. Ames: Iowa State University Press, 1988. 468p. CORE

Comprehensive volume on genetics and breeding. Selected, but key references included.

Sprague, G. F., and J. W. Dudley, eds. *Corn and Corn Improvement*. 3d ed. Madison, Wis.: American Society of Agronomy, 1988. 986p. (*Agronomy* no. 18) CORE

Frequently cited, definitive work on corn, including genetic engineering. Intended as reference source for researchers and students. Includes extensive bibliographical references.

Legumes

Allen, O. N., and Ethel K. Allen. *The Leguminosae, a Source Book of Characteristics, Uses, and Nodulation*. Madison, Wis.: University of Wisconsin Press, 1981. 812p. CORE

Global census of nodulation data of 50% of genera and 20% of species of legumes. Includes index of common names and sixty-seven pages of references.

Hebblethwaite, P.D., ed. *The Faba Bean (Vicia faba L.): A Basis for Improvement*. London and Boston: Butterworths, 1983. 573p.

An examination and collection of research on faba by thirty-two authors from nine countries. Divided into five parts: background, physiology, and breeding; husbandry; pests; diseases, and; harvesting and post-harvest handling. Includes references.

Hebblethwaite, P.D., ed. *The Pea Crop: A Basis for Improvement*. London and Boston: Butterworths, 1985. 486p.

Proceedings of the 40th Easter School in Agricultural Science Papers cover the areas of genetics, breeding, agronomy, physiology, utilization, and marketing. Problems of poor standing ability, low yield, yield instability, and soilborne disease are discussed.

International Crops Research Institute for the Semi-Arid Tropics. *Chickpea in the Nineties: Proceedings of the 2d International Workshop on Chickpea Improvement, ICRISAT Center, India, 1989*. Patancheru, India: ICRISAT, 1990. 403p.
Updated information on work on chickpeas since 1979. Thirty-one papers covering chickpea status and potential, utilization, genetic resources and enhancement of germplasm, physiology and agronomy, pathology, insect pests, breeding strategies and approaches to crop improvement, and transfer and exchange of technology. Twenty additional abstracts on advances in improvement in all the principal chickpea growing areas worldwide. Abstracts in French and English.

Lock, Michael. *Legumes of Africa: A Checklist*. Richmond, England: Royal Botanic Gardens, Kew, 1989. 619p.
First volume published as part of the International Legume Database and Information Service Project. Lists nomenclature, characteristics, habitat, economic importance, and geography. Indexed by country, with 567 references.

Nene, Y. L., Susan D. Hall, and V. K. Sheila. *The Pigeonpea*. Wallingford, U.K.: CAB International, 1990. 490p. CORE
Written as a standard reference on pigeonpea. Covers all aspects of production and utilization. Excellent bibliographies.

Poehlman, John Milton. *The Mungbean*. Boulder: Westview Press, 1991. 375p.
Comprehensive review of knowledge and research information on the production, improvement, and utilization of mungbean. Research covered deals only with importance as an economic crop plant. Also includes some information on blackgram. Extensive references.

Schoonhoven, A. van, and O. Voysest, eds. *Common Beans: Research for Crop Improvement*. Oxon: C.A.B. International in association with CIAT, 1991. 980p.
A comprehensive work on Phaseolus emphasizing improved production through breeding and changing agronomic processes. Intended for use as reference source for researchers globally.

Singh, S. R., and K. O. Rachie, eds. *Cowpea Research, Production, and Utilization*. Chichester and New York: Wiley, 1985. 460p. CORE
Reports from a congress on developments in research on cowpea. Covers origin, genetics, physiology, breeding, pathology, entomology, agronomy, and nutrition. Seventy pages of references.

Singh, S. R., K. O. Rachie, and K. E. Dashiell, eds. *Soybeans for the Tropics: Research, Production, and Utilization*. Chichester and New York: Wiley, 1987. 230p. CORE
Based on papers presented at the Tropical Soybean Workshop held at IITA in 1985. Covers all tropical areas. Divided into three parts: crop improvement, cultivation, and protection; regional production and research; and potential and methods of use. Over thirty pages of references.

Webb, C., and G. Hawtin, eds. *Lentils*. Farnham Royal, England: Commonwealth

Agricultural Bureaux and International Center for Agricultural Research in the Dry Areas, 1981. 216p. CORE

Only English language reference work for researchers concerned with increasing production of lentils. Covers production, trade, and uses, origin, taxonomy, and domestication, morphology and growth patterns, genetic resources, genetics and breeding methodology, adaptation to environments, agronomy, mechanization of harvesting, weed and their control, improved nitrogen fixation, diseases, insects and other pests, and nutritional value and quality.

Wilcox, J. R., ed. *Soybeans: Improvement, Production, and Uses*. 2d ed. Madison, Wis.: American Society of Agronomy, 1987. 888p. (*Agronomy* no. 16) CORE

Summarizes knowledge to-date on soybean as a plant, a crop, and on the utilization of soybean products. Emphasizes research since 1973, covering genetics, breeding, taxonomy, physiology, diseases, and pests. Extensive bibliographical reference follow each chapter.

Rice

Bajaj, Y. P. S., ed. *Rice*. Berlin and New York: Springer-Verlag, 1991. 645p. (*Biotechnology in Agriculture and Forestry* no. 14) CORE

Divided into eight sections and written by eighty-three worldwide experts, this volume explores the state of the art in rice biotechnology. Emphasis has been given to the production of haploids and the release of cultivars, regeneration of plants from protoplasts, hybridization, and somaclonal variation for the induction and conservation of genetic variability.

de Datta, Surajit K. *Principles and Practices of Rice Production*. Malabar, Fla.: Krieger, 1987. 618p. CORE

Intended as an aid to improvement of research and application of research for purposes of increasing yield. Directed toward production in Asia. Well illustrated with good bibliographies.

International Rice Research Institute. *Progress in Rainfed Lowland Rice: Proceedings of the International Rice Research Conference, Bubaneswar, India, 1986*. Los Banos, Philippines: International Rice Research Institute, 1986. 446p.

Papers presented at a conference on improvement of lowland rice. Frequently cited, includes an index of varieties and bibliographical references.

International Rice Research Institute. *Progress in Upland Rice Research: Proceedings of the 2nd International Upland Rice Conference, Jakarta, Indonesia, 1986*. Manila, Philippines: International Rice Research Institute, 1986. 567p. CORE

Papers on the improvement and management of African-type rice. Includes bibliographies.

Sorghum

Doggett, Hugh. *Sorghum*. 2d ed. Burnt Mill, England: Longman Scientific, co-published in the United States with J. Wiley, 1988. 512p. (*Tropical Agriculture Series*) CORE

Overview of all aspects of sorghum production and improvement with over forty pages of references.

Wheat

Bajaj, Y. P. S., ed. *Wheat*. Berlin and New York: Springer-Verlag, 1990. 687p. (*Biotechnology in Agriculture and Forestry* no. 13) CORE
Treatise on *in vitro* propagation and biotechnology of wheat. Includes bibliographical references.

Heyne, E. G., ed. *Wheat and Wheat Improvement*. 2d ed. Madison, Wis.: American Society of Agronomy, 1987. 765p. (*Agronomy* no. 13) CORE
Comprehensive volume on all aspects of wheat production with a focus on advances in the last twenty years. Global in scope.

Other Crops

Blackburn, Frank. *Sugar-Cane*. London and New York: Longman, 1984. 414p. CORE
Definitive volume on sugarcane with over twenty pages of references. Covers botany, environment, varieties, cultivation, pests and diseases, and processing.

Cassava in Tropical Africa: A Reference Manual. Ibadan, Nigeria: International Institute of Tropical Agriculture, 1990. 176 p.
Intended as both an instructional and a reference tool, the volume covers the production and post-harvest technology of cassava. The first part describes production constraints including diseases and pests, weeds, soils and agronomic factors, and socioeconomic considerations. In part two, morphology, physiology, and breeding are related to yield and disease resistance. Part three covers post-harvest technology and part four describes cassava research. A bibliography of recommended reading is included.

Coste, René. *Coffee; The Plant and the Product*. London: Macmillan, 1992. 328 p.
An updated replacement for Coste's *Le Cafeier* published in 1968 emphasizing the progress resulting from 20 years of agronomic, technological, chemical, and physiological research. Designed as a practical guide for coffee growers. Includes references to the principal studies conducted on coffee since 1968. (Translation of *Caféiers et Cafés*, Maisonneuve et Larose, 1989)

Hartley, C. W. S. *The Oil Palm (Elaeis guineensis Jacq.)*. 3d ed. London: Longman Scientific; and New York: Wiley, 1988. 761p. (*Tropical Agriculture Series*) CORE
Comprehensive volume on an increasingly important economic crop. The history of the industry, botany of the oil palm, climate and soils required, factors affecting growth, flowering, and yield, selection and breeding, germination and storage of seed, intercropping, diseases and pests, and aspects of plantation maintenance are covered. References follow each section.

Kohel, R. J., and C. F. Lewis, eds. *Cotton*. Madison, Wis.: American Society for Agronomy, 1984. 605p. (*Agronomy* no. 24) CORE
Compendium of scientific knowledge on cotton to-date with a United States focus. Includes references.

Mauney, Jack R., and James McD. Stewart, eds. *Cotton Physiology*. Memphis, Tenn.: Cotton Foundation, 1986. 786p. (*Cotton Foundation Reference Book Series* no. 1)
Part of the Cotton Foundation Reference Book Series, this volume tells "more

than just 'what to do' and 'when to do it.'" Describes "the cotton plant's fruiting and vegetative development as related to environmental conditions, cultural practices, etc." Includes over 2,200 references.

Robbelen, Gerhard, R. Keith Downey, and Amram Ashri, eds. *Oil Crops of the World: Their Breeding and Utilization.* New York: McGraw-Hill, 1989. 553p. CORE

Excellent reference book on the importance, chemical nature, use, and genetics and breeding of oil crops. Covers the seventeen most important crops and touches on some new annuals. Includes bibliographies.

Stover, R. H., and N. W. Simmonds. *Bananas*. 3d ed. Harlow, England: Longman Scientific and Technical, 1987. 468p. (*Tropical Agriculture Series*) CORE

This edition updates the chapters on botany, classification, and cultivars. References to core sources from previous editions included in bibliographies though not necessarily cited in text. [B557]

Tsunoda, S., K. Hinata, and C. Gomez-Campo, eds. *Brassica Crops and Their Wild Allies: Biology and Breeding*. Tokyo: Japan Scientific Societies Press, 1980. 354p. CORE

Older but useful volume on the systematics, domestication, breeding, and conservation of Brassica. Includes bibliographical references and data tables.

Webster, C. C., and W. J. Baulkwill, eds. *Rubber*. Burnt Mill, England: Longman Scientific and Technical, 1989. 614p. (*Tropical Agriculture Series*) CORE

In addition to covering the scientific aspects of rubber such as physiology, genetics, and breeding, the volume includes appendices of the literature on rubber to 1945, clone identification and nomenclature, and standards for rubber. Also includes references to current literature.

Wrigley, Gordon. *Coffee*. Harlow, England: Longman Scientific and Technical, 1988. 639p. (*Tropical Agriculture Series*) CORE

Text of single authorship with editorial contributions emphasizing improvement and production of coffee. Twenty-six pages of references.

Pasture Crops

Crowder, L. V., and H. R. Chheda. *Tropical Grassland Husbandry*. London and New York: Longman, 1982. 562p. (*Tropical Agriculture Series*) CORE

Comprehensive treatment of grasses in the tropics, this volume of the *Tropical Agriculture Series,* covers the tropical environment, the botany and systematics of forages, growth and development, distribution, introduction, utilization and evaluation, improvement, seed production, and management practices. Good bibliographies after each chapter.

Hanson, A. A., D. K. Barnes, and R. R. Hill, eds. *Alfalfa and Alfalfa Improvement*. Madison, Wis.: American Society of Agronomy, 1988. 1084p. (*Agronomy* no. 29) CORE

With a focus toward North America, this compendious volume encompasses all aspects of alfalfa production and improvement, including seeds. Extensive bibliographies included.

Heath, Maurice E., Robert F. Barnes, and Darrel S. Metcalfe, eds. *Forages: The Science of Grassland Agriculture*. 4th ed. Ames: Iowa State University Press, 1985. 643p. CORE

The work of 110 authors, this major text has been significantly reorganized, with some chapters combined, other dropped or added. Appendices of common and botanical names and silo capacities are included. [B799]

International Board for Plant Genetic Resources. *Forage and Browse Plants for Arid and Semi-Arid Africa*. Rome: International Board for Plant Genetic Resources, 1984. 293p.

Taken from the database SEPASAT (Survey of Economic Plants for Arid and Semi-Arid Tropics) at Kew. A selection of about 550 of the 5,000 species included in the database. Chosen on the basis of palatability, each species is listed and indexed by Latin name, with English, French or native language name given if known. Each is described, and its ecology, distribution, potential for improvement, and agronomy are given. Seed collections for each are also included.

Jones, Michael M. B., and Alec Lazenby, eds. *The Grass Crop: The Physiological Basis of Production*. London and New York: Chapman and Hall, 1988. 369p.

Describes the ecophysiology of temperate grasses. Serves as a reference source for researchers to information on the influence of environment and management practices on grass productivity.

Skerman, P. J., D. G. Cameron, and F. Riveros. *Tropical Forage Legumes*. 2d ed., rev. and exp. Rome: FAO, 1988. 692p. (*FAO Plant Production and Protection Series* no. 2) CORE

This volume is a synthesis of existing knowledge on forage legumes in the tropics. Covers the role of legumes in agriculture, land use for nutrition, management, irrigation, weed control, seed production, breeding, and composition. Includes a catalog of pasture and browse species. Appendices of analysis tables, sources of seeds, and common names given. Illustrated with sixty pages of references. [B803]

Skerman, P. J., and F. Riveros. *Tropical Grasses*. Rome: Food and Agriculture Organization of the United Nations, 1990. 832p. (*FAO Plant Production and Protection Series* no. 23) CORE

Companion volume to *Tropical Forage Legumes,* a source of ready information on grasses of economic importance. Emphasis is on those species used for soil conservation, stabilization and utilization of saline areas, and grazing under plantation crops, and those which grow in high aluminum-content soils. Also includes a section on pest grasses. Breeding of improved grasses is not covered, but known cultivars are listed.

Soderstrom, Thomas R., ed. *Grass Systematics and Evolution*. Washington, D.C.: Smithsonian Institution Press, 1987. 473p.

Papers from the International Symposium on Grass Systematics and Evolution held in Washington in July, 1986, presenting a synthesis of the current knowledge on systematics and evolution of grasses. Treats the importance of grasses as an economic crops. Seventy pages of references.

Stubbendieck, James L., Stephan L. Hatch, and Charles H. Butterfield. *North American Range Plants*. 4th ed. Lincoln: University of Nebraska Press, 1992. 493p.

Identifies the 200 most important range plants in North America based on abundance, desirability, or noxious properties. Descriptions include characteristics for identification, magnified illustrations, distribution maps. Nomenclature, life span, origin, season of growth, growth form, floral, fruit and vegetation characteristics, forage value, and uses are given for each.

Watson, Leslie, and Michel J. Dallwitz. *The Grass Genera of the World.* Oxon: CAB International, 1992. 1038p.
Definitive reference work with detailed descriptions of 785 genera of grasses with information on nomenclature, general morphology, leaf anatomy and physiology, biochemistry, haploid and 2cDNA values, fruit and embryo structure, seedling form, cytology, intergeneric hybrids, phytogeography and distribution, ecology, pathogens, classification and economic aspects.

Hydroponics

Resh, Howard M. *Hydroponic Food Production: A Definitive Guidebook of Soilless Food Growing Methods.* 4th ed. Santa Barbara, Calif.: Woodbridge Press, 1989. 462p.
An expanded version of the earlier edition, covering new systems and designs, media, and cultures. [B671]

Sholto Douglas, James. *Advanced Guide to Hydroponics (Soilless Culture).* New ed. London: Pelham Books, 1985. 368p.
Treats research on hydroponics done since previous edition, including new techniques and designs in an added section. New illustrations and tables are included as well. [B673]

Organic and Alternative Methods

Akobundo, I. O., and A. E. Deutsch., eds. *No-Tillage Crop Production in the Tropics.* Corvallis, Or.: International Plant Protection Center, Oregon State University for the West African and International Weed Science Societies, 1983. 235p. CORE
Papers focusing on production improvement through the use of alternative methods of agriculture. Covers problems specific to tropical climates. Good bibliographies.

Francis, Charles A., ed. *Multiple Cropping Systems.* New York: Macmillan; and London: Collier Macmillan, 1986. 383p. CORE
A critical review and synthesis of the international literature on multiple cropping, specifically intended for researchers with limited access to printed resources. Biological, social, and economic aspects of multiple cropping are treated, with good bibliographies for each chapter.

Lampkin, Nicolas. *Organic Farming.* Ipswich, U.K.: Farming Press, 1990. 701p.
Comprehensive coverage of soils, nutrition, fertilizers, rotation, weed and pest management, effects of livestock, grass and fodder crops, horticultural crops, marketing and economics, and issues.

Mulongoy, K., M. Gueye, and D.S.C. Spencer, eds. *Biological Nitrogen Fixation*

and Sustainability of Tropical Agriculture. Chichester and New York: Wiley, 1992. 488 p.
Proceedings of the fourth International Conference of the African Association for Biological Nitrogen Fixation (AABNF) held in 1990. Contains an overview of 20 years of biological nitrogen fixation research in Africa and defines sustainable agriculture and its measurement. Four parts cover nitrogen-fixing systems, biotechnology and modelling, measurement and socioeconomic impact, and sustainability.

Sattler, Friedrich and Eckard v. Wistinghausen. *Bio-Dynamic Farming Practice*. 1st English ed. Stourbridge: Bio-Dynamic Agricultural Association, 1992. 333 p. [Translation of the original 1989 German ed.]
A how-to book on combining traditional farming methods with up-to-date management, crop and animal practices to achieve an ecologically sound farming system. Includes basic information on soils, manures, and tillage followed by extensive treatment of crop husbandry practices, with coverage of specific economic crops. References to additional literature sources included.

Sprague, Milton A., and Glover B. Triplett, eds. *No-Tillage and Surface Tillage Agriculture: The Tillage Revolution*. New York: Wiley, 1986. 467p. CORE
Divided into four parts: environmental resources and their management; characteristics of major cropping systems; pest control management; economics and summary evaluation of surface tillage systems. Global in scope, covering all climatic regions. Includes bibliographies.

Poisonous Plants

Cheeke, Peter R., and Lee R. Shull. *Natural Toxicants in Feeds and Poisonous Plants*. Westport, Conn.: AVI, 1985. 492p.
Intended for use by both animal and plant scientists, this volume covers botanical characteristics of crop plants, range and pasture weeds from around the world, chemical nature and metabolism of toxicants, and pathology induced by particular toxicants.

James, L. F., ed. *Poisonous Plants: Proceedings of the Third International Symposium*. Ames: Iowa State University Press, 1992. 661p.
International in scope, looks at poisonous plants which are of economic importance in the harvesting of forages as well as livestock production. Chapters cover identification, analysis, toxins, and mechanisms of intoxication for poisonous plants in general and for locoweeds, Swainsonia and legumes, pyrrolizidine alkaloids, and lupines, Phomopsis and Thermopsis specifically.

Keeler, Richard F., and Anthony T. Tu, eds. *Plant and Fungal Toxins*. New York, N.Y.: M. Dekker, 1983. 934p. (*Handbook of Natural Toxins* no. 1)
Has a greater emphasis on toxicology than its predecessor *Effects of Poisonous Plants on Livestock* by Keeler (N.Y., Academic Press, 1978). Information is given on the chemistry of toxins, their sources, gross and histopathologic effects, and mechanisms of action. Toxins are grouped to illustrate effects on the cardiovascular and pulmonary systems, carcinogenic effects, reproductive effects, psychic or neurotoxic effects, gastrointestinal and hepatic effects, effects on species

interaction, and usefulness in medicine. Indexed by author and subject. [B807]

Roth, Lutz, Max Daunderer, and Kurt Kormann. *Giftpflanze—Pflanzengifte: Vorkommen, Wirkung, Therapie*. 2d ed. Landsberg: Ecomed, 1984. 1,030p. [Translated title: Poisonous Plants—Plant Poisons]

One of very few recent, complete handbooks on poisonous plants of the world. Includes phytogeography, chemistry, symptoms and effects, antidotes, severity of poisoning for species. Also includes poisonous mushrooms and plant poisons, with an index of symptoms and illnesses. Illustrated for purposes of identification. Keyword index with formulae and non-German names. Section on plant poisons gives complete chemistry. References listed with nearly every plant, mushroom, and poison.

Turner, Nancy J., and Adam F. Szczawinski. *Common Poisonous Plants and Mushrooms of North America*. Portland, Or.: Timber Press, 1991. 311p.

More than an identification guide, this book provides easy reference to all plants, marine and land, cultivated and wild, and mushrooms which can potentially cause physical harm. Mushrooms are listed alphabetically by scientific name and plants by prevalent common name. Plants are grouped by the environment in which they are most commonly found. Includes references.

Vahrmeijer, J. *Poisonous Plants of Southern Africa That Cause Stock Losses*. Cape Town: Tafelberg, 1981. 168p.

Written for agriculturists, veterinarians, botanists and farmers, emphasis is on the location and identification of poisonous plants and the symptoms they produce.

Seed Science

Fehr, Walter R., and Henry H. Hadley, eds. *Hybridization of Crop Plants*. Rev. ed. Madison, Wis.: American Society of Agronomy and Crop Science Society of America, 1982. 765p. CORE

Serves as a reference on principles and procedures used to obtain hybrid seed of self- and cross-pollinated crop plants. Intended for use by both teachers and scientists. Does not cover breeding and genetics.

Hebblethwaite, P. D., ed. *Seed Production*. London and Boston: Butterworths, 1980. 694p. CORE

Listed by crop type, covers seed production, quality, and vigor for herbage plants, sugar beets, cereals, soy and plantation crops, and vegetables.

"International Rules for Seed Testing 1985." *Seed Science and Technology* 13 (2) (1985): 300–513.

With amendments adopted in 1986 and 1989, the 1985 rules supersede those adopted in 1976. The 1986 amendments were published on adhesive paper to be applied to the pages of the International Rules . . . volume. The 1989 amendments were published as a separate volume, with pages intended to be cut up and applied to the 1985 volume. New sections on purity (App. D and E) have been added as part of the 1989 amendments. [B852]

Kelly, A. Fenwick. *Seed Production of Agricultural Crops*. Harlow, England: Longman Scientific and Technical, 1988. 227p.

Intended to aid the improvement of seed production, combines the principles of seed production with the practical requirements. Written for seed breeders. Includes some references.

Mayer, A. M., and A. Poljakoff-Mayber. *The Germination of Seeds*. 4th ed. Oxford and New York: Pergamon, 1989. 270p. CORE

A 30% rewrite of the text of the 3d edition with an additional 300 references attesting to the influence of techniques of molecular biology and membrane research. References for deleted sections have been retained. Chapter on seed technology and unconventional ways of propagation has been added. [B856]

Murray, David R., ed. *Seed Physiology*. Sydney and Orlando: Academic, 1984. 2 vols. CORE

With a focus on development, Vol. 1 ("Development") covers the nutrition of the developing seed. Vol. 2 ("Germination and Reserve Mobilization") discusses accumulation of reserve materials, their location and synthesis.

Plant Varieties and Seeds. Vol. 1. (1988)–. Oxford: Blackwell Scientific for the National Institute of Agricultural Botany. 3 nos. yearly.

Research papers and reviews from international sources on all aspects of variety and seed improvement. Emphasizes practical applications rather than theoretical research extending across agricultural, vegetable, and ornamental species.

Sgaravatti, E., ed. *World List of Seed Sources = Liste Mondiale de Sources de Semences = Lista Mundial de Fuentes de Semillas*. 3d ed. Rome: Food and Agriculture Organization of the United Nations, 1986. 425p.

Addresses of approximately 7,000 plant breeding stations and seed producers in 150 countries, with a heavy emphasis on crops.

Srivastava, J. P., and L. T. Simarski, eds. *Seed Production Technology*. Rev. ed. Aleppo, Syria: International Center for Agricultural Research in the Dry Areas, 1986. 287p.

A guide to techniques and methods of seed production in the Middle East and North Africa. Describes the physiology, anatomy, testing, and viability of seeds in general and cereals, forages, chickpea, and lentils specifically.

Genetics, Cytology, and Breeding

Bhojwani, Sant S., Vibha Dhawan, and Renu Arora. *Plant Tissue Culture: A Classified Bibliography*. Amsterdam and New York: Elsevier, 1986. 789p.

Compilation of literature on all aspects of plant tissue culture of higher plants, including gymnosperms. About 12,000 citations with full title, classified under twenty-six subject headings. Relevant books and conference proceedings are included in a separate chapter. Indexed by plant name. Supplemented by an additional 406 page volume, *Plant Tissue Culture: A Classified Bibliography, 1985–1989*, published by Elsevier in 1990. The chapter on gymnosperm gametophyte has been supplanted by a new chapter on genetic engineering, which had previously been combined with somatic hybridization. Supplement includes citations to about 6,000 papers and 100 books.

Blum, Abraham. *Plant Breeding for Stress Environments*. Boca Raton, Fla.: CRC Press, 1988. 223p. CORE

A definitive work on breeding for adverse conditions including drought, heat, cold, salinity, and high mineral content. Contains thirty pages of references.

Fehr, Walter R., ed. *Principles of Cultivar Development*. New York and London: Macmillan, 1987. 2 vols. CORE

A comprehensive treatise on the methods presently in use by breeders for cultivar development. Does not deal extensively with new or future techniques using cellular and molecular biology. Provides sufficient information to evaluate best technique for a particular situation. Vol. 2 covers the seventeen species representing the major cultivars grown commercially.

Gupta, P. K., and T. Tsuchiya, eds. *Chromosome Engineering in Plants: Genetics, Breeding, Evolution*. Amsterdam and New York: Elsevier, 1991. 2 vols. (*Developments in Plant Genetics and Breeding* no. 2A-B)

A collection of review articles on the cumulated research in plant cytogenetics. Written as a reference work, Part A covers general topics and cereals and millets.

Janick, Jules. ed. *Plant Breeding Reviews*. Vol. 1 (1983–). Westport, Conn.: AVI. Annual.

A review "journal" established to consolidate all aspects of plant breeding related to crop improvement. Emphasis is on major agronomic, horticultural, and forest crops, but minor crops will also be covered. Topics dealing with the theory of breeding systems and methodologies, testing, and evaluation as well as matters pertaining to the profession will also be included. Vol. 7 is entirely dedicated to the U.S. National Germplasm System.

Kalloo, G., and J. B. Chowdhury, eds. *Distant Hybridization of Crop Plants*. Berlin and New York: Springer-Verlag, 1992. 271p.

"An overview of basic and applied aspects of hybridization of distantly related species and genera of crop plants" and the importance of the transfer of attributes for crop improvement. Covers crossability relation and barriers to crossability, unilateral incompatibility, sterility in the hybrids, cytogenetics, and biotechnology. References follow each chapter.

King, Robert C., and William D. Stansfield. *A Dictionary of Genetics*. 4th ed. New York: Oxford University Press, 1990. 406p.

Now includes 7,100 definitions. Appendices cover classification, domesticated species, chronology of the field with an index of scientists and a bibliography of classical papers in genetics, a periodicals list, list of multijournal publishers, and foreign words in scientific titles. [B887]

Mayo, Oliver. *The Theory of Plant Breeding*. 2d ed. Oxford: Clarendon; and New York: Oxford University Press, 1987. 334p. CORE

A synthesis of ideas from population and quantitative genetics, this volume attempts to provide a theoretical basis for plant, specifically crop breeding. Covers the impact and progress of genetic engineering and the techniques of recombinant DNA. Reviewed as a "comprehensive, stimulating reference on plant breeding for the professional." Contains over forty pages of references and a glossary.

Neyra, Carlos A., ed. *Biochemical Basis of Plant Breeding*. Boca Raton, Fla.: CRC Press, 1985–1986. 2 vols.

"A comprehensive survey of progress and current knowledge of those biochemical processes with great potential for the development of superior plant genotypes: photosynthesis, photorespiration, starch biosynthesis, nitrate assimilation, biological nitrogen fixation, and protein synthesis." The first three processes are covered in Vol. 1, "Carbon Metabolism," and the remainder in Vol. 2., "Nitrogen Metabolism." Extensive references follow each chapter.

The Plant Cell. Vol. 1. (1989)–. Rockville, Md.: American Society of Plant Physiologists. Monthly.

Covers meeting reports of the Society. Research articles include many which are crop specific. Available positions are advertized.

Plant Cell Reports. Vol. 1. (1981)–. Berlin and New York: Springer-Verlag. Six issues yearly.

"Original, short communications dealing with new advances concerning all aspects of research and technology in plant cell science, plant cell culture and molecular biology including biochemistry, genetics, cytology, physiology, phytopathology, plant regeneration, genetic manipulation and nucleic acid research." Many articles cover specific crops.

Poehlman, John Milton. *Breeding Field Crops*. 3d ed. Westport, Conn.: AVI, 1987. 724p. CORE

With the retained structure and subject coverage of the earlier edition, many chapters of this latest version have been extensively revised and new chapters on quantitative genetics, hybrid breeding, and plant cell and tissue culture have been added. Still a valuable basic resource to both student and practitioner. [B960]

Rieger, R., A. Michaelis, and M. M. Green. *Glossary of Genetics: Classical and Molecular*. 5th ed. Berlin and New York: Springer-Verlag, 1991. 533p.

Revision of the 1976 edition through elimination of little-used or obsolete terms and inclusion of the new vocabulary of molecular genetics. Includes cross references and over forty pages of bibliographic citations. [B889]

Sybenga, J. *Cytogenetics in Plant Breeding*. Berlin and New York: Springer-Verlag, 1992. 469p. (*Monographs in Theoretical and Applied Genetics* no. 17)

A basic, scholarly text on chromosome structure and function and methods in chromosome analysis and engineering. Introductory chapters cover meiosis, mitosis, and karyotype analysis. Remaining sections describe in detail methods of diagnosis of abnormalities, gene transfer between forms and species, gene dose manipulation (duplications, deficiencies, haploidy, polyploidy), and manipulation of genetic systems (allopolyploidization of autopolyploids, permanent translocation, heterozygosity, hybrid breeding, apomixis). Also contains an extensive system of cross references, detailed index, and numerous diagrams.

Vasil, Indra K, ed. *Cell Culture and Somatic Cell Genetics of Plants*. Vol. 1. (1984)–. Orlando, Fla.: Academic Press. CORE

Series intended for use by researchers, published volumes cover: laboratory procedures and applications; cell growth, nutrition, cytodifferentiation, and cryopreservation; plant regeneration and genetic variability; cell culture in phytochemistry; phytochemicals in plant cell cultures; molecular biology of plant nuclear genes; molecular biology of plastids; photosynthetic apparatus; scale-up and automation in-plant propagation. Crop plants are well represented.

Biotechnology and in vitro *Propagation*

Directory of Plant Biotechnology Companies in USA. Athens: FORE, 1991+. Annual.

Alphabetical listing of companies with addresses, phone and fax numbers, and description of research. Indexed by keyword and state.

Atkinson, Bernard, and Ferda Mavituna. *Biochemical Engineering and Biotechnology Handbook*. 2d ed. New York, N.Y.: Stockton Press, 1991. 1271p.

A clear and detailed reference source to basic concepts and data in plant biotechnology. Profusely illustrated and easy to use laboratory handbook. References included.

Bajaj, Y. P. S., ed. *Crops I-II*. Berlin and New York: Springer-Verlag, 1986–88. 2 vols. (*Biotechnology in Agriculture and Forestry* no. 2 and 6)

Comprehensive survey of the literature on crop biotechnology. Vol. 1 comprises thirty-three chapters dealing with cereals, vegetables, legumes and tubers, and some potential crops such as triticale, hordecale, buckwheat, winged bean, and amaranth. Vol. 2 comprises thirty-one chapters dealing with fruits, some additional vegetables, grasses, and pasture crops. Extensive references.

Bajaj, Y. P. S., ed. *Somaclonal Variation in Crop Improvement I*. Berlin and New York: Springer-Verlag, 1990. 685p. (*Biotechnology in Agriculture and Forestry* no. 11)

Early chapters describe aspects of somaclonal variation, including genetic molecular basis, gene amplification, mosaics and chimeras, the role of the environment, and variability of tolerance to stresses. Chapters detailing the pros and cons of somaclonal variation in cereals, vegetables, fruits, and ornamentals follow.

Collins, Glenn B., and Joseph G. Petolino, eds. *Applications of Genetic Engineering to Crop Improvement*. Boston: M. Nijhoff/W. Junk, 1984. 604p. CORE

Assesses the relationship between conventional plant breeding and genetic engineering. Identifies six areas in crop improvement which lend themselves to cellular and molecular approaches by emphasizing techniques which complement existing procedures: nitrogen fixation; photosynthesis; seed quality; stress tolerance; disease resistance, and; secondary products. Covers whole plant, cellular, and molecular approaches. Over 2400 references.

Thottappilly, G. et al., eds. *Biotechnology: Enhancing Research on Tropical Crops in Africa*. Wageningen, The Netherlands: CTA; Ibadan, Nigeria: International Institute of Tropical Agriculture, 1992. 364p.

Reviewed and revised papers from a workshop held in Ibadan in November 1990 to consider ways in which biotechnology can improve agriculture in Africa. Forty-eight papers presented by scientists from twenty-four countries (sixteen African) are included in the volume in seven sections: (1) need for biotechnology research in Africa; (2) enhancing the genetic base; (3) cell and tissue culture; (4) controlled gene manipulation; (5) using molecular markers; (6) other selected applications of biotechnology, and; (7) policy issues. References follow each chapter.

Walker, John M., and Michael Cox. *The Language of Biotechnology: A Dictionary of Terms*. Washington, D.C.: American Chemical Society, 1988. 255p.
Definitions with illustrations of routinely used terms in various areas of biotechnology.

Withers, Lyndsey A., and P. G. Alderson, eds. *Plant Tissue Culture and Its Agricultural Applications*. London and Boston: Butterworths, 1986. 526p.
Proceedings of the 41st Easter School in Agricultural Science. Reviews the significant areas of advancement in plant tissue culture and presents current research findings. Covers anther and embryo culture, protoplast technology, and transformation. Three sections consider morphogenesis and clonal propagation, plant health and germplasm conservation and storage, and genetic improvement. Includes references.

Plant Patents

Crespi, R. S. *Patenting in the Biological Sciences: A Practical Guide for Research Scientists in Biotechnology and the Pharmaceutical and Agrochemical Industries*. Chichester and New York: J. Wiley, 1982. 211 p.
Written with emphasis on and examples from British patent law, a nonetheless valuable volume designd to provide scientists with background on the law and practice of patents in order to communicate with law professionals. Chapters cover patent disclosure, the structure of the patent specification, categories of patentable invention, conditions for patentability, the mechanism of patenting, inventorship, ownership and contractual relationships, patenting strategies, and enforcement and exploitation of patents. In addition, two chapters deal specifically with chemical, microbiological and gentic engineering patents. Reprints from European and U.S. law are included in the appendices.

Crespi, R.S. *Patents: A Basic Guide to Patenting in Biotechnology*. Cambridge and New York: Cambridge University Press, 1988. 191 p.
A companion to Crespi's earlier volume (see above) with emphasis on biotechnological inventions. Presented in question and answer format, this volume provides practical interpretations of patent law and practice rather than citing statutes and cases. Guidelines to Japanese patent law are included, but emphasis is on European, especially British law. A list of suggested titles is given.

Plant Patent Directory. Washington, D.C.: American Association of Nurserymen, 1990. 617p.
Patents granted by the U.S. Patent and Trademark Office between 1931 and 1989. Divided into five parts: a numerical listing of patents; an alphabetical listing of plants; an alphabetical listing of denomination; a patent listing by originator; a patent listing by assignee. Name variations make access problematic. Amendments are not included. Supersedes *Plant Patents with Common Names . . .*
[B917]

Plant Varieties Journal. Vol. 1. (1988)–. Canberra, A.C.T.: Registrar of Plant Variety Rights, Bureau of Rural Science. Quarterly.
Official journal of the Australian Plant Variety Rights Office. Covers general

plant variety information and publishes notices of application, granting, and revocation of rights to plant variety registration in Australia by national and international institutions, corporations, and individuals.

Plant Variety Protection Office Official Journal. Vol. 7, no. 2. (1989)–. Beltsville, Md.: U.S. Dept. of Agriculture, Agricultural Marketing Service. Quarterly.
New title of *Official Journal of the Plant Variety Protection Office*. Includes cumulative, updated indexes by crop type. [B966]

Registered Field Crop Varieties, 1926–1981. Madison, Wis.: Crop Science Society of America, 1982. 57p.
Cultivars of crops registered as of Jan. 1, 1982. Variety name, variety number, year of registration, originating institution, agency, or organization, and description are provided for forty-six categories of 2,088 varieties. References to registration articles published in *Journal of the American Society of Agronomy*, *Agronomy Journal*, and *Crop Science* are also included.

B. Crop Protection

Phyllis Reich

The theory and practice of crop protection have changed since the publication of Blanchard and Farrell. Producers of agricultural chemicals have retired some pesticides and introduced many others. The most significant change, however, is the emphasis on nonchemical control of crop pests and diseases. Biological control and integrated pest management crop protection measures require a deeper knowledge of the biological mechanisms of plant pests and pathogens. They also require an understanding of the physical, chemical, and biological interactions of the disease agents in the ecosystem. The literature reflects these changes.

Although this update to Blanchard and Farrell lists biological control and integrated pest management publications under a separate heading, this information is also included in many of the titles in other sections. It has become an integral and inseparable part of crop protection literature. This is a selected listing covering only titles published in 1980 or later.

Literature Guides

Alebeek, F. A. N. *Integrated Pest Management: A Catalogue of Training and Extension Materials for Projects in Tropical and Subtropical Regions*. Wageningen: Agricultural University, Technical Centre for Agricultural and Rural Cooperation, 1989. 305p.
This catalogue details handbooks, guides, field manuals, brochures, slide sets, films, videos and tape recordings appropriate to education and training for integrated pest management in tropics and subtropics.

Gilbert, P. and C. J. Hamilton. *Entomology: A Guide to Information Sources*. 2d ed. London: Mansell, 1990. 259p.
Covers books, monographs, review articles, 344 journal titles and eighty newsletters. An introduction treats the history, early literature and insects in art which is followed by sections on: the naming and identification of insects; specimens and collections; the literature of entomology; searching and locating literature; keeping up with current events; entomologists and their organizations; and miscellaneous services, which includes translations, services and guides.

Hamilton, C. J. *Pest Management: A Directory of Information Sources. Vol. 1. Crop Protection*. Wallingford: CAB International, 1991. 352p.
A comprehensive annotated guide to the literature of all aspects of crop protection. Contents cover: bibliographic sources (including books, reviews, primary and secondary journals, newsletters and electronic databases). Provides directories of relevant libraries and information centers, consultants, professional organizations, suppliers of pesticide materials, and regulatory bodies. Vols 2 and 3 will be concerned with animal husbandry and public health.

Harris, K. M. and P. R. Scott, eds. *Crop Protection Information: An International Perspective*; Proceedings of the International Crop Protection Information Workshop. Wallingford: CAB International, 1989. 321p.
The objectives of the workshop were: "(1) To review present systems of making crop protection information available to users, emphasizing the requirements of developing countries; (2) To identify deficiencies in present methods of dissemination of crop protection information and opportunities rising from new communication techniques; and (3) To develop proposals to ensure that information reaches users in a timely manner." Appended is an exhaustive survey of electronic databases appropriate to crop protection and the recommendations from the working groups addressing informational needs in five areas: crops; organisms; plant quarantine; integrated pest management; and pesticides.

Rossman, A. Y., M. E. Palm, and L. J. Spielman. *A Literature Guide for the Identification of Plant Pathogenic Fungi*. St. Paul, Minn.: American Phytopathological Society, 1987. 252p.
A guide intended to direct users to sources of information for the identification of plant pathogenic fungi. The guide is divided into two sections: "Important References to Major Groups of Plant Pathogenic Fungi," contains references for placing fungi in the higher taxonomic levels. This is followed by a section on "Genera of Plant Pathogenic Fungi" which considers 607 genera, organized alphabetically by genus.

Directories

Hall, G. S. and D. L. Hawksworth, compilers. *International Mycological Directory*. 2d ed. Wallingford, U.K.: Publ. by CAB International on behalf of the International Mycological Association, and CAB International Mycological Institute, 1990. 163p.
A directory of sources of information, reference materials and organizations.

Half the book is a listing of organizations and institutions with summary data and information. Second half is indexes covering these topics: Awards and Honours; Courses; Culture Collections; Databases; Herbaria; Special Interests. Includes a listing of journals, periodicals and newsletters issued by government agencies or societies and associations. Compiled for and distributed at the IMA's fourth congress in 1990.

Harvey, L. T. *Pesticide Directory: A Guide to Producers and Products, Regulators, Researchers and Associations in the United States*. Fresno, Calif.: Thomson Publications, 1992. 161p.

A directory of basic manufacturers and formulators, national, regional and state organizations and aerial applicator associations. Includes a listing of relevant magazines and newsletters, and a directory of poison control centers.

Harvey, L. T. *A Guide to Agricultural Spray Adjuvants Used in the United States*. 4th ed. Fresno: Thomson Publications, 1992. 208p.

Lists the majority of spray adjuvants. Includes a brief description, formulation and application rates. Divides into seven sections according to the type of adjuvant.

International Pesticide Directory. London: McDonald Publications, 1981– . Annual.

In 3 parts: Part I lists pesticide suppliers, their addresses and products; Part II is an alphabetical list of pesticides with their active ingredients, a brief description of their uses and names of manufacturers and suppliers; Part III is an alphabetical list of active ingredients on which the pesticides are based.

Julien, M. H. *Biological Control of Weeds: A World Catalogue of Agents and Their Targets*. 3d ed. Wallingford: CAB International, 1992. 186p. CORE

Arranged in four lists: (1) exotic invertebrates and fungi that have been released for biocontrol of weeds and their target weeds; (2) exotic vertebrates (primarily fish) for weed control; (3) organisms manipulated for use as weed control agents; (4) organisms released as biocontrol agents now occurring in countries other than those in which they were released. These lists include information on: weeds and their control, with the weeds listed under plant and family names; control agents listed alphabetically under their target weeds; status of the agents and the degree of control; research organizations; and literature references.

Kidd, Hamish, Douglas Hartley, and George Ekstrom. *World Directory of Pesticide Control Organisations*. Nottingham: Royal Society of Chemistry, 1989. 311p.

A directory of international and national authorities in 132 countries concerned with pesticides and their control. Some 1,200 contact addresses are provided.

Lisansky, S. G., Alison Robinson, and J. Coombs. *The Worldwide Directory of Agrobiologicals*. 2d ed. Newbury, Berkshire: CPL Press, 1991. 438p.

Formerly, "The Green Growers Guide." The book is a comprehensive guide to over 850 crop protection products derived from natural sources. In three sections: (1) An introduction containing information on consumer attitudes on pesticides and organic products, an update on legislation relating to labelling and pesticides, and a summary treatment of the companies in the industry. (2) A product specification section under the headings insect control; fungi and disease control; weed control; silage inoculants; and rhizobium (soil inoculants). (3) Appendixes which include an active ingredient index, a company index of manufacturers and distributors giving the names of the products with which they are involved, and a company listing with addresses.

Thomson, W. T. *A Worldwide Guide to Beneficial Animals (Insects/Mites/Nematodes) Used for Pest Control Purposes*. Fresno, Calif.: Thomson Pub., 1992. 92p.

Use of insects, mites and nematodes for the control of insect pests of agricultural and horticultural crops, and ornamental plants. Includes a list of commercial suppliers in North America, Europe, Australia and Israel; description of packaging; a list of the pests controlled and recommendations for the use of the controlling agent.

Warrell, E., ed. *Crop Protection Directory, 1988–89*. London: Elaine Warrell, 1988. 256p.

A directory of sources of information, products and services in the United Kingdom. Divided into sections covering registration and legislation, producers, distributors, contractors, safety, biological and integrated control, advisory training, trade associations and databases. Provides contact names, addresses and telephone numbers.

Dictionaries and Terminology

Bojnansky, V. and A. Fargasova. *Dictionary of Plant Virology, in Five Languages, English, Russian, German, French and Spanish*. New York: Elsevier, 1991. 472p.

The general part of this dictionary contains 4,611 English entries followed by the Russian, German, French and Spanish equivalents. The specialized section is divided into two parts: Section A consists of the names of the virus and viroid diseases of higher plants (Bryophyta, Pteridophyta and Spermatophyta). The names of mycoplasmas, rickettsia and viruslike diseases with genetic, physiological or unidentified origin are presented in Section B.

Fiala, I. and F. Fevre. *Dictionnaire des Agents Pathogenes des Plantes Cultivees: Latin, Francais, Anglais*. Paris: INRA, 1992. 136p.

A multilingual dictionary listing 300 viruses, 300 fungi and 84 bacteria responsible for crop diseases. The species are presented alphabetically in Latin, French and English.

Gjaerum, H. B. *Nordic Names of Plant Diseases and Pathogens: Bacteria and Fungi*. Copenhagen: Det kgl. Danske Landhusholdningsselskab, Nordiske Jorddbrugsforskeres Forening, 1985. 547p.

Lists bacterial and fungal diseases on cultivated plants in Nordic countries. The main register is arranged alphabetically by host plants. Bacterial and fungal diseases are arranged separately in alphabetical order by scientific name, with common names.

Halliday, P. *A Dictionary of Plant Pathology*. Cambridge: Cambridge University Press, 1989. 369p.

Includes over 8,000 entries. Covers fungi from 460 genera, 675 viruses, bacteria, spiroplasmata, nematodes and viroids. Each pathogen is briefly described with supporting bibliographic references. Also includes crops and their diseases, disease names, fungicide names, taxonomic groups, toxins, vectors and biographies of some past plant pathologists.

Hull, R., F. Brown, and C. Payne. *Virology: Directory and Dictionary of Animal, Bacterial and Plant Viruses*. New York: Stockton Press, 1989. 325p.

A dictionary of names of viruses and their higher order taxa, as well as terms in use in the literature of virology. In many cases, definitions are supported by references to reviews and articles. Appended are lists of insect species in which specific viral infections have been recorded. Included also is a table of phage isolates from bacteria, cyanobacteria and mycoplasmas.

Important Weeds of the World: Scientific and Common Names, Synonyms and WSSA Approved Computer Codes. 3d ed. Leverkusen: Agrochemicals Division of Bayer AG, 1983. 711p.

Includes a listing by plant family names and indexes according to the Latin, English, German, French and Spanish names.

Jacobson, Martin. *Glossary of Plant-Derived Insect Deterrents*. Boca Raton, Fla.: CRC Press, 1990. 213p.

Lists plants found to be effective as insect feeding deterrents, as well as those which were tested and found to lack deterrent properties. Some 1,500 plant species from 175 families are treated. Arranged in four sections: (1) Introduction, (2) Methods for antifeedant bioassay, (3) Biological test results for cryptogams (less than a page long) and phanerogams, and (4) Scientific and common names of referenced insects. Includes an extensive bibliography and a detailed index.

Names of British Plant Diseases and Their Causes: A List of English and European Names and the Scientific Names of the Causal Organisms. 5th ed. Kew, Eng.: Commonwealth Mycological Institute, 1984. 76p. (*Phytopathological Paper* no. 28)

Hosts are listed alphabetically according to their common English names. Diseases are listed alphabetically under each host by their common names, followed by the scientific name of its causal organism.

Reynolds, I. F., ed. *Pesticides, Synonyms and Chemical Names*. 8th ed. Woden, Australia: Commonwealth Department of Health, 1987. 334p.

Divided into color coded sections: Section I consists of an alphabetical listing of common names, trade names, trivial names and code number of pesticides. Trade names used in Australia are listed first, followed by overseas trade names; Section II is an index of chemical names of pesticides which have "Recommended Common Names" and some pesticides used in Australia which do not have recommended common names. In these cases a trade name is given; Section III outlines the system of organic chemical nomenclature used in the publication.

Thesaurus of Agricultural Organisms, Pests, Weeds, and Diseases, compiled and edited by Derwent Publications Ltd., with the assistance of CIBA-GEIGY SA. London: Chapman and Hall, 1990. 2 vols.

Originally produced as a set of manuals between 1974 and 1977 for use with Derwent's PESTDOC database to standardize the use of the common and Latin names of agricultural organisms. The current publication primarily covers weeds, pests and disease organisms. It also includes the names of crop species and a number of non-target invertebrates, vertebrates, plants and microorganisms. There is a "Main Entry" for each organism consisting of the Latin name, one or two "Higher Taxa," Latin name synonyms, and the common names in English, French, and German when these are known. An index of inverted genus species names is appended.

Weed Science Society of America, Subcommittee on Standardization of Common and Botanical Names of Weeds. "Composite List of Weeds," *Weed Science* 32 (Supplement 2) (1984). 137p.

"The names of 1,934 weed species of current or potential importance in the United States and Canada are arranged alphabetically in three lists: by scientific name, by WASSA-approved common name, and by the five-letter "Bayer code." Many common woody plants or plants of generally weedy habit which are not pests are not included." [B826]

Williams, G. H., and K. Hunyadi. *Dictionary of Weeds of Eastern Europe: Their Common Names and Importance*. Amsterdam: Elsevier, 1987. 479p.

A dictionary in Latin, Albanian, Bulgarian, Czech, German, English, Greek, Hungarian, Polish, Romanian, Russian, Serbo-Croat and Slovak.

Williams, G. H. *Elsevier's Dictionary of Weeds of Western Europe: Their Common Names and Importance*. Amsterdam: Elsevier Scientific, 1982. 320p.

A dictionary in Latin, Danish, German, English, Spanish, Finnish, French, Icelandic, Italian, Dutch, Norwegian, Portuguese and Swedish.

Williams, G. H., and van der Zweep, eds. *Interdisciplinary Dictionary of Weed Science: Dansk, Deutsch, English, Espanol, Francais, Italino, Nederlands, Portugues*. Wageningen, The Netherlands: PUDOC, 1990. 546p.

Brings together standard terms in weed science as well as terms from plant physiology, chemistry, soil science, toxicology and engineering. Contains about 3,000 terms in Danish, German, English, Spanish, French, Italian, Dutch, and Portuguese.

Wood, A. M. *Insects of Economic Importance: A Checklist of Preferred Names*. Wallingford: CAB International, 1989. 150p.

A checklist of the scientific names of the most important genera and species of arthropods in agriculture, horticulture, veterinary medicine and forestry. Covers mites, ticks, spiders, scorpions and insects. The family, order and authority are given after each scientific name. Cross references from the non-preferred name to the preferred name are provided.

Manuals and Handbooks, General

Attwood, P. J., ed. *Crop Protection Handbook—Cereals*. Croydon: BCPC Publications, 1985. 229p.

Discusses weeds, diseases and pests in cereals and their chemical and non-chemical control. Includes twenty-eight tables. Appended are diagrams of important growth stages of cereals; the decimal code for cereal growth stages; and a list of proprietary names.

Bohmont, B. L. *The Standard Pesticide User's Guide*. Rev. ed. Englewood Cliffs, N.J.: Prentice-Hall, 1990. 498p.

Covers insect pests, plant disease agents, vertebrate pests and weeds. Chapters include: Pesticide laws, liability and record keeping; The Pesticide label; Pesticide safety; Pesticides and environmental considerations; Pesticide formulations and adjuvants; Pesticide application equipment; Pesticide equipment calibration; Pesticide calculations and useful formulas; Pesticide transportation, storage, decontamination and disposal; and Integrated pest management. Appended are the following tables: pesticide information telephone numbers; restricted-use pesticides; United States and Canadian Pesticide Control Offices; Trade name/common name cross references; EPA field reentry standards; toxicity classification for pesticide; stability information; cold weather handling. Includes also an index of common pesticide classifications, names and trade names.

Conner, J. D., L. S. Ebner, C. A. O'Connor, et al. *Pesticide Regulation Handbook*. 3d ed. New York: Executive Enterprises, 1991. 540p.

A comprehensive presentation of United States pesticide regulations. Covers the preparation and submission of data; registration procedures; pesticide tolerances and food additive regulations; federal ground water regulations; exports and imports; state regulations; storage regulations; the Federal Insecticide, Fungicide and Rodenticide Act; a list of acronyms and references to legal cases.

Johnston, A., and C. Booth, eds. *Plant Pathologist's Pocketbook*. 2d ed. Farnham Royal: Commonwealth Agricultural Bureaux, 1983. 439p. CORE

Treats fungal, bacterial and viral and diseases as well as mycoplasma-like organisms as plant pathogens. Chapters on: non-infectious disorders; air pollution; biocontrol of fungal plant pathogens; plant quarantine; insect and other arthropod pests; weeds; parasitic higher plants; fungicides; crop loss assessment; and post-harvest losses. Includes a list of fungicides and bactericides, a glossary of plant pathology terms and a list of useful addresses.

Matthews, G. A. *Pesticide Application Methods*. 2d ed. Harlow, Eng.: Longman Scientific and Technical, 1992. 405p. CORE

This handbook of pesticide application techniques provides detailed information on targets, formulation, sprayers and other equipment. [C331]

Nyvall, R. F. *Field Crop Diseases Handbook*. 2d ed. New York: Van Nostrand Reinhold, 1989. 817p. CORE

Each chapter treats a disease. Diseases of alfalfa, barley, buckwheat, corn, cotton, field beans, flax, millet, oats, peanut, rapeseed and mustard, red clover, rice, rye, safflower, sorghum, sugarbeets, sugarcane, sunflowers, tobacco, wheat, wild rice are considered. For each disease, the causal agent, distribution, symptoms and control are given.

Pimentel, D. *CRC Handbook of Pest Management in Agriculture*. 2d ed. Boca Raton, Fla.: CRC Press, 1991. 3 vols. CORE

Volume 1 covers estimated losses of crops from plant pathogens and weeds, cultural environmental controls of pests in crops, and pesticide use in agriculture. Volume 2 includes chapters on biological and quarantine controls, pesticide application methods, resistance to pesticides, and biological control of insect pests, plant pathogens and weeds. Volume 3 covers pests on particular crops.

Powell, Charles C., and Richard K. Lindquist. *Ball Pest and Disease Manual.* Geneva: Ball Pub., 1992. 332p.
A reference manual for the diagnosis and management of pests and diseases of flower and foliage crops. Contents include: diagnosing problems, identifying diseases, identifying insect and mite pests, controlling infectious diseases, using fungicides, using pesticides, using integrated pest management and writing plant protection programs. Appended are a list of useful addresses, conversion tables and an index of host plants, diseases and pests.

Scopes, N., and L. Stables, eds. *Pest and Disease Control Handbook.* Surrey: BCPC, 1989. 732p. CORE
A comprehensive treatments of the pests and diseases of crops grown in the United Kingdom. Discusses pesticide usage and application including storage and disposal, and principles of insecticide and fungicide evaluation. Considers pests and diseases of twelve groups of crops: cereals; oilseed; rape; brassica seed crops; field beans; grass and fodder crops; potatoes; sugar beet; vegetables; fruit and hops; protected crops; and ornamentals and turf. Under each crop, the biology and method of the causal organism is provided.

Smith, I. M., J. Dunez, R. A. Lelliott, et al. *European Handbook of Plant Diseases.* Oxford: Blackwell Scientific, 1988. 583p. CORE
Provides a comprehensive treatment of the diseases of cultivated plants and major forest and amenity trees of Europe. Chapters are by major systematic groups of pathogens, with emphasis on the biology of the pathogen, the host, host specialization, the geographical distribution, the epidemiology, economic impact and control. Entries for pathogens are divided into: (1) name, synonym and anamorph name; (2) basic description of the pathogen (3) host index.

United States Department of Agriculture. *Guidelines for the Control of Plant Diseases and Nematodes.* Washington, D.C.: U.S. Govt. Print. Off., 1986. 274p. (*USDA Agriculture Handbook* no. 656) CORE
Intended for plant pathologists, nematologists, and research and extension workers. Considers the uses of chemicals, as well as alternative measures for the control of plant diseases and nematodes on crops. Includes tables of major crop diseases and suggestions for their control, a seed treatment list and a fungicide index.

Weiser, J. *Biological Control of Vectors: Manual for Collecting, Field Determination and Handling of Biofactors for Control of Vectors.* Published on behalf of the World Health Organization. New York: John Wiley, 1991. 189p.
Intended for field workers to assist with the identification and handling of new pathogens, parasites and predators of arthropod vectors. Considers the major groups of vector pathogens, parasitoids and parasites and predators. Includes three bibliographies: one on handbooks, with thirty-seven references; another on WHO/VBC publications of twenty-one references; and a bibliography of seventy-one references of selected literature.

Westcott's Plant Disease Handbook. 5th ed. Revised by R. K. Horst. New York: Van Nostrand Reinhold, 1990. 953p. CORE
A classic work, revised to reflect advances in the discipline. Chemical and

pesticide regulations have been updated. This edition also includes new host plants and some taxonomic changes in bacteria, fungi and mistletoes. [B1026]

Pesticides

AGROW. *Biological Crop Protection*. Richmond, Surrey: PJB Publications, 1989. 109p.
Provides information on biological control agents: their markets and the companies which produce them. The book is divided into three sections: Microbial pest control; predators and parasites in crop protection; and pheromonal crop protection.

Altman, J. *Pesticide Interactions in Crop Production: Beneficial and Deleterious Effects*. Boca Raton, Fla.: CRC Press, 1993. 624p.
"A broad-based book covering all aspects of how pesticides affect crops. Focuses on the physical, chemical, biological and ecological interactions of pesticides on crops. Considers the effects of pesticides on the environment and on crop pests. Describes the effects of pesticides on soil symbionts."

Cairns, Thomas. *Comprehensive Analytical Profiles of Important Pesticides*. Boca Raton, Fla.: CRC Press, 1992. 288p.
Information for each pesticide includes formulation and uses; properties; analytical methods and toxicological data; fish and wildlife toxicity studies; and food and feed tolerances. Covers insecticides, fungicides, miticides, herbicides, plant growth regulators and fumigants.

Canada. Agriculture Research Branch. *Guide to the Chemicals Used in Crop Protection*. 7th ed. 1982. 595p. (Publication 1093) [C077]

"In the preparation of the Seventh Edition, 121 compounds listed in the Sixth Edition have been omitted (and listed separately) and 199 new compounds have been added, while the more recent information of some of the compounds in the previous edition has been included." Provides chemical name, structure, alternative names, history, physical properties, chemical properties, biological properties, formulations, analysis and residue information. Includes a list of manufacturers and addresses.

Flick, W. W. *Agricultural Chemical Products*. Park Ridge, N.J.: Noyes Publications, 1988. 327p.
Describes over 750 agricultural chemical products currently in use. These include pesticides, herbicides, insecticides, fungicides, plant growth regulators, animal repellents, nematicides, seed treatments, composting aids, micronutrients and specialty fertilizers. The data presented are selected from manufacturers' descriptions. Products with EPA registration numbers indicate that the U.S.Environmental Protection Agency has approved the product label. Products are presented alphabetically by company. Includes a trade name index and a list of suppliers' addresses.

Flick, E. W. *Fungicides, Biocides and Preservatives for Industrial and Agricultural Applications*. Park Ridge, N.J.: Noyes Publications, 1987. 284p.

Describes about 350 chemical compounds currently available. The data represent selections from manufacturers' descriptions. All products have EPA registration numbers, and are presented by company, with the companies listed alphabetically. Includes a listing of suppliers' addresses, trade name and chemical name indexes.

Hartley, Douglas, and Hamish Kidd, eds. *Agrochemicals Handbook*. 3d ed. Cambridge: Royal Society of Chemistry Information Services, 1991. Looseleaf.

A collection of datasheets on pesticide chemicals arranged alphabetically. Primarily British products include herbicides, fungicides, insecticides, rodenticides, nematicides, acaricides, mollusicides, bactericides, and pest attractants and repellents. Datasheets give chemical structure and chemical, physical, analytical, use and toxicity data. Activity lists "comprise fifteen listings of the common names of all active ingredients possessing each type of biological activity (e.g., herbicides, fungicides insecticides)." Includes also a listing of manufacturers and chemical family and CAS registry numbers indexes.

James, D. R., and Hamish Kidd, eds. *Pesticide Index: An Index of Chemical Common and Trade Names of Pesticides and Related Crop-Protection Products*. 2d ed. Cambridge: Royal Society of Chemistry, 1991. 280p.

A quick reference guide to common, chemical and trade pesticide names and their manufacturers. Covers herbicides, insecticides and fungicides, including 800 active ingredients, arranged alphabetically by the active ingredient and by trade name. The *Chemical Abstracts* registry number is given for each active ingredient.

Jourdain, D. and E. Hermouet, eds. *Regional Agro-Pesticide Index*. Bangkok: International Co-operation Centre of Agricultural Research for Development (CIRAD), 1990. 3 vols.

Lists 670 pesticide-active ingredients and more than 7,000 products in fourteen countries in Asia, the Pacific and Africa. Active ingredients are classified according to their target pests. Includes insecticides, fungicides, herbicides, molluscicides, nematacides, rodenticides and some others. Entries give chemical family, manufacturer, uses and physical properties.

MSDS Reference for Crop Protection Chemicals. 4th ed. New York: Chemical and Pharmaceutical Press, 1989+. Annual with quarterly supplements.

A comprehensive guide. Materials safety datasheets for each pesticide provide product identification data, the manufacturer, hazardous ingredients, hazard rating, physical properties, and shipping and safety information. Provides brand name, manufacturer and common and chemical name indexes. Includes information on compliance criteria, hazardous chemical inventory and accidental release reporting tables, tabulation of products with extremely hazardous substances, DOT information table and a list of poison control centers.

Plant Health Guide: Complete Product Listings for More than 47 Crops. Willoughby, Ohio: Meister Publishing, 1992. 178p.

"Summarizes and provides comparison between disease control products regarding formulations, rates, diseases controlled, restrictions, days to harvest, reentry and use remarks." Covers registered fungicides, bactericides, antibiotics and nematicides. Organized alphabetically by crop.

Thomson, W. T. *Agricultural Chemicals*. Books I-IV. Fresno, Calif.: Thomson Publications, 1989– 1991. 4 vols.

Entries for each book give chemical and common names, uses, chemical structure, origin, toxicity, formulations, phytotoxicity, application rates, diseases controlled by the pesticide, related mixtures and compounds. Provides conversion tables, manufactures' addresses and glossaries. Book I: Insecticides, acaricides and ovicides, 1989 revision.; Book II: Herbicides, 1993 revision.; Book III: Fumigants, growth regulators, repellents and rodenticides, 1991–1992 revision.; Book IV: Fungicides, 1991 revision.

The U.K. Pesticide Guide. Wallingford: CAB International, 1988– .

An annual guide to available products, with indications on how to use them in the U.K. The pesticides are listed alphabetically by the active ingredient, using British Standards Institution common names. Information on uses, efficacy, crop safety and product precautions is provided. A Crop Guide section lists suggested pesticides for each type of crop.

U.S. Environmental Protection Agency. *Pesticide Fact Handbook*. Park Ridge, N.J.: Noyes, 1988–. 2 vols.

Contains 217 pesticide factsheets issued by the U.S. Environmental Protection Agency. "The Fact Sheets are issued if one of the following regulatory actions occurs: (1) a Registration Standard is issued, (2) a significantly different use pattern has been registered, (3) a new chemical is registered, or (4) a Special Review determination document has been issued. Each factsheet includes: the description of the chemical (including generic, common and trade names), use paterns and formulations, summary of science findings, and chemical characteristics.

Worthing, Charles, and R. J. Hance, eds. *The Pesticide Manual: A World Compendium*. 9th ed. Farnham, Eng.: British Crop Protection Manual, 1991. 1141p. CORE

The main sections of the book are preceded by a glossary of scientific and common names of weeds, pests and diseases referred to in the text, a bibliography, a list of names and addresses of firms, and notes on common names, chemical nomenclature and structures. The two main sections are: Part I lists pesticides in use or under development. It includes chemicals and microbial agents, animal ecotoparasites, plant growth regulators, pest repellents, synergists, herbicide safeners and some timber preservatives. Each entry covers the nomenclature, development properties, formulations, uses, toxicology and a reference to methods of analysis. Part II lists superseded compounds. There are four indexes: (1) CAS registry numbers, (2)molecular formulae, (3) standard codes used to identify pesticides, and (4) chemical, common and trivial names and trademarks.

Crop Diseases, General

Agrios, G. N. *Plant Pathology*. 3d ed. San Diego, Calif.: Academic Press, 1988. 803p. CORE

A fundamental and well written text. Contents: Introductions to plant pathology; Parasitism and disease development; How pathogens attack plants; Effects of pathogens on plant physiological functions; How plants defend themselves against pathogens; Genetics of plant disease; Effect of environment on development of infectious plant diseases; Plant disease epidemiology; Control of plant diseases; Environmental factors that cause plant diseases; Plant diseases caused by fungi, prokaryotes, parasitic higher plants, viruses, nematodes, and flagellate protozoa. Includes also information on the application of biotechnology in plant pathology.

Clark, C. A., and J. W. Moyer. *Compendium of Sweet Potato Diseases*. St. Paul, Minn.: APS Press, 1988. 75p. CORE

Considers bacterial, fungal, nematode, and viral diseases. Discusses diseases caused by environmental stresses, nutritional factors, toxicities and those of unknown etiology. Includes a glossary, seventy-four color photographs and fifty-three black and white photographs and illustrations.

Cook, R. J., and R. L. Veseth. *Wheat Health Management*. St. Paul, Minn.: APS Press, 1991. 152p. CORE

A practical guide for growers in North America. Contents include: How wheat manages its health in the wild; limiting effects of the physical and chemical environment; limiting effects of pests and disease; economic implications of wheat health; weed, pest and disease control; pre-planting health management; health management at planting; post-planting health management. Includes ninety-four color and eighty-three black and white photographs.

Ellis, M. A., R. H. Converse, and R. N. Willams, et al. *Compendium of Raspberry and Blackberry Diseases and Insects*. St. Paul, Minn.: APS Press, 1991. 122p.

Contents: Diseases caused by fungi, viruses and virus-like agents, arthropod pests of Rubus and abiotic factors. Treats also the effects of cultural practices on disease. Includes a glossary and over 140 color photographs.

Fredderiksen, R. A. *Compendium of Sorghum Diseases*. St. Paul, Minn.: APS Press, 1986. 98p. CORE

Treats diseases caused by bacteria, fungi, viruses and virus-like organisms, nematodes, parasitic plants and abiotic factors. Includes beneficial microorganisms, insect pests and disease management. Contains a glossary, 118 color photographs and forty-four black and white photographs and illustrations.

Gair, R., J. E. E. Jenkins, and E. Lester. *Cereal Pests and Diseases*. 4th ed. Ipswich, Eng.: Farming Press, 1987. 268p.

Designed as a guide for agriculturalists and farmers to the diseases, pests and disorders of cereals grown in the United Kingdom. Covers damage, symptoms, the life histories of the pests and diseases, crop losses and control methods. Each crop; wheat, barley, oats, rye and maize is treated in one chapter dealing with pests and another with diseases. Contains fifty-seven plates.

Gunn, J. S., ed. *Crop Protection Handbook; Potatoes*. Farnham, Eng.: British Crop Protection Council, 1992. 192p.

Examine the broad spectrum of potato diseases. Has specific chapters on viral, fungal, and bacterial diseases, as well as general pests. Also covers disease vectors, nematodes and weeds, as well as diseases in storage and in seed crops. Emphasis is given to integrated control and, in particular, cultivation methods for crop protection.

Hagedorn, D. J. *Compendium of Pea Diseases*. St. Paul, Minn.: APS Press, 1984. 73p. CORE

Covers diseases caused by bacteria, fungi, viruses, nematodes, as well as abiotic diseases. Includes a guide to the identification of pea diseases in the field, a glossary, 110 color photographs, and thirteen black and white photographs and illustrations.

Hall, R. *Compendium of Bean Diseases*. St. Paul, Minn.: APS Press, 1991. 102p. CORE

Contents include fungal, bacterial, viral, nematode and mycoplasma-like diseases. Provides descriptions of symptoms, causes, cycles and control practices. Illustrated with 171 color photographs and thirty-eight black and white photographs and illustrations.

Harris, K. F., and K. Maramorosch, eds. *Vectors of Plant Pathogens*. New York: Academic Press, 1980. 467p. CORE

A comprehensive work on major vectors. Covers insects and invertebrates as carriers of plant disease.

Hill, D. S., and J. M. Waller. *Pests and Diseases of Tropical Crops*. London: Longman, 1982–88. 2 vols. CORE

Designed to permit a preliminary identification of the most important pests and diseases of major crops, based on characteristics or symptoms. Lists pests and diseases under each of the major forty crops, giving descriptions, life history, distribution and control of selected pests and diseases for each crop. Includes distribution maps, scientific, common and family names of the pathogen.

Hoffman, G. M., F. Nienhaus, and F. Schonbeck, et.al. *Compendium of Pest and Disease Management*. 2d ed. Berlin: Paul Parey, 1985. 488p.

Treats diseases, weeds and pests. Covers symptoms, methods of crop protection and integrated control systems. Includes color photographs of disease symptoms.

Jones, A. L. and H. S. Aldwinckle. *Compendium of Apple and Pear Diseases*. St. Paul, Minn.: APS Press, 1990. 100p.

Treats diseases caused by fungi, bacteria, mycoplasmas, nematodes, bacteria and viruses, as well as abiotic diseases, such as environmental, genetic, nutritional and physiological disorders. Contains a glossary and 170 color photographs and twenty-six black and white photographs and illustrations.

Jones, J. B., J. P. Jones, and R. E. Stall, et. al. *Compendium of Tomato Diseases*. St. Paul, Minn.: APS Press, 1991. 100p. CORE

Considers the botany, culture and diseases of tomatoes. Covers diseases caused by fungi, bacteria, viruses, viroids, mycoplasma-like organisms, nematodes, arthropods, as well as physiological, genetic and nutritional diseases. Illustrated with 157 color photographs.

Kolte, S. J. *Diseases of Annual Edible Oilseed Crops. Vol. I. Peanut Diseases. Vol. II. Rapeseed-Mustard and Sesame Diseases*. Boca Raton, Fla.: CRC Press, 1984–85. 2 vols. CORE

After briefly discussing the botany, origin and distribution of peanuts, rapeseed-mustard and sesame seed, the diseases of these oilseeds are considered. They include fungal, bacterial, viral, mycoplasma, parasitic nematode and phanerogamic and non-parasitic diseases. Covers the geographic distribution, economic importance, symptoms, host ranges, survival, factors affecting infection, and the development of the disease and its control.

Maas, J. L. *Compendium of Strawberry Diseases*. St. Paul, Minn.: APS Press, 1984. 159p. CORE

Treats non-infectious diseases, arthropod and mollusk pests, and bacterial, fungal, viral nematode and insect-vectored diseases. Includes a glossary, 148 color photographs and 156 black and white photographs and illustrations.

Mathre, D. E. *Compendium of Barley Diseases*. St. Paul, Minn.: APS Press, 1982. 94p. CORE

Treats bacterial, fungal, viral, nematode and animal pest diseases and disorders; also includes disorders caused by environmental stress. Contains 116 color photographs and fifty-five black and white photos and illustrations.

McGee, Denis C. *Maize Diseases: A Reference Source for Seed Technologists*. St. Paul, Minn.: APS Press, 1988. 150p.

Considers 150 diseases of maize caused by fungi, bacteria, mycoplasmas and viruses. Diseases are grouped according to seedborne and seed transmission aspects of pathogens. Information for each disease includes the disease names, pathogen, symptoms, economic importance, disease distribution, host plants, other than maize, variability, control, seedborne aspect, effect on seed quality, pathogen transmission, seed treatments and seed health tests. Includes also some key references to the literature.

McGee, Denis C. *Soybean Diseases: A Reference Source for Seed Technologists*. St. Paul, Minn.: APS Press, 1991. 151p. CORE

Considers all diseases of soybeans caused by fungi, bacteria, mycoplasmas and viruses. Diseases are grouped according to seedborne and seed transmission aspects of pathogens. Information for each disease includes the disease names, pathogen, symptoms, economic importance, disease distribution, other hosts, variability, control, seedborne aspect, effect on seed quality, pathogen transmission, seed treatments and seed health tests. Includes also some key references to the literature.

Mukhopadhyay, A. N. *CRC Handbook on Diseases of Sugar Beet*. Boca Raton, Fla.: CRC Press, 1987. 2 vols.

Considers in seven chapters the diseases caused by fungi, bacteria, mycoplasma, viruses, nematodes and those of noninfectious origin. Covers the history, symptoms, and causes of the diseases, as well as the control measures used in different countries.

Pearson, R. C., and A. C. Goheen. *Compendium of Grape Diseases*. St. Paul, Minn.: APS Press, 1988. 121p.

Considers diseases caused by fungi, bacteria and bacteria-like organisms, viruses and virus-like organisms, nematodes, mites and insects and abiotic factors. Includes effects of cultural practices on disease, selection of planting materials, a glossary, 188 color photographs, and thirty black and white photos and illustrations.

Porter, D. M., D. H. Smith, and R. Rodriguez-Kabana. *Compendium of Peanut Diseases*. St. Paul, Minn.: APS Press, 1984. 93p. CORE

Considers foliar, soilborne, nematode, virus and abiotic diseases. Treats also of beneficial organisms, insects and mites and dodder. Includes a glossary, 157 color photographs and eighty-nine black and white photos and illustrations.

Raychaudhuri, S. P., and J. P. Verma, eds. *Review of Tropical Plant Pathology*. New Delhi, India: Today and Tomorrow's Printers, 1984–1989. 6 vols. CORE
Volume 1. Covers diseases of cereals, maize, millets, rice, and wheat; Vol. 2. Fruit diseases; Vol. 3. Vegetable diseases; Vol. 4. Diseases of plantation crops and forest trees; Vol. 5. Diseases of fibre and oilseed crops; and Vol. 6. Techniques and plant quarantine.

Ricaud, C. et al., eds. *Diseases of Sugarcane: Major Diseases*. Amsterdam: Elsevier, 1989. 399p. CORE
Contributions from authorities in the field grouped under the headings of bacterial diseases, fungal diseases and virus diseases. Includes information on sugarcane quarantine and the world distribution of sugarcane diseases. Contains some color plates.

Rich, A. E. *Potato Diseases*. New York: Academic Press, 1983. 238p. CORE
Diseases are discussed in these sections: bacterial diseases; fungal diseases; virus, viroid and mycoplasma diseases; insect and nematode diseases and non-infectious diseases. Includes also a section on seed potato certification. Appended are a glossary and bibliography.

Richardson, M. J. *Supplement to an Annotated List of Seed-Borne Diseases*. 3d ed. Zurich, Switzerland: International Seed Testing Association, 1981–83. 2 vols. (*Phytopathological Papers* no. 23, suppl. 1–2) CORE
Supplements to: M. J. Richardson, *An Annotated List of Seed-Borne Diseases*. 3d ed. Kew, 3 Eng.: Commonwealth Mycological Institute, 1979, which included references to seed-borne pathogens up to December, 1976. Supplement 1 extends the references to December 1979, supplement 2 to December 1982. Provides a list of Commonwealth Mycology Institute Maps, CMI descriptions of plant pathogenic fungi and bacteria; CMI/ABB descriptions of plant viruses and CIH descriptions of plant parasitic nematodes issued subsequent to the third edition.

Rowe, R. C. *Potato Health Management*. St. Paul, Minn.: APS Press, 1993.
A practical guide which brings together information required for a complete potato crop protection program from pre-planting to postharvest. Covers management of physiological disorders, weeds, insect pests, aphids, leafhoppers, viruses and viruslike organisms, soilborne pathogens and nematodes.

Shaw, H. D., and G. B. Lucas. *Compendium of Tobacco Diseases*. St. Paul, Minn.: APS Press, 1991. 68p.
Covers diseases caused by fungi, bacteria, mycoplasma-like organisms, nematodes, viruses, parasitic higher plants. Also treats non-infectious or abiotic diseases, environmental disorders, nutritional problems, integrated pest management and genetic improvement. Includes a glossary and 145 color photographs.

Sherf, A. F., and A. A. MacNab. *Vegetable Diseases and Their Control*. 2d ed. New York: John Wiley, 1986. 728p. CORE
Chapters cover diseases of asparagus and artichoke, snap and dry beans, lima bean, beet, pea, carrot, celery, corn, crucifer, cucurbits, lettuce, onion, garlic, eggplant, leek and shallots, Capsicum, spinach, sweet potato, potato and some

minor crops. Diseases are listed by causal agents, with information on symptoms, cause, disease cycle and control. Covers bacteria, fungi, viruses, mycoplasmas, nematodes and abiotic agents. Appended are a list of references and a glossary.

Shurtleff, M. C. *Compendium of Corn Diseases*. 2d ed. St. Paul, Minn.: APS Press, 1980. 117p. CORE

Considers disease development and control, infectious diseases, and non-infectious or abiotic diseases, and hosts for maize pathogens. Includes a glossary, a guide to nutrient deficiency symptoms, 76 color photographs and 103 back and white photos and illustrations.

Singh, U., and J. Kumar, eds. *Plant Diseases of International Importance*. Englewood Cliffs, N.J.: Prentice-Hall, 1992–1993. 4 vols. CORE

Volume 1. Diseases of cereals and pulses; Vol. 2. Diseases of vegetable and oil seed crops; Vol. 3. Diseases of fruit crops; and Vol. 4. Diseases of sugar, forest and plantation crops. A comprehensive treatment. The history, distribution (including a distribution map), symptoms, etiology, epidemiology, disease cycle and control are given for each disease discussed. Each chapter contains extensive bibliographic references.

Stutevill, D. L., and D. C. Erwin. *Compendium of Alfalfa Diseases*. 2d ed. St. Paul, Minn.: APS Press, 1990. 104p. CORE

Covers diseases caused by bacteria, mycoplasma-like organisms, fungi, nematodes, viruses, parasitic flowering plants, insects, and abiotic agents, such as nutritional deficiencies, herbicide injury and toxicities. Includes a glossary and 132 color photographs and sixty-six black and white illustrations.

Thurston, H. D. *Tropical Plant Diseases*. St. Paul, Minn.: APS Press, 1984. 208p. CORE

This book treats fifteen important tropical crop groups: cereals; root and tuber crops; food grain legumes; bananas and plantains; beverage crops; tropical fruits and nuts; sugar cane; vegetables; oil crops; rubber; fiber crops; forages; spices; tobacco; and drug crops. Following a brief introduction to each crop group, the major diseases of the crops are described. Includes selected bibliographic references and a chapter on the international agencies involved with tropical plant diseases.

Watkins, G. M. *Compendium of Cotton Diseases*. St. Paul, Minn.: APS Press, 1981. 95p. CORE

Considers infectious diseases, abiotic diseases and crop management for disease control. Includes a guide to the identification of diseases in the field, a glossary and fifty-nine color photographs, and forty-five black and white photos and illustrations.

Whiteside, J. O., S. M. Garnsey, and L. W. Timmer. *Compendium of Citrus Diseases*. St. Paul, Minn.: APS Press, 1988. 105p.

Treats bacterial, fungal, nematode, virus and virus-like diseases. Considers also flowering parasites, mineral deficiencies and toxicities, chemical injuries, pest injuries. Includes a guide to the identification of diseases and disease groups, a glossary and 171 color photographs and thirty-three black and white photos and illustrations.

Whitney, E. D., and J. E. Duffus. *Compendium of Beet Diseases and Insects*. St. Paul, Minn.: APS Press, 1986. 107p. CORE
Covers biotic diseases and disorders, major insects and arthropods, abiotic disorders, and disease diagnosis. Includes a glossary and 243 color photographs and thirty-six black and white photographs and illustrations.

Wiese, M. V. *Compendium of Wheat Diseases*. 2d ed. St. Paul, Minn.: APS Press, 1987. 124p. CORE
Considers diseases caused by bacteria, mycoplasmas, fungi, nematodes, virus and virus-like agents, parasitic plants and abiotic factors. Includes a glossary, seventy-four color photographs and 115 black and white photographs and illustrations. [B1041]

Williams, R. D., ed. *Crop Protection Handbook, Grass and Clover Swards*. Croydon, Eng.: BCPC Publications, 1984. 104p.
Covers in seven chapters these aspects: the significance of weeds, pests and diseases in grass and clover crops; integrated control of weeds, insects and pathogens; control during the establishment phase; management of established grassland; controlling weeds in permanent swards; pests of established grasses and legumes; and diseases of established swards. Appended are: details on the identification of broad-leaved weeds and crop and weed grasses; and safety precaution information.

Zadoks, J. C., and F. H. Rijsdijk. *Agro-Ecological Atlas of Cereal Growing in Europe. Vol. 3. Atlas of Diseases and Pests in Europe*. Wageningen: PUDOC, 1984. 169p. CORE
The distribution and extent of damage caused by pests and diseases of barley, maize, oats, rye and wheat are displayed in sixty-nine maps, each giving the host and disease or pest agent. Crop damage is illustrated by way of two types of maps: maps showing mean annual damage, and those showing isoderms (in which points with equal mean damage are joined by lines). The period covered is primarily 1965–70. Vol. 1 *Agro-Climatic Atlas of Europe;* Vol. 2 *Atlas of the Cereal-Growing Areas in Europe*.

Zillinsky, F. J. *Common Diseases of Small Grain Cereals: A Guide to Identification*. Mexico: Centro Internacional de Mejoramiento de Maiz y Trigo, 1983. 141p. CORE
Intended as a guide to assist in the identification of pathogens of cereal crops. Diseases of bread wheat, durum wheat, barley, oats and triticale are the main focus. Covers the rusts, Helminthosporium diseases, Septoria complex and septoria-like diseases, smuts and bunts, fusarium, root and crown diseases, miscellaneous leaf and head and bacterial diseases. Limited to the use of simple techniques in identifying the most common disease organisms.

Bacterial Diseases

Bradbury, J. F. *Guide to Plant Pathogenic Bacteria*. Farnham Royal, Eng.: CAB International, 1986. 332p. CORE
A comprehensive reference work in two sections: I. treats the nomenclature,

synonymy, characteristics, type cultures, host ranges, symptom and geographical distribution of some 2,000 bacterial names; II. consists of a host-pathogen index of about 2,500 host plants. Includes a bibliography.

Lelliott, R. A., and D. E. Stead. *Methods for the Diagnosis of Bacterial Diseases of Plants*. Oxford: Blackwell Scientific Publications, 1987. 216p.
Aimed at agricultural extension services, this book deals with the selection and treatment of specimens, their isolation procedure, culture and the diagnostic strategy and procedures. It considers the disease symptoms and their expression, host tests, media and methods. Includes a host-pathogen index and a bibliography.

Saettler, A. W., N. W. Schaad, and D. A. Roth, eds. *Detection of Bacteria in Seed and Other Plant Material*. St. Paul, Minn.: APS Press, 1989. 122p.
The detection of bacteria is considered in four sections: review of methods; detection of plant pathogenic bacteria in grain legumes and cereals; detection of plant pathogenic bacteria in vegetables; detection of plant pathogenic bacteria in vegetatively-propagated crops.

Fungi and Fungal Diseases

Batra, L. R. *World Species of Monilinia (Fungi): Their Biology, Biosystematics and Control*. Berlin: J. Cramer, 1991. 246p.
Treats Monilinia and its anamorph, a taxon with some forty phytopathogenic species. In addition to the systematics and life histories of Monilinia, the book considers host-pathogen interactions, symptoms, losses, host specificity and control measures for eight economically important species.

Bailey, J. A., and M. J. Jeger, eds. *Colletotrichum: Biology, Pathology and Control*. Wallingford: CAB International, 1992. 388p.
A comprehensive treatment of a plant pathogenic fungal genus of great economic importance. Coverage includes taxonomy, cellular and molecular biology, epidemiology, field pathology, host resistance, and biological and chemical control.

Colem, G. T., and H. C. Hoch, eds. *The Fungal Spore and Disease Initiation in Plants and Animals*. New York: Plenum Press, 1991. 555p.
Treats the early phases of fungi pathogenesis in plants and animals. Discusses fungal-host interactions as opportunities for research, spore attachment and invasion, host response, and fungal spore products.

Farr, D. F, G. F. Bills, G. P. Chamuris, and A. Y. Rossman. *Fungi on Plants and Plant Products in the United States*. St. Paul, Minn.: APA Press, 1989. 1,252p. CORE
A monumental work on fungi on plants and plant products in the United States with a brief summary of the diseases they cause. This book revises and updates the information on fungi in the *USDA Handbook 165* (1960). It covers 78,000 unique fungus-host combinations and 13,000 accepted fungal names listed by their scientific names. Arranged into three sections: the host-fungus list in which host families are listed alphabetically by genus, with the fungi for each host; the fungus list; and a bibliography of 4,030 references. Includes a host index; host common name index; fungus index; and fungal author names.

Lenne, J. H. *A World List of Fungal Diseases of Tropical Pasture Species*. Wallingford, Eng.: CAB International, 1990. 162p. (*Phytopathological Papers* no. 31) CORE
Diseases recorded on 429 species of thirty-seven genera of legumes; and on 667 species of thirty-one genera of grasses. Includes a comprehensive bibliography of more than 300 references.

Mycoplasmas and Prokaryotes

Maramorosch, K., and S. P. Raychaudhuri. *Mycoplasma Diseases of Crops: Basic and Applied Aspects*. New York: Springer Verlag, 1988. 456p.
An authoritative work bringing together basic and applied knowledge on several plant mycoplasma diseases of crops. In three parts: Part I discusses detection, characterization and cultivation of plant mycoplasmas; Part II treats interactions between insect vectors, plant mycoplasmas and viruses; Part III considers mycoplasma diseases of rice, potato, corn, citrus and other plants, as well as current control technologies.

Singh, R. S. *Plant Pathogens: The Prokaryotes*. New Delhi, India: IBH Publishing Co., 1989. 216p.
Discusses the morphology, taxonomy, reproduction, biology and survival of plant pathogenic prokaryotes. Includes a bibliography.

Nematodes

Brown, R. H. and B. R. Kerry, eds. *Principles and Practice of Nematode Control in Crops*. Sydney, Australia: Academic Press, 1987. 447p. CORE
This book is comprised of contributions from several authorities. It covers the principles of nematode control, chemical control, resistance and tolerance, interaction with other organisms, biological control, quarantine, control strategies in high-value crops, control strategies in low-value crops, and control strategies in subsistence agriculture.

Dropkin, V. H. *Introduction to Plant Nematology*. 2d ed. New York: John Wiley, 1989. 304p. CORE
Discusses the twenty-five major genera of plant parasitic nematodes, their morphology, biology, pathology and control. Provides an in-depth coverage of the relationship between nematodes and plants. Considers the distribution, population dynamics, community structure and ecology of plant parasitic nematodes. Addresses biological control, sanitation, crop rotation, cultural practices and nematicides as control measures.

Luc, M., R. A. Sikora, and J. Bridge, eds. *Plant Parasitic Nematodes in Subtropical and Tropical Agriculture*. Wallingford, Eng.: CAB International, 1990. 629p. CORE
Considers nematode morphology and methods for assessing the effect of nematodes on crop growth in general as well the effect on particular crops or group of crops. Discusses the effect of tropical climates on nematode distribution. In-

cludes color plates, a nematode list and a list of the nematode genra and species cited.

Nickle, W. R., ed. *Manual of Agricultural Nematology*. New York: M. Dekker, 1991. 1035p. CORE
Contributions from authorities in the field describing the major plant nematode groups and their chacteristics. Discusses DNA sequencing and probes for identifying nematodes. Considers the use of insect parasitic nematodes for the control of insects. Contains 2000 references to the literature.

Nickle, W. R. *Plant and Insect Nematodes*. New York: M. Dekker, 1984. 925p. CORE
Covers nematode parasites of soybeans, corn, rice, cotton, potatoes, tobacco, alfalfa, cereals, grasses, vegetable crops, peanuts, citrus, peach and other tree crops, grapes and other small fruit, sugar beet and sugarcane. Includes contributions on the importance of agricultural plant nematology and the history, importance and development of insect nematology. Contains many illustrations and photographs.

Insect and Other Arthropod Diseases

Alford, D. V. *A Colour Atlas of Fruit Pests: Their Recognition, Biology and Control*. London: Wolfe, 1984. 320p.
Addresses the recognition, biology and control of fruit and hop pests. Considers earwigs and thrips, aphids, hoppers, psyllids, mites, miscellaneous pests, parasites and predators and pesticides.

Blackman, R. L. and V. F. Eastop. *Aphids on the World's Crops; An Identification and Information Guide*. Chichester, Eng.: John Wiley, 1984. 466p. CORE
Provides concise information on the identity, economic importance, geographical distribution, host-plant range and life cycle of aphids on agricultural and horticultural crop worldwide. Over 200 crops are listed by their scientific names, followed by their English names. Aphids found on each crop are listed. This is followed by a systematic treatment of aphid genera, their recognition, host-plants, distribution and biology.

Croft, B. A. *Arthropod Biological Control Agents and Pesticides*. New York: Wiley, 1989. 723p.
A comprehensive treatment. Chapters include: pesticide susceptibility; pesticide selectivity; pesticide resistance; and pesticide resistance management.

Dent, D. *Insect Pest Management*. Oxon, Eng.: CAB International, 1991. 604p. CORE
An in-depth coverage of sampling, monitoring and forecasting; yield loss assessment; chemical, biological and cultural control; host-plant resistance; and evaluation of integrated pest management. Treats primarily crop protection, but includes also some examples from medical and veterinary entomology.

Hill, D. S. *Agricultural Insect Pests of Temperate Regions and Their Control*. Cambridge: Cambridge University Press, 1987. 659p.
Discusses pest ecology, principles of pest control, pest damage to crops, biolog-

ical control and pesticides and pesticide application. Treats also the recognition, biology, plant damage, distribution and control of insects and mites in eleven taxa. Annotated lists of major and minor pests are presented under host crops, with indications of the pest distribution and the plant damage.

Hill, D. S. *Agricultural Insect Pests of the Tropics and Their Control*. 2d ed. Cambridge: Cambridge University Press, 1983. 746p. CORE

Intended for students and professional workers concerned with the identification and control of insect and mite pests in tropical regions. Some 310 species are treated in detail. Includes chemical and biological control methods. [C092]

Insect Control Guide. Willoughby, Ohio: Meister Publication, 1992. 447p.

Arranged by crop, with the pests and the controlling agents listed under each crop. Where appropriate, the controlling agents are grouped under the headings of chemical control or biocontrols. Includes indexes to crops, insecticides, biocontrol agents/beneficial organisms and pests. Contains also a directory of manufacturers.

McKinlay, R. G., ed. *Vegetable Crop Pests*. London: Macmillan Press, 1992. 406p.

A comprehensive work on the major invertebrate pests of economic importance on vegetable crops in temperate regions. Information for each pest includes geographical distribution, description, life cycle, plant damage and control, both chemical and non-chemical with emphasis on the non-chemical. Includes a glossary.

Ortega, C. A. *Insect Pests of Maize: A Guide for Field Identification*. Mexico, D.F.: Centro Internacional de Mejoramiento de Maiz y Trigo, 1987. 106p.

A field guide to arthropod pests and their insect parasites and predators. Provides a key for the identification of corn pests, insect pest names, and color photographs.

Steedman, A., ed. *Locust Handbook*. 3d ed. Chatham, Eng.: Natural Resources Institute, 1990. 204p.

Information on biology, distribution and movement of locusts and grasshoppers. Includes distribution maps.

Wilson, M. R. and M. Claridge. *Handbook for the Identification of Leafhoppers and Planthoppers of Rice*. Wallingford, Eng.: CAB International, 1991. 142p.

Intended for use by specialists and field workers. Provides keys for the identification of over seventy leafhopper and planthopper species in the major rice growing regions of the world. Treats the diseases they transmit, their identification, biology, distribution and diagnosis. Richly illustrated. Includes a comprehensive and wide-ranging bibliography.

World Crop Pests. Amsterdam: Elsevier Science Publishers, 1985–. Currently 5 vols.

A continuing series of books providing a very thorough coverage of the world crop pests. Aimed at entomologists, extension specialists, pest control specialists, naturalists and environmentalists. Each volume is comprised of contributions from recognized specialists. All the volumes are divided into three parts: (1) a detailed treatise of the pest organism; (2) an analysis of its interactions with agents for biological control; (3) a discussion of the status and future trends of pest control in the various crops. Volumes to date: Vol. 1 A-B. Spider mites: Their biology, natural enemies and control; Vol. 2 A-C. Aphids: Their biology natural enemies and control; Vol. 3 A-B. Fruit flies: Their biology, natural enemies and control; Vol. 4 A-B. Armored scale insects: Biology, natural enemies and control; Vol. 5. Tortricid pests: Their biology, natural enemies and control.

Virus Diseases

Brunt, A. A., K. Crabtree, and A. Gibbs, eds. *Viruses of Tropical Plants: Descriptions and Lists from the VIDE Database*. Wallingford, Eng.: CAB International, 1990. 707p.

Provides descriptions of v

iruses isolated from tropical plants. It also contains information on virus groups to assist with identifying novel viruses. This book uses the VIDE (Virus Identification Data Exchange) database to describe and identify viruses. "Information on over 500 characters is currently sought for each virus . . . and is stored in the VIDE database." Section 2 lists known natural hosts of plant viruses in tropical plants, first in an alphabetical listing of infected plant species and then as an alphabetical listing by families. Section 3 gives the virus descriptions.

Edwardson, J. R. and R. G. Christie. *CRC Handbook of Viruses Infecting Legumes*. Boca Raton, Fla.: CRC Press, 1991. 680p. CORE

Table 1 presents plant virus groups and some of their characteristics; table 2 treats legumes naturally infected with viruses; and table 3 treats viruses naturally infecting legumes. Gives for each virus, the virus properties and inclusions, the hosts and control. Includes many references to the literature.

Fraenkel-Conrat, H. *The Viruses: Catalogue, Characterization, and Classification*. New York: Plenum Press, 1985. 266p.

A comprehensive catalog to assist with the identification of viruses of vertebrates, invertebrates, plants and prokaryotes. In three sections. Section II deals with plant viruses. Section III lists phages and their hosts.

Mandahar, C. L., ed. *Plant Viruses*. Boca Raton, Fla.: CRC Press, 1989. 2 vols. Volume I. Structure and Replication. Volume II. Pathology.

The structure and replication of plant viruses, viroids, satellites of plant viruses and spiroplasma-like organisms are considered in Volume I. Volume II discusses the diseased plant, viruses as pathogens and disease control.

Matthews, R. E. F. *Diagnosis of Plant Virus Diseases*. Boca Raton, Fla.: CRC Press, 1993. 347p. CORE

A summary of methods and procedures for the diagnosis of plant diseases caused by viruses and viroids. Includes references to the literature.

Matthews, R. E. F. *Plant Virology*. 3d ed. San Diego, Calif.: Academic Press, 1991. 835p. CORE

A comprehensive text on plant virology. Discuses methods for isolation, assay, detection and diagnosis. Treats the replication of viruses, viroids, satellite viruses and satellite RNAs. Considers also transmission, movement and host range, disease symptoms and effects on metabolism, disease induction and economic importance and control. Includes a 120 page bibliography. [B1160]

Plant Virus Diseases of Horticultural Crops in the Tropics and Subtropics. Taipei: Food and Fertilizer Technology Center for the Asian and Pacific Region, 1986. 193p.

Comprises twenty papers presented at the FFTC seminar. Includes an overview of the plant viral situations in Malaysia, Thailand and Korea. Nine papers are devoted to vegetables and eight to fruit crops. Most papers are from Taiwan (7) or the United States (6) with seven countries represented. Papers are brief.

Walkey, D. G. A. *Applied Plant Virology*. 2d ed. London: Chapman and Hall, 1991. 338p. CORE

Examines plant virus classification, virus symptoms, mechanical transmission, virus isolation, purification, identification, epidemiology, control measures, control through resistant cultivars and production of virus-free plants. Includes a glossary.

Weeds

Akobundu, I. O. *Weed Science in the Tropics: Principles and Practices*. New York: J. Wiley, 1987. 522p. CORE

Considers weed biology, ecology, management, and use of herbicides. Emphasis is on problems and control in cereals, legumes, fibre, vegetable and plantation crops. Includes a glossary and a list of common herbicides and trade names.

Aldrich, R. J. *Weed-Crop Ecology: Principles in Weed Management*. North Scituate: Breton Publishers, 1984. 465p. CORE

Considers weed behavior and management in the context of ecology and agricultural management. Treats weed taxonomy and evolution, crop ecology and the role of weed in allelopathy. Discusses the mode of action of herbicides, biological, cultural and chemical weed control. Provides lists of common and scientific names of weeds and chemical names of herbicides.

Ashton, F. M., and T. J. Monaco, eds. *Weed Science: Principles and Practice*. 3d ed. New York: Wiley, 1991. 466p. CORE

In three sections covering weed biology, weed management, herbicide properties, herbicides as they relate to the plant and the soil, herbicide registration, formulation and application. This third edition stresses environmental impact of herbicides and integrated management.

Fletcher, W. W. *Recent Advances in Weed Research*. Farnham Royal, Eng.: Commonwealth Agricultural Bureaux, 1983. 266p. CORE

Presents the major developments in weed research prior to 1982. Discusses weed distribution, the development of new herbicides and plant growth regulators for weed control, biological control of weeds, herbicide application and the mode of action of herbicides.

Gwynne, D. C., and R. B. Murray. *Weed Biology and Control in Agriculture and Horticulture*. London: Batsford Academic and Educational, 1985. 258p.

Discusses weed origin, distribution and seed biology. Considers weed control methods, herbicide formulation and application. Provides a guide to the identification of weed seedlings. Includes a glossary of weed names and a selected bibliography.

Hance, R. J., and K. Holly, eds. *Weed Control Handbook: Principles.* 8th ed. Oxford: Blackwell Scientific Publications, 1990. 582p. CORE
Includes basic practices as well as principles. Chapters by authorities in the field cover weed biology, evolution of weed control, herbicides and their properties, herbicide formulation and application, herbicides in plants and the soil, herbicide resistance and evaluation, weed control in cereals, vegetable crops, fruit and other perennial crops, grasslands and turf. [C126]

Holm, LeRoy, and J. V. Pancho. *A Geographical Atlas of World Weeds.* New York: John Wiley and Sons, 1979. 391p. CORE
"A first approximation of the weed species in the world and their distribution," ranked according to their importance in each area. The Latin and up to three of the most common names are given for each species. [B830]

Lorenzi, H. J., and L. S. Jeffery. *Weeds of the United States and Their Control.* New York: Van Nostrand, 1987. 355p. CORE
Gives descriptions, habitats and control of some 300 weeds in the United States. Provides distribution maps for each weed. Includes tables on methods of herbicide application and lists of herbicides registered for use in the United states and their manufacturers.

TeBeest, D. O., ed. *Microbial Control of Weeds.* New York: Chapman and Hall, 1991. 284p.
The book reviews the current status with contributions from authorities. In five sections: Biological control of weeds; Host-parasite interactions; Genetic manipulation of plant pathogens; Application technology and economic aspects of biological control.

Weed Science Society of America, Herbicide Handbook Committee. *Herbicide Handbook of the Weed Science Society of America.* 6th ed. Champaign, Ill.: Weed Science Society of America, 1989. 301p. CORE
The latest edition of this classic contains thirty-one new compounds. It treats 148 herbicides and some other agricultural chemicals, such as safeners, additives and plant growth regulators. Herbicides are listed alphabetically by their WASSA common names. Entries for each chemical provide information on nomenclature, chemical and physical properties, herbicide uses, use precautions, physiological and biochemical behavior, behavior in soil, toxicological properties, and synthesis and analytical methods. Herbicides that are no longer of commercial interest in North America are listed in a separate section under the heading, "Chemicals Presented in Previous Editions." Includes a list of companies, their products and addresses. Updated every five years. [C065]

Nutritional Diseases

Bergmann, Werner, ed. *Nutritional Disorders of Plants: Development, Visual and Analytical Diagnosis.* Jena and New York: Gustav Fischer, 1992. 742p. Has a color atlas of 386 p.
Contains 945 color pictures of deficiency and toxicity symptoms of plants subject to nutrient deficiencies. Includes chapters on plant analysis and the evaluation

of analytical data by "computerized nutrient element ranges." Text in English, French and Spanish.

Biological Control, Integrated Pest Management, and Allelopathy

Biocontrol News and Information. London: Commonwealth Agricultural Bureaux, 1980– . Quarterly.

An abstracting service for some 2,500 journal articles, reports, conferences and books. Presents a few review papers and news items, such as editorials, announcements, books, conference reports, forthcoming meetings. In addition to biological control, covers integrated control and biology, taxonomy and ecology as relate to biological control.

Burn, A. J., T. H. Coaker, and P. C. Jepson, eds. *Integrated Pest Management*. London: Academic Press, 1987. 474p. CORE

Discusses forecasting and monitoring; cultural methods, the plant and the crop; effectiveness of native natural enemies, chemical methods; genetic control; and planning an integrated pest management system. This is followed by chapters on integrated pest management in horticultural and cereal crops, sugarbeet, fruit and hops, forests, protected crops and stored products.

Cook, R. J., and K. F. Baker. *The Nature and Practice of Biological Control of Plant Pathogens*. St.Paul, Minn.: APS Press, 1983. 539p. CORE

Discusses biological control as it relates to the pathogen, the host, soil ecosystem, the introduction of antagonists and agricultural practices. Lists forty-five antagonists with descriptions of their role in biological control, as well as their taxonomy and morphology. Contains seventy illustrations and a bibliography.

Grainge, M., and S. Ahmed. *Handbook of Plants with Pest-Control Properties*. New York: Wiley, 1988. 470p. CORE

Lists 2,400 plant species with pest control potential divided into three sections: Section I is an alphabetical listing of pest-control species. Section II provides information on the plant characteristics, descriptions of the active materials, methods of preparation, method of application, environmental conditions in use, and additional economic value of the plant. Section III lists alphabetically pests that are controlled by the plant. Includes 1,398 bibliographic references.

Harley, K. L. S., and I. W. Forna. *Biological Control of Weeds: A Handbook for Practioners and Students*. Melbourne, Australia: Inkata Press, 1992. 74p.

A brief "how to book" covering the basic concepts of biocontrol by natural enemies, selection of weeds for biocontrol, biocontrol agents, techniques for the assessment of host specificity and shipping and quarantine procedures.

Leslie, A. R., and G. W. Cuperus, eds. *Successful Implementation of Integrated Pest Management for Agricultural Crops*. Boca Raton, Fla.: Lewis Publishers, 1993. 193p.

Contributions from a symposium organized by the Environmental Chemistry Division of the American Chemical Society which present successful cases of integrated pest management in managing insect pests, diseases and weeds on corn, soybeans, cotton, vegetables, wheat, peanuts and stored grains.

Lomer, C. J., and C. Prior, eds. *Biological Control of Locusts and Grasshoppers.* Wallingford, U.K.: CAB International in association with International Institute of Tropical Agriculture, 1992. 394p.
Proceedings of a workshop at IITA. Section I contains summaries of the activities of the organizations concerned with locust and grasshopper control. "The remaining sections cover technical aspects of locust and grasshopper control; (II) agents for biocontrol; (III) exploration and chacterization; (IV) mass production, application and formulation; and (V) biology, ecology, field experimentation and environmental impact." A summary of the discussions is given in the last section. Presenters are from all continents.

Mandava, Bhushan. *CRC Handbook of Natural Pesticides: Methods.* Boca Raton, Fla.: CRC Press, 1985–90. 6 vols. CORE
This series provides a comprehensive interdisciplinary review of naturally occurring pesticides for insect control. Contents: Vol. 1. Theory, practice and detection; Vol. 2. Isolation and identification; Vol. 3. Insect growth regulators; Vol. 4. Phermones; Vol. 5. Microbial insecticides; Vol. 6. Insect attractants and repellents.

Mukerji, K. G., and K. L. Garg, eds. *Biocontrol of Plant Diseases.* Boca Raton, Fla.: CRC Press, 1988. 2 vols. CORE
Chapters contributed by authorities in the field are devoted to separate aspects of biocontrol: fungal diseases, nematodes, insects and weeds, biocontrol concepts and practices, and agents of biological control and their application. Discusses plant breeding strategies for biocontrol.

Ohlendorf, B., and M. L. Flint, eds. *UC IPM Management Guidelines.* Oakland, Calif.: University of California, Integrated Pest Management Education and Publications, 1990. 2 vols. Looseleaf. (*Publication* no. 3339)
Presents pest management guidelines for twenty major crops in California, including fruit and vegetables, lucerne and cotton. For each crop, information is given on the biology, morphology and control of arthropods, pathogens and weeds. Lists pesticides appropriate for each pest, with recommendations for concentrations. Covers also cultural and biological control methods.

Rice, Elroy L. *Allelopathy.* 2d ed. Orlando, Fla.: Academic Press, 1984. 422p. CORE
Discusses the role of allelopathy in preventing seed decay, the chemical nature of allelopathins and their behavior in plants and the significance of allelopathy in agriculture, forestry, plant pathology and ecosystems. Presents a list of the phyla of plants with allelopathic species. Includes over 1,000 references to the literature. [B189–190]

Rice, Elroy L. *Pest Control with Nature's Chemicals: Allelochemics and Pheromones in Gardening and Agriculture.* Norman, Okla.: University of Oklahoma Press, 1983. 224p.
Considers the effects of crop residues, inhibitory compounds in rice paddies and plant excretions on crops. Discusses allelopathy between crops and weeds and the role of allelopathy in biological weed control. Considers the importance of allelopathy in plant diseases. Treats interactions between organisms: plants to nematodes, plants to insects, and insect to insect. Includes an extensive bibliography. Few illustrations.

Rizvi, S. J. H., and V. Rizvi, eds. *Allelopathy: Basic and Applied Aspects*. London and New York: Chapman and Hall, 1992. 480p.
Contributions from over forty authorities in the field beginning with an overview of the subject, followed by theoretical and conceptual topics. Some specific plant studies and research with allelochemicals are described. Includes new information on allelochemical applications to nematode management.

Tests and Analytical Methods

Cairns, T. and J. Sherma, eds. *Emerging Strategies for Pesticide Analysis*. Boca Raton, Fla.: CRC Press, 1992. 352p.
Reports on analytical technologies in the field of pesticide residue analysis. Covers extraction and cleanup, and alternative analytical approaches to conventional detection methods.

Hickey, K. D., ed. *Methods for Evaluating Pesticides for Control of Plant Pathogens*. St. Paul, Minn.: APS Press, 1986. 312p.
Details pesticide evaluation procedures for the laboratory and greenhouse. Describes field tests for fruit and nut crops; vegetable crops; field crops; ornamentals and turfgrasses; seed treatments; tree injections; and soil treatments. Covers also nematicide test procedures. Includes sixty-one illustrations.

U.S. Environmental Protection Agency. *Manual of Chemical Methods for Pesticides and Devices*. 2d ed. Arlington, Va.: AOAC International, 1992. 1 vol. Looseleaf.
"A compendium of chemical methods for the analysis of pesticides in technical materials and commercial pesticide formulations. The manual contains 287 methods that have been contributed by federal and state agencies and private industries." Includes a cross-reference list and a bibliography.

Wilcox, W. F., ed. *Biological and Cultural Tests for Control of Plant Diseases*. St. Paul, Minn.: APS Press, 1986– . Annual.
An annual series providing brief evaluations of biological and cultural control measures for plant diseases. Each volume contains reports based on replicated disease control tests, as well as tests for these plant groups: fruits and nuts; vegetables; corn and sorghum; soybeans; small grains; field crops; turfgrass; ornamentals and trees; citrus and tropical.

Quarantine

Kahn, R., ed. *Plant Protection and Quarantine*. Boca Raton, Fla., 1988–1989. 3 vols. CORE
Vol. 1. Biological concepts; describes the biological concepts and characteristics of twelve groups of pests, pathogens and weeds. These are related to plant quarantine programs. Vol. 2. Selected pests and pathogens of significance; reviews selected pests. Vol. 3. Special topics; considers the development of safety precautions, guidelines and protocols. Discusses the natural enemies of weeds, insects, mites and ticks, plant pathogens, nematodes, snails and other invertebrates.

FAO Digest of Plant Quarantine Regulations. Rome: Food and Agricultural Organization, 1990. Looseleaf.
Abstracts of the rules of regulations for seventy countries relating to the importation of plant materials. The requirements are organized under various commodity classes for each country. These include the administrative authority, legislation, main provisions, general prohibitions and restrictions, and prohibitions and restrictions for commodity classes.

Safety

Watterson, Andrew. *Pesticide Users' Health and Safety Handbook: An International Guide*. Aldershot Hants: Gower Technical, 1988. 504p.
Datasheets are presented on the health and safety aspects of over 200 pesticides. Each data sheet includes the common, chemical and trade names, synonyms, chemical formula, the RN CAS Rn number, uses, toxicity, regulatory control and standards, and use precautions.

11. Primary United States Historical Literature of Crop Science, 1850–1949

DONNA HERENDEEN
Center for Research Literature, Chicago

WALLACE C. OLSEN
Cornell University

This chapter deals with crops, their improvement, breeding, introduction, pathology, and protection including their entomological and other pests. Research on both crops and crop insects have their roots in their parent worlds of plants and entomology. In the United States prior to 1875, the literature concerned with agricultural crops was imbedded in the general agricultural literature. Knowledge of farming practices or controlling rusts or determining which seeds to plant was incorporated in agricultural periodicals or journals when not passed by word of mouth. This pattern of development of crops literature is the same for farm animals and other rural literature. As the land became more settled, knowledge became more routinized. Before more scientific approaches were introduced, the practitioner used popular periodicals and newspapers for information.

A. Early American Crop Science Literature

One of the major events influencing the development of crop science was the establishment of the U.S. Agricultural Society, founded in 1852 and given much credit for organizing early United States agricultural interests. Its members were the main force behind the congressional acts supporting agricultural education and research. They lobbied to create the U.S. Department of Agriculture which was established in 1862. Other major Congressional acts followed the same year: The Homestead Act of 1862 opened vast areas to settlement and cultivation, causing a boom in agriculture; the Morrill Land Grant College Act (1862) followed by the Hatch Experiment

Station Act (1887) created the academic environment for crop science education and research.[1]

Several individuals such as Orange Judd, Liberty Hyde Bailey, U. P. Hedrick, and Marshall P. Wilder heavily influenced the development of the crop science literature of this period. They are significant because their work, unlike that of the many specialized workers, had broad influence over the entire developing field of crop science.

Orange Judd has been described as

> the most successful editor of the 1850's . . . . Judd and his contemporaries meant to protect readers from ideas that wouldn't work. That is why the *American Agriculturalist* denounced 'humbugs' and the *Country Gentleman* printed fertilizer analyses. The enlarged audience was supposed to rely on good professors and informed editors. No longer collaborators in agricultural study, the readers were to be consumers of information vended by experts. With that, agricultural journalism had assumed much of its present character.[2]

A brief biography of Orange Judd was published in 1927 by William E. Ogilvie.[3]

Liberty Hyde Bailey (1858–1954) was a noted horticulturalist, teacher, writer and editor whose activities were wide ranging. He was President or Chairman of the American Association for the Advancement of Science, American Country Life Association, American Society for Horticultural Science, Botanical Society of America, Association of Agricultural College Experiment Stations, American Pomological Society, and New York State Horticultural Society.[4] His contribution to the horticultural and agricultural literature is impressive; he edited 117 titles by ninety-nine authors between 1890 and 1940.[5] His *Standard Cyclopedia of Horticulture* ran through several editions from 1900 through 1935 and is still in use.

U. P. Hedrick (1879–1951) was a prolific agricultural writer whose many collaborative publications received wide distribution.[6] He is most noted for *A History of Horticulture in America to 1860*. In addition, he published a series of seven immense volumes on the fruits of New York,

1. Wayne D. Rasmussen, ed., *Readings in the History of American Agriculture* (Urbana: University of Illinois Press, 1960).
2. Donald B. Marti, "Agricultural Journalism and the Diffusion of Knowledge: The First Half-Century in America," *Agricultural History* 54 (1) (1980): 36–37.
3. William E. Ogilvie, *Pioneer Agricultural Journalists* (Chicago, Ill.: Arthur G. Leonard, 1927), pp. 29–37.
4. Philip Dorf, *Liberty Hyde Bailey: An Informal Biography* (Ithaca, N.Y.: Cornell University Press, 1959), p. 241.
5. Dorf, *Liberty Hyde Bailey*, p. 244.
6. Paul W. Gates, "Ulysses Prentiss Hedrick, Horticulturist and Historian," *New York History* 47 (3) (July 1966): 219–247.

including *The Cherries of New York; The Grapes of New York*, and *The Peaches of New York*.

Marshall P. Wilder (1798–1886), founder of the American Pomological Society, worked on fruit nomenclature. He endowed the Wilder Medal, considered the most highly coveted medal in American pomology. His creation of the American Pomological Society's *Code of Nomenclature* is still an important reference for fruits grown in the United States.[7]

Plant Breeding and Introduction

During 1850–1949 crop improvement consisted of two major activities, plant breeding and plant introduction. Both have a long history reaching back into prehistoric times, but during the nineteenth century they were modernized.

Modern plant breeding emerged due to the influence of three major works: Charles Darwin's *Origin of Species* in 1859, the rediscovery of Gregor Mendel's genetic principles in 1900, and Hugo de Vries' publication on the theory of mutation in 1901–03.[8] Together they provided the intellectual tools to create scientific plant breeding programs.

In the early 1900s Erwin F. Smith, W. A. Orton, and E. L. Rivers were the first in the United States working on breeding plants that would be resistant to disease. Many methods were developed while trying to breed plants resistant to cotton wilt. From 1925 until 1953 most research utilized and adapted these plant breeding methods. Plant breeding and genetics was radically changed by J. D. Watson and F. H. C. Crick, who constructed a model of the structure of DNA (deoxyribose nucleic acid), the fundamental molecule of inheritance. Watson and Crick published their discovery as "Molecular Structure of Nucleic Acids" in 1953.[9] This knowledge provided the foundation for the field of molecular genetics.

The development of hybrid corn is one of the outstanding contributions of plant breeding made by William James Beal, who in 1876 performed the

7. D. V. Fisher and W. H. Upshall, eds., *History of Fruit Growing and Handling in United States of America and Canada 1860–1972* (University Park, Pa.: American Pomological Society, 1976), pp. 328–338.

8. de Vries' ideas were originally published in *Die Mutationstheorie: Versuche und Beobachtungen über die Entstehung von Arten im Pflanzenreich* (Leipzig: Veit and Comp., 1901–03), translated as *The Mutation Theory: Experiments and Observations on the Origin of Species in the Vegetable Kingdom* (Chicago: Open Court, 1909–10); and in *Species and Varieties, Their Origin by Mutation; Lectures Delivered at the University of California by Hugo de Vries* (Chicago: Open Court, 1905).

9. (a) J. D. Watson and H. C. Crick, "Molecular Structure of Nucleic Acids: A Structure for Deoxyribose Nucleic Acid," *Nature* 171 (4356) (1953): 737–738. (b) Robert Olby, *The Path to The Double Helix* (London: Macmillan Press Ltd., 1974).

first controlled crosses of corn to increase yield. In the early development of crop hybrids there were two research program; one lead by George Harrison Shull (1874–1954) and the other by Edward M. East (1879–1938) and Donald F. Jones (1890–1963). Each group independently worked on systems for growing modern hybrid seed corn. As a result of a paper given by Shull in 1908, East and Shull became aware of each other's work and agreed "that neither of them would ever raise the issue of priority of discovery."[10] In 1919, East and Jones published the landmark *Inbreeding and Outbreeding: Their Genetic and Sociological Significance*. Earlier influential works by East were *The Genetics of the Genus Nicotiana* (1928), *Heterozygosis in Evolution and in Plant Breeding* (1919), and *Inheritance in Maize* (1911).

At this time Barbara McClintock started working in maize genetics and began publishing on cytogenetics and maize breeding in 1926. Most of McClintock's publications appeared in *Genetics, Proceedings of the National Academy of Sciences USA,* and the *Carnegie Institution of Washington Year Book*. She received the Nobel Prize for Physiology or Medicine for her work on "jumping genes" in 1983. Many of the ideas for which she received the Nobel were published in her paper "The Origin and Behavior of Mutable Loci in Maize" in 1950. Her work was ahead of its time and not fully appreciated until much later in the 1970s.[11]

Plant introduction has been no less important than plant breeding. Howard Hyland has emphasized its importance: "One of the basic reasons for the present status of the United States as a world leader in agriculture production lies in the utilization of exotic plant germplasm for crop improvement."[12]

Extensive efforts in plant introduction did not begin before the establishment of the USDA section on Seed and Plant Introduction in 1898. Before that time uncoordinated introductions were made by private explorations, immigrants, shippers and traders, and government officials. Plant introductions between 1898 and 1930 contributed more to crop improvement and production than at any other time in United States agricultural history. David Fairchild was the force behind the establishment and increasing activity of the USDA Seed and Plant Introduction Section. He was an active

10. Grant G. Cannon, *Great Men of Modern Agriculture* (New York: Macmillan, 1963), p. 223.

11. Nina Federoff and David Botstein, *The Dynamic Genome: Barbara McClintock's Ideas in the Century of Genetics* (Plainview, N.Y.: Cold Spring Harbor Laboratory Press, 1992).

12. Howard L. Hyland, "History of Plant Introduction in the United States," in C. W. Yeatman et al., eds., *Plant Genetic Resources: A Conservation Imperative* (Boulder, Colo.: Westview Press, 1985 [*American Association for the Advancement of Science Selected Symposium* no. 87]), p. 5.

introducer and recruited and encouraged numerous others. He became a national celebrity when his autobiography, *The World Was My Garden: Travels of a Plant Explorer*, became a nonfiction best-seller in 1938.[13]

Frank N. Meyer (1875–1918) is one of the most famous of the economic plant explorers. His biography outlines his accomplishments, his expeditions, comments from his contemporaries, and an incomplete list of his plant introductions and germplasm. The Meyer Medal was created in his honor, and the biography includes a list of recipients.[14]

Another explorer, Wilson Popenoe (1892–1975), specialized in South American crops and is most associated with avocado introductions. Frederic Rosengarten's *Wilson Popenoe: Agricultural Explorer, Educator, and Friend of Latin America* covers the life and travels of Popenoe and lists sixty-eight of his principle publications from 1911 to 1971.[15]

While working on grain crop improvement, Mark A. Carleton (1866–1925), discovered a wheat type being grown by Russian Mennonites in Kansas that was more resistant to rust and could endure the harsh conditions of the North American plains. Through his own persistence, and with little encouragement from his colleagues, he traveled to Russia and Central Asia and found the resistant wheat strains. He introduced them to the United States, where they have become the foundation of the United States durum wheat industry.[16] Carleton published *The Small Grains* (1916) as part of Liberty H. Bailey's *Rural Textbook Series*. Carleton's other major writings appeared in publications of the U.S. Department of Agriculture, Bureau of Plant Industry.

Agricultural Entomology

Agricultural entomology in the United States was first supported by societies and later by state and federal governments. The first entomological society to form in the United States was the Entomological Society of Pennsylvania in 1842. Publications of the early North American entomological societies contained mostly taxonomic works. The Brooklyn Entomological Society was established in 1872; its *Bulletin* contained a few articles and

13. Joseph Ewan, *A Short History of Botany in the United States* (New York: Hafner Publishing, 1969), p. 21.

14. Isabel S. Cunningham, *Frank N. Meyer, Plant Hunter in Asia* (Ames: Iowa State University Press, 1984).

15. Frederic Rosengarten, Jr., *Wilson Popenoe: Agricultural Explorer, Educator, and Friend of Latin America* (Lawai, Kaui, Hawaii: National Tropical Botanic Garden, 1991).

16. Edward J. Dies, *Titans of the Soil: Great Builders of Agriculture* (Chapel Hill, N.C.: University of North Carolina Press, 1949), pp. 141–49.

references to insect interactions with plants and galls. When the *Bulletin* ceased, it was replaced by *Entomologica Americana* and a second periodical, *Entomological News*, which began publication in 1895. From its beginning, *Entomological News* dedicated space to applied aspects with papers on subjects such as coconut pests (vol. 1–2) and resistance to cyanide (vol. 3–4).

The Entomological Club of the American Association for the Advancement of Science was established in 1875 and had its meeting activities published in the *Canadian Entomologist*. The *Canadian Entomologist* was for the most part taxonomic in orientation, but included information on insects injurious to crop plants and on agricultural research. The Entomological Club of the AAAS is considered the "mother society" to the American Association of Economic Entomologist (est. 1889) and the Entomological Society of America (est. 1906).[17] The Entomological Club was discontinued after the formation of the American Association of Economic Entomologists.

Unlike most earlier entomology associations, the American Association of Economic Entomologists and its publications focused on the applied rather than the taxonomic aspects of entomology. The American Association of Economic Entomologists frequently held its meetings in conjunction with the meetings of the Association of Agricultural Colleges and Experiment Stations, and members were limited to those entomologists connected with the U.S. Department of Agriculture or State Experiment Stations. This restriction was later modified to allow entomologists from colleges and universities. The American Association initially published its proceedings in *Insect Life* or in the *Bulletins* of the U.S. Department of Agriculture, but in 1908 it created the *Journal of Economic Entomology* as its official publication. The American Association of Economic Entomologists was active in promoting legislation and improvement in research and instruction.

Many entomologists organized regionally, with many of these local organizations becoming branches or affiliates of the American Association of Economic Entomologists. The Pacific Slope Association of Economic Entomologists was organized in 1909 and had its proceedings published in the *Journal of Economic Entomology*. The Cotton States Entomologists was organized in 1908. The New York Entomological Society was organized in 1892 and immediately began the *Journal of the New York Entomological Society*. The Florida Entomological Society, which organized in 1916, was influential in that state and began *The Florida Buggist* in 1917. The Georgia

17. Herbert Osborn, *Fragments of Entomological History, Including Some Personal Recollections of Men and Events* (Columbus, Ohio: Published by the Author, 1937), vol. 1, p. 124.

Entomological Society began *Georgia Entomologist* in 1942.[18] Many more societies appeared later.

The Entomological Society of America was established in 1906 in "response to an evident need for an organization of an international character which would provide for a large number of amateur and professional entomologists who were not included in the American Association of Economic Entomologists on account of the distinctive economic character of that society."[19] J. H. Comstock was the first President; its *Annals* began later.

Government support for economic entomology began when Marshall P. Wilder organized the U.S. Agricultural Society in 1852 to lobby for the establishment of land-grant colleges and the creation of a national department of agriculture. The Morrill Act of 1862 granted land to states for agricultural and industrial colleges, opening a new path for academic support of economic entomology.

The first official entomological course was created by Henry Goadby in 1859 at Michigan State College, and the first course in agricultural entomology was given by Manley Miles at the same institution four years later. A. J. Cook had the first continuous course in entomology in 1867, followed by A. C. Burrill in Illinois (1868), C. H. Fernald in Maine (1871), J. H. Comstock at Cornell (1874), and E. H. Popenoe in Kansas (1879).[20] After Cook at Michigan State College, the second program in agricultural entomology was at Cornell University.[21]

A number of individuals are considered to be among the first professional agricultural entomologists in the United States: A. J. Cook (1842–1916), Asa Fitch (1809–1879), Townend Glover (1813–1883), Thaddeus W. Harris (1795–1856), William LeBaron (1814–1876), J. A. Lintner (1822–1898), A. S. Packard, Jr. (1839–1905), Charles Valentine Riley (1843–1895), William Saunders (1835–1914), Cyrus Thomas (1825–1910), and Benjamin D. Walsh (1808–1869). Sorensen provides a detailed account of the rise and development of agricultural entomology in the United States for this time period.[22]

Thaddeus William Harris was "the most significant teacher of agricultural entomologists" working in the period before 1850. He began publishing in agricultural entomology in 1828 and "stands apart as the original prototype of the profession."[23] One of his valued publications is *A Treatise on Some of*

18. Ibid., pp. 136–139.
19. Ibid., p. 128.
20. Ibid., pp. 93–120.
21. Conner Sorensen, "The Rise of Government Sponsored Applied Entomology, 1840–1870," *Agricultural History* 62 (2) (1988): 112–114.
22. Ibid.
23. Ibid., pp. 99, 100.

the Insects Injurious to Vegetation.[24] Good biographical sketches and references on Harris and later entomologists can be found in *American Entomologists.*[25]

The next generation of agricultural entomologists included Walsh, Fitch, Glover and LeBaron. Walsh edited the *Practical Entomologist* and, with C. V. Riley, *American Entomologist.* Walsh was State Entomologist of Illinois and published such works as *Insects Injurious to Vegetation in Illinois* in the *Transactions of the Illinois State Agricultural Society* betweeen 1861 and 1864. In 1854 Asa Fitch became involved with entomological investigations for the State of New York. He produced fourteen reports all published in the *Transactions of the New York State Agricultural Society* from 1855 to 1872.[26] A recent biography, *Asa Fitch and the Emergence of American Entomology*, includes an entomological bibliography and a catalog of taxonomic names and type specimens.[27]

Walsh was replaced in 1870 as State Entomologist of Illinois by William LeBaron, who wrote on economic entomology for farm newspapers and journals.[28] He became the editor of *Prairie Farmer* in 1865. As State Entomologist many of his works appeared in his annual reports, e.g., the *Outlines of Entomology, Published in Connection with the Author's Annual Reports upon Injurious Insects. Part First, Including the Order of Coleoptera.*[29]

A biography of Townend Glover written by Charles R. Dodge in 1888 includes a bibliography of Glover's publications.[30] His writings were almost exclusively confined to reports published in the *Annals of the Patent Office* and by United States Department of Agriculture. In 1854 Glover was an entomologist for the Bureau of Agriculture in the U. S. Patent Office. He left the Patent Office and later in 1863 was appointed as the first U.S. Entomologist for the newly created Department of Agriculture. Glover was a skillful artist and his publications were beautifully illustrated. The last of

24. Thaddeus W. Harris, *A Treatise on Some of the Insects Injurious to Vegetation*, new ed., (New York: Orange Judd, 1890). (Also published 1862 and 1884)

25. Arnold Mallis, *American Entomologists* (New Brunswick, N.J.: Rutgers Universtiy Press, 1971).

26. Ibid., pp. 35–43.

27. Jeffrey K. Barnes, *Asa Fitch and the Emergence of American Entomology, With an Entomological Bibliography and a Catalog of Taxonomic Names and Type Specimens* (Albany: University of the State of New York, State Education Dept., Biological Survey, 1988 [*New York State Museum Bulletin* no. 461]).

28. Mallis, *American Entomologists*, pp. 49–50.

29. *Fourth Annual Report on the Noxious and Beneficial Insects of the State of Illinois* (Springfield, Ill.: State Journal Steam Print, 1874).

30. Charles R. Dodge, "The Life and Entomological Work of the Late Townend Glover: First Entomologist of the U.S. Dept. of Agriculture," *U.S. Dept. of Agriculture, Division of Entomology Bulletin* 18 (1888).

his long line of monographs was *Illustrations of North American Entomology in the Orders of Coleoptera, Orthoptera, Neuroptera, Hymenoptera, Lepidoptera, Homoptera and Diptera* (Washington, D.C., 1878) which had 273 plates; only twelve copies were printed. Illness forced Townend Glover to retire in 1878; he was replaced by C. V. Riley.

Cyrus Thomas and Joseph Albert Lintner published heavily in the agricultural press. Thomas wrote on economic entomology for the *Prairie Farmer, Rural New Yorker, American Agriculturist, Farmer's Review*, and others. Thomas succeeded Walsh and LeBaron as State Entomologist of Illinois. Lintner wrote 900 articles and thirteen reports, and for twenty-five years was the entomological editor of *Country Gentleman*. His *Entomological Contributions* were published by the New York State Cabinet of Natural History.[31]

A later generation of influential economic entomologists included Charles Valentine Riley, A. S. Packard, W. Saunders, and A. J. Cook. According to one historian ". . . the early publications of the agricultural colleges and experiment stations were for the most part compilations from the writings of Fitch, Riley, Walsh, LeBaron, Thomas, Packard and A. J. Cook."[32]

C. V. Riley stands out in this group. "Riley was undoubtedly one of the greatest and most controversial figures in the history of American Entomology."[33] Riley had over 2,400 publications, published two journals, *American Entomologist* and *Insect Life*, and was the founder and first president of the Entomological Society of Washington. He was also involved in the founding of the Federal Entomological Service, which became part of the Entomological Society of America in 1953.[34] Riley became USDA Entomologist in 1878, but because of an argument with his superior he resigned after nine months. He was reappointed in 1881, and a division of entomology was created in the USDA.

Alpheus S. Packard's influence on economic entomology was through his textbooks and numerous students. His *Guide to the Study of Insects and a Treatise on Those Injurious and Beneficial to Crops* was in its ninth edition by 1889. In 1867, Packard helped found the *American Naturalist* and was its editor for many years. He also published *Insects Injurious to Forest and Shade Trees* (1881) and *Text-Book of Entomology*.[35]

31. Mallis, *American Entomologists*, pp. 50–54.

32. L. O. Howard, "The Rise of Applied Entomology in the United States," *Agricultural History* 3 (3) (1929): 137.

33. Mallis, *American Entomologists*, pp. 69.

34. Judith J. Ho and Willie Yuille, eds., *The Papers of Charles Valentine Riley* (Beltsville, Md.: National Agricultural Library, 1990 [USDA, National Agriculture Library, *Bibliographies and Literature of Agriculture* no. 92]).

35. Mallis, *American Entomologists*, pp. 296–302.

William Saunders was the first Canadian agricultural entomologist. With C. J. S. Bethune in 1868 he founded and edited the *Canadian Entomologist*, and they were the sole contributors to the first two numbers.[36] *Canadian Entomologist* was considered a North American publication rather than a Canadian one and contained numerous submissions from entomologists in the United States.[37]

Albert J. Cook was responsible for introducing Paris green for the control of the codling moth in 1880; he was also one of the founders of the American Association of Economic Entomologists. In addition, he was responsible for securing the financial backing for the *Journal of Entomology*.[38]

Plant Pathology

In his discussion of the beginnings of plant pathology in North America, John A. Stevenson blames the importation of new crops for a steady increase in plant pests and disease problems, noting that relatively little was done to address the problem before 1870.[39] The first movement by the government to address the problem was in 1871, when Thomas Taylor was hired by the U.S. Department of Agriculture to work on plant disease research.

Early government involvement in plant pathology began in the U.S. Department of Agriculture. The USDA established a Mycology Section in 1885 (renamed Vegetable Pathology two years later) with F. Lamson-Scribner as head and the first federal plant pathologist. He focused on grape diseases and was involved in testing the Bordeux mixture. Lamson-Scribner's *Fungus Diseases of the Grape, and Other Plants and Their Treatment* (1894) was the first American phytopathological handbook.[40]

Lamson-Scribner's successor in 1888 was Beverly T. Galloway, who was active in plant disease research. Galloway influenced the development of phytopathological literature when he took over editorship of the *Journal of Mycology*, which changed from a taxonomy-oriented journal to include information on plant pathology. The changes resulted in the *Journal of Mycology* becoming the first American phytopathology journal.[41]

Erwin F. Smith worked on the USDA staff with both Lamson-Scribner

36. Mallis, *American Entomologists*, p. 109.
37. Osborn, *Fragments of Entomological History*, vol. 1, pp. 133–134.
38. Mallis, *American Entomologists*, pp. 138–141.
39. John A. Stevenson, "The Beginnings of Plant Pathology in North America," in C. S. Holton et al., eds., *Plant Pathology, Problems and Progress 1908–1958* (Madison: University of Wisconsin Press, 1959), pp. 14–23.
40. Ibid., p. 20.
41. Ibid.

and Galloway. He is well known for his extensive work on viral diseases, specifically peach yellows. Smith later concentrated on bacterial diseases and is viewed as the founder of that area of phytopathology.[42] He published *Bacteria in Relation to Plant Diseases* (1905) and investigated over one hundred bacterial diseases during his career.[43]

Research in plant pathology and plant breeding meet in the area of disease resistance. The practice of breeding plants to be resistant to disease was extensively used in the United States by W. A. Orton. In the early 1900s, he used plant breeding to develop cotton and other crops to be resistant to wilt diseases.

Before the establishment of pathology programs in universities, the earliest course work in plant pathology was given at the University of Illinois. Beginning in 1873 Thomas J. Burrill taught plant pathology as part of a botany course at that university. Burrill was also the first to establish the connection between bacteria and plant disease in a paper on pear blight in 1877.[44]

The Hatch Act and the Morrill Act were highly influential in the development of phytopathology education and research. As John A. Stevenson notes, "It is impossible to overemphasize the importance of this series of Congressional enactments which established the agricultural colleges and experiment stations, thus making a unique provision for agricultural education and research."[45]

The establishment of the land-grant colleges helped establish departments of plant pathology. The first was in 1903 at the University of California, Berkeley, with R. E. Smith as chief. Plant pathology departments were established in 1907 at Cornell University, with Herbert Hice Whetzel as head, and at the University of Minnesota under E. M. Freeman. The next department at a land-grant university was at the University of Wisconsin in 1909. Along with the specific departments came societies and specialized publications. The American Society of Plant Pathology began in 1909 and two years later introduced its journal *Phytopathology*.[46]

The USDA provided early support for research in nematology. Nathan Cobb was employed by the U.S. Department of Agriculture in 1907 to work in nematology, which he described as "the last great organic group worthy of a separate branch of biological science comparable with entomol-

42. Ibid., p. 21.

43. C. Lee Campbell, "Erwin Frink Smith—Pioneer Plant Pathologist," *Annual Review of Phytopathology* 21 (1983): 25.

44. Stevenson, "Beginnings of Plant Pathology," p. 19.

45. Stevenson, "The Beginnings of Plant Pathology" p. 21.

46. S. E. A. McCallan, "The American Phytopathological Society—The First Fifty Years," in Holton, *Plant Pathology*, pp. 24–31.

ogy"[47] Cobb was one of the authors of the United States Plant Quarantine Act of 1912, which helped bring under control the influx of plant diseases and pests into the United States. In the 1920s the Bureau of Plant Industry created a Division of Nematology with Cobb as director.[48]

Pesticides

The theory of pesticides originated with Marshall Ward, who in 1880 suggested that plants could be protected from disease by chemicals applied before the plant became infected. One of the earliest modern chemicals was the Bordeaux mixture, a fungicide composed of a mixture of copper sulfate and lime, first used by Alexis Millardet in 1882. "The modern era of fungicide research may be dated from Millardet's discovery of Bordeaux mixture in the vineyards of Medoc, France, during the severe epiphytotoic of downy mildew on grape in 1879–1882."[49] The Bordeaux mixture was the major defense against fungal disease for the next seventy-five years; it wasn't replaced until 1956 when an effective oil mixture was introduced.[50]

An early insecticide, Paris green, was first used in 1868 but by 1910 lead arsenate and other substances were also used as insecticides.[51] Because poor quality products were increasingly being used in the United States, the Insecticide Act was passed in 1910 to insure the marketing of legitimate and effective products.

By 1916 synthetic insecticides such as PDB (paradichlorobenzene) developed by USDA entomologist E. B. Blakeslee, came into use. DDT was developed and introduced for use during World War II after being tested by the state agricultural experiment stations.[52]

An understanding of the impact of insecticides developed slowly. In 1940, before DDT was released, there were doubts about its safety. In 1949 the USDA and other agencies began studying the possible effects of insecticide use and establishing limits on the amounts of residue left in food. Congressional hearings followed, and in 1954 Congressman A. L. Miller forced passage of an amendment to the Food, Drug and Cosmetic Act of

47. R. N. Huettel and A. M. Golden, "Nathan Augustus Cobb: The Father of Nematology in the United States," *Annual Review of Phytopathology* 29 (1991): 22.

48. Ibid., p. 24.

49. George L. McNew, "Landmarks During a Century of Progress in Use of Chemicals to Control Plant Diseases," in Holton, *Plant Pathology*, p. 42.

50. Ibid., p. 43.

51. John H. Perkins, *Insects, Experts, and the Insecticide Crisis: The Quest for New Pest Management Strategies* (New York: Plenum Press, 1982), p. 3.

52. Ibid., p. 5.

1938. The amendment required a manufacturer to demonstrate the safety of a product, and it set legal amounts of insecticide allowable in food products. Not until the publication of Rachel Carson's *Silent Spring* (1962) were the effects of insecticides on wildlife and the environment addressed seriously.[53]

B. Historical Literature for Preservation

The objective of this historical chapter is to identify the crops literature worthy of long-term preservation. The need for preservation stems from the changes in the process of papermaking that occurred in the 1860s when rag was replaced by wood fiber. The result was an acidic paper that self-destructs over time by turning yellow and becoming brittle. The paper tears easily, and entire pages fall away or pulverize to dust. By contrast, books from earlier periods with a high rag content are still pliable with little evidence of brittleness, yellowing, or disintegration.

The historical preservation concentrates on the crop science literature published in the United States between 1850 and 1950. Three distinct types of publications important to crop science were identified: (a) monographs, (b) popular and trade periodicals, and (c) scholarly and professional journals. It was not the intention of this project to identify every title published between 1850 and 1950 in the United States. The aim was to identify those publications which could be considered worthy of preservation, or the core of historically valuable published literature.

In the analysis and evaluation processes several categories of publications were excluded from consideration:

(1) Due to their age and rarity, all agricultural monographs published in the United States before 1850 were excluded since they have already been declared of high value and worthy of preservation. The same applies to journals and other publications related to agriculture issued before 1850.

(2) Federal, state, and land-grant publications in series were also excluded since most have been microfilmed or are actively being considered for preservation. The material issued by these three publishing agents are of such value that there appeared little need to evaluate individual series titles.

(3) Foreign publications including those from Canada and the United Kingdom were excluded unless the subject was important to the development of the literature of crop science in the United States.

53. Perkins, *Insects, Experts and the Insecticide Crisis*, pp. 29–55.

The immense literature remaining was evaluated by different methods to ascertain relative value for future historical use. The literature in the following lists must be viewed as but a portion of all the potential crops literature of historical importance. These lists provide the primary or core material in any preservation operation on a national scale.

C. The Historical Monographs

The books were identified using citations gleaned from landmark or overview works published before 1970. Some of these source documents were chosen by Mary Van Buren, former librarian of the New York State Experiment Station in Geneva, N.Y., in consultation with the faculty of the Station and with the director of the Core Agricultural Literature Project. These source documents were extensively augmented by the Core Agricultural Literature Project staff. The following list provided 80% of the monograph titles which were entered into composite lists for peer evaluation.

Source Documents for Identification of Historical Crop Science Monographs

Ahlgren, Gilbert H. *Forage Crops*. 1st. ed. New York, Toronto and London; McGraw-Hill Book Co., Inc., 1949. 418p.

Allard, R. W. *Principles of Plant Breeding*. New York and London; John Wiley and Sons,Inc., 1960.

Bailey, L. H. *The Standard Cyclopedia of Horticulture*. New York and London; Macmillan, 1914. Vol. 3, pp. 150–1562.

Batchelor, L. D., and H. J. Webber, eds. *Citrus Industry. Volume II. Production of the Crop*. 1st ed. Berkeley and Los Angeles, Calif.; University of California Press, 1948.

Chandler, William H. *Evergreen Orchards*. 2d ed., rev. Philadelphia, Pa.; Lea and Febiger, 1958. 525p. (Reprinted 1964.)

Chupp, Charles, and Adam F. Sherf. *Vegetable Diseases and Their Control*. New York; The Ronald Press Co., 1960. 693p.

Coats, Alice M. *The Quest for Plants: A History of the Horticultural Explorers*. London; Studio Vista, Ltd., 1969. 400p.

Esau, Katherine. *Plant Anatomy*. New York; John Wiley and Sons, Inc., 1953. 735p.

Foster, Adriance S., and E. M. Gifford. *Comparative Morphology of Vascular Plants*. San Francisco, Calif.; W. H. Freeman and Co., 1959. 555p.

Gardner, V. R. *Principles of Horticultural Production*. East Lansing, Mich.; Michigan State University Press, 1966. 583p.

Gill, N. T., and K. C. Vear. *Agricultural Botany*. 2d ed. London; Gerald Duckworth and Co., Ltd., 1966. 637p.

Great Britain. Ministry of Agriculture and Fisheries. *Books on Agriculture and Horticulture*. London; Her Majesty's Stationery Office, 1954.

Great Britain. Ministry of Agriculture and Fisheries. *A Selected and Classified List of Books in English Relating to Agriculture, Horticulture, etc. in the Library of the Ministry of Agriculture*. London; His Majesty's Stationery Office, 1939. 89p. (*Bulletin* no. 78)

Harvey-Gibson, R. J. *Outlines of the History of Botany*. London; A. and C. Black, Ltd., 1919. 274p.

Hill, Albert F. *Economic Botany: A Textbook of Useful Plants and Plant Products*. 2d ed. New York, Toronto and London; McGraw-Hill Book Co., Inc., 1952. 560p.

Holman, Richard and Wilfred W. Robbins. *Textbook of General Botany*. 2d ed. New York; J. Wiley and Sons, 1924. 590p.

Janick, Jules, ed. *Classic Papers in Horticultural Science*. Englewood Cliffs, N.J.; Prentice-Hall, 1989. 585p.

Janick, Jules, et al. *Science and Introduction to World Crops*. 3d ed. San Francisco, Calif.; W. H. Freeman and Co., 1981. 868p.

Lawrence, George H. M. *Taxonomy of Vascular Plants*. New York; The Macmillan Co., 1951. 823p.

Lawrence, George H. M., and Kenneth F. Baker. *History of Botany*; Two Papers Presented at a Symposium, William Andrews Clark Memorial Library, Dec. 1963. Los Angeles, Calif.; University of California, and Pittsburgh, Pa.; Carnegie Institute of Technology, 1965. 70p.

Mabberley, D. J. *The Plant Book: A Portable Dictionary of the Higher Plants . . .* arranged largely on the principles of editions 1–6 (1896/1897–1931) of Willis' *A Dictionary of Flowering Plants and Ferns*. Cambridge University Press, 1987. 706p. (1st ed., 1896.)

Metcalfe, C. R., and L. Chalk et al. *Anatomy of the Dicotyledons Leaves, Stem and Wood in Relation to Taxonomy with Notes on Economic Uses*. Oxford; Clarendon Press, 1950. 2 vols. 1500p.

Nieuwhof, M. *Cole Crops: Botany, Cultivation, and Utilization*. London; Leonard Hill Books, 1969. 353p.

Nonnecke, Ib Libner. *Vegetable Production*. New York; Van Nostrand Reinhold, 1989. 657p.

Salaman, Radcliffe N. *The History and Social Influence of the Potato*. Cambridge University Press, 1949. 685p.

Schilletter, Julian C. and Harry W. Richey. *Textbook of General Horticulture*. 1st ed. New York and London; McGraw-Hill Book Co., Inc., 1940. 367p.

Shoemaker, James S. *General Horticulture*. 2d ed. Chicago, Philadelphia and New York; J. B. Lippincott Co., 1956. 464p.

Synge, P. M., and E. Napier, eds. *Fruit, Present and Future*. London; Royal Horticultural Society, 1966. 2 vols.

Thompson, Homer C. *Vegetable Crops*. 4th ed. New York, Toronto and London; McGraw-Hill Book Co., 1949. 611p.

Tindall, H. D. *Fruits and Vegetables in West Africa*. Rome; Food and Agriculture Organizaiton of the United Nations, 1965. 259p.
Webber, Ronald. *The Early Horticulturists*. Newton Abbot, U.K.; David and Charles (Holdings) Ltd., 1968. 224p.
Weber, Herbert J. and Leon D. Batchelor. *The Citrus Industry; Volume I. History, Botany and Breeding*. Berkeley and Los Angeles, Calif.; University of California Press, 1948. 1028p.
Whittle, Tyler. *The Plant Hunters: Being an Examination of Collecting with an Account of the Careers & the Methods of a Number of Those Who Have Searched the World for Wild Plants*. Philadelphia, New York and London; Chilton Book Company, 1970. 281p.
Wilsie, Carroll P. *Crop Adaptation and Distribution*. San Francisco and London; W. H. Freeman and Co., 1962. 448p.
Wilson, Harold K. *Grain Crops*. 2d ed. New York, Toronto and London; McGraw-HillBook Co., Inc., 1955. 396p.

Approximately 2,200 monographs were identified by this method and entered into a master list. The monograph list was then culled to remove publications under fifty pages in length, those considered to have no application to United States crops and practices, and those strongly oriented to ornamentals and landscape gardening. After an analysis of the titles and their subject concentrations, an additional source was used to provide coverage in important subject areas. This was the *Dictionary Catalog of the National Agricultural Library, 1862–1965*, where select subject headings were surveyed and approximately 400 additional titles gathered. A final list of about 1,600 monograph titles was submitted to crop scientists for evaluation and ranking.

With such a sizeable list of monographs to be evaluated, finding reviewers with adequate background and knowledge was a prodigious task. With the aid of older Cornell faculty members in crop and plant breeding, pathology, and protection, reviewers were found who could evaluate and rank the titles. The Mann Library is very appreciative of the assistance of these five reviewers.

Reviewers of Historical Crop Science Monographs

R. F. Carlson
Dept. of Horticulture
Michigan State University

Arthur A. Muka
Dept. of Entomology
Cornell University

Edward H. Smith
Dept. of Entomology
Cornell University

D. G. Woolley
Iowa State University

Charles B. Heiser, Jr.
Dept. of Biology
Indiana University, Bloomington

Lists were sent to ten additional reviewers but were not returned. Reviewers were encouraged to suggest additional titles for incorporation into the list, although no more than fifteen titles were recommended. Reviewers were asked to rank the titles keeping these criteria in mind:

(1) The work had an important influence in the field, whether as a work by an important author, as a work frequently referred to by later authors, as a major compilation or compendium of the time, or as a foundation book on which later efforts were built.
(2) The work was the first of its kind published or a work recording major advances in the field.
(3) The work embodies an historical record of changes in the field.
(4) The title is a superior work of a leader in the crop or horticultural sciences.
(5) The work includes unusual or valuable etchings, prints, or illustrations.
(6) The title has survived in a very limited number of copies.

A three-point scale was used for reviewers' rankings:

(1) A very important historical title worthy of preservation
(2) Worth preserving but of secondary importance
(3) A title of marginal historical value

The final rankings were divided into two levels, the first level indicating higher priority. Over a third of the titles on the original list were eliminated from the final ranking. The resulting monograph list has 1,071 titles, of which 246 (23%) are first priority for preservation.

Primary Core Historical Monographs Worthy of Preservation, 1850–1949 (N = 1,071)

A

Second Abbott, Francis B. Hand-Book of Small Fruits. Chicago, Ill.; 1889?

Second Adams, Richard L. Field Manual for Sugar Beet Growers; A Practical Handbook for Agriculturists, Field Men and Growers. Chicago, Ill.; Beet Sugar Gazette Co., 1913. 134p.

First Adriance, Guy W., and F. R. Brinson. Propagation of Horticultural Plants. 1st ed. New York and London; McGraw-Hill, 1939. 314p. (2d ed., New York, 1955. 298p.)

Second Ahlgren, G. H. Forage Crops. 1st ed. New York, Toronto, London; McGraw-Hill Book Co., Inc., 1949. 418p.

Second Aiken, George D. Pioneering with Fruits and Berries. New York; Stephen Daye Press, 1936. 94p.

Second Alexander, E. R., et al. Southern Field-Crop Enterprises, Including Soil Management . . . ed. by Kary C. Davis. Philadelphia and Chicago; J. B. Lippincott Co., 1928. 550p. (New Edition, 1937. 574p.)

Second Alford, George H. How to Prosper in Boll Weevil Territory. Chicago, Ill.; International Harvester Co. of New Jersey, Agricultural Extenision Department, 1914. 32p.

First Allen, C. L. Cabbage, Cauliflower and Allied Vegetables from Seed to Harvest. New York; Orange Judd Co., 1901. 127p.

Second Allen, Lewis F. Rural Architecture; Being a Complete Description of Farmhouses, Cottages, and Outbuildings, Comprising Wood Houses, Workshops, Tool Houses, Carriage and Wagon Houses, Stables, Smoke and Ash Houses, Ice-houses, Apiary or Bee House, Poultry Houses, Rabbitry, Dovecote, Piggery, Barns and Sheds for Cattle, etc., Together with Lawns, Pleasure Grounds and Parks; the Flower, Fruit and Vegetable Garden; also, Useful and Ornamental Domestic Animals for the Country Resident, etc; Also, the Nest Method of Conducting Water into Cattle Yards and Houses. Beautifully illustrated. New York; C. M. Saxton, 1863. 378p.

Second Allen, Walter F. English Walnuts; What You Need to Know About Planting, Cultivating and Harvesting This Most Delicious of Nuts. Lawrenceville, N.J.; W. F. Allen, 1912. 29p.

Second American Agriculturist. Onions. How to Raise Them Profitably . . . New York; Orange Judd Co., 1859. 31p.

First Ames, O. Economic Annuals and Human Cultures. Cambridge; Botanical Museum of Harvard University, 1938.

Second Anderson, E. Introgressive Hybridizaiton. New York; John Wiley and Sons, 1949.

Second Anderson, Oscar G. and F. C. Roth. Insecticides and Fungicides, Spraying and Dusting Equipment: A Laboratory Manual with Supplementary Text Material. New York; J. Wiley and Sons, Inc., 1923. 349p.

First Arber, Agnes. The Gramineae; the Study of Cereal, Bamboo, and Grass. Cambridge; Cambridge University Press, 1934. 480p.

Second Arthur, Joseph C. Manual of the Rusts in the United States and Canada. New York; John Wiley and Sons, 1934. (Also published in 1934 by Purdue Research Foundation, Lafayette, Ind. Reissued by New York; Hafner Pub. Co., 1962.)

Second Arthur, Joseph C., et al. The Plant Rusts (Uredinales). New York; J. Wiley and Sons, London; Chapman and Hall, Ltd., 1929. 446p.

Second Ashmead, William H. Orange Insects; A Treatise on the Injurious and Beneficial Insects Found on the Orange Trees of Florida. Jacksonville, Fla.; Ashmead Bros., 1880. 78p.

Second Auchter, E. C., and H. B. Knapp. Orchard and Small Fruit Culture. New York and London; Wiley, Chapman and Hall, 1937. 627p. (Later ed., New York; John Wiley and Sons, 1949.)

Second Avery, G. S., et al. Hormones and Horticulture. New York; McGraw-Hill Co., 1947.

B

First Babcock, Ernest B., and R. E. Clausen. Genetics in Relation to Agriculture. 1st ed. New York; McGraw-Hill Book Co., 1918. 675p. (2d ed., 1927. 673p.)

Second Baerg, William J. Introduction to Applied Entomology. Russellville, Ark.; Russellville Printing Co., 1937. 72p. (2d ed., Minneapolis, Minn.; Burgess Pub. Co., 1942. 146p.)

Second Bagenal, N. B., ed. Fruit Growing: Modern Cultural Methods. London; Ward, Lock and Co., Ltd., 1939. 399p. (Rev. ed., 1945. 416p.)

Second Bailey, Alton E., ed. Cottonseed and Cottonseed Products: Their Chemistry and Chemical Technology. New York; Interscience Publishers, 1948. 936p.

First Bailey, C. H. The Chemistry of Wheat Flour. New York; Reinhold Pub. Corporation, 1925. 324p.

First Bailey, L. H. The Apple Tree. New York; Macmillan Co., 1922. 117p.

Second Bailey, L. H. Cyclopedia of American Horticulture; Comprising Suggestions for Cultivation of Horticultural Plants, Descriptions of the Species of Fruits, Vegetables, Flowers and Ornamental Plants Sold in the United States and Canada, Together with Geographical and Biographical Sketches. Assisted by Wilhelm Miller, illustrated with over two thousand original engravings. New York; The Macmillan Co., 1900–1902. 4 vols. 2016p. (4th ed. (bound in 6 vols with additional preface, extra plates and a synopsis of the vegetable kingdom) New York; Doubleday, Page and Co., 1906. 6th ed., 1909.)

First Bailey, L. H. Hortus. New York; The Macmillan Co., 1941. (1935)

Second Bailey, L. H. Manual of Cultivated Plants Most Commonly Grown in the Continental United States and Canada, by L. H Bailey and the Staff of the Bailey Hortorium at Cornell University. Rev. ed., completely restudied. New York; Macmillan Co., 1949. 1116p.

Second Bailey, L. H. Manual of Cultivated Plants; a Flora for the Identification of the Most Common or Significant Species of Plants Grown in the Continental United States and Canada, For Food, Ornament, Utility, and General Interest, Both in the Open and Under Glass. New York; Macmillan, [1924]. 851p. (2d ed., London; Macmillan Co. 1949.)

First Bailey, L. H. The Standard Cyclopedia of Horticulture. New York and London; Macmillan, 1914. 6 vols. (Later published, New York and London; Macmillan, 1947. 3 vols.)

Second Bailey, L. H. The Survival of the Unlike; A Collection of Evolution Essays Suggested by the Study of Domestic Plants. New York; Macmillan, 1896. 515p. (5th ed., 1906. 515p.)

Second Baker, J. O. The Complete Market Gardener. London; Gifford, 1949.

Second Balls, William L. Studies of Quality in Cotton. London; Macmillan and Co., Ltd, 1928. 376p.

Second Barker, Ralph E. Small Fruits. New York; Rinehart, 1954. 90p.

Second Barnhart, Floyd. Cotton. 1st ed. Caruthersville, Mo.; 1949. 279p. (2d ed., 1949. 5th ed., 1956.)

First Barron, A. F. British Apples. London; Macmillan, 1884. 248p.

Second Barske, Charlotte. King Cotton, The Story of Cotton With a Moving Picture of Build . . . New York; 1938. 23p.

Second Bartholomew, Elam. Handbook of the North American Uredinales (Rust Flora). Stockton, Kans.; H. L. Covert, 1928. 193p. (2d ed., 1933.)

Second Barton, L. V., and W. Crocker. Twenty Years of Seed Research at Boyce Thompson Institute for Plant Research. London; Faber and Faber, 1948. 148p.

Second Bartrum, Edward. The Book of Pears and Plums . . . with Chapters on Cherries and Mulberries. London and New York; J. Lane, 1903. 96p.

Second Batchelor, L. D., and H. J. Webber, ed. The Citrus Industry. 1st ed. Berkeley and Los Angeles; University of California Press, 1948.

Second Bates, Frank A. How to Make Orchards Profitable. Boston; The Ball Publishing Co., 1912. 123p.

Second Bates, G. Weed Control. London; Spon, 1948. 235p.

Second Bateson, William Mendel's Principles of Heredity, by W. Bateson. Cambridge, U.K.; University Press, 1909. (Later ed., 1913. 413p.)

Second Bateson, William. The Methods and Scope of Genetics. Cambridge, U.K.; University Press, 1908. 49p.

Second Bateson, William. Problems of Genetics, by William Bateson. New Haven, Conn.; Yates University Press, 1913. 258p.

Second Bateson, William. Materials for the Study of Variation, Treated with Special Regard to Discontinuity in the Origin of Species. London, New York; Macmillan, 1894. 598p.

Second Baur, Erwin. On the Etiology of Infectious Variegation. Originally publ.: "Zur Aetiologie der infectiösen Panachierung." in Berichte der Deutschen Botanischen Gesellschaft. 22 (1904): 453–460. Ithaca, N.Y.; American Phytopathological Society, 1942. [*Phytopathological Classics* no. 7]

Second Bawden, Frederick C. Plant Diseases. London; Thomas Nelson and Sons, 1948. 206p.

Second Bayliss, William M. Principles of General Physiology. London and New York; Longmans, Green and Co., 1915. 850p. (5th ed., 1960.)

Second Beal, W. J. Grasses of North America. New York; Henry Holt and Co., 1887. 2 vols.

Second Bealby, J. T. How to Make an Orchard in British Columbia. A Handbook for Beginners. New York; The Macmillan Co., 1912. 86p.

Second Bean, W. J. Trees and Shrubs Hardy in the British Isles. 1st ed. London; J. Murray, 1914–1933. Vols. I-II have date, 1914; vol. III, 1933. 3 vols. (3d ed., London; J. Murray, 1921. 2 vols. 8th ed., rev., George Taylor, ed., London; J. Murray 1970–81. 4 vols.)

Second Beardsley, Josephine B. From Wheat to Flour. Chicago, Ill.; Wheat Flour Institute, 1937. 38p.

Second Beattie, W. R. Celery Culture. New York; Orange Judd Co., 1907. 143p.

Second Beaven, E. S. Barley. Edinburgh; James Thin, 1947.

Second Beck, F. V. Field Seed Industry in the United States; an Analysis of the Production, Consumption and Prices of Leguminous and Grass Seeds. Madison; University of Wisconsin Press, 1944.

Second Bedford, Duke of, and S. Pickering. Science and Fruit Growing: Being an Account of the Results Obtained at the Woburn Experimental Fruit Farm Since Its Foundation in 1894. London; Macmillan and Co., Ltd., 1919. 348p.

Second Beijerinck, M. W. *Concerning a Contagium Vivum Fluidum as a Cause of the Spot-Disease of Tobacco Leaves*, English translation by J. Johnson (Ithaca, N.Y.: American Phytopathological Society, 1942 [*Phytopathological Classics* no. 7]). (Originally published as "Ueber ein Contagium Vivum Fluidum als Ursache der Fleckenkrankheit der Tabaksblätter" in *Verhandelingen der Koninklyke Akademie van Wettenschappen te Amsterdam* 65 (2) (1898): 3–21.)

Second Bengtson, Nels A., and D. Griffith. The Wheat Industry; For Use in Schools. New York; Macmillan Co., 1915. 341p.

First Bennett, Frederick T. Outlines of Fungi and Plant Diseases for Students and Practitioners of Agriculture and Horticulture. London; Macmillan and Co., 1924. 254p.

Second Berkeley, M. J. Observations, Botanical and Physiological, on the Potato Murrain . . . together with selections from Berkeley's Vegetable Pathology. East Lansing, Mich.; American Phytopathological Society, 1948. 108 p. The Observations . . . were published originally in the Journal of the Horticultural Society of London (1846) l: 9–34; Vegetable Pathology appeared in 173 articles in the Gardeners' Chronical, between 7 Jan. 1854 and 3 Oct. 1857. [*Phytopathological Classics* no. 8]

Second Bewley, William F. Diseases of Glasshouse Plants. London; E. Benn, Ltd., 1923. 208p.

Second Bews, J. W. The World's Grasses: Their Differentiation, Distribution, Economics, and Ecology. London and New York; Longmans, Green and Co., 1929. 408p.

Second Biggle, Jacob. Biggle Berry Book; Small Fruit Facts From Bud to Box, Conserved into Understandable Form. Philadelphia; Wilmer Atkinson Co., 1911. 144p. (On cover: Biggle Farm Library. 5th ed., 1913.)

Second Biggle, Jacob. Biggle Garden Book; Vegetables, Small Fruits and Flowers for Pleasure and Profit. Philadelphia; W. Atkinson Co., 1908. 184p. (On cover: Biggle Farm Library. 3d ed., 1912.)

First Biggle, Jacob. Biggle Orchard Book; Fruit and Orchard Gleanings from Bough to Basket, Gathered and Packed into Book Form. Philadelphia; W. Atkinson Co., 1906. 144p. (On cover; Biggle Farm Library; 2d ed., 1908. 3d ed., 1911.)

Second Bigwood, George. Cotton. London; Constable and Co., Ltd., 1918. 203p.

First Black, J. J. The Cultivation of the Peach and the Pear on the Delaware and Chesapeake Peninsula: With a Chapter on Quince Culture and the Culture of Some of the Nut-Bearing Trees. Wilmington, Del.; The James and Webb Co., 1886. 397p. (New York; Orange Judd Company, 1887.)

First Blacknall, O. W. New and Enlarged Manual on Practical Strawberry and Berry Fruit Culture, also of Grapes, Asparagus, Rhubarb, etc. Kittrell, N.C.; O. W. Blacknall, 1900. 118p.

First Bonavia, Emanuel. The Cultivated Oranges and Lemons, Etc. of India and Ceylon; With Researches into their Origin and the Derivation of their Names, and Other Useful Information, with an Atlas of Illustrations.

London; W. H. Allen, 1888–1890. 394p. (Later Printing, Dehra Dun; Bishen Singh Mahendra Pal Singh and Periodical Experts. 1973.)

Second Bosson, Charles P. Observations on the Potato, and Remedy for the Potato Plague. Boston; Published by E. L. Pratt, 1846. 118p.

Second Bourcart, Emmanuel. Insecticides, Fungicides, and Weedkillers: A Practical Manual on the Diseases of Plants and Their Remedies, for the Use of Manufacturing Chemists, Agriculturists, Arboriculturists and Horticulturists, translated from the French, rev. and adapted to British Standards and Practice, by Donald Grant. London; Scott, Greenwood and Son, 1913. 431p. (2d English ed., rev. and enl. by Thomas R. Burton. Lodon; E. Benn, 1926. 431p.)

Second Bowdidge, Elizabeth. The Soya Bean; Its History, Cultivation (in England) and Uses . . . forward by Sir John T. Davies. London; Oxford University Press, H. Milford, 1935. 83p.

Second Boy Scouts of America. Citrus Fruit Culture. New York; Boy Scouts of America, 1931. 68p.

Second Boy Scouts of America. Cotton Farming. New York; Boy Scouts of America, 1931. 57p.

First Boyd, John R. M. The Farmers' Alfalfa Guide. Facts Based Upon Experience in the Field . . . Indianapolis, Ind.; 1913. 33p. (Columbus, Ohio; J. R. M. Boyd, 1914. 47p. Rev. ed., Pittsburg, Pa.; Farmers' Alfalfa Bureau, 1917. 46p.)

Second Boys, Charles V. Weeds, Weeds, Weeds. London; Wightman, 1937. 71p.

Second Boysen-Jensen, P. Growth Hormones in Plants. 1st ed. New York; McGraw-Hill Book Co., 1936.

Second Braungart, Dale C. An Introduction to Plant Biology. St. Louis; Mosby, 1942. 411p.

Second Brefeld, Oscar. Investigations in the General Field of Mycology . . . trans. by Frances Dorrance. Wilkes-Barre, Pa.; Heinrich Schoningh. 59p.

Second Brenchley, Winifred E. Weeds of Farm Land. London; Longmans, Green and Co., 1920. 239p.

Second Brett, W. Vegetable Growing. London; Pearson, 1948. 192p.

Second Bridgeman, Thomas. Fruit Gardening. Containing Complete Practical Directions for the Selection, Propagation and Cultivation of All Kinds of Fruit. Philadelphia, Pa.; H. T. Coates and Co., 1901. 211p.

Second Bridgeman, Thomas. Kitchen-Gardening; Containing Complete Practical Directions for the Planting and Cultivation of All Kinds of Vegetables. Philadelphia; Henry T. Coates and Co., 152p.

First Brill, F. Farm-Gardening and Seed Growing. New York; Orange Judd, 1872. 151p. (Later ed., 1884. 166p.)

Second Brooks, C. P. Cotton; Its Uses, Varieties, Fibre Structure, Cultivation and Preparation For Market and as an Article of Commerce, Also the Manufacture of Cotton Seed Oil and Cotton Seed Meal and Fertilizer. New York; Spon and Chamberlain, London; E. and F. N. Spon, Ltd., 1898. 384p.

First Brooks, Frederick T. Plant Diseases. London and New York; Oxford University Press, 1928. 386p. (2d ed., 1953. 457p.)

First Brooks, Reid M., and H. P. Olmo. Register of New Fruit and Nut Varieties,

1920–1950. Berkeley and Los Angeles, Calif.; University of California Press, 1952. 206p. (2d ed., 1972. 708p.)

First Brown, Bliss S. Modern Fruit Marketing; A Complete Treatise Covering Harvesting, Packing, Storing, Transporting and Selling of Fruit. New York; Orange Judd Co., 1916. 283p.

Second Brown, H. D., and C. S. Hutchison. Vegetable Science. Chicago and London; Lippincott, 1949.

First Brown, Harry B. Cotton; History, Species, Varieties, Morphology, Breeding, Culture, Diseases, Marketing, and Uses. 1st ed. New York; McGraw-Hill Book Co., 1927. 517p. (2d ed., 1938. 592p. 3d ed., 1958. 566p.)

Second Bruck, Werner F. Plant Diseases . . . trans. by J. R. Ainsworth-Davis. London, Glasgow and Bombay; 1912. 153p.

Second Brues, Charles T. Insects and Human Welfare; an Account of the More Important Relations of Insects to the Health of Man, to Agriculture, and to Forestry. Cambridge; Harvard University Press, 1920. 104p. (Rev. ed, 1947. 154p.)

Second Bruner, Lawrence. A Preliminary Introduction to the Study of Entomology. Together with a Chapter on Remedies, or Methods that can be Used in Fighting Injurious Insects; Insect Enemies of the Apple Tree and its Fruit, and the Insect Enemies of Small Grains. Lincoln, Nebr.; J. North and Co., Printers, 1894. 322p.

First Budd, J. L. A Horticultural Handbook. De Moines, Iowa; Wallace Pub. Co., March, 1900. 160p.

Second Budd, J. L., and N. E. Hansen. American Horticultural Manual. New York; John Wiley and Sons, 1902. 417p. (Part I Comprising the leading principles and practices connected with the propagation, culture, and improvement of fruits, nuts, ornamental trees, shrubs, and plants in the United States and Canada)

Second Buell, Jonathan S. The Cider Makers' Manual. A Practical Hand-Book, Which Embodies Treatises on the Apple; Construction of Cider Mills, Cider-Presses, Seed-Washers, and Cider Mill Machinery in General; Cider Making; Fermentation; Improved Processes in Refining Cider, and Its Conversion into Wine and Champagne; Vinegar Manipulation by the Slow and Quick Processes; Imitation Ciders; Various Kinds of Surrogate Wines; Summer Beverages; Fancy Vinegars, etc . . . Rev. ed with additions. Buffalo; Hans, Nauert and Co., 1874. 183p.

Second Bull, Henry G., ed. The Apple and Pear as Vintage Fruits. The Technical Descriptions of the Fruit are for the Most part by Robert Hogg. Hereford; Jakeman and Carver, 1886. 247p. ("This work was issued under the auspices of the Woolhope naturalists' field club, and forms a condensed and cheap edition of the information in The Herefordshire Pomona on vintage fruits." - Pref.)

Second Bull, Henry G., ed. The Herefordshire Pomona, Containing Coloured Figures and Descriptions of the Most Esteemed Kinds of Apples and Pears. With Illustrations Drawn and Coloured From Nature by Miss Ellis and Miss Bull. Hereford, U.K.; Jakeman and Carver, 1876–1885. 2 vols.

First Buller, Arthur H. R. Essays on Wheat, Including the Discovery and Introduction of Marquis Wheat, the Early History of Wheat-Growing in Western Canada, The Origin of Red Bobs and Kitchener, and the Wild Wheat of Palestine. New York; The Macmillan Co., 1919. 339p.

Second Bunyard, Edward A. A Handbook of Hardy Fruits More Commonly Grown in Great Britain; Apples and Pears. London; Murray, 1920. 2 vols. 204p.

Second Bunyard, Edward A. A Handbook of Hardy Fruits More Commonly Grown in Great Britain; Stone and Bush Fruits, Nuts, etc. London; Murray, 1925. 258p.

Second Bunyard, George. Apples and Pears. London and Edinburgh; T. C. and E. C. Jack, 1911. 115p.

Second Burbidge, A. W. Horticulture. Stamford, 1877.

Second Burbidge, Frederick W. Cultivated Plants; Their Propagation and Improvement. Edinburgh and London; W. Blackwood and Sons, 1877. 618p.

Second Burkett, Charles W., and Clarence H. Poe. Cotton, Its Cultivation, Marketing, Manufacture, and the Problems of the Cotton World. New York; Doubleday, Page and Co., 1906. 331p.

Second Burlison, William L. Fighting Illinois Weeds. Springfield, Ill.; Illinois Farmer's Institute, 1941. 61p.

Second Burlison, William L., and A. W. Nolan. Farm Crop Projects. New York; Macmillan Co., 1930. 458p.

Second Burpee, W. Atlee. Vegetables for the Home Garden. Philadelphia; W. Atlee Burpee and Co., 1896. 127p. (3d ed., 1898. Rev. ed., 1912.)

Second Burr, Fearing. The Field and Garden Vegetables of America: Containing Full Descriptions of Nearly Eleven Hundred Species and Varieties . . . Boston; Crosby and Nichols, 1863. 674p. (Later ed., Boston; J. E. Tilton and Co., 1865. 667p.)

Second Burritt, M. C. Apple Growing. New York; Outing Pub. Co., 1912. 177p.

First Burton, W. G. The Potato; A Survey of Its History and of Factors Influencing Its Yield, Nutritive Value and Storage. London; Chapman and Hall, 1948. 319p.

First Bush, &. Son and Meissner. Illustrated Descriptive Catalogue of American Grape Vines; A Grape Growers' Manual. 3d ed. St. Louis; R. P. Studley and Co., 1883. 133p. (4th ed. St. Louis; R. P. Studley and Co., 1895. 108p.) (1st ed. in French: Les Vignes Américaines . . . Montepiher, C. Coulet, 1876. 130p.)

Second Bush, Carroll E. Nut Grower's Handbook; a Practical Guide to the Sucessful Propagation, Planting, Cultivation, Harvesting and Marketing of Nuts. New York; Orange Judd Pub. Co., Inc., 1941. 189p. (Rev. ed. 1951. 199p.)

Second Bush, R. Fruit Growing Outdoors. London; Faber and Faber Ltd., 1946. 518p. (1st ed. pub. in 3 vol. by Penguin Books in 1942.)

Second Bush, R. Fruit Salad; Jottings from a Fruit Grower's Diary. London; Cassell, [1947]. 165p.

Second Bush, R. Harvesting and Storing Garden Fruit. London; Faber and Faber, 1947. 162p.

Second Bush, R. Tree Fruit Growing. Vol. I Apples, Vol. II Pears, Quinces and Stone Fruits. Harmondsworth Middlesex; Penguin Books, 1943. 2 vols. v. 1, 167p.; v. 2, 158p.

Second Butler, Edwin J. Fungi and Diseases in Plants Calcutta and Simla: Thacker, Spink and Co., 1918. 547p.

Second Butler, Edwin J., and S. G. Jones. Plant Pathology. London; Macmillan, 1949. 979p.

C

Second California Walnut Growers Association. The California Walnut. Los Angeles, 1919. 70p.

Second California Walnut Growers Association. All The Answers About Walnuts. Los Angeles, Calif.; California Walnut Growers' Assoc., 1937. 31p.

Second Cameron, James. Culture of Alfalfa or Lucerne, From Fifteen Years' Experience. Beaver City, Nebr.; Merwin Pub. Co., 1900. 32p.

First Camp, A. F. Citrus Industry of Florida. Tallahassee; Florida Dept. Agric., in Cooperation with Univ. of Florida, [1945]. 198p.

First Carleton, M. A. The Small Grains. New York; The Macmillan Co., 1916. 699p. (Also issued 1924.)

First Carman, Elbert S. The New Potato Culture; As Developed by the Trench System, by the Judicious Use of Chemical Fertilizers, and by the Experiments Carried on at the Rural Grounds During the Past Fifteen Years. New York; The Rural Pub. Co., 1891. 165p. (2d ed., rev., 1893. 100p.)

First Cary, Augustus G. Cranberry Culture on a Western Plan, With Valuable Items and Recipes from Growers and Members of the Wisconsin Cranberry Association. Cincinnati, Ohio; Rasall and Co., 1891. 77p.

Second Case, H. C. M., and P. E. Johnston. Principles of Farm Management. Philadelphia; J. B. Lippincott Co., 1953.

Second Chadwick, L. The Cultivator's Hand Book on Universal or Planetary Law of the Plants, Sun, Moon and Signs. Chicago; R. R. Donnelly and Sons Co., Printers, 1895. 110p.

Second Chandler, W. H. Fruit Growing. Boston and New York; Houghton Mifflin, [1925]. 777p. (Published under the editorial supervision of George A. Works . . . and William S. Taylor)

Second Chandler, W. H. North American Orchards. Philadelphia, Pa.; Lea and Febiger, 1928. 516p.

Second Chapman, V. J. Seaweeds and Their Uses. London; Methuen, 1950. 287p.

Second Chatterton and Son. Growing, Harvesting and Marketing of Domestic Beans . . . comp. by B. A. Stickle. Lansing, Mich.; Chatterton and Son, 1928. 33p.

Second Cheney, R. H. Coffee: A Monograph of the Economic Species of th Genus Coffea L. New York; New York University Press, 1925. 244p.

Second Chester, Kenneth S. The Nature and Prevention of the Cereal Rusts as Exemplified in the Leaf Rust of Wheat. Waltham, Mass.; Chronica Botanica Co., 1947. 269p.

Second Chester, Kenneth S. Nature and Prevention of Plant Diseases. Philadelphia; Blakiston Co., 1942. 584p. (2d ed., 1947. 525p.)

Second Chicago and Northwestern Railway Company. Alfalfa; By a Practical Producer of This Great Money Corp of the West and Northwest; Its Adaptation to the Agricultural Conditions of the Territory Reached by the Northwestern Line, and Its Influence Upon Cattle, Hog and Dairy Interests.

Chicago, Ill.; Industrial Dept., Chicago Northwestern Railway, 1911? 36p.

Second Child, David L. The Culture of the Beet, and Manufacture of Beet Sugar. Boston; Weeks, Jordan and Co., Northampton, J. H. Butler, 1840. 136p.

Second Childers, Norman Franklin. Fruit Science: Orchard and Small Fruit Management. Philadelphia, Pa.; Lippincott, 1949. 630p.

Second Childers, W. H. Fruit Science. Chicago; J. B. Lippincott, 1949.

First Chorlton, William. The American Grape Grower's Guide: Intended Especially for the American Climate. Being a Practical Treatise on the Cultivation of the Grape-Vine in Each Department of Hot House, Cold Grapery, Retarding House, and Outdoor Culture. With Plans for the Construction of the Requisite Buildings, and Giving Best Methods of Heating the Same. every Department Being Fully Illustrated. New York; Orange Judd and Co., [1852]. 204p.

First Chorlton, William. Chorlton's Grape Growers Guide. A Hand-Book of the Cultivation of the Exotic Grape. New ed. with descriptions of the later exotic grapes . . . by George Thurber. New York; Orange Judd Co., 1883. 208p. (New ed., 1887. 211p. Published in 1856 under title: The American Grape Grower's Guide.)

Second Chupp, Charles. Manual of Vegetable-Garden Diseases. New York; Macmillan Co., 1925. 647p.

Second Church, Ella Rodman. The Home Garden. New York; D. Appleton and Co., 1881. 121p. (In Appleton's Home Book Series)

First Claassen, Hermann. Beet-Sugar Manufacture . . . authorized trans. from 2d German ed., by William T. Hall and George William Rolfe. 1st ed. New York; J. Wiley and Sons, 1906. 280p. (2d ed., 1910. 343p.)

Second Clausen, C. P. Entomophagous Insects. New York; McGraw-Hill, 1940. 688p.

First Clements, Frederic E. Plant Physiology and Ecology. New York; H. Holt and Co., 1907. 315p.

First Coburn, F. D. The Book of Alfalfa; History, Cultivation and Merits. Its Uses as a Forage and Fretilizer . . . New York; Orange Judd Pub. Co., Inc., 1906. 336p. (New Rev. ed., 1907 and 1912.)

First Coburn, Foster D. Alfalfa, Lucerne, Spanish Trefoil, Chilian Clover, Brazilian Clover, French Clover, Medic, Purple Medic (Medicago sativa) Practical Information on Its Production, Qualities, Worth, and Uses, Especially in the United States and Canada. New York; O. Judd Co., 1901. 163p.

Second Cockerham, Kriby L. A Manual for Spraying. New York; Macmillan Co., 1923. 87p.

Second Coit, J. Eliot. Carob Culture. San Diego, California; Walter Rittenhouse, 1949.

Second Coit, J. Eliot. Citrus Fruits; An Account of the Citrus Fruit Industry, With Special Reference to California Requirements and Practices and Similar Conditions. New York; Macmillan Co., 1915. 520p.

Second Collings, Gilbeart H. The Production of Cotton. New York; J. Wiley and Sons, Inc., London; Chapman and Hall, Ltd., 1926. 256p.

Second Comin, Donald. Onion Production. New York; Orange Judd Pub. Co., Inc., 1946. 186p.

First	Compton, D. A. The Prize Essay on the Cultivation of the Potato. New York; O. Judd and Co., 1870. 40p.
First	Condit, Ira J. The Fig. Waltham, Mass.; Chronica Botanica Co., 1947. 222p. (A New Series of Plant Science Books No. 19)
Second	Conradi, Albert F., and W. A. Thomas. Farm Spies; How the Boys Investigated Field Crop Insects. New York; Macmillan Co., 1916. 165p.
Second	Cook, Melville T. The Diseases of the Tropical Plants. London; Macmillan and Co., Ltd., 1913. 317p.
Second	Cook, Melville T. Viruses and Virus Diseases of Plants. Minneapolis, Minn.; Burgess Pub. Co., 1947. 244p.
Second	Cooke, Matthew. Injurious Insects of the Orchard, Vineyard, Field, Garden, Conservatory, Household, Storehouse, Domestic Animals, etc., with Remedies for Their Extermination. Sacramento, Calif.; H. S. Crocker and Co., Printers, 1883. 472p.
Second	Cooke, Matthew. Insects, Injurious and Beneficial; Their Natural History and Classification, For the Use of Fruit Growers, Vine Growers, Farmers, Gardeners and Schools. Sacramento, Calif.; H. S. Crocker and Co., 1883. 156p.
Second	Cooke, Mordecai C. Fungoid Pests of Cultivated Plants. London; Spottiswoode and Co., Ltd., 1906. 278p. (Reprinted from the Journal of the Royal Horticultural Society, vols. 27–29.
Second	Cooke, Mordecai C. Rust, Smut, Mildew, and Mould, an Introduction to the Study of Microscopic Fungi. 2d ed. London; R. Hardwicke, 1870. 242p. (5th ed., rev. and enlarged, London; W. H. Allen and Co., 1886. 262p.)
Second	Cooper, J. C. Walnut Growing in Oregon. Portland, Oreg.; Passenger Dept., Oregon Railroad and Navigation Co., Southern Pacific Co. Lines in Oregon, 1910. 62p. (Also pub. in 1912, 64p.)
Second	Copeland, Edwin B. Rice. London; Macmillan and Co., Ltd., 1924. 352p.
Second	Copeland, Melvin T. The Cotton Manufacturing Industry of the United States. Cambridge; Harvard University Press, 1912
Second	Copley, G. H. Fruit Growing. London; Crowther, 1946. 130p.
Second	Corbett, Lee Cleveland. Garden Farming. Boston; Gill and Co., 1913. 473p.
First	Coulter, John M. Fundamentals of Plant-Breeding. New York and Chicago; D. Appleton and Co., 1914. 346p.
Second	Coulter, John M. Practical Science; An Address Delivered on the Occasion of the Dedication of Plant Industry Hall, at the Univ. Farm, on Tuesday, June Tenth, Nineteen Hundred Thirteen. Lincoln, Neb.; The Univ. Press, 1913. 12p.
Second	Coulter, John M., and Merle C. Coulter. Plant Genetics. Chicago, Ill.; The Univ. of Chicago Press, 1918. 214p.
Second	Coulter, Merle C. Outline of Genetics, With Special Reference to Plant Material. Chicago, Ill.; The Univ. of Chicago Press, 1923. 211p.
Second	Cox, Alonzo. Cotton Markets and Cotton Merchandising. 2d ed. Austin, Texas; Hemphill's, 1949. 183p. (1st ed. Cotton Markets and Marketing. Austin, Texas; University of Texas, 1945. 135p.)
Second	Cox, Joseph F., and Lyman Jackson. Field Crops and Land Use. New York; J. Wiley and Sons, London; Chapman and Hall, Ltd., 1942. 473p.

Second Cox, Joseph F., and C. R. Megee. Alfalfa. New York; J. Wiley and Sons, Inc., London; Chapman and Hall, Ltd., 1928. 101p.

Second Crabb, A. Richard. The Hybrid-Corn Makers: Prophets of Plenty. New Brunswick, N.J.; Rutgers University Press, 1947. 331p.

First Crafts, A. S. et al. Water in the Physiology of Plants. Waltham, Mass.; Chronica Botanica, 1949.

First Crane, M. B. The Genetics of Garden Plants, by M. B. Crane and W. J. Lawrence. With a Foreword by Sir Daniel Hall. 1st ed. London; Macmillan and Co., Ltd., 1934. 236p. (2d ed., 1938. 4th ed., 1952.)

Second Crawford, M. Strawberry Culture. Cuyahoga Falls, Ohio; M. Crawford Co., 1902. 62p.

First Crawford, Morris De Camp. The Heritage of Cotton, The Fibre of Two Worlds and Many Ages. New York and London; G. P. Putnam's Sons, 1924. 244p. (Later printings, 1931 and New York; Fairchild, 1948. 294p.)

First Crocker, William. Growth of Plants; Twenty Years' Research at Boyce Thompson Institute. New York; Reinhold, 1940. 459p.

Second Croff, G. W. The Lychee and the Lungan. New York; Orange Judd Co., 1921.

Second Crossweller, W. T. The Gardeners' Company, 1605–1907. 1908.

Second Crowther, D. S. Fruit for Small Gardens. London; Collingridge, [1949]. 152p.

Second Croy, Homer. Corn Country. 1st ed. New York; Duell, Sloan and Pearce, 1947. 325p.

First Crozier, A. A. The Cauliflower. Ann Arbor, Mich.; Peter Henderson and Co., 1884. 400p.

Second Cruess, William V. Commercial Fruit and Vegetable Products; A Textbook for Student, Investigator and Manufacturer. New York; McGraw-Hill Book Co., 1924. 530p.

Second Cunningham, Gordon H. Fungous Diseases of Fruit-Tree in New Zealand and Their Remedial Treatment. Auckland, N.Z.; New Zealand Fruitgrowers' Federation Ltd., 1925. 382p.

Second Curtiss, Daniel S. Wheat Culture. How to Double the Yield and Increase the Profits. New York; Orange Judd Co., 1880. 72p. (4th ed, N.Y., 1898.)

Second Cutting, Hiram A. An Address Upon Farm Pests, Including Insects, Fungi, and Animalcules, Delivered at the Meeting of the New Hampshire Board of Agriculture. Manchester, N.H.; Printed by J. B. Clarke, 1879. 75p.

D

Second Daker, J. S. Early Vegetables Under Glass. London; Cassell, 1936. (5th ed., 1948. 171p.)

Second Dakota Improved Seed Co. Alfalfa in the Northwest. 2d ed. Mitchell, S.Dak.; 1912. 58p.

Second Dallimore, W. The Pruning of Trees and Shrubs. London; Dalau and Co., Ltd., 1926.

Second Dana, William B. Cotton From Seed to Loom. A Handbook of Facts for the Daily Use of Producer, Merchant and Consumer. New York; W. B. Dana and Co., 1878. 291p.

Second Darby, William D. Cotton, The Universal Fiber; A Survey of the Cotton Industry From the Raw Material to the Finished Product, Including Descriptions of Manufacturing and Marketing Methods and a Dictionary of Cotton Goods. New York; Dry Good Economist, 1922. 68p. (Revised ed., 1932. 63p.)

First Darlington, C. D. Chromosomes and Plant Breeding. London; Macmillan, 1932.

Second Darlington, C. D., and K. Mather. Elements of Genetics. New York; The Macmillan Co., 1949.

Second Darlington, E. D., and L. M. Moll. How and What to Grow in a Kitchen Garden of One Acre. Philadelphia; W. Atlee Burpee, 1888. 198p. (8th ed. 1895.)

First Darrow, G. M. The Strawberry; History, Breeding, and Physiology. 1st ed. New York; Holt, Rinehart and Winston, [1966]. 446p.

First Darwin, Charles. Different Forms of Flowers on Plants of the Same Species. London; John Murray, 1877. 352p.

First Darwin, Charles. The Effects of Cross and Self Fertilization in the Vegetable Kingdom. London; J. Murray, 1876. 482p.

First Darwin, Charles with F. Darwin. The Power of Movement in Plants. London; John Murray, 1880. 529p.

First Darwin, Charles. The Variation of Animals and Plants Under Domestication. 1st ed. London; J. Murray, 1868. 2 vols. (2d rev., ed., New York; D. Appleton, 1890.)

First Darwin, F. The Life and Letters of Charles Darwin. New York; D. Appleton and Co., 1887. 2 vols.

First Daubenmire, R. F. Plants and Environment; a Textbook of Plant Autecology. New York and London; Wiley, Chapman and Hall, 1947. 424p.

Second Davis, George W. A Treatise on the Culture of the Orange, Together with a Description of Some of the Best Varieties of the Fruit, Gathering, Curing and Preparing the Fruit for Shipment and Market. Jacksonville, Fla.; C. W. Dacosta, 1881. 60p.

Second Davis, Kary C. Field-Crop Enterprises, Including Soil Management. Chicago and Philadelphia; J. B. Lippincott Co., 1928. 528p. (Other editions, 1937 and 1946.)

Second Day, J. W., D. Cummins, and A. I. Root. Tomato Culture. Medina, Ohio; A. I. Rood, 1892. 135p.

First De Bary, A. Comparative Morphology and Biology of the Fungi Mycetozoa and Bacteria . . . authorized English translation by Henry E. F. Garmsey, rev. by Isaac Bayley Balfour. Oxford; Clarendon Press, 1887. 525p.

First De Candolle, A. Origin of Cultivated Plants. New York; D. Appleton, 1885. 468p. (The International scientific series, v. 48 Original French edition, 1883. First American edition, 1885. Second American edition, New York; D. Appleton, 1886. Reprint of the 2nd edition, 1886. Appleton-Centry-Crofts, Inc., 1895. The International Scientific Series, v. 49, New York Hafner Pub. Co., 1959.)

First De Coin, Robert L. History and Cultivation of Cotton and Tobacco. London; Chapman and Hall, 1864. 306p.

First De Kruif, Paul H. Hunger Fighters. New York; Harcourt, Brace and Co., 1928. 377p.

First de Vries, Hugo. The Mutation Theory: Experiments and Observations on the Origin of Species in the Vegetable Kingdom . . . trans. by J. B. Farmer and A. D. Darbishire. Chicago; Open Court, 1909–1910.

First de Vries, Hugo. Plant-Breeding; Comments on the Experiments of Wilson and Burbank. Chicago; The Open Court Pub. Co., 1907. 360p.

Second de Vries, Hugo. Species and Varieties, Their Origin by Mutation; Lectures Delivered at the University of California by Hugo de Vries. Chicago: The Open Court Pub. Co., 1905. 847p.

Second Dearing, William G. Corn Culture; New Methods Versus Old. Louisville, Ky.; J. P. Morton and Co., Inc., 1913. 105p.

First Deerr, Noel. The History of Sugar. London; Chapman and Hall, Ltd, 1950. 2 vols.

Second Dethier, Vincent G. Chemical Insect Attractants and Repellents. Philadelphia, Pa.; Blakiston, 1947. 289p.

First Dewey, Dellon M. The Nurseryman's Pocket Specimen Book, Colored From Nature. Fruits, Flowers, Ornamental Trees, Shrubs, Roses, etc . . . Rochester, N.Y.; D. M. Dewey, 1872. 54p.

Second Dice, L. R. The Biotic Provinces of North America. Ann Arbor; Univ. of Michigan Press, 1943. 78p.

Second Dickinson, Albert, Co. Dickinson's Alfalfa Facts. Useful Guide For Growing Alfalfa. 2d ed. Chicago, Ill.; 1913. 48p.

Second Dickson, J. G. Diseases of Field Crops. New York; McGraw-Hill Book Co., Inc., 1947. 429p.

Second Dickson, J. G. Outline of Diseases of Cereal and Forage Crop Plants of the Northern Part of the United States. Minneapolis; Burgess Pub. Co., 1940.

Second Dictionary of Gardening; a Practical and Scientific Encyclopedia of Horticulture . . . edited by Fred J. Chittendon, assisted by specialists. Oxford; Clarendon Press, 1951. 4 vols. 2316p.

Second Dies, Edward J. Soybeans, Gold From The Soil. New York; The Macmillan Co., 1942. 122p. (Rev. ed., 1943.)

Second Dillion-Weston, William A. R., and R. E. Taylor. The Plant in Health and Disease. London; Crosby Lockwood, 1948. 164p.

Second Dixon, Henry H. Practical Plant Biology; A Course of Elementary Lectures on the General Morphology and Physiology of Plants. London and New York; Longmans, Green and Co., 1922. 291p. (2d ed., Dublin; Hodges, 1943. 337p.)

Second Dixon, Henry H. Transpiration and the Ascent of Sap in Plants. London; Macmillan, 1914. 216p.

Second Dobell, C. Antony van Leewenhoek and His "Little Animals." London; 1932. (Reprinted New York; Dover Publicaitons, [1960].)

First Dobzhansky, Theodosius G. Genetics and the Origin of Species. New York; Columbia Univ. Press, 1937. 364p. (2d ed., 1941.)

Second Dodge, B. O., and H. W. Rickett. Diseases and Pests of Ornamental Plants. Rev. ed. New York; Ronald Press, 1948.

Second Dominguez, Zeferino. The Modern Cultivation of Corn. San Antonio, Tex.; Dominguez Corn Book Publishing Co., 1914. 351p.

First Dondlinger, Peter Tracy. The Book of Wheat; An Economic History and Practical Manual of the Wheat Industry. New York; Orange Judd Co., 1908. 369p.

Second Douglas, Charles E. Rice, Its Cultivation and Preparation. London; Pitman and Sons, Ltd., 1924. 143p.

Second Douglass, Benjamin Wallace. Orchard and Garden. Indianapolis, Ind.; 1918. 360p.

Second Dowling, Reginald N. Sugar Beet and Beet Sugar. London; Benn Ltd, 1928. 277p.

Second Dowling, Reginald N. Sugar Beet From Field to Factory. London; E. Benn Ltd., 1925. 72p.

Second Downing, Andrew J. The Fruits and Fruit Trees of America; or The Culture, Propagation, and Management, in the Garden and Orchard, of Fruit Trees Generally . . . New York and London; Wiley and Putnam, 1845. 594p. (2d rev. and corr., with large additions, New York; Wiley, 1890. 1098p.)

Second Dowson, W. J. Manual of Bacterial Plant Diseases. New York; Macmillan, 1949. 183p. (2d ed. with title: Plant Diseases Due to Bacteria. Cambridge; Cambridge University Press, 1957. 231p.)

First Drain, B. D. Essentials of Systematic Pomology. New York; John Wiley and Sons, 1925. 284p.

Second Dreer, Henry A. Dreer's Open-Air Vegetables. Philadelphia; Henry A. Dreer, 1897. 148p.

Second Dreer, Henry A. Dreer's Vegetables Under Glass. Philadelphia; Henry A. Dreer, 1896. 99p.

First Duggar, Benjamin M. Fungous Diseases of Plants; With Chapters on Physiology, Culture Methods and Technique. Boston, Mass. and New York; Ginn, 1909. 508p.

Second Duncan, F. Martin. Insect Pests and Plant Diseases in the Vegetable and Fruit Garden. London; Constable and Co., Ltd., 1919. 95p.

Second Duncan, F. Martin. Our Insect Friends and Foes. London; Methuen and Co., Ltd., 1911. 296p.

Second Dunkin, H. The Pruning of Hardy Fruit Trees. 2d ed. London and New York; Dent, 1947. 82p.

Second Duran-Reynals, M. L. The Fever Bark Tree. New York; Doubleday and Co., Inc., 1946. 275p.

Second Dutton, R. The English Garden. New York; Scribner, 1938. 122p.

Second Dwyer, T. J. Guide to Hardy Fruits and Ornamentals. Cornwall, N.Y.; T. J. Dwyer and Son, 1903. 125p.

Second Dygert, H. Arthur. Crops that Pay; Pecans, Pomelos, Ginseng. Philadelphia, Pa.; 1903. 60p.

E

Second Ealand, Charles A. Insects and Man; An Account of the More Important Harmful and Benefical Insects, Their Habits and Life-Histories, Being an Introduction to Economic Entomology for Students and General Readers. New York; The Century Co., 1915. 343p.

Second Earle, Franklin S. Sugar Cane and Its Culture. New York; J. Wiley and Sons, Inc., London; Chapman and Hall, Ltd., 1928. 355p.

Second East, E. M., and H. K. Hayes. Heterozygosis in Evolution and in Plant Breeding. Washington, D.C.; Governtment Printing Office, 1912. 58p.

Second East, E. M., and H. K. Hayes. Inheritance in Maize. New Haven, Conn.; Connecticut Agricultural Experiment Station, 1911. 142p.

First East, E. M., and D. F. Jones. Inbreeding and Outbreeding. Philadelphia; Lippincott Co., 1919. 285p.

First Edgar, William C. The Story of a Grain of Wheat. New York; D. Appleton and Co., 1903. 195p. (Also printed: London; Newnes, 1903. New York, 1915. 195p. New York and London; D. Appleton and Co., 1925. 232p. 1903 ed. reprinted Freeport, N.Y.; Books for Libraries Press. 1972.)

First Egleston, Nathaniel H. Hand-Book of Tree-Planting; or Why to Plant, Where to Plant, What to Plant, How to Plant. New York; D. Appleton and Co., 1884. 126p.

First Eisen, Gustav. The Raisin Industry. San Francisco; H. S. Crocker and Co., 1890. 223p.

First Elliott, Charlotte H. Manual of Bacterial Plant Pathogens. Baltimore; Williams and Wilkins Co., 1930. 349p.

Second Elliott, F. R. Hand-Book for Fruit Growers. New ed. Rochester; Rochester Lithographing Co., 1876. 144p. (Rochester; D. M. Dewey, 1876. 128p.)

First Elliott, F. R. Elliott's Fruit Book: or the American Fruit-Growers' Guide. New York; C. M. Saxton, 1854. 503p. (Subsequently issued under title: The Western Fruit-Book. 4th ed., 1859. 528p.)

First Elliott, William R. Practical and Comprehensive Treatise on Fruit and Floral Culture, and a Few Hints on Landscape Gardening. Pittsburgh; Jos. Horne and Co., 18--? 100p.

Second Ellis, C., and M. W. Swaney. Soilless Growth of Plants. New York; Reinhold Publishing Corporation; London; Chapman and Hall, 1938. 155p.

Second Emerson, Edward R. The Story of the Vine. New York and London; G. P. Putnam's Sons, The Knickerbocker Press, 1902. 252p.

Second Emerson, William D. History and Incidents of Indian Corn, and Its Culture. Cincinnati, Ohio; Wrightson and Co., Printers, 1878. 464p.

First Enfield, Edward. Indian Corn; Its Value, Culture, and Uses. New York; D. Appleton and Co., 1866. 308p.

First Eriksson, Jacob. Fungous Diseases of Plants in Agriculture, Horticulture and Forestry, translated from the German by William Goodwin. 2d ed. London; Bailiere, Tindall and Cox, 1930. 526p. (1st ed., in 2 books Fungoid Diseases of Agricultural Plants, 1929, and Die Pilzksankhanten der Garten -und Parkgenachse, 1928.)

Second Ernle, Lord. English Farming, Past and Present. London and New York; Longmans, Green and Co., 1912. 504p. (5th ed., London; Longmans, 1936.)

Second Essig, E. O. A History of Entomology. New York; Macmillan, 1931. 1029p.

Second Evershed, Henry. Improvement of the Plants of the Farm. London; Printed by W. Clowes and Sons, Ltd., 1884. 39p.

F

Second Fabricius, Johann Christian. Attempt at a Dissertation on the Diseases of Plants, translation by Mrs. Margaret Kolpin Ravn of "Forsog Tilen Afhandling om Planternes Sygdomme," published in Det Kongelige Norske Videnskabers Selskabe Skrifter (1774) 5: 431–492. [*Phytopathological Classics* no. 1] Translation published in 1926, 66p.

Second Fairchild, David. The World Was My Garden; Travels of a Plant Explorer. New York and London; Scribner's Sons Ltd., 1938. 494p.

Second Falconer, William. Mushrooms: How to Grow Them. New York; Orange Judd Co., 1891. 169p.

Second Farmer, Lawrence J. Fall-bearing Strawberry Secrets Gathered From Personal Experience and Now Disclosed for the First Time. Edited by Walter E. Andrews. Philadelphia, Pa.; 1912. 62p.

Second Farmer, Lawrence J. Farmer on the Strawberry. A Series of Papers on the Subject of Strawberry Culture. Pulaski, N.Y.; Democrat Print, 1891. 53p. (1912; 94p.)

Second Farmers and Manufacturers Beet Sugar Association. The Story of Beet Sugar From the Seed to the Sack. Bay City, Mich.; Farmers and Manufacturers Beet Sugar Association, 1933. 14p. (5th ed., Saginaw, Mich. 1945.)

Second Farrer, Reginald J. On the Eaves of the World. London; E. Arnold, 1917. 2 vols.

Second Farrer, Reginald J. The Rainbow Bridge. New York; Longmans, Green and Co., 1921. 383p.

Second Farrington, E. I. Ernest H. Wilson, Plant Hunter. Boston; The Stratford Co., 1931. 197p.

Second Faulkner, R. P. The First Principles of Horticulture. London; Pitman, 1947. 122p.

Second Favor, E. H. The Fruit-Grower's Guide-book. St. Joseph, Mo.; The Fruit-Grower, 1911. 285p.

First Fawcett, H. S. Citrus Diseases and Their Control. New York; McGraw-Hill, 1926. 656p.

Second Fay, Ivan G. A Workbook in Field Crops and Soils. Danville, Ill.; Interstate, 1938. 193p.

Second Fergus, Ernest N., and C. Hammonds. Field Crops Management . . . ed. by R. W. Gregory. Philadelphia and New York; J. B. Lippincott Co., 1942. 600p. (Revised 1949. Later title, Field Crops, 1958, based on authors' Field Crops Management.)

Second Fernald, Henry T. Applied Entomology; An Introductory Text-Book of Insects in Their Relations to Man. 1st ed. New York; McGraw-Hill Book Co., Inc., 1921. 386p. (2d ed, 1926. 395p. 5th ed, 1955. 385p.)

Second Fernald, M. L., and A. C. Kinsey. Edible Wild Plants of Eastern North America. Cornwall-on-Hudson; Idlewild Press, 1943. 452p.

Second Ferree, Christian J., and J. T. Tussaud. The Soya Bean and the New Soya Flour. Rev. trans. from Dutch. London; W. Heinemann, Ltd., 1929. 79p.

Second Ferris, Gordon F. The Principles of Systematic Entomology. Stanford, Calif.; Stanford University Press, 1928. 169p.

First Field, Thomas W. Pear Culture; A Manual for the Propagation, Planting, Cultivation, and Management of the Pear Tree . . . New York; A. O. Moore, 1863. 286p.

Second Findlay, Archibald. The Potato: Its History and Culture, with Descriptive List of Varieties Raised and Introduced. Cupar-Fife; Printed by A. Westwood and Son, 1905. 88p.

Second Fish, A. C. The Profits of Orange Culture in Southern California. 2d ed. Los Angeles; 1890. 25p.

Second Fish, D. T. The Hardy-Fruit Book. London; Gill, 1882? 2 vols.

First Fisher, R. A. The Genetical Theory of Natural Selection. Oxford; Clarendon Press, 1930.

Second Fiske, G. Burnap, Compiler. Prize Gardening. New York; Orange Judd Co., 1901. 307p.

First Fitz, James. The Southern Apple and Peach Culturist . . . Richmond, Va.; J. W. Randolph and English, 1872. 336p.

First Flagg, William J. Hand-Book of the Sulphur-Cure, As Applicable to the Vine Disease in America, and Disease of Apple and Other Fruit Trees. New York; Harper and Bros., 1870. 99p.

Second Flamont, Adolphe. A Practical Treatise on Olive Culture, Oil Making and Olive Pickling. San Francisco; Louis Gregoire and Co., Booksellers, 1887. 76p.

Second Fletcher, F. J. Market Nursery Work: A Series of Six Books on the Cultivation of Crops for Market, Volume V: Orchard Fruit Tree Culture . . . London; Benn Brothers Ltd., 1922. 73p.

Second Fletcher, Stevenson W. How to Make a Fruit Garden. New York; Doubleday, Page and Co., 1906. 283p.

First Fletcher, Stevenson W. The Strawberry in North America; History, Origin, Botany, and Breeding. New York; Macmillan Co., 1917. 234p.

Second Flint, Charles Louis. Grasses and Forage Plants; A Practical Treatise Comprising Their Natural History; Comparative Nutritive Value, etc. New York; G. P. Putnam and Co.; Boston; William White, Printer, 1857. 236p. (Later ed., Boston, 1879.)

Second Flint, Wesley P., and C. L. Metcalf. Insects, Man's Chief Competitors. Baltimore, Md.; Williams and Wilkins Co. in cooperation with the Century of Progress Exposition, 1932. 133p.

Second Fogg, John M. Weeds of Lawn and Garden; a Handbook for Eastern Temperate North America. Philadelphia; Univ. of Pennsylvania Press, 1945. 215p.

Second Foister, C. E. The Relation of Weather to Plant Diseases. London; H. M. Stationary Office, 1929. 50p.

Second Folmer, Henry D. Alfalfa on Wildwood Farm and How to Succeed With It. Columbus, Ohio; Nitschke Bros., Printers, 1911. 105p.

Second Fontana, Felice. Observations on the Rust of Grain, translation by Pascal Pompey Pirone, of Fontana's Osservazioni sopra la Ruggine del Grano. Lucca, Italy; Jacob Giusti, 1767. Translation published in 1932, 40p. [*Phytopathological Classics* no. 1]

Second Forbes, Stephen A. The Kind of Economic Entomology Which the Farmer Ought to Know . . . Read at the Illinois State Farmers' Institute at Decatur, Ill. Feb. 24, 1904. Bloomington, Ill.; Pantagraph Printing and Stationery Co., 1904. 16p.

First Ford, E. B. Mendelism and Evolution. New York; L. MacVeagh, The Dial Press, 1931. 116p. (8th ed., London; Methuen Co.; New York; Wiley, 1965. 122p.)

First Franck, J. Photosynthesis in Plants. Ames; Iowa State College Press, 1949. 500p.

First Fraser, Samuel. American Fruits, Their Propagation, Cultivation, Harvesting and Distribution. New York; Orange Judd, 1924. 888p.

Second Fraser, Samuel. A Practical Treatise on the Potato. New York; Orange Judd Co., 1905. 185p.

Second Fraser, Samuel. The Strawberry. New York; Orange Judd Pub. Co., Inc., 1926. 120p.

Second Fred, Edwin B., Ira L. Baldwin, and Elizabeth McCoy. Root Nodule Bacteria and Leguminous Plants. Madison, Wis.; University of Wisconsin, 1932. 343p. (University of Wisconsin Studies in Science No. 5)

Second Free, M. Gardening. New York; Harcourt, Brace and Co., 1937.

Second Freeman, Edward M. Minnesota Plant Diseases. St. Paul, Minn., 1905. 432p.

Second Fremlin, R. The Potato in Farm and Garden; Embracing Every Phase of Its Cultivation, with Chapters on Disease and Special Cultures. London and New York; G. Routledge and Sons, 1883. 178p.

Second French, Allen. The Book of Vegetables and Garden Herbs. New York; The Macmillan Co., 1907. 312p.

Second Fruits and Gardens. Fruits and Gardens New Spray Manual, the Fruit Grower's Hand Book. rev. ed. Zeeland, Mich.; 1928. 120p.

Second Fryer, John C. F., and F. T. Brooks. Insect and Fungus Pests of the Farm. London; E. Benn Ltd, 1928. 198p.

Second Fryer, Percival J. Insect Pests and Fungus Diseases of Fruit and Hops; A Complete Manual for Growers. Cambridge; Cambridge University Press, 1920. 728p.

Second Fuller, Andrew S. The Illustrated Strawberry Culturist. New York; Orange Judd Co., 1887. 59p.

Second Fuller, Andrew S. The Nut Culturist. New York; Orange Judd Co., 1896. 289p.

Second Fuller, Andrew S. The Propagation of Plants. New York; Orange Judd Co., 1887. 349p.

First Fuller, Andrew S. The Small Fruit Culturist. New York; Orange Judd Co., 1867. 276p. (New ed., 1881. 187p. New ed., with an appendix, 1887. 197p. 3d ed., 1897, 298p.)

Second Fulton, James A. Peach Culture. New York; O. Judd and Co., 1870. 190p. (New, rev., and greatly enl. ed., 1889. 204p.)

Second Funk, Eugene D. Corn Disease Investigations . . . 3d ed., rev. Springfield, Ill.; Illinois Farmers' Institute, 1927. 23p.

G

First Gallesio, Georges. Orange Culture; A Treatise on the Citrus Family. Jacksonville, Florida; Charles H. Walton, 1876. 65p. (Trans. from the French for The Florida Agriculturist)

First Gardner, Victor R., F. C. Bradford, and H. D. Hooker. The Fundamentals of Fruit Production. New York; 1922. 686p. (3d ed., 1951. 739p.)

Second Gardner, Victor R. The Cherry and Its Culture. New York; Orange Judd Pub. Co., 1930. 128p. (Revised ed., New York; Orange Judd Pub. Co., 1946. 146p.)

First Gardner, Victor R., F. C. Bradford, and H. D. Hooker. Orcharding. 1st ed. New York; McGraw-Hill Book Co., Inc., 1927. 311p.

Second	Garey, Thomas A. Orange Culture in California . . . With an Appendix on Grape Culture by L. J. Rose. San Francisco, Calif., Pub. for T. A. Garey by Pacific Rural Press; 1882. 227p.
Second	Garner, Robert J. The Grafter's Handbook. London; Faber and Faber, 1947. 223p.
Second	Garner, W. W. The Production of Tobacco. Philadelphia; The Blakiston Company, 1946. 516p.
Second	Gause, G. F. The Struggle for Existence. Baltimore; Williams and Wilkins, 1934.
Second	Gilbert, Arthur W., Mortier F. Barrus, and Daniel Dean. The Potato. New York; Macmillan Co., 1917. 318p.
Second	Gilroy, Clinton G. The History of Silk, Cotton, Linen, Wool, and Other Fibrous Substances, Including Observations on Spinning, Dyeing and Weaving . . . New York; C. M. Saxton, New London, Conn.; E.R. Fellows, 1853. 464p.
Second	Glasscock, H. Good Control of Plant Diseases. London; English University Press, 1949. 180p.
Second	Globe Nurseries. A Guide to Fruit Growing. Britsol, Tenn.; Davis, 1906. 36p.
First	Goff, E. S. Lessons in Commercial Fruit Growing. Madison Wis.; University Cooperative Association, 1902. 221p.
Second	Goff, E. S. Principles of Plant Culture. Madison, Wis.; Published by Author, 1897. 276p. (3d ed., rev., Madison, Wis.; University Cooperative Co., 1906. 303p. (Preface by Frederick Cranefield).)
Second	Goldschmidt, Richard B. The Material Basis of Evolution. New Haven; Yale Univ. Press; London; H. Milford, Oxford Unviersity Press, 1940. 436p.
Second	Good, R. Plants and Human Economics. Cambridge; University Press, 1933.
Second	Goodspeed, W. E. The California Walnut. Los Angeles, Calif.; California Walnut Grower's Association, 1919. 72p.
Second	Gordon, G. The Wasted Orchards of England. London; Gardners' Magazine, 1896. 123p.
Second	Goudiss, Charles H. The Invaluable Apple; Fruit That is Food and Medicine for Men. New York; Priv. Printed by the People's Home Journal, 1921. 18p.
Second	Gourley, Joseph H. Orchard Management. New York and London; Harper and Bros., 1925. 247p.
Second	Govv, E. S. A Syllabus of Horticulture. Madison, Wis.; State Journal Printing Co., 1891. 110p.
First	Graber, Laurence F. Alfalfa; A Handbook For the Alfalfa Grower and Student. Madison, Wis.; L. F. Graber, 1918. 76p.
Second	Grainger, John. Virus Diseases of Plants. London; H. Milford, Oxford University Press, 1934. 104p.
Second	Granger, C. A. Growing of Sugar Beets. Lehi, Utah; n.d. 15p.
First	Grant, E. B. Beet-Root Sugar and Cultivation of the Beet. Boston; Lee and Shepard, 1867. 158p.
Second	Gray, George D. All About the Soya Bean, In Agriculture, Industry and Commerce . . . with an Introductory Chapter by James L. North. London; J. Bale, Sons and Danielsson, Ltd., 1936. 140p.

Second Great Britain. Ministry of Agriculture and Fisheries. A Selected and Classified List of Books in English Relating to Agriculture, Horticulture, etc. in the Library of the Ministry of Agriculture. London; His Majesty's Stationery Office, 1939. (Bulletin No. 78)

First Great Western Sugar Company. Methods of Analysis and Laboratory Control of the Great Western Sugar Company . . . by the Chemical Department of the Great Western Sugar Company. 1st ed. Denver, Colo.; The Great Western Sugar Company, 1920. 259p.

First Great Western Sugar Company. Technology of Beet Sugar Manufacture; a Textbook Describing the Theory and Practice of the Process of Manufacture of Beet Sugar. 1st ed. Denver, Colo.; The Great Western Sugar Company, 1920. 324p.

Second Green, Charles A. Green's Four Books. Devoted to: 1. How We Made the Old Farm Pay. 2.Peach Culture. 3. How to Propagate Fruit Plants, Vines and Trees. 4. General Fruit Instuctor. Rochester, N.Y.; 1895. 142p.

Second Green, Charles A. Green's Six Books: Devoted to Apple Culture; Pear Culture; Plum and Cherry Culture; Raspberry and Blackberry Culture; Grape Culture; Strawberry, Currant, Gooseberry and Persimmon Culture. Rochester, N.Y.; 1896. 142p.

Second Green, Charles A. How to Propagate and Grow Fruit. Rochester, N.Y.; 1885. 80p. (Rochester, NY,1885, Union and Adv. Co's. Print. 64 pp.)

Second Green, Joseph R. The Soluble Ferments and Fermentation. Cambridge; University Press, 1899.

First Green, Samuel B. Vegetable Gardening. St. Paul, Minn.; Pub. by Author; Webb Pub. Co., Agents, 1896. 224p. (2d ed., rev., 1899. 12th ed., rev., 1912.)

Second Green, Samuel B. Amateur Fruit Growing. Minneapolis; Farm, Stock and Home Pub. Co., 1894. 132p. (St. Paul, Minn.; Webb Publishing Company, 1905. 138p.)

Second Green, Samuel B. Popular Fruit Growing. St. Paul, Minn.; Webb Pub. Co., 1909. 298p. (4th ed., 1912. 328p.)

Second Greening, Charles E. An Important Message to the Farmer; How to Save the Farm Orchard From Destruction. Monroe, Mich.; 1917. 12p.

First Gregg, Thomas. A Hand-Book of Fruit-Culture. New York; Fowler and Wells, 1857. 163p.

Second Gregg, Thomas. How to Raise Fruits. New York; S. R. Wells and Co., 1877. 183p.

Second Gregory, Charles T., and J. J. Davis. Common Garden Pests, What They Are and How to Control Them. Des Moines, Iowa; Better Homes and Gardens, 1928. 150p.

Second Gregory, James J. H. Cabbages and Cauliflowers: How to Grow Them. Boston; Cashman Keating and Co., Printers, 1889. 88p. (Rev. ed., Marblehead, Mass; J. J. H. Gregory, 1907. 93p.)

First Gregory, James J. H. Carrots, Mangold-Wurtzels and Sugar Beets: How to Raise Them, How to Keep Them, and How to Feed Them. Marblehead, Mass.; N. A. Lindley and Co., 1877. 61p. (Boston; J. J. H. Gregory, 1882. J. J. Arakelyan, 1900. 65p.)

Second Greiner, T. Celery for Profit: An Expose of Modern Methods in Celery Growing. Philadelphia; W. Atlee Burpee and Co., Spring, 1893. 85p. (9th ed., 1898.)

Second Greiner, T. The Garden Book for Practical Farmers. Philadelphia; The Farmer Co., 1901. 190p. (Part I. Published as No. 1, Vol III, of the Practical Farmer's Library. Part II, Published as No. 4, Vol. III, of the Practical Farmer's Library.)

First Greiner, T. How to Make the Garden Pay. Philadelphia; Wm. Henry Maule, 1890. 272p. (2d ed., rev., 1894. 319p.)

Second Greiner, T. Money in Potatoes; 400 Bushels to the Acre as A Field Crop. A Complete Instructor for the Potato Grower . . . Philadelphia; Franklin News Company, 1885. 53p.

First Greiner, T. The New Onion Culture; A Story for Young and Old, Which Tells How to Grow 2,000 Bushels of Fine Bulbs on One Acre . . . Buffalo, N.Y.; Haas and Klein, Printers, 1891. 62p. (Rewritten and greatly enlarged, New York; Orange Judd Company, 1903. 114p. A Complete Guide in Growing Onions for Profit, 1911.)

Second Greiner, T. Onions for Profit: An Expose of Modern Methods in Onion Growing. Philadelphia; W. Atlee Burpee and Co., 1893. 104p.

Second Greiner, T. The Young Market-Gardener: Beginner's Guide. La Salle, New York; Sprint, 1896. 119p. (T. Greiner's Garden Series No. 2); Buffalo, N.Y.; J. W. Klein Printing, Co., 1896.)

First Greiner, T., and C. H. Arlie. How to Grow Onions; With Notes on Varieties. Philadelphia; W. Atlee Burpeee and Co., 1888. 71p.

Second Griffiths, A. B. Diseases of Crops and Their Remedies; a Handbook of Economic Biology for Farmers and Students. London; 1896. 174p.

Second Grove, William B. British Stem- and Leaf-Fungi (Coelomycetes); a Contribution to Our Knowledge of the Fungi Imperfecti Belonging to the Sphaeropsidales and the Melanconiales. Cambridge, U.K.; Cambridge Univ. Press, 1935. 2 vols.

Second Growers Guide, . . . a Compilation of Useful Information for the Grower. Nashville, Tenn.; The Grower's Guide Co., 1898. 416p.

First Grubb, Eugene H., and W. S. Guilford. The Potato: A Compilation of Information From Every Available Source. Garden City, N.Y.; Doubleday, Page and Co., 1912. 545p.

Second Grubb, N. H. Cherries. London; Crosby Lockwood, 1949. 186p.

Second Grundy, Fred. A Fortune in Two Acres. New York; The Rural Pub. Co., 1893. (The Rural Library Vol. 1, no. 24.)

Second Guenther, Ernest, et al. The Essential Oils; Vol. 1. History, Origin in Plants, Production, Analysis; Vol. 2. Constituents of Essential Oils; Vol. 3. Individual Essential Oils of the Plant Families: Rutaceae and Labiatae; Vol. 4. Individual Essential Oils of the Plant Families: Gramineae, Lauraceae, Burseraceae, Myrtaceae, Umbelliferae, and Geraniaceae; Vol. 5. Individual Essential Oils of the Plant Families, pt. 2, Rosaceae, Myristicaceae, etc . . . ; and Vol. 6. Individual Essential Oils of the Plant Families Ericaceae, Betulaceae, etc . . . New York; D. Van Nostrand Co., 1948–52. 6 vols.

Second Gurney, C. W. Northwestern Pomology; A Treatise on the Growing and Care of Trees, Fruits and Flowers in the Northwestern States. Concord, Neb.; The Author, 1894. 293p.

H

Second Haas, Paul, and T. G. Hill. The Chemistry of Plant Products. London and New York; Longmans, Green and Co., 1913. 401p. (3d ed., 1921.)

First Haberlandt, Gottlieb. Physiological Plant Anatomy, (Physiologische Pflanzenanatomie) . . . trans. from the 4th German ed. by Montagu Drummond. 1st print. London; Macmillan and Co., Ltd., 1914. 777p. (2d reprint ed., New Delhi, India; Today and Tomorrow's Printers and Publishers, 1990.)

Second Haldane, R. C. Subtropical Cultivations and Climates. London; William Blackwood and Sons, 1886. 308p.

Second Hall, Archibald J. Cotton-Cellulose; Its Chemistry and Technology. London; E. Benn Ltd., 1924. 228p.

Second Hall, Bolton. The Garden Yard; A Handbook of Intensive Farming. Philadelphia; D. McKay, 1909. 321p.

Second Hall, D. M. A Practical Handbook on the Culture of Small Fruits, and Guide to Success in Raising the Various Small Fruits for Home Use and for Market. Bangor, Maine; Dirigo Rural Printing Establishment, 1881. 104p.

Second Hall, W. An Architect of Nature: Being the Autobiography of Luther Burbank with Biographical Sketch. London; Watts and Co., 1939. 139p.

Second Halsted, Byron D. Injurious Garden Insects. New York; Phillips and Hunt, Cincinnati; Walden and Stowe, 1883. 16p.

Second Hanson, Val T. Practical Tree Surgery with a Short Course in Apple Culture. Appleton, Wis.; C. C. Nelson Pub. Co., 1932. 101p.

Second Harcourt, Helen. Florida Fruits, and How to Raise Them. Louisville, Ky.; John P. Morton and Co., 1886. 347p.

Second Hard, M. E. The Mushroom. Columbus, Ohio; Ohio Library Co., 1908. 609p.

Second Hardenburg, F. V. Potato Production. Ithaca, N.Y.; Comstock Pub. Co., 1949.

Second Harding, A. R. Ginseng and Other Medicinal Plants. Columbus, Ohio; A. R. Harding Pub. Co., 1908. 301p.

First Harland, S. C. The Genetics of Cotton. London; Jonathan Cape, Ltd., 1939.

Second Harris, Thaddeus W. A Treatise on Some of the Insects Injurious to Vegetation. New York: Orange Judd Co., 1862. 640p. (3d ed., 1890. 662p.)

First Harshberger, John W. A Text-Book of Mycology and Plant Pathology. Philadelphia; P. Blakiston's Son and Co., 1917. 779p.

Second Hartig, Robert. Text-Book of the Diseases of Trees, Rev. and ed., translated by William Somerville. London and New York: Macmillan and Co., 1894. 331p.

Second Harwood, W. S. New Creations in Plant Life: an Authorative Account of the Life and Work of Luther Burbank. New York and London; The Macmillan Co., 1905. 368p. (2d ed., rev. and enl., 1907. 430p. Reprinted, 1941.)

Second Hawkes, John G. Potato Collecting Expeditions in Mexico and South America. Cambridge, Eng.; School of Agriculture, 1941–44. 2 vols.

Second Hawks, Ellison. Pioneers of Plant Study. New York; Macmillan Co., 1928. 288p. (This book was originally planned, and some parts of it written, in collaboration with the late G. S. Boulger.)

Second Hay, Roy. Gardener's Chance; From War Production to Peace Possibilities. London; Putnam, [1946]. 139p.

Second Hayes, Herbert K., and Ralph J. Garber. Breeding Crop Plants. 1st ed. New York; McGraw-Hill Book Co., 1921. 328p. (2d ed., 1927. 438p.)

Second Hayes, Herbert K., and Forrest R. Immer. Methods of Plant Breeding. 1st ed. New York and London; McGraw-Hill Book Co., 1942. 432p. (2d ed., 1955. 551p.)

Second Hayne, Ralph A. More and Better Potatoes to the Acre. Chicago, Ill.; International Harvester Co., Agricultural Extension Dept., 1920. 62p.

First Hayward, Herman E. The Structure of Economic Plants. New York; Macmillan Co., 1938. 674p. (Reprint, New York; Stechert-Hafner, 1967. (Historiae Naturalis Classica no. 54.)

Second Hazlitt, W. Carew. Gleanings in Old Garden Literature. New York; G. J. Coombes, 1887. 263p.

Second Heald, Frederick D. Introduction to Plant Pathology. 1st ed. New York and London; McGraw-Hill Book Co., Inc., 1937. 579p. (2d ed., 1943. 603p.)

First Heald, Frederick D. F. Manual of Plant Diseases. 1st ed. New York and London; McGraw-Hill, 1926. 891p. (2d ed., 1933. 953p.)

Second Hedges, Isaac A. Sugar Canes and Their Products, Culture and Manufacture. rev. and enl. ed. St. Louis, Mo.; The Author, 1881. 90p.

First Hedrick, U. P. Cyclopedia of Hardy Fruits. New York; Macmillan Co., 1922. 370p. (2nd ed., 1938. 402p.)

Second Hedrick, U. P. Fruits for the Home Garden. London and New York; Oxford University Press, 1944. 171p.

Second Hedrick, U. P. Grapes and Wines from Home Vineyards. London and New York; Oxford University Press, 1945. 326p.

First Hedrick, U. P. A Laboratory Manual in Systematic Pomology, an Effort to Place Before the Students of Pomology in the Michigan Agricultural College a Means by Which an Intimate and Accurate Knowledge of Pomology may be Acquired. East Lansing, Mich., 1903. 91p.

First Hedrick, U. P. Manual of American Grape-Growing. New York; Macmillan Co., 1919. 458p. (Rev. ed., New York; The Macmillan Co., 1924. 458p.)

First Hedrick, U. P., ed. Sturtevant's Notes on Edible Plants. Albany, N.Y.; J. B. Lyon Co., State Printers, 1919.

Second Hedrick, U. P., N. B. Booth, O. M. Taylor, and R. Wellington, et al. The Grapes of New York. Albany, N.Y.; J. B. Lyon Co., State Printers, 1908. 564p.

Second Hedrick, U. P., G. H. Howe, O. M. Taylor, and Alwin Berger, et al. The Small Fruits of New York. Albany, N.Y.; J. B. Lyon Co., State Printers, 1925. 614p.

Second Hedrick, U. P., G. H. Howe, O. M. Taylor, E. H. Francis, and H. B. Tukey. The Pears of New York. Albany, N.Y.; J. B. Lyon Co., State Printers, 1921. 636p.

Second Hedrick, U. P., G. H. Howe, O. M. Taylor, and C. B. Tubergen. The Peaches of New York. Albany, N.Y.; J. B. Lyon Co., State Printers, 1917. 541p.

Second Hedrick, U. P., R. Wellington, O. M. Taylor, and M. J. Dorsey. The Plums of New York. Albany, N.Y.; J. B. Lyon Co., State Printers, 1911. 616p.

Second Henderson, Peter. Garden and Farm Topics. New York; Peter Henderson and Co., 1884. 244p.

Second Henderson, Peter. Henderson's Handbook of Plants. New York; Peter Henderson and Co., 1881. 411p.

Second Henderson, Peter &. Co. Henderson's Bulb Culture. New York; Peter Henderson and Co., 1904. 68p.

First Henrici, Arthur T. Molds, Yeasts and Actinomycetes; A Handbook for Students of Bacteriology. Minneapolis, Minn.; Burgess-Roseberry Co., 1929. 159p. (2d ed., New York, London; Wiley, Chapman and Hall, 1947.)

Second Henslow, George. The Origin of Floral Structures through Insect and Other Agencies. New York; Appleton, 1888. 349p. (U.K. ed., London; Kegan Paul, Trench, 1888. 2d ed., 1893.)

Second Heriot, Thomas H. P. The Manufacture of Sugar From the Cane and Beet. London and New York; Longmans, Green and Co., 1920. 426p.

Second Herrick, Glenn W. Insects of Economic Importance: Outlines of Lectures in Economic Entomology. Ithaca, N.Y.; Carpenter and Co., 1915. 138p. (Rev. ed., New York; Macmillan, 1920. 172p.)

First Herrmann, H. French Method of Intensive Cultivation and Asparagus Forcing; a Treatise on the French Method of Gardening. Louisville, Ky.; 1910. 50p.

First Hesler, Lexemuel R., and H. H. Whetzel. A Manual of Fruit Diseases. New York; Macmillan Co., 1917. 462p. (Later printing 1920)

Second Hess, Katherine P. Textile Fibers and Their Uses. Philadelphia; J. B. Lippincott Co., 1931. 354p. (6th ed., Chicago; Lippincott, 1958. 549p.)

Second Hexamer, F. M. Asparagus; Its Culture for Home Use and for Market. New York; Orange Judd Co., 1901. 168p.

Second Heyne, E. B. Catalogue of European Vines; With Their Synonyms and Brief Descriptions. San Francisco; Dewey and Co., 1881. 63p.

First Hill, Albert F. Economic Botany; A Textbook of Useful Plants and Plant Products. 1st ed. New York; McGraw-Hill Book Co., 1937. 592p.

Second Hillman, Joseph. The Cultivation of Cotton; A Short Treatise Specially Bearing on Fertilization and the Control of the Ravages of the Boll Weevil. New York; W.S. Myers, 1905. 25p.

First Hills, William H. Small Fruits. Boston; Cupples, Upham and Co., 1886. 138p.

Second Hoare, A. H. The English Grass Orchard and the Principles of Fruit Growing. London; E. Benn, Ltd., 1928. 227p.

Second Hoare, A. H. Fruit Culture. London; Nelson, 1948. 347p.

Second Hoare, A. H. Vegetable Crops for Market. London; C. Lockwood and Sons, 1937. 198p. (3d ed., London; Crosby Lockwood, 1949. 336p.

Second Hoare, Arthur H. Commercial Apple Growing. London; M. Hopkinson Ltd, 1937. 245p. (2d ed, London; Bodley Head, 1948. 288p.)

Second Hogg, J. The Vegetable Garden. New York; Dick and Fitzgerald, 1877. 137p. (Cover has the legend, Dick's Garden Hand-Books. The Vegetable Garden)

Second Hogg, Robert. The Fruit Manual. 1st ed. London; Cottage Gardener Office, 1860. 280p. (2d ed., 1861. 5th ed., 1884.)

First Holden, Perry G. The ABC of Corn Culture. Springfield, Ohio; The Simmons Publishing Co., 1906. 92p.

Second Holden, Perry G. Corn Secrets. 1st ed. Philadelphia, Pa.; Wilmer Atkinson Co., 1910. 79p.

Second Holden, Perry G. Growing Prize Corn. Philadelphia, Pa.; Wilmer Atkinson Co., 1914. 47p.

Second Holden, Perry G. Successful Corn Culture . . . Des Moines, Iowa; 1907. 84p. (10th ed., rev. and enl., Des Moines, Iowa; Successful Farming Pub. Co., 1912. 86p.)

First Hollister, E. J. Livingston's Celery Book. Columbus, Ohio; A. W. Livingston's Sons, 1898. 96p.

Second Holmes, F. O. Handbook of Phytopathogenic Viruses. Minneapolis, Minn.; Burgess Pub. Co., 1939. 390p.

First Holmes, Francis S. The Southern Farmer and Market-Gardener: Being a Compilation of Useful Articles on These Subjects From the Most Apporoved Writers . . . Charleston, S.C.; Burges and James, 1842. 244p. (New ed., Charleston, S.C.; Wm. R. Babcock, 1852. 249p.)

Second Hook, Wallace. A Primer of Agriculture. 1st ed. Packwood, Idaho; W. A. Hook, 1912. 64p.

Second Hooper, Edward J. Hooper's Western Fruit Book; A Compendious Collection of Facts, From the Notes and Experience of Successful Fruit Culturists, Arranged for Practical Use in the Orchard and Garden. Cincinnati; Moore, Wilstach, Keys & Co., 1857. 333p. 3d rev. ed., 18--; 355p.

Second Horsfall, James G. Fungicides and Their Action. Waltham, Mass.; Chronica Botanica Co., 1945. 239p.

First Horvath, Arthemy A. The Soybean Industry. New York; The Chemical Publishing Co. of New York, Inc., 1938. 221p.

Second Howard, Albert. An Agricultural Testament. London and New York; Oxford University Press, 1940. 253p. (Continuation of an earlier book: The Waste Products of Agriculture, 1931.)

Second Howard, Leland O. The Insect Menace. New York and London; The Century Co., 1931. 347p.

Second Howard, W. L., and E. H. Favor. The Home Garden. St. Joseph, Mo.; The Fruit-Grower Co., 1905. 58p. (Brother Jonathan Series, No. 4)

Second Howe, Walter. The Garden; As Considered in Literature by Certain Polite Writers; with a Critical Essay. New York and London; G. P. Putnam's Sons, 1890. 309p.

Second Howes, Frank N. Nuts; Their Production and Everyday Uses. London; Faber and Faber, Ltd., 1948. 264p.

Second Howes, Frank N. Vegetable Gums and Resins. Waltham, Mass.; Chronica Botanica Co., 1949. 188p.

Second Hubbard, H. G. Insects Affecting the Orange. Washington, D.C.; Govt. Print. Off., 1885. 227p.

First Hume, H. Harold. Citrus Fruits and Their Culture. Jacksonville, Fla.; The H. and W. Drew Co., 1904. 597p. (3d. ed. rev., New York; Orange Judd Company, 1909. 587p.)

Second Hume, H. Harold. The Cultivation of Citrus Fruits. New York; Macmillan Co., 1926. 561p. (Rev. ed. titled: Citrus Fruits, 1957. 444p.)

First Hume, H. Harold. The Pecan and Its Culture. Petersburg, Va.; The American Fruit and Nut Journal, 1906. 159p. (2d. ed., Glen St. Mary, Fla.; Published by the author, 1910. 159p. 3d. ed., rev., 1912. 195p.)

Second Humphreys, Phebe Westcott. The Practical Book of Garden Architecture. Philadelphia and London; J. B. Lippincott Co., 1914. 330p.

Second Hunt, Thomas F. The Cereals in America. Reprinted with Corrections. New York; Orange Judd Co., 1905. 421p. (1st printing 1904.)

Second Hunter, Herbert. Oats: Their Varieties and Characteristics; A Practical Handbook for Farmers, Seedsmen, and Students. London, Eng.; Ernest Benn, Ltd., 1924. 131p.

First Hunter, Herbert, and H. Martin Leake. Recent Advances in Agricultural Plant Breeding. London, Eng.; J. and A. Churchill, 1933. 361p.

First Hurst, B. F. The Fruit Grower's Guide. Boise, Idaho.; 1905. 144p. (International Library of Technology. Correspondence course in fruit growing. 3 vols. 1912, 1913, 1914. Scranton, Pa.)

First Husmann, George. Grape Culture and Wine-Making in Californa. San Francisco; Payot, Upham and Co., 1888. 380p.

Second Husmann, George. American Grape Growing and Wine Making. New York; Orange Judd Co., 1880. 243p. (New ed., 1883. 310p. 4th ed., rev., 1896. 269p.)

Second Husmann, George. The Cultivation of the Native Grape, and Manufacture of American Wines. New York; G. E. and F. W. Woodward, 1866. 192p. (4th ed., rev., New York; Orange Judd Company, 1896. 169p.)

Second Huston, Tom, Peanut Company, Columbus, Ga. Peanuts; Culture and Marketing of the White Spanish Variety in the Southeastern States . . . comp. by Bob Barry and Grady Porter. Columbus, Ga.; Tom Huston Peanut Company, 1930? 28p. (3d ed., Columbus, Ga.; 1932. 32p.)

Second Hutcheson, T. B., T. K. Wolfe, and M. S. Kipps. The Production of Field Crops; a Textbook of Agronomy. New York and London; McGraw-Hill Book Co., 1936. (Later eds. by Kipps.)

Second Hutchins, W. T. All About Sweet Peas. Philadelphia; W. Atlee Burpee and Co., 1892. 25p. (2d ed., A complete epitome of the literature of this fragrant annual, 1894. 131p. Rev. ed., 1894. 131p.)

Second Hutchins, W. T. Sweet Peas Up-to-Date. Philadelphia; W. Atlee Burpee and Co., 1897. 72p.

Second Huxley, J. Evolution—The Modern Synthesis. New York; Harpers, 1942.

Second Hylander, C. J., and O. B. Stanley. Plants and Man. Philadelphia; The Blakiston Co., 1941.

Second Hyslop, J. A. Losses Occasioned by Insects, Mites and Ticks in the United States. Washington, D.C.; U.S. Bureau of Entomology and Plant Quarantine, Division of Insect Pest Survey and Information, 1938. 57p. (Rev. ed. of An Estimate of the Damage by Some of the More Important Insect Pests in the United States. Washington, 1930.)

I

Second Illustrated Pear Culturist by an Amateur. New York; C. M. Saxton and Co., New London; Starr and Co., 1858. 106p.

First Iltis, Hugo. Life of Mendel. London; George Allen and Unwin, 1932. (Translated by Eden and Cedar Paul from Gregor Johann Mendel; Leben, Werk und Wirking. Julius Springer, Berlin. 1924. Reprinted in 1966.)

Second Ingalls, Walter F. Soy Beans. Cooperstown, N.Y.; The Arthur H. Crist Co., 1912. 36p.

Second Ingold, C. T. Spore Discharge in Land Plants. Oxford; Clarendon Press, 1939.

Second International Apple Shippers' Association. The Story of National Apple Week, Past and Present . . . Rochester, N.Y.; 1883? 52p.

Second Isely, Dwight. Methods of Insect Control. Fayetteville, Ark.?; 1938. (2d ed, rev., Minneapolis, Minn.; Burgess Publishing Co., 1941. 4th ed, rev., 1947.)

Second Ivanowski, Dmitrii. Concerning the Mosaic Disease of the Tobacco Plant. Originally publ.: "Ueber die Mosaikkrankheit der Tabakspflanze." in St. Petersb. Acad. Imp. Sci. Bul. 35 (nouv. ser. i. e. ser. 4, v. 3) (Sept. 1892): 67–70. Ithaca, NY; American Phytopathological Society, 1942. [*Phytopathological Classics* no. 7]

J

Second J. I. Case Company. How to Produce High Protein Hay. Racine, Wis., 1946. 15p.

Second Jacks, G. V., and R. O. Whyte. The Rape of the Earth. London; Faber and Faber, Ltd., 1939.

Second Jackson, A. V. Secrets of Mushroom Growing Simply Explained, by the Largest Grower in America. Chicago; 1906. 44p. (5th ed., Boston; Hooper Printing Company, 1913. 68p.)

Second Jacques, D. H. New Illustrated Rural Manuals. New York; Fowler and Wells, 1859. (Comprising: The House, The Garden, The Farm, Domestic Animals. Complete in one Volume)

First Jacques, George. A Practical Treatise on the Management of Fruit Trees. Worcester; Erastus N. Tucker, 1849. 256p. (Adapted to northern states, New York; Edward Livermore, 1856.)

Second James, W. O., and A. R. Clapham. The Biology of Flowers. Oxford; Clarendon Press, 1935.

Second James, W. O. An Introduction to Plant Physiology. Oxford; Clarendon Press, 1936.

Second Jamieson, G. S. Vegetable Fats and Oils, the Chemistry, Production and Utilization of Vegetable Fats and Oils for Edible, Medicinal and Technical Purposes. New York; Chemical Catalog Co., 1932. 444p. (2d ed., New York; Reinhold Pub. Corp., 1943.)

Second Jennings, H. S. Genetics. New York; W. W. Norton, 1935. 373p.

Second Jepson, Willis L. A Flora of the Economic Plants of California. University of California; Associated Students Store, 1924. 223p.

Second Jepson, Willis L. The Trees of California. University of California; Associated Students Store, 1909.

First Johansen, Donald A. Plant Microtechnique. New York and London, Eng.; McGraw-Hill, 1940. 523p.

Second Johnson and Stokes. Farm Gardening. Philadelphia; Johnson and Stokes, 1898. 124p.

Second Johnson, Charles. The Seed Grower. A Practical Treatise on Growing Vegetable and Flower Seeds and Bulbs for the Market. Marietta, Pa.; 1906. 191p.

Second Johnson, George William. Cottage Gardeners' Dictionary. London; W. S. Orr, 1852. 927p.

First Johnson, Mark W. How to Plant and What to Do with the Crops. New York; Orange Judd Co., 1886. 89p.

First Johnson, Samuel W. How Crops Grow. New York; O. Judd Co., 1869. 394p. (Reprinted 1879, 1887. Rev. and enl., 1890, 1895, 1900, 1908. 416p.)

Second Johnson, William H. Cotton and Its Production. London; Macmillan and Co., Ltd., 1926. 536p.

First Jones, B. W. The Peanut Plant, Its Cultivation and Uses . . . New York; O. Judd Co., 1885. 69p. (Reprinted 1896.)

Second Jones, Donald F. Selective Fertilization. Chicago, Ill.; The Univ. of Chicago Press, 1928. 163p.

Second Jones, Henry Albert, and Samuel Leonard Emsweller. The Vegetable Industry. 1st ed. New York; McGraw-Hill, 1931. 431p. (McGraw-Hill vocational texts)

First Jones, Henry Albert, and Joseph Tooker Rosa. Truck Crop Plants. 1st ed. New York; McGraw-Hill Book Company, Inc., 1928. 538p.

Second Jones, Llewellyn, and Scard Frederic I. The Manufacture of Cane Sugar. London; E. Stanford, 1909. 454p. (2d ed., rev., London; Duckworth and Co., 1921. 481p.)

Second Jones, S. G. Introduction to Floral Mechanism. London and Glasgow; 1939. 274p.

Second Jones, William N. Plant Chimaeras and Graft Hybrids. London; Methuen and Co., Ltd., 1934. 136p.

Second Jordan, David Starr, and Vernon L. Kellogg. The Scientific Aspects of Luther Burbank's Work. San Francisco, Calif.; A. M. Robertson, 1909. 115p.

Second Jordan, Ralph W. Onions. St. Paul, Minn.; Webb Pub. Co., 1915. 95p.

Second Jordan, Samuel M. Making Corn Pay. Springfield, Mass.; The Phelps Pub. Co., 1913. 75p. (Practical Farm Library)

K

Second Kahn, Allen R. Sugar; A Popular Treatise. 4th ed. Los Angeles, Calif.; U. S. Sugar Publications Co., 1921. 78p.

Second Kains, Maurice G. Culinary Herbs. New York; Orange Judd Co., 1912. 143p.

First Kains, Maurice G. Ginseng. New York; Orange Judd Co., 1899. 53p. (New ed., rev., 1903. Rev., 1902. 144p.)

Second Kains, Maurice G. Modern Guide to Successful Gardening. New York; Greenberg, 1934. 370p.

Second Kains, Maurice G., and L. M. McQuesten. Propagation of Plants; A Complete Guide for Professional and Amateur Growers . . . New York; Orange Judd Pub. Co., 1938. 555p. (New ed., New York; Orange Judd, 1950.) (1st ed. by Kains only, 1916.)

First Kansas State Horticultural Society. The Apple . . . What it is. How to Grow it. Its Commercial and Economic Importance. How to Utilize it . . . comp. and rev. by the Kansas State Horticultural Society, William H. Barnes, Secretary. Topeka, Kans.; J. S. Parks, Printer, 1898. 229p.

First Kansas State Horticultural Society. The Cherry in Kansas, with a Chapter on the Apricot and the Nectarine . . . comp. and rev. for the Kansas State Horticultural Society by William H. Barnes, Secretary. Topeka, Kans.; Issued by the State, 1900. 128p.

First Kansas State Horticultural Society. The Grape in Kansas . . . comp. and rev. for the Kansas State Horticultural Society, by Willam H. Barnes, Secretary. Topeka, Kans.; Issued by the State, 1901. 139p.

Second Kansas State Horticultural Society. The Peach . . . How to Grow Your Trees. How to Plant and Care for Them. How to Fight its Enemies. How to Gather, Pack, and Market. How to Enjoy it in the Home . . . comp. by the Kansas State Horticultural Society, William H. Barnes, Secretary. Topeka, Kans.; W. Y. Morgan, State Printer, 1899. 159p.

Second Kansas State Horticultural Society. The Plum in Kansas, with a Chapter on the Prune. How to Grow Them . . . comp. and rev. for the Kansas State Horticultural Society by William H. Barnes, Secretary. Topeka, Kans.; 1900. 159p.

Second Karling, John S. Plasmodiophorales. New York; The Author, 1942. 144p.

Second Karling, John S. The Simple Holocarpic Biflagellate Phycomycites, Including a Complete Host Index and Bibliography. 1st ed. New York; The Author, 1942. 123p.

Second Kavina, K. Atlas of Fungi. London; Lincolns-Prager, 1947. 31p.

First Kearney, Thomas H. Cotton; History, Botany, and Genetics. Washington; American Genetic Association, 1931. 47p.

Second Kellogg, Vernon L., and Rennie W. Doane. Elementary Textbook of Economic Zoology and Entomology. New York; H. Holt and Co., 1915. 532p.

First Kendall, R. C. Cotton and Common Sense. A Treatise on Perennial Cotton (Gossypium arboreum); Its Commercial Value as Compared with Herbaceous Cotton, The Feasibility of Its Culture in Northern Latitudes . . . , New York; Mapes and Lockwood, 1862. 32p.

Second Kerr, Thomas. A Practical Treatise on the Cultivation of Sugar Cane, and Manufacture of Sugar. London; J. J. Griffin and Co., Glasgow; R. Griffin and Co., 1851. 139p.

Second Keystone Pecan Company, Manheim, Penn. The Paper Shell Pecan . . . Lancaster, Pa.; Press of Examiner Printing House, 1915. 48p. (Also Printed; 1918. 56p., 1919. 64p., New Era Printing Co.; 1921. 52p.)

Second Kidder, Alfred F. Corn . . . Baton Rouge; 1914. 80p.

Second Kidner, Alfred W. Asparagus. London; Faber and Faber, Ltd., 1947. 168p.

Second King, Eleanor, and Wellmer Pessels. Insect People. 1st ed. New York and London; Harper and Brothers, 1937. 63p. (Facts and pictures about the lives and habits of garden insects.)

Second King, Franklin H. Farmers of Forty Centuries; or, Permanent Agriculture in China, Korea, and Japan. Madison, Wis.; Mrs. F. H. King, 1911. 441p. (Later Printing, Emmanus, Penn.; Rodale Press. 1973.)

Second Klages, K. H. W. Ecological and Crop Geography. New York; Macmillan Co., 1941.

First Klippart, John H. The Wheat Plant: Its Origin, Culture, Growth, Development, Composition, Varieties, Diseases, etc . . . Together with a Few

Remarks on Indian Corn, Its Culture, etc. Cincinnati, Ohio; Moore, Wilstach, Keys and Co., 1860. 706p.

Second Klose, Nelson. America's Crop Heritage; The History of Foreign Plant Introduction by the Federal Government. Ames, Iowa; Iowa State College Press, 1950. 156p.

Second Knapp, A. W. The Cocoa and Chocolate Industry. London; Sir Isaac Pitmann and Sons, 1930.

Second Knapp, George R. How to Grow Strawberries. Greenfield, Mass.; H. D. Watson Co., 1886. 63p.

Second Knapp, Halsey B., and E. C. Auchter. Growing Tree and Small Fruits. New York; J. Wiley and Sons, Inc.; London; Chapman and Hall, Ltd., 1929. 510p. (2d ed., New York; 1941. 615p.)

Second Knight, Paul. Problems of Insect Study . . . Ann Arbor, Mich.; Edwards, 1933. 118p. (Reprinted 1935. 2d ed., 1939. 132p.)

First Knott, James E. Vegetable Growing. Philadelphia; Lea and Febiger, 1930. 352p. (5th ed., 1955. 358p.)

Second Knowlton, J. M. Our Hardy Grapes. New York; Coutant and Baker, 1863. 96p.

First Knuth, Paul E. Handbook of Flower Pollination: Based Upon Hermann Muller's Work 'The Fertilization of Flowers By Insects' . . . trans. by J. R. Ainsworth Davis. Oxford; Clarendon Press, 1906–1909. 3 vols. (Later printings; 1908, 1909)

Second Krieger, Louis C. C. The Mushroom Handbook. New York and London; Macmillan, 1936. 512p.

Second Kuster, Ernst. Pathological Plant Anatomy, authorized translation by Frances Dorrance. 1915. 258p.

L

Second Lacy, T. Jay. Fruit Culture for the Gulf States, South of Latitude 32. Alexandria, La.; Press of Town Talk, 1888. 50p.

Second Lance, Inc., Charlotte, N.C. The Peanut World; The Story of Peanuts and Some Delicious Peanut Foods. Carlotte, N.C.; 1946. 32p.

Second Landreth, Burnet. 999 Queries, With Answers Upon Agricultural and Horticultural Subjects. Philadelphia; David Landreth and Sons, Press of MacCalla and Co., 1895. 200p.

First Landreth, Burnet. Market-Gardening and Farm Notes. New York; Orange Judd Co., 1893. 215p.

Second Lansdell, Joseph. Grapes, Peaches, Nectarines and Melons . . . rev. and modernized by A. J. Macself. London; Collingridge, 1948. 143p.

Second Large, E. C. The Advance of the Fungi. London; Jonathan Cape; New York; H. Holt and Co., 1940. 488p.

Second Lawrence, William H. Apple Growing. Orenco, Oreg.; H. V. Meade, 1913. 31p. (Written specially for use in the Pacific horticultural correspondence school, Portland, Oregon.)

Second Lawrence, William J. C. Practical Plant Breeding. London; Allen and Unwin, 1937. 155p. (2d ed., 1948. 3d ed., 1951. 166p. 3d ed., rev., 1957. 165p.)

Second Lawson, William. A New Orchard and Garden . . . reprinted from the third edition with a preface by Elanour Sinclair Rohde. Whereunto is newly added the Art of Propagating Plants With the True Ordering of All Manner of Fruits, In Their Gathering, Carrying Home, and Preservation. London; Cresset Press, Ltd., 1927. 116p.

First Leach, Julian G. Insect Transmission of Plant Diseases. 1st ed. New York; McGraw-Hill, 1940. 615p.

Second Leavens, George D. Corn, The Foundation of Profitable Farming. New York; The Coe-Mortimer Co., 1915. 80p.

Second Leavens, George D. Potatoes: A Money Crop. How to Grow, Fertilize, Spray and Harvest Them at a Profit. 5th ed? New York; The Coe-Mortimer Co., 1916. 45p.

Second Lectures on Plant Pathology and Physiology in Relation to Man; A Series of Lectures Given at the Mayo Foundation and The Universities of Minnesota, Iowa, Wisconsin, the Des Moines Academy of Medicine, Iowa, and Iowa State College 1926–1927. Philadelphia and London; W. B. Saunders Co., 1928. 207p.

First Lelong, B. M. Culture of the Citrus in California. Sacramento; A. J. Johnston, Superintendent State Printing, 1900. 206p. (Revised by State Board of Horticulture, 1902, Published by State Board of Horticulture, 169p.)

Second Leonard, Warren H., and Andrew G. Clark. Field Plot Technique. 1st ed. Minneapolis, Minn.; Burgess Pub. Co., 1939. 271p. (2d ed by LeClerg, Erwin L.; H. L. Leonard and A. G. Clark, 1962. 373p.)

First Levitt, J. Frost Killing and Hardiness of Plants. Minneapolis; Burgess Pub. Co., 1941. 211p.

Second Link, George K. K. Readings in Phytopathology; Ancient Greek Phytopathology. 1st ed. Chicago; Univ. of Chicago Bookstore, 1941. 18p.

Second Lloyd, J. W. Productive Vegetable Growing. Philadelphia; J. B. Lippincott Co., 1914. 339p.

Second Lodeman, Ernest G. The Spraying of Plants; A Succinct Account of the History, Principles and Practice of the Application of Liquids and Powders to Plants, for the Purpose of Destroying Insects and Fungi. New York and London; Macmillan and Co., 1897. 399p.

Second Loeb, Jacques. Regeneration from a Physico-Chemical Viewpoint. 1st ed. New York; McGraw-Hill, 1924. 143p.

Second Long, Harold C. Weed Suppression by Fertilizers and Chemicals. London; Caledonian Press, 1934. 57p. (2d ed. titled: Suppression of Weeds by Fertilizers and Chemicals. London; Lockwood, 1945. 87p.)

Second Loomis, Walter E., and C. A. Shull. Experiments in Plant Physiology. New York and London; McGraw-Hill, 1939.

Second Loomis, Walter E. and C. A. Shull. Methods in Plant Physiology. New York; McGraw-Hill Book Co., 1937. 472p.

Second Lorette, L. The Lorette System of Pruning . . . translated by W. R. Dykes. London; M. Hopkinson and Co., Ltd., 1925. 166p. (Rev. ed., Emmaus, Penn.; Rodale Press, 1946. 239p.)

Second Louisiana State Rice Milling Co. Rice and How to Cook It. With Illustrations of the Primitive and Modern Methods of Planting, Cultivation and

Milling. New Orleans, La.; Louisiana State Rice Milling Co., Inc., 1919. 47p.

First — Lowther, Granville, and William Worthington, eds. The Encyclopedia of Practical Horticulture. North Yakima, Washington; The Encyclopedia of Horticulture Corporation, 1914. 3 vols. 2037p.

Second — Lucas, I. B. Dwarf Fruit Trees for Home Gardens. New York; De La Mare and Co., 1946. 123p.

Second — Lundegardh, H. Environment and Plant Development. London; Edward Arnold and Co., 1931. (Translated from the German, Jena; Klima und Boden, 1925.)

Second — Lupton, J. M. Cabbage and Cauliflower for Profit. Philadelphia; W. Atlee Burpee and Co., 1894. 122p.

Second — Lyman, Joseph B. Cotton Culture . . . With an Additional Chapter on Cotton Seed and Its Uses by J. R. Sypher. New York; O. Judd and Co., 1868. 190p.

First — Lyon, W. S. Gardening in California. Los Angeles, Calif.; Geo. Rice and Sons, 1897. 156p. (Published by the author, 180 pp.)

M

Second — MacCurdy, R. M. The History of the California Fruit Growers Exchange. Los Angeles; Published by the Exchange, 1925. 106p.

Second — MacGerald, Willis, ed. Practical Farming and Gardening. Chicago and New York; Rand, McNally and Co., 1902. 500p. (Contents: Edgerton, J. J. Modern Ideas in Soil Treatment and Tillage; Edgerton, J. J. Field Crops; Their Adaptations and Economic Relations With Specific Cultural Directions. Erwin, A. T. Vegetable Garden and Trucking Crops. Taft, L. R. Fruit Culture and Forestry. etc . . .)

Second — MacMillan, R. C. Planting for Plenty. New York; Faber and Faber, 1947. 112p.

Second — Maksimov, N. A. A Plant in Relation to Water; A Study of the Physiological Basis of Drought Resistance . . . trans. by R. H. Yapp. New York; Macmillan Co., 1929. 451p. (London; Allen and Unwin, 1929.)

Second — Maksimov, N. A. Plant Physiology. New York; McGraw-Hill Book Co., Inc., 1938.

Second — Malden, Walter J. The Potato in Field and Garden. London; W. A. May, 1895. 217p.

Second — Mantell, C. L. The Water Soluble Gums. New York; Reinhold Pub. Corp., 1947. 279p.

Second — Manville, A. H. Practical Orange Culture, Including the Culture of the Orange, Lemon, Lime, and Other Citrus Fruits, as Grown in Florida. Jacksonville, Florida; Ashmead Bros., 1883. 116p.

Second — Markham, Ernest. Raspberries and Kindred Fruits; How to Obtain Fresh Supplies Daily From June to November, with Chapters on the Loganberry, Hybrid Berries and Giant Blackberries. London; Macmillan and Co., Ltd., 1936. 67p.

Second — Marsh, H. T. The Apple Industry. Cumberland, Md.; F. Mertens' Sons, 1910. 18p.

Second Martin, J. H., and W. H. Leonard. Principles of Field Crop Production. New York; Macmillan, 1949. 1176p. (2d ed., 1967.)

Second Martin, Hubert. The Scientific Principles of Plant Protection. London; E. Arnold and Co., 1928. 316p. (6th ed. titled: The Scientific Principles of Crop Protection. London; E. Arnold; New York; Crane, Russak, 1973. 423p.)

Second Martin, J. N. Botany with Agricultural Applications. New York; John Wiley and Sons, Inc., 1920.

First Marvin, A. T. The Olive: Its Culture in Theory and Practice. San Francisco; Payot, Upham and Co., 1888. 146p.

Second Mason, Albert F. Spraying, Dusting, and Fumigating of Plants; A Popular Handbook on Crop Protection. New York; Macmillan Co., 1928. 539p.

Second Massachusetts Horticultural Society. History of the Massachusetts Horticultural Society, 1829–1878. Portrait of H. A. S. Dearborn. Boston; Massachusetts Horticultural Society, 1880. 545p.

First Massee, George E. Disease of Cultivated Plants and Trees. London; Duckworth and Co., 1910. 602p. (U.S. ed., New York; Macmillan Co., 1914. 602p.) (Replaced: Textbook on Plant Diseases, 1907.

Second Massee, George E. Mildews, Rusts and Smuts; A Synopsis of the Families, Peronosporaceae, Erysiphaceae, Uredinaceae and Ustilaginaceae. London; Dalau and Co., 1913. 229p.

Second Massee, George E. A Text-Book of Plant Diseases Caused by Cryptogamic Parasites. London; Duckworth and Co., New York; Macmillan Co., 1899. 458p. (3d ed., 1907. 472p.)

Second Matchette, William H. Potatoes; How to Grow More and Better Potatoes, A Guide for the Business Farmer. Waterloo, Ia.; The Galloway Bros.-Bowman Co., 1913. 47p.

Second Mather, K. Biometrical Genetics. New York; Dover Publications, 1949.

Second Matsuura, Hajime. A Bibliographical Monograph on Plant Genetics (Genetic Analysis) 1900–1925. Tokyo; Tokio Imperial University, 1929. 499p. (2d ed., rev. and enl., Sapporo; Hokkaido Imperial University, 1933. 787p.)

Second Matthews, Joseph Merritt. The Textile Fibers, Their Physical, Microscopical and Chemical Properties. 4th ed., rewritten and enl. New York; J. Wiley and Sons, Inc., 1924. 1053p. (Later eds. under: Maursberger, H. R., ed. Matthew's Textile Fibers. 5th ed. New York; John Wiley and Sons, Inc., 1947.)

Second Maxson, Asa C. Insects and Diseases of the Sugar Beet. Fort Collins, Colo.; Beet Sugar Development Foundation, 1948. 425p.

Second Maxson, Asa C. Principal Insect Ememies of the Sugar Beet in the Territories Served by the Great Western Sugar Company. Denver, Colo.; Great Western Sugar Co., 1920. 137p.

Second Maxwell, Francis. Modern Milling of Sugar Cane. London; N. Rodger, 1932. 423p.

Second Mayer, Adolf. *Concerning the Mosaic Disesase of Tobacco*, English translation by J. Johnson (Ithaca, N.Y.: American Phytopathological Society, 1942 [*Phytopathological Classics* no. 7]). (Originally published as "Ueber die Mosaikkrankheit des Tabaks" in *Landwirtschaftlichen Versuchs-Stationen* 32 (1886): 451–467.)

First Maynard, Samuel T. The Practical Fruit Grower. Springfield, Mass.; The Phelps Pub. Co., 1886. 108p. (American Agriculture, No. 1., 1898. 128p. New York; Orange Judd Company, 1909.)

First Maynard, Samuel T. Successful Fruit Culture. New York; Orange Judd Co., 1905. 274p.

Second McAlpine, Daniel. Fungus Diseases of Stone-Fruit Trees in Australia and Their Treatment. Melbourne; R. S. Brain, Govt. Printer, 1902. 165p.

Second McAlpine, Daniel. The Rusts of Australia; Their Structure, Nature, and Classification. Melbourne; R. S. Brain, Govt. Printer, 1906. 349p.

Second McCollom, William C. Vines and How to Grow Them. Garden City, New York; Doubleday, Page and Co., 1911. 315p.

Second McConnell, Primrose. Crops, Their Characteristics and Cultivation . . . London, Paris, New York, etc.; Cassell and Co., Ltd., 1908. 115p.

Second McCubbin, Walter A. Fungi and Human Affairs, with Special Reference to Plant Diseases. Yonkers-On-Hudson, N.Y. and Chicago; World Book Co., 1924. 111p.

Second McCullough (J. Chas) Seed Co. Farm and Grass Seed Manual. 1st ed. Cincinnati; 1948. 126p. (2d ed., 1950. 148p.)

Second McEwen, George. The Culture of the Peach and Nectarine . . . ed. and enl. by John Cox. London; Groombridge and Sons, 1859. 52p.

First McIntosh, T. P. The Potato, Its History, Varieties, Culture and Diseases. London and Edinburgh; Oliver and Boyd, 1927. 264p.

Second McIntyre, A. R. Curare. Chicago; University of Chicago Press, 1947. 240p.

First McLaurin, John. The Model Potato: An Exposition of the Proper Cultivation of the Potato; the Causes of Its Diseases, or "Rotting"; The Remedy Therefore; Its Renewal, Preservation, Productiveness, and Cooking. New York; S. R. Wells, 1872. 102p.

Second Medsger, Oliver P. Edible Wild Plants . . . with introd. by Ernest Thompson Seton. New York; Macmillan Co., 1939. 323p.

Second Meech, W. W. Quince Culture. New York; The American Garden, 1888. 143p. (Orange Judd Company, 1888. Rev. ed., Orange Judd Company, 1896. 180p.)

Second Meeker, Ezra. Hop Culture in the United States: Being a Practical Treatise on Hop Growing in Washington Territory, From the Cutting to the Bale . . . Puyallup, Wash.; E. Meeker and Co., 1883. 170p.

Second Meisel, Max. A Bibliography of American Natural History. The Pioneer Century, 1769–1865. Brooklyn, N.Y.; 1927. 3 vols.

Second Melander, Axel L. The Practical Control of Apple Diseases and Pests. How to Identify the Pests and Diseases . . . , Portland, Oreg.; 1912. 41p. (Written specially for use in the Pacific horticultural correspondence school, Portland, Oregon.)

Second Melhus, Irving E., and G. C. Kent. Elements of Plant Pathology. New York; The Macmillan Co., 1939. 493p.

Second Menand, L. Autobiography, and Recollections of Incidents Connected with Horticultural Affairs, etc., from 1807 up to This Day, 1892 . . . Albany, N.Y.; Weed, Parsons and Co., 1892. 200p. (2d ed., Cohoes, N.Y.; L'Independant Printing Office, 1898. 350p. From 1807 up to this day 1898.)

Second Mendel, G. Experiments in Plant-Hybridization. Cambridge, Mass.; Harvard University Press, 1925. 353p. (Translation from Verh. Naturf. Ver. in Brunn. Abhandlungen IV, 1866 in Royal Horticultural Society Journal, v. 26, 1901.)

Second Messick, Henry L. Citrus Tree Culture; Facts on Citrus Trees and Their Diseases. La Verne, Calif.; La Verne Leader, 1919. 64p.

First Metcalf, Clell L., and W. P. Flint. Destructive and Useful Insects; Their Habits and Control. 1st ed. New York; McGraw Hill Book Co., Inc., 1928. 918p. (4th ed., McGraw-Hill, 1962. 1087p.)

Second Meyer, Bernard S., and Donald B. Anderson. Plant Physiology. New York; D. Van Nostrand, 1939.

Second Michels, Charles A. Systematic Laboratory Studies of Field Crops and Weeds. Philadelphia; Lea and Febiger, 1928. 298p.

First Milburn, Thomas. Fungoid Diseases of Farm and Garden Crops . . . pref. by E. A. Bessey. London and New York; Longmans, Green and Co., 1915. 118p.

First Millard, Hannah. Grapes and Grape Vines of California. San Francisco; Edward Bosqui and Co., 1877.

Second Millardet, Pierre M. A. The Discovery of Bordeaux Mixture; Three Papers: I. Treatment of Mildew and Rot; II. Treatment of Mildew with Copper Sulphate and Lime Mixture; III. Concerning the History of the Treatment of Mildew and Copper Sulphate, translation by Felix John Schneiderhan, from Millardet papers. Ithaca, N.Y.; The Cayuga Press, 1933, 25p. [*Phytopathological Classics* no. 3]

Second Miller, Andrew. Wheat and Its Products; a Brief Account of the Principal Cereal: Where it is Grown, and the Modern Method of Producing Wheaten Flour. London and New York; Sir I. Pitman and Sons, Ltd, 1916. 134p.

Second Miller, Edith M., et al. Some Great Commodities. Garden City, N.Y.; Doubleday, Page and Co., 1922. 287p.

First Miller, Edwin C. Plant Physiology, with Reference to the Green Plant. 1st ed. New York and London; McGraw-Hill Book Co., Inc., 1931. 900p. (2d ed., 1938. 1201p.)

Second Miller, T. B. Farm and Garden Compendium; Agriculture, Horticulture, Floricutlure. Philadelphia; 1893. 85p.

Second Millspaugh, Charles F. American Medicinal Plants; An Illustrated and Descriptive Guide to American Plants Used as Homeopathic Remedies. The Work is Illustrated from Drawings of Each Plant In Situ by the Author. New York; Boericke and Tafel, 1884–1887. 2 vols.

Second Mitchell, J. W., and J. H. Holland. A Textbook in Tropical Agriculture. London; The Macmillan Co., 1929.

Second Mitchell, John W., and Paul C. Marth. Growth Regulators for Garden, Field and Orchard. Chicago; Univ. of Chicago Press, 1947. 129p.

Second Mitzky and Co., C. Our Native Grape. Rochester; W. W. Morrison, 1893. 218p.

Second Moeller-Krause, Werner. Practical Handbook for Beet-Sugar Chemists; Rapid Methods of Technico-Chemical Analyses of the Products and By-Products and of Material Used in the Manufacture of Beet Sugar. Easton, Pa.; The Chemical Publishing Co., London; Williams and Norgate, 1914. 132p.

Second — Mohl, Hugo von. Principles of the Anatomy and Physiology of the Vegetable Cell . . . trans. by Arthur Henfry. London; J. Van Voorst, 1852. 158p.

Second — Molegode, Walter. Maize or Indian Corn. Kandy, Miller and Co., Printers, 1920. 7p.

Second — Montgomery, Edward G. Productive Farm Crops. Philadelphia and London; J. B. Lippincott Co., 1916. 501p. (2d ed., rev. 1918. 501p.; 5th ed., rev., 1938. 521p.)

First — Moore, T. W. Treatise and Handbook of Orange Culture in Florida. 1st ed. Jacksonville, Fla.; Sun and Press Job Rooms, 1877. 73p. (2d ed., rev., New York; E. R. Pelton and Co.; Jacksonville, Fla., Ashmead Bros., 1881. 184p. 3d ed., 1883. 184p. 4th ed., rev., New York; E. R. Pelton and Co.; Jacksonville, Horace Drew, 1886. 184p. Treatise and Handbook of Orange Culture in Florida, Louisiana and California. 4th ed., rev., New York; E. R. Pelton and Co., 1892. 189 pp.)

First — Moore, Shepard Wells. Practical Orcharding on Rough Lands. Akron, Ohio; The New Werner Co., 1911. 289p. (Cincinnati, Ohio; Steward and Kidd Co., 1911.)

Second — Moreland, W. S. A Practical Guide to Successful Farming. Garden City, New York; Halcyon House, 1943.

Second — Morris, Robert T. Nut Growing. New York; Macmillan Co., 1921. 236p.

First — Morse, J. E. The New Rhubarb Culture. New York; Orange Judd Co., 1901. 130p.

Second — Morse, Lester L. Field Notes on Lettuce. 2d ed., rev. San Francisco, Calif.; C. C. Morse and Co., 1923. 76p.

Second — Morse, Lester L. Field Notes on Onions. San Francisco, Calif.; C. C. Morse and Co., 1923. 48p.

Second — Mortimer, W. Golden. History of Coca, The Devine Plant of the Incas: with an Introductory Account of the Incas, and of the Andean Indians of To-day. New York; J. H. Vail, 1901. 576p. (Reprint of the 1901 ed., Comp. and unabridged, San Francisco, Calif.; And/or Press, 1974.)

Second — Morton, K., and J. Morton. Fifty Tropical Fruits of Nassau. Coral Gables, Fla.; Text House, Inc., 1946. 113p.

Second — Mosley, Frederick O. Fungoid and Insect Pests and Their Control. Reading, U.K.; The Author, 1918.

Second — Muench, Friedrich. School for American Grape Culture. St. Louis; Conrad Witter, 1865. 139p.

Second — Muenscher, Walter C. Weeds. New York; Macmillan Co., 1935. 577p. (2d ed., New York; Macmillan Co., 1955. 560p.)

Second — Mundkur, B. B. Fungi and Plant Disease. London; Macmillan, 1949. 246p.

First — Munson, Thomas V. Foundations of American Grape Culture. Denison, TX; T. V. Munson and Son, 1909. 252p. (Also printed in New York; Orange Judd Company, 1909. Reprinted 25 years later)

Second — Murke, Franz. Condensed Description of the Manufacture of Beet Sugar. New York; J. Wiley and Sons, Inc., 1921. 175p.

First — Murneek, A. E., and R. O. Whyte. Vernalization and Photoperiodism; A Symposium. Waltham, Mass.; Chronica Botanica, 1948. 196p.

Second — Murray, Andrew. Economic Entomology. London; Chapman and Hall, 1877. 433p.

Second Mutual Orange Distributors, Redlands, Calif. A Manual for Citrus Growers . . . ed. by Chas. W. Horn et al. 1st ed. Redlands, Calif.; Mutual Orange Distributors, 1937. 102p. (2d ed, rev. and enl., 1943. 138p. 3d. ed, rev., 1947. 141p.)

Second Myers, Allen O. Alfalfa, "The Grass" in Ohio; Where How and Why to Grow It. Columbus, Ohio; The F. J. Heer Printing Co., 1907. 187p.

Second Myrick, Herbert, ed. The Book of Corn; A Complete Treatise Upon the Culture, Marketing and Uses of Maize in America and Elsewhere, for Farmers, Dealers, Manufacturers and Others . . . New York; O. Judd Co., 1903. 368p. (2d ed., rev., 1904. 372p.)

Second Myrick, Herbert. The Hop; Its Culture and Cure, Marketing and Manufacture . . . New York and Springfield, Mass.; Orange Judd Co., 1914. 299p.

N

Second Neck, S. S. The Present and Future Productions of Florida. Ocala, Fla.; Banner Steam Printing House, 1888. 134p. (Subject Oranges)

Second Nelson, A. Principles of Agricultural Botany. London; T. Nelson, 1946. 556p.

Second New Treatise on the Culture, Management and Insects, Relating to the Pear Tree. New York; 1858. 67p.

Second Newland, H. O. The Planting, Cultivation and Expression of Coconuts, Kernels, Cacao and Edible Vegetable Oils and Seeds of Commerce. A Pracitical Handbook for Planter, Financier, Scientists, and Others. London; C. Griffin and Co., Ltd., 1919. 111p. (Griffin's Technological Hand-Books.)

Second Newman, James S. Southern Gardener's Practical Manual. Harrisburg, Pa.; Published by the Author, Mount Pleasant Press, J. Horace McFarland Co., 1906. 220p.

Second Newman, L. H. Plant Breeding in Scandinavia. Ottawa; Canadian Seed Grower's Association, 1912.

Second Newsham, John C. The Potato Book. London; C. A. Pearson, Ltd., 1917. 92p.

Second Nichols, Floyd B. Making Money on Farm Crops. St Joseph, Mo.; The Fruit-Grower and Farmer, 1913. 279p.

Second Nicholson, G. The Illustrated Dictionary of Gardening; A Practical and Scientific Encyclopedia of Horticulture for Gardeners and Botanists . . . edited by George Nicholson; assisted by Professor J. W. H. Trail . . . and J. Garrett. London, New York; L. U. Gill, J. Penman, 1887–1889. 8 vols. (Oxford; Clarendon Press, 1951. 4 vols. 2315p. 2d ed., Oxford; Clarendon Press, 1956. 4 vols. 2316p.)

Second Nickerson, W. J., ed. Biology of the Pathogenic Fungi. Waltham, Mass.; Chronica Botanica, 1947. 236p.

First Nicol, Hugh. Biological Control of Insects. Harmondsworth, U.K.; Penguin Books, 1943. 174p.

Second Nicol, J. Life and Adventures. Edinburgh; W. Blackwood, 1822. 215p. (Later American Edition, New York; Farrar and Rinhard. 1936.)

First Nixon, E. L. The Principles of Potato Production. New York; Orange Judd Pub. Co., Inc., 1931. 123p.

Second Nolan, Aretas W., and James H. Greene. Corn Growing; A Manual for Corn Clubs. Chicago and New York; Row, Peterson and Co., 1917. 80p.

Second Nordin, J. Green. The Sweet Potato, How to Grow and Keep It. Russellville, Ark.; Courier-Democrat Print., 1912. 50p.

Second Northcote, Rosalind. The Book of Herbs. London and New York; Lane, 1903. 212p.

Second Northrup, King and Co. Onion Culture; Planting, Cultivation, Harvesting and Marketing . . . Minneapolis, Minn.; Northrup, King and Co., 1915. 24p.

Second The Nurseryman's Directory; Reference Book of Nurserymen, Florists, Seedsmen, Tree Dealers, etc., for the United States, 1883. Alphabetically Arranged by States and Post Offices. Galena, Ill.; D. W. Scott and Co., [1883]. 328p. (1883, 328p.)

O

First O'Kane, Walter C. Injurious Insects; How to Recognize and Control Them. New York; The Macmillan Co., 1912. 414p.

First Oemler, A. Truck-Farming at the South. New York; Orange Judd Co., 1883. 270p. (New, rev. ed., 1888. 1900, 274p.)

Second Oliver, F. W. Makers of British Botany; a Collection of Biographies of Living Botanists. Cambridge; University Press, 1913. 332p.

Second Oliver, Francis W., ed. The Exploitation of Plants. London; J. M. Dent and Sons, Ltd., 1917. 170p.

Second Oliver, George W. Plant Culture. New York; A. T. De La Mare Printing and Pub. Co., Ltd., 1900. 193p. (2d ed., 1909. 308p. 3d ed., 1912. 312p.)

Second Ontario. Dept. of Agriculture. The Fruits of Ontario. Toronto; L. K. Cameron, 1907. 275p. (2d ed., 1914. 320p.)

First Oosting, H. J. The Study of Plant Communities; An Introduction to Plant Ecology. San Francisco; Freeman, 1948.

First Ormerod, Eleanor A. Report of Observations of Injurious Insects and Common Farm Pests During the Year [1877]-1900: With Methods of Prevention and Remedy. London; T. P. Newman, 1877–1901.

First Ormerod, Eleanor A. A Text-Book of Agricultural Entomology: Being a Guide to Methods of Insect Life and Means of Prevention of Insect Ravage for the Use of Agriculturists and Agricultural Students. 2d ed. London; Simpkin, Marshall, Hamilton, Kent and Co., 1892. 238p.

First Osborn, Herbert. Agricultural Entomology for Students, Farmers, Fruit-growers and Gardeners. Philadelphia and New York; Lea and Febiger, 1916. 347p.

First Osborn, Herbert. Meadow and Pasture Insects. Columbus, Ohio; The Educator's Press, 1939. 288p.

Second Osborne, Thomas B. The Proteins of the Wheat Kernel. Washington, D.C.; Carnegie Institution, 1907. 119p.

Second Osborne, Thomas B. The Vegetable Proteins . . . London and New York; Longmans, 1909. 125p. (2d ed., 1924. 154p.)

First Ott, E. Cellulose and Cellulose Derivatives. New York; Interscience Publishers, 1943. 1176p. (2d ed. by E. Ott, H. M. Spurlin and M. W. Grafflin, 1954–55. 3 vols.)

First Owens, Charles E. Principles of Plant Pathology. New York; J. Wiley and Sons, Inc., London; Chapman and Hall, Ltd., 1928. 629p. (Originally issued at Ann Arbor, Mich.; Edwards Bros., 1924, in mimeograph.)

P

First Pabor, William E., compiler. Fruit Culture in Colorado. Denver; W. E. Pabor, Publisher, 1883. 82p.

Second Pacific Coast Borax Co. Boron in Agriculture. New York; Pacific Coast Borax Co., 1939. 24p. (Rev. ed., 1944. 62p.)

Second Packard, A. S. Guide to the Study of Insects and a Treatise on Those Injurious and Beneficial to Crops. Salem, Mass.; Naturalist's Book Agency, 1869. 702p. (9th ed., New York; H. Holt and Co. and Boston; Estes and Lauriat, 1889. 715p.)

First Paddock, Wendell, and Orville B. Whipple. Fruit-Growing in Arid Regions. New York; The Macmillan Co., 1910. 395p. (Rural Science Series, edited by L. H. Bailey)

Second Palmer, Julius A., Jr. About Mushrooms. Boston; Lee and Shepard, 1894. 100p.

First Pammel, Louis H. Weeds of the Farm and Garden. New York; Orange Judd Co., 1911. 281p.

First Pardee, Richard G. A Complete Manual for the Cultivation of the Strawberry: with a Description of the Best Varieties. Also, Notices of the Raspberry, Blackberry, Currant, Gooseberry, and Grape; with Directions for their Cultivation and Selection of the Best Varieties. New York; C. M. Saxton, 1854. 144p.

Second Parker, Bertha M., and Robert E. Gregg. Insect Friends and Enemies. Evanston, Ill. and New York; Row, Peterson and Co., 1941. 35p.

First Parker, H. The Hop Industry. London; P. S. King and Son, Ltd., 1934. 327p.

Second Parry, John R. Nuts for Profit. Parry, N.J.; Pub. by Author, 1897. 157p.

Second Parry, William. Fifty Years Among the Small Fruits: Telling What and How to Plant. Parry, N.J.; 1885. 64p.

Second Passe, Crispijn van de. Hortus Floridus . . . trans. by Spencer Savage. London; Cresset Press, 1928–1929. (Text translated from the Latin)

Second Patterson, Cecil F. Hardy Fruits with Special Reference to Their Culture in Western Canada. Saskatoon, Sask.; 1935. 321p.

Second Peairs, Leonard M. Insect Pests of Farm, Garden, and Orchard. 4th ed. New York; J. Wiley, London; Chapman and Hall, 1941. 549p. ("Dr. E. Dwight Sanderson, author of the first edition of this work and co-author of the second and third editions, has withdrawn from participation in the work and, in accordance with his wish, his name has been omitted from the title page of this edition." - Pref.)

Second Pearson, Robert H. The Book of Garden Pests. London and New York; Lane Co., 1908. 214p.

Second Peavey, Charles T. Grain. Chicago; C. T. Peavey, 1928. 122p.

First Peck, Charles H. Mushrooms and Their Use. Cambridge, Mass.; Cambridge Botanical Supply Co., 1897. 80p.

Second Pedersen, J., and G. H. Howard. How to Grow Cabbages and Cauliflowers Most Profitably. Philadelphia, Pa.; W. Atlee Burpee and Co., 1888. 85p.

First Peek, S. W. The Nursery and Orchard. Atlanta, Ga.; Jas. P. Harrison and Co., 1885. 208p.

First Percival, John. Agricultural Botany: Theoretical and Practical. London; Duckworth and Co., 1900. 798p. (4th ed., New York; H. Holt, 1910. 828p. 8th ed., London; Duckworth and Co., 1936. 839p.)

First Percival, John. The Wheat Plant; a Monograph. London; Duckworth and Co., 1921. 463p.

Second Perrine, Henry E. A True Story of Some Eventful Years in Grandpa's Life. Buffalo, N.Y.; Press of E. H. Hutchinson, 1885. 303p.

Second Perry, Josephine. The Cotton Industry. 1st ed. New York and Toronto; Longmans, Green and Co., 1943. 128p.

Second Petherbridge, Frederick R. Fungoid and Insect Pests of the Farm. Cambridge; Cambridge University Press, 1916. 174p. (2d ed., 1923. 177p.)

First Phange, N. M. G. Citrus Culture for Profit. Jacksonville, Fla.; Wilson and Toomer, 1913. 80p.

Second Philo, Edgar W. Farming With Alfalfa Bacteria Culture. 1st ed. Elmira, N.Y.; Philo Press, 1916. 78p.

First Pickering, Charles M. D. Chronological History of Plants: Man's Record of His Own Existence Illustrated Through Their Names, Uses, and Companionship. Boston; 1879. 1222p.

Second Pink, James Potatoes and How to Grow Them. 2d ed. London; Crossley Lockwood, 1879. 95p.

Second Piper, Charles V. Forage Plants and Their Culture. New York; Macmillan Co., 1914. 618p. (Rev. ed., 1942. 671p.)

First Piper, Charles V., and William J. Morse. The Soybean. New York; McGraw-Hill, 1923. 329p.

Second Pool, R. J. Marching with the Grasses. Lincoln; University of Nebraska Press, 1948. 210p.

Second Poole, Mrs Hester M. Fruits, and How to Use them. New York; Fowler and Wells, 1890. 242p.

Second Popenoe, Paul B. Date Growing in the Old World and New. Altadena, Calif.; West India Gardens, 1913. 316p.

First Popenoe, Wilson. Manual of Tropical and Subtropical Fruits, Excluding the Banana, Coconut, Pineapple, Citrus Fruits, Olive, and Fig. New York; The Macmillan Co., 1920. 474p. (The Rural Manuals, ed. by L. H. Bailey)

Second Powell, Edward P. Hedges, Windbreaks, Shelters, and Live Fences. New York; Orange Judd Co., 1900. 322p.

Second Powell, Edward P. The Orchard and Fruit Garden. New York; McClure, Phillips and Co., 1905. 321p. (New York; Doubleday, Page and Co., 1910. 322p.)

First Powell, Fred W. The Bureau of Plant Industry: Its History, Activities and Organization. Baltimore, Md.; Johns Hopkins Press, 1927. 121p.

Second Power, Richard Anderson. Know Your Weeds. Viroqua, Wis; 1937. 235p.

Second Prange, Nettie M. G. Citrus Culture for Profit; Practical Directions by

Wilson and Toomer Fertilizer Co. St. Augustine, Fla.; The Record Co., 1911. 83p.

Second Preyer, Hugo. Ten Years' Practical Experience in Grape and Small Fruit Culture. Canton, Ohio; Bascom and Caxton, Printers, 1875. 77p.

Second Price, E. M. The Walnut . . . a Comprehensive Treatise on How to Grow it. Sacramento, Calif.; The Jos. M. Anderson Co., 1910. 68p.

First Price, R. H. Sweet Potato Culture for Profit. Dallas, Texas; Texas Farm and Ranch Pub. Co., 1896. 110p.

Second Prévost, Benedict. Memoir on the Immediate Cause of Bunt or Smut of Wheat, and of Several Other Diseases of Plants, and on Preventives of Bunt; translated from the French by George Keitt of Prévost's Memoire sur La Cause Immediate de La Carie ou Charbon des Bles . . . C'est une Botanique a faire, que celle des Plantes Microscopiques in Senebier, Physiol. Veg. vol. II, p. 291. Menasha, Wisc.; American Phytopathological Society, 1939. 95p. [*Phytopathological Classics* no. 6]

Second Purdy, A. M. Small-Fruit Instructor. South Bend, Ind.; Purdy and Hance, 1869. 32p. (New ed., Palmyra, N.Y.; Published by the Author, 1887. 128p.)

First Putnam, Henry. Touches on Agriculture; Including a Treatise on the Preservation of the Apple-Tree, Together with Family Recipes, experiments on Insects, etc . . . Portland, Me.; A. W. Thayer, Printer, 1824. 43p.

Second Pyre, W. Clover, The Great Cash Money Crop and All About It: A Manual of Clover Culture. Waterloo, Iowa; The Galloway Bros., Bowman Co., 1913. 102p.

Q

First Quayle, Henry J. Insects of Citrus and Other Subtropical Fruits. Ithaca, N.Y.; Comstock Pub. Co., Inc., 1938. 583p.

First Quinn, P. T. Pear Culture for Profit. New York; The Tribune Association, 1869. 136p. (New ed., rev., New York; Orange Judd Company, 1883. 136p.)

R

Second Raber, Oran L. Principles of Plant Physiology. New York; The Macmillan Co., 1928. 377p. (Rev. ed., 1933 and 1938)

First Rabinowitch, E. I. Photosynthesis and Related Processes. New York; Interscience, 1945.

First Ramsbottom, J. A Handbook of the Larger British Fungi. London; British Museum (Natural History), 1923.

Second Rather, Howard C. Field Crops. 1st ed. New York and London; McGraw-Hill Book Co., Inc., 1942. 454p. (2d ed., 1951. 446p.)

First Rawson, Herbert. Success in Market-Gardening, and Vegetable Growers' Manual. Boston; Published by the Author, 1887. 208p. (7th ed., rev., Boston; Published by the Author, 1892. 140 pp.)

Second Raymond, O. G. Commercial Apple Orchards. Cumberland, Md.; F. Mertens' Sons, 1910. 12p.

Second Redgrove, H. Stanley. Spices and Condiments. London; Sir I. Pitman and Sons, Ltd., 1933. 361p.

First	Rehder, Alfred. A Manual of Cultivated Trees and Shrubs. New York; The Macmillan Co., 1927. 930p. (2d ed., New York; The Macmillan Co., 1940.)
Second	Reynolds, P. K. The Banana. Boston and New York; Houghton Mifflin, 1927. 181p.
Second	Rhoads, S. N., ed. Botanica Neglecta. Philadelphia; Priv. Print., 1916. 55p.
Second	Riker, Albert J., and Regina S. Riker. Introduction to Research on Plant Diseases; A Guide to the Principles and Practice for Studying Various Plant-Disease Problems. St. Louis and Chicago; J. S. Swift Co., 1936. 117p.
First	Risien, E. E. Pecan Culture for Western Texas. San Saba, Tex.; Published by the author, 1904. 55p.
First	Rixford, E. H. The Wine Press and the Cellar. San Francisco, New York; Payot Upham and Co., D. Van Nostrand, 1883. 240p.
Second	Robbins, Wilfred W. Botany of Crop Plants. 3d ed., rev. Philadelphia, London; Blakiston, American Book Supply, 1931.
Second	Robbins, Wilfred W. and Francis Ramaley. Plants Useful to Man. Philadelphia, Pa.; P. Blakiston's Son and Co., Inc., 1933. 428p.
First	Robbins, Wilfred W., et al. Weed Control, A Textbook and Manual. 1st ed. New York and London; McGraw- Hill Book Co., 1942. 543p.
Second	Robert, J. C. The Story of Tobacco in America. New York; Alfred A. Knopf, Inc., 1949. 296p.
First	Roberts, Herbert F. Plant Hybridization Before Mendel. Princeton, N.J.; Princeton University Press, 1929. 374p.
Second	Robinson, Douglas H., and S. G. Jary. Agricultural Entomology. London; Duckworth, 1929. 314p.
First	Rockwell, F. F. The Gardener's Pocket Manual. New York; McBride, Nast and Co., 1914. 90p.
Second	Rockwell, F. F. Gardening Indoors and Under Glass. New York; McBride, Nast and Co., 1912. 210p.
Second	Rockwell, F. F. Making a Garden of Small Fruits. New York; McBride, Nast and Co., 1914. 56p. (House and Garden Making Books)
Second	Rodale, Jerome I. Pay Dirt: Farming and Gardening with Composts. New York; Devin-Adair Co., 1945. 242p.
Second	Rodrian, Richard. Plant Life and Fruit Protection Against Destroying Insects. Denver, Colo.; Bradford-Robinson, 1943. 14p.
Second	Roe, E. P. The Home Acre. New York; Dodd, Mead and Co., 1889. 252p.
Second	Roe, E. P. A Manual on the Culture of Small Fruits. Newburgh, N.Y.; Journal Printing Establishment, 1877. 82p.
First	Roe, E. P. Play and Profit in My Garden. New York; Dodd and Mead, 1873. 349p. (New ed., New York; Orange Judd Company, 1893.)
First	Roe, E. P. Success with Small Fruits. New York; Dodd, Mead and Co., 1880. 313p. (The Illustrated quarto new edition; Preface dated 1886), (Different form 1881. 388p.)
Second	Roebuck, Arnold. Insect Pests and Fungous Diseases of Farm Crops. London; Benn Bros., Ltd., 1923. 55p.
Second	Roeding, G. C. The Smyrna Fig at Home and Abroad. Fresno, Calif.; Published by the Author, 1903. 87p.

Second	Roeding, George C. Roeding's Fruit Grower's Guide. Fresno, Calif.; 1919. 97p.
Second	Roessle, T. How to Cultivate and Preserve Celery. New York; C. M. Saxton, Barker and Co., 1860. 102p.
Second	Rogers, C. Trodden Glorey. Santa Barbara, Calif.; W. Hebberd, [1949]. 129p.
Second	Rogers, E. A. Practical Potato Culture. Philadelphia; J. B. Haines, 1913. 126p.
Second	Rogers, J. E. The Book of Useful Plants. Garden City, N.Y.; Doubleday, Page and Co., 1913. 374p. (The Garden Library)
Second	Rogers, W. S. Garden Planning. Garden City, N.Y.; Doubleday, Page and Co., 1911. 423p.
First	Rolfs, P. H. Vegetable-Growing in the South for Northern Markets. Richmond; The Southern Planter Pub. Co., 1896. 255p.
Second	Roper, William N., ed. The Peanut and Its Culture. Petersburg, Va.; American Nut Journal, 1905. 62p.
Second	Rowles, William F. Every Man's Book of Garden Difficulties. New York; Geo. H. Doran,
First	Russell, Edward J. Plant Nutrition and Crop Production. Berkeley, Calif.; University of California Press, 1926. 115p.
Second	Russell, Edward J. Soil Conditions and Plant Growth. London and New York; Longmans, Green, 1912. 168p. (11th ed. with title: Russell's Soil Conditions and Plant Growth, by E. Walter Russell. Burnt Mill, Engl.; Longman Scientific and Technical, and New York; Wiley, 1988. 991p.)
Second	Ruston, Edwin. Floral Talks. New York; W. N. Swett and Co., 1892. 96p. (The Central Square Series, no.34)
Second	Rutter, John. The Culture and Diseases of the Peach. Harrisburg, Pa.; Every Saturday Night Office, 1880. 95p.

S

Second	Sakharotrest (The United Sugar Industry of the Soviet Union). The Selection, Breeding and Culture of Sugar Beet Seeds in the Soviet Union . . . New York; Amtorg Trading Corporation, 1927. 54p.
First	Salaman, Redcliffe N. Potato Varieties. Cambridge, U.K.; Cambridge Univ. Press, 1926. 378p.
Second	Sampson, Arthur W. Native American Forage Plants. New York; John Wiley and Sons., Inc., 1924.
Second	San Pedro, Los Angeles and Salt Lake Railroad Co. Story of the Orange; A History of the Golden Fruit Which Has Traveled Through Every Semi-Tropic Clime to Find the Perfect Home in the Sun-Kissed Valleys of Southern California. Los Angeles, Calif.; 1913. 24p.
Second	Sanders, Thomas W. The Book of the Potato . . . London; W. H. and L. Collingridge, 1905. 222p.
Second	Sanders, Thomas W. The Encylopaedia of Gardening, A Dictionary of Cultivated Plants, etc., Giving an Epitome of the Culture of All Kinds Generally Grown in Gardens in This Country, Together With a Complete List of Their Common or Popular Names. London; W. H. and L. Collingridge, 1869. 434p. (First Appeared in serial form in Amateur Gardening . . . from Nov. 15, 1890 until Aug. 10, 1895. 13th ed.,

London, Collingridge, 1908. 446p. Published later under the title: Sanders' Encyclopaedia of Gardening . . . revised by A. J. Macself. 22d ed., London; Collingridge, New York; Transaltlantic Arts, 1952.)

Second Sanders, Thomas W. Fruit and its Cultivation in Garden and Orchard. London; W. H. and L. Collingridge, 1940. 288p.

Second Sanders, Thomas W. Fruit Foes; A Description of the Various Insect, Animal and Fungal Pests That Attack Fruit Trees, with Remedies for Their Prevention and Eradication. London; W. H. and L. Collingridge, 1921. 111p.

Second Sanders, Thomas W. Garden Foes. London; W. H. and L. Collingridge, Ltd., 1929? (3 vols. in one. I. Flower Foes. II. Fruit Foes. III. Vegetable Foes.)

Second Sanders, Thomas W., and J. Lansdell. Grapes: Peaches: Melons: and How to Grow Them. A Handbook Dealing With Their History, Culture, Management and Propagation. and Melons . . . London; Collingridge, 1924. 150p.

Second Sanderson, Ezra D. Insect Pests of Farm, Garden and Orchard. 1st ed. New York; J. Wiley and Sons, 1912. 684p. (3d ed., rev., 1931. 568p.)

Second Sanderson, Ezra D. Insects Injurious to Staple Crops. 1st ed. New York; Wiley & Sons, London; Chapman & Hall, 1902. 295p. (Reprint. 1911. 295p.)

Second Sanford, Albert H. The Story of Agriculture in the United States. Boston; D. C. Heath and Co., 1916. 394p.

Second Saunders, L. H. Vegetable Growing in the Tropics. London; Oxford University Press, 1940.

First Saunders, William. Insects Injurious to Fruits. Philadelphia; J. B. Lippincott and Co., 1883. 436p.

First Sawyer, Joseph Dillaway. How to Make a Country Place. New York; Orange Judd Co., 1914. 430p.

Second Saylor, Henry H. The Book of Annuals. New York; McBride, Nast and Co., 1913. 127p.

Second Scherer, James A. B. Cotton as a World Power; A Study in the Economic Interpretation of History. New York; Fredrick A. Stokes Co., 1916. 452p.

Second Schilletter, J. C., and H. W. Richey. Textbook of General Horticulture. 1st ed. New York and London; McGraw-Hill Book Co., Inc., 1940.

Second Scribner, Frank L. Fungus Diseases of the Grape and Other Plants and Their Treatment. Little Silver, N.J.; J. T. Lovett Co., 1890. 134p.

Second Schuur, Peter J. How to Grow Celery Anywhere. Kalamazoo, Mich.; Union Seed Co., 1896. 112p.

Second Seabrook, William P. Modern Fruit Growing. London; The Lockwood Press, 1918. 172p. (8th ed., Benn, 1949.)

First Sears, Fred C. Productive Orcharding. Philadelphia; J. B. Lippincott Co., 1914. 315p.

Second Sears, Fred C. Productive Small Fruit Culture . . . Philadelphia, Pa.; J. B. Lippincott Co., 1920. 368p. (2d ed., 1925. 368p.)

Second Sempers, Frank W. Injurious Insects and the Use of Insecticides; A New Descriptive Manual on Noxious Insects, with Methods for Their Repression. Philadelphia; W. A. Burpee and Co., 1894. 216p. (3d ed., 1896.)

Second Sevey, Glenn C. Bean Culture. New York; Orange Judd Co., 1907. 130p.

Second Sevey, Glenn C. Peas and Pea Culture. New York; Orange Judd Co., 1911. 92p.

Second Sewell, Cornelius V. V. Common Sense Gardens. New York; The Grafton Press, Publishers, 1906. 396p.

Second Seymour, Edward L. D., ed. Farm Knowledge. Garden City, N.Y.; Doubleday, Page and Co., 1918. 4 vols. (Rev. ed., 1919. 4 vols.)

First Shamel, Archibald D. Manual of Corn Judging. Urbana, Ill.; The Author, 1902. 35p. (2d ed., New York; O. Judd Co., 1903. 72p.)

Second Shamel, Archibald D. The Study of Farm Crops . . . No. 1–12; Sept. 1900–Aug. 1901. Taylorville, Ill.; C. M. Parker, 1900–1901. 12 pieces.

Second Shaw, Ellen Eddy. Gardening and Farming. Garden City, N.Y.; Doubleday, Page and Co., 1911. 376p. (The Children's Library of Work and Play)

Second Shaw, Thomas. Clovers and How To Grow Them. New York; Judd, 1906. 349p. (Reprinted 1912.)

Second Shaw, Thomas. Weeds and How to Eradicate Them. Toronto; J. E. Bryant Co., Ltd., 1893. 208p. (3d ed. rev., St. Paul, Minn.; Webb Pub. Co., 1911. 236p.)

Second Sheehan, James. Your Plants. New York; Orange Judd Co., 1893. 79p.

Second Shell Chemicals Limited, London. Better Fruit, Vegetables and Flowers. London; Shell Chemicals, Dist. by Simpkin Marshall, 1941. 64p.

Second Sherman, J. V., and S. L. Sherman. The New Fibers. New York; D. Van Nostrand Co., Inc., 1946. 537p.

Second Sherrington, Charles S. The Endeavor of Jean Fernel, with a List of the Editions of his Writings. Cambridge, U.K.; University Press, 1946. 223p.

First Shields, O. D., Compiler. A Western Book for Western Planters. Loveland, Colo.; O. D. Shields, 1905. 147p.

Second Shirley, Jasper N. Success With Alfalfa At One-Tenth the Usual Cost. Indianapolis, Ind.; 1920. 52p.

Second Shoemaker, J. S. Small Fruit production. New York; 1928. 123p.

Second Shoemaker, J. S. Small-Fruit Culture; A Text for Instruction and Reference Work, and a Guide to Field Practice. Philadelphia; Blakiston, 1934. 434p. (2d ed, 1948.)

Second Shoesmith, Vernon M. The Study of Corn. New York; O. Judd Co., 1910. 96p.

Second Simmonds, Peter Lund. Tropical Agriculture: A Treatise on the Culture, Preparation, Commerce and Consumption of the Principal Products of the Vegetable Kingdom. London and New York; E. and F. N. Spon, 1877. 315p. (New ed., 1889.)

Second Simpson, J. Quick Fruit Culture. 1900.

Second Sinn, J. F. The Growing of Gold. Facts About Growing Alfalfa, The Practical Gold Mine for the Farmer of Today. Clarinda, Idaho; A. A. Berry Seed Co., 1913. 50p.

Second Skinner, Hubert M., and A. L. McCredie, eds. Library of Agriculture. Volume IV. Horticulture and Truck Farming. Chicago; Cree Pub. Co., 1912. 539p.

Second Slade, Daniel Denison. The Evolution of Horticulture in New England. New York and London; Putnam's Sons, 1895. 180p.

Second Slingerland, Mark V. Our Insect Enemies in 1903 . . . Ithaca, N.Y.?; 1904? 6p.

Second Smith, Datus C. 19 Years Clover Growing in North Dakota . . . Experiences at Cloverlea Farm. Blanchard, N. Dak.; Cloverlea Seed Co., 1916. 22p.

First Smith, Erwin F. Bacteria in Relation to Plant Disease. Washington; Carnegie Inst., 1905–1914. 3 vols.

First Smith, Erwin F. An Introduction to Bacterial Diseases of Plants. Philadelphia and London; W. B. Saunders Co., 1920. 688p.

Second Smith, Grace M., compil. Lecture Notes for Alfalfa Charts . . . , from the lectures by Professor P. G. Holden and his Associates. Chicago, Ill.; International Harvester Co. of New Jersey, Agricultural Extension Dept., 1914. 29p.

Second Smith, Grace M. Studies in Alfalfa; A Teacher's Handbook Designed Especially for Use in the Corn Belt States. Chicago, Ill.; International Harvester Co. of New Jersey, Agricultural Extension Dept., 1914. 32p.

Second Smith, J. W. Agricultural Meteorology; the Effect of Weather on Crops. New York; The Macmillan Co., 1920. 304p.

Second Smith, John. A Dictionary of Popular Names of the Plants which Furnish the Natural and Acquired Wants of Man, in all matters of Domestic and General Economy, their History, Products and Uses. London; 1882.

First Smith, John B. Economic Entomology for the Farmer and Fruit-Grower, and For Use as a Text-Book in Agricultural Schools and Colleges. Philadelphia, Pa.; J. B. Lippincott Co., 1896. 481p. (2d ed., rev., 1906. 475p.)

First Smith, John B. Our Insect Friends and Enemies; The Relation of Insects to Man, to Other Animals, to One Another, and to Plants, with a Chapter on the War Against Insects. Philadelphia, Pa. and London; J. B. Lippincott Co., 1909. 314p.

First Smith, Kenneth M. Virus Diseases of Farm and Garden Crops. Worcester; Littlebury and Co., 1945.

Second Smith, Kenneth M. Recent Advances in the Study of Plant Viruses. London; J. and A. Churchill, 1933. 423p.

Second Smith, Kenneth M. A Textbook of Agricultural Entomology. Cambridge; University Press, 1931. 285p. (2d ed., 1948. 289p.)

Second Smith, Kenneth M. A Textbook of Plant Virus Diseases. Philadelphia; Blakiston's Sons and Co., 1937. 615p. (3d ed., New York; Academic Press, 1972. 684p.)

Second Smith, Kenneth M. The Virus; Life's Enemy. Cambridge, U.K.; Cambridge University Press, 1941. 176p. (1948, reprinted with appendix.)

Second Smith, Thomas. French Gardening. London; Joseph Fels and Utopia Press, 1909. 128p.

Second Smith, Thomas. The Profitable Culture of Vegetables, for Market Gardeners, Small Holders, and Others. London and New York; Longmans, Green and Co., 1911. 452p. (Revised by W. E. Shewell-Cooper, 1937.)

Second Smith, William C. How to Grow One Hundred Bushels of Corn per Acre on Worn Soils. Delphi, Ind.; Smith Pub. Co., 1910. 111p. (2d ed., rev. and enl., Cincinnati; Stewart and Kidd Co., 1912. 188p.)

First Smith, Worthington G. Diseases of Field and Garden Crops; Chiefly Such as are Caused by Fungi. London; Macmillan and Co., 1884. 353p.

Second Smock, R. M., and A. M. Neubert. Apples and Apple Products. New York and London; Interscience, 1950.

Second Snodgrass, Robert E. Principles of Insect Morphology. New York and London; McGraw-Hill, 1935. 667p.

First Snowden, Joseph D. The Cultivated Races of Sorghum. London, Eng.; Adlard and Son, Ltd., 1936. 273p.

Second Snyder, Harry. Bread; A Collection of Popular Papers on Wheat, Flour and Bread. New York; The Macmillan Co., 1930. 293p.

Second Somerville, William. Farm and Garden Insects. London and New York; Macmillan, 1897. 127p.

Second South Orchards Company, Mobile, Ala. The Paper-Shell Pecan and the Satsuma Orange; Their Development, Native Soils, Where They Bear Best, Money Made Growing Them, How Bearing Orchards Can Be Bought, How they Can Be Paid For, What They Will Pay . . . 4th ed. Mobile, Ala. and Chicago, Ill.; South Orchards Co., 1912. 51p.

Second Southwick, Lawrence. Dwarf Fruit Trees . . . ed. by E. Robinson. New York; Macmillan Co., 1948. 126p.

First Spalding, William A. The Orange: Its Culture in California with a Brief Discussion of the Lemon, Lime, and Other Citrus Fruits. Riverside, Calif.; Press and Horticulturist, 1885. 97p.

Second Spencer, Edwin R. Just Weeds. New York; C. Scribner's Sons, 1940. 317p. (New expanded ed., 1957. 333p.)

Second Spencer, Guilford L. A Hand-Book for Chemists of Beet-Sugar Houses and Seed-Culture Farms. Containing Selected Methods of Analysis, Sugarhouse Control, Reference Tables, etc . . . 1st ed. New York; J. Wiley and Sons, 1897. 475p.

Second Spinden, H. J. Tobacco Is American: the Story of Tobacco before the Coming of the White Man. New York; New York Public Library, 1950. 120p.

Second Spooner, Alden. The Cultivation of American Grape Vines, and Making of Wine. Brooklyn; A. Spooner and Co., 1846. 96p. (2d ed., Brooklyn; E. B. Spooner; New York, A. O. Moore, 1858.)

Second Spurway, C. H. Soil Fertility Diagnosis and Control. Ann Arbor, Mich.; J. W. Edwards, 1948.

Second Stadler, L. J. Some Observations on Gene Variability and Spontaneous Mutation. Michigan State College; Spragg Memorial Lectures in Plant Breeding, 1942.

First Stakman, Elvin C. A Study in Cereal Rusts: Physiological Races . . . St. Paul, Minn.; University Farm, 1914. 56p.

Second Staley, A. E., Manufacturing Co. The Wonder Bean. Decatur, Ill.; 1947. 30p.

Second Standard Oil Company. A Fruit Growers Handbook and Record. Chicago, Ill.; Standard Oil Co. (Indiana), 1936. 112p. (Supplement, Chicago, 1937.)

Second	Stanford, Ernest E. Economic Plants. New York; Appleton-Century Co., 1934. 571p.
Second	Stanhope, H. A. Potatoes. London; Agricultural and Horticultural Assoc., Ltd., 1907. 20p.
Second	Stebbins, Cyril A. The Principles of Agriculture through the School and the Home Garden. New York; The Macmillan Company, 1913. 380p.
Second	Stedman, J. M. A Treatise on Spraying . . . St. Joseph, Mo.; The Fruit-Grower Co., 1905. 123p. ("Brother Jonathan" Series, no.2)
First	Steinhaus, E. A. Insect Microbiology. Ithaca, N.Y.; Comstock Pub. Co., 1946. 763p.
Second	Sterling Chemical Company (Cambridge, Mass.). Sterlingworth Bug Book and Agricultural Spray Guide no. 18. Cambridge, Mass.; Sterling Chemical Co., 1918. 72p.
First	Stevens, Frank L. The Fungi Which Cause Plant Diseases. New York; The Macmillan Co., 1913. 754p.
First	Stevens, Frank L. Plant Disease Fungi. New York; Macmillan Co., 1925. 469p.
First	Stevens, Frank L., and J. G. Hall. Diseases of Economic Plants. New York; Macmillan Co., 1910. 513p. (Rev. ed., 1921. 507p. Later rev. ed., 1922. 507p.)
Second	Steward, Henry. Irrigation for the Farm, Garden and Orchard. New York; Orange Judd Co., 1883. 264p. (Rev., 1886. 276p.)
Second	Stewart, Homer L. Celery Growing and Marketing a Success. Tecumseh, Mich.; The Blade Printing and Paper Company, 1891. 151p.
Second	Stewart, Homer L. The Pecan, and How to Grow it. Chicago; Woman's Temperance Pub. Associaiton, 1893. 90p.
Second	Stone, W. B. and Co. The Pecan Business, From Planting the Nuts to Gathering the Nuts. Thomasville, Ga.; 1915?, 32p.
Second	Stoney, John. Pruning and Planting Fruit Trees and How to Establish a Fruit-Cage. London; Lindsay Drummond, 1944. 58p.
Second	Stringfellow, H. M. The New Horticulture. Galveston, Texas; Published by the Author, 1896. 216p. (Trans. into German by Friedrich Wannieck, as Der Neue Gartenbau, Frankfurt, 1901. New rev. ed., Dallas, Tex.; Farm and Ranch Publishing Company, 1906. 146p.)
First	Strong, W. C. Culture of the Grape. Boston; J. E. Tilton and Co., 1866. 355p.
First	Stuart, W. The Potato; Its Culture, Uses, History and Classification. Philadelphia and London; J. B. Lippincott Co., 1923. 518p. (2d ed., rev., 1927. 3d ed., rev., 1928. 4th ed., rev., 1937.)
Second	Sturrock, D. Tropical Fruits for Southern Florida and Cuba. Jamaica Plain; Atkins Institution of the Arnold Arboretum of Harvard University, 1940.
Second	Sturtevant, A. H., and G. W. Beadle. An Introduction to Genetics. Philadelphia and London; W. B. Saunders Co., 1939. 391p.
First	Sturtevant, Edward L. Maize; An Attempt at Classification . . . Geneva, N.Y., December, 1883. Rochester, N.Y.; Democrat and Chronicle Print., 1884. 9p. (Printed for Private Distribution Only.)
Second	Sugar Research Foundation, Inc., New York. Sugar. An Illustrated Story

of the Production and Processing of a Natural Food and Useful Chemical. New York; 1947? 36p. (Reprinted 1948.)

Second Sutton and Sons, Reading, U.K. The Culture of Vegetables and Flowers from Seeds and Roots. London; Hamilton, Adams and Co., 1884. 308p. (9th ed., London; Simpkin, Marshall, Hamilton, Kent, 1919. 427p. Later ed. 1936.)

Second Sutton, L. N. The Cool Greenhouse. New York and London; Putnam, 1935.

Second Swanson, Charles O. Wheat and Flour Quality. Minneapolis, Minn.; Burgess Pub. Co., 1938. 227p.

Second Swanson, William W. Wheat. New York; The Macmillan Co., 1930. 520p.

First Sweetman, Harvey L. The Biological Control of Insects; With a Chapter on Weed Control. Ithaca, N.Y.; Comstock Pub. Co., Inc., 1936. 461p.

Second Synnestvedt, Paul. Fighting Insect Plagues; A Treatise on the Evils Arising From Improper Treatment of Soils, With Some Suggestions as to How to Correct Them. Philadelphia, Pa.; International Printing Co., 1924. 85p.

T

Second Tabor, Grace, and Gardner Teall. The Garden Primer. New York; McBride Winston and Co., 1910. 118p. (Philadelphia; John C. Winston and Co., 1910. New ed., rev., McBride, Nast and Co., 1911. 164p.)

Second Taft, L. R. Greenhouse Construction. New York; Orange Judd Co., 1894. 208p.

Second Taft, L. R. Greenhouse Management. New York; Orange Judd Co., 1898. 382p.

Second Talbert, T. J., and A. E. Murneek. Fruit Crops: Principles and Practices of Orchard and Small Fruit Culture. Philadelphia, Pa.; 1939. 345p.

Second Talbert, Thomas J. General Horticulture. Philadelphia, Pa.; 1946. 452p.

Second Tansley, A. G., and T. F. Chipp, eds. Aims and Methods in the Study of Vegetation. London; British Empire Vegetation Committee and Crown Agents for the Colonies, 1926.

First Taubenhaus, J. J., and E. W. Mally. The Culture and Diseases of the Onion. New York; Dutton, 1923. 286p.

Second Taubenhaus, J. J. The Culture and Diseases of the Sweet Potato. New York; E. P. Dutton and Co., 1923. 286p.

Second Taubenhaus, J. J. Diseases of Greenhouse Crops and Their Control. New York; E. P. Dutton, 1920. 429p.

Second Taubenhaus, J. J. Diseases of Truck Crops and Their Control. New York; E. P. Dutton and Co., 1918. 396p.

Second Tawell, G. H. Good Market Gardening. English Univ. Press, 1948.

Second Taylor, G. O. The Propagation of Hardy Trees and Shrubs. Oxford; Dulau, 1947.

Second Taylor, Harold V. The Apples of England. London; Crosby Lockwood, 1936. 266p. (3d ed., 1948.)

Second Taylor, Harold V. The Plums of England. London; Crosby Lockwood, 1949. 151p.

Second Taylor, N. Practical Encyclopedia of Gardening. Garden City, N.Y.; Garden City Pub. Co., 1942.

First Teat, John Thomas. The Farmer's Garden, and Its Management. Cardington, Ohio; 1896–98. 107p.

Second Ten Eyck, Albert M. Wheat; A Practical Discussion of the Raising, Marketing, Handling and Use of the Wheat Crop, Relating Largely to the Great Plains Region of the United States and Canada. 1st ed. Lincoln, Neb.; Campbell Soil Culture Pub. Co., 1914. 194p.

First Terry, T. B. The ABC of Potato Culture . . . Medina, Ohio; A. I. Root, 1885. 42p. (2d ed., rev., 1893. 212p. 3d ed., rev., 1901.)

Second Terry, T. B. The ABC of Strawberry Culture. Medina, Ohio; 1902. 235p. (2d ed. How to Grow Strawberries)

First Terry, T. B., and A. I. Root. How to Grow Strawberries. Medina, Ohio; A. I. Root, 1890. 144p.

First Theobald, Frederic V. Report on Economic Zoology, 1896–1922. London and Asford, U.K.; Kent, 1896–1923. 19 nos.

Second Theobald, Frederick V. The Insect and Other Allied Pests of Orchard Bush and Hothouse Fruits and Their Prevention and Treatment. Wye; The Author, 1909. 550p.

Second Thomas, Arthur G. Wheat From the Fields of Boaz. New York; Nelson and Phillips, 1878. 285p.

Second Thomas, Harry H. The Book of the Apple . . . Together with Chapters on the History and Cookery of the Apple, and on the Preparation of Cider. London and New York; J. Lane, 1902. 112p.

Second Thomas, Theodore. Our Mountain Garden. New York; The Macmillan Co., 1904. 212p.

Second Thompson, C. R. Good Fruit Farming. London; English Univ. Press, 1949. 296p.

Second Thompson, C. R. The Pruning of Apples and Pears by Renewal Methods. London; Faber and Faber, 1941.

Second Thompson, Fred S. Rhubarb or Pie-Plant Culture. Milwaukee, Wis.; J. N. Yewdale and Sons Co., 1894. 76p.

Second Thompson, H. C. Asparagus Production. New York; Orange Judd, 1942. 124p.

Second Thompson, H. C. Vegetable Crops. New York; McGraw-Hill, 1923.

Second Thompson, Robert. Gardener's Assistant: Practical and Scientific; a Guide to the Formation and Managment of the Kitchen, Fruit, and Flower Garden, and the cultivation of Conservatory, Green-House, and Stove Plants, with a Copius Calendar of Gardening Operations. London, etc.; Blackie and Son, 1859. 774p. (Revised by William Watson, Six Volumes, London, 1901.)

First Thompson, W. W. A Plain and Simple Treatise on Growing, Gathering and General Management of the le Conte and Kieffer Pears, and Other Fruits of the South. Clarksville, Tenn.; W. P. Titus, 1889. 66p.

Second Thornton, A. W. The Suburbanite's Handbook of Dwarf Tree Culture . . . Bellingham, Wash.; Press of S. B. Irish and Co., 1901. 115p.

Second Thorp, Francis Newton. An American Fruit Farm; Its Selection and Management . . . New York; G. P. Putnam's Sons, 1915. 348p.

Second Tillet, M. Dissertation on the Cause of the Corruption and Smutting of the

Kernels of Wheat in the Head and on the Means of Preventing These Untoward Circumstances . . . Translated from the French by Harry B. Humphrey of Tillet's Dissertation sur La Cause qui Corrompt et Noirict Les Graines de Bled dans Les Epis . . . Bordeaux, Pierre Brun, 1755. Ithaca, N.Y.; American Phytopathological Society, 1937. 191p. [*Phytopathological Classics* no. 5] Title page of original French dissertation is photographical reproduced in the translation.

First — Tillinghast, Isaac F. A Manual of Vegetable Plants. Factoryville, Pa.; Tillinghast Bros., 1878. 102p. (La Plume, Penn., 1881, I. F. Tillinghast, 101 pp.)

Second — Todd, John A. The World's Cotton Crops. London; A. and C. Black, Ltd., 1915. 460p. (Later printing, 1923.)

Second — Todd, Sereno Edwards. The Apple Culturist. New York; Harper and Bros., 1871. 334p.

Second — Tracy, Will W. Tomato Culture. New York; Orange Judd Co., 1907. 150p.

Second — Treat, Mary. Injurious Insects of the Farm and Garden. New York; Orange Judd Co., 1882. 288p.

Second — Trelease, William. The Genus Phoradendron. A Monographic Revision. Urbana, Ill.; The University, 1916. 224p.

First — Trimble, Isaac P. A Treatise on the Insect Enemies of Fruit and Fruit Trees; The Curculio and the Apple Moth. New York; W. Wood and Co., 1865. 139p.

First — Tritschler, Charles H., and W. D. Buchanan. A Practical Treatise on How to Grow Flowers, Fruits, Vegetables, Shrubbery, Evergreens, Shade Trees, Ornamental Trees. Nashville, Tenn.; McQuiddy Printing Co., 1910. 167p.

First — Troop, James. Melon Culture. New York; Orange Judd Co., 1911. 105p.

Second — Trowbridge, J. M. The Cider Makers' Hand Book. New York; Orange Judd Co., 1890. 119p.

Second — Tschiffely, A. F. Southern Cross to North Pole Star. London; Heinemann, 1933.

First — Tubeuf, Karl. Diseases of Plants Induced by Cryptogamic Parasites; Introduction to the Study of Pathogenic Fungi, Slime Fungi, Bacteria and Algae . . . English ed. by William G. Smith. London and New York; Longmans, Green and Co., 1897. 598p.

Second — Tukey, Harold Bradford. The Pear and Its Culture. New York; Orange Judd Pub. Co., 1928. 125p.

First — Tukey, Harold Bradford, ed. Plant Regulators in Agriculture. New York; Wiley, 1947. 269p.

Second — Turner, William. Fruits and Vegetables Under Glass. New York; A. T. De La Mare Co., Ltd., 1912. 255p.

U

Second — Underwood, L. The Garden and Its Accessories. Boston; Little, Brown and Co., 1907. 215p.

Second — Urban, Abram Linwood. The Voice of the Garden. Author's ed. Philadelphia; T. Meehan and Sons, 1912. 93p.

Second — Utter, Delbert. Making Special Crops Pay. Springfield, Mass.; The Phelps Pub. Co., 1913. 60p.

V

Second Van Deman, Henry E. Orange and Grape Fruit Culture in Louisiana . . . Baltimore and New York; Printed by Munder-Thomas Press, 1914. 20p.

First Van Meter, Ralph A. Bush Fruit Production. New York; Orange Judd, 1928. 123p.

Second Van Ornam, F. B. Potatoes for Profit. 3d ed. Philadelphia; W. Atlee Burpee and Co., 1896. 84p.

Second Van Welzer, A. C. Fig Culture. Houston, Texas; J. V. Dealy Co., 1901. 218p.

Second Veitch, J. H. Hortus Veitchii; a History of the Rise and Progress of the Nurseries of Messrs. James Veitch and Sons, Together with an Account of the Botanical Collectors and Hybridists Employed by Them and a List of the Most Remarkable of their Introductions. London; James Veitch and Sons, Ltd., 1906. 542p.

Second Verdoorn, Frans, ed. Plants and Plant science in Latin Americia. Waltham, Mass.; Chronica Botanica Co., 1945. 384p. (A New Series of Plant Science Books No. 16)

Second Verrill, A. H. Foods America Gave the World. Boston; L. C. Page and Company, 1937.

Second Viala, Pierre, and L. Ravaz. American Vines (Resistant Stock) . . . complete translation of 2nd ed. by Raymond Dubois and W. Percey Wilkinson . . . rev. by P. Viala. Melbourne; R. S. Brain, Govt. Printer, 1901. 297p. (U.S. ed., different traslation, later date: Viala, Pierre and L. Ravaz . . . , translation of 2d ed. by Raymond Dubois and E. H. Twight. San Francisco, Calif.; Press of Freygang-Leary Co., 1903. 299p.)

First Vilmorin-Andrieux et Cie. The Vegetable Garden. Illustrations, Descriptions, and Culture of the Garden Vegetables of Cold and Temperate Climates. London; J. Murray, 1885. (English Edition. Published under the Direction of W. Robinson, Editor of "The Garden" 3d ed. Murray, 1920. Translation of Vilmorin, Audrieux and Cie: Les Plantes Potageres)

Second Virginia-Carolina Chemical Co. The New Corn Culture. Richmond, Va.; Virginia-Carolina Chemical Co., 1915. 32p.

Second Von Loesecke, H. W. Bananas; Chemistry, Physiology, Technology. New York; Interscience Publishers, 1949. 189p. (2d rev. ed., 1950. 189p.)

Second Vyvyan, M. C. Fruit Fall and Its Control by Synthetic Substances. East Malling, U.K.; Imperial Bureau of Horticulture and Plantation Crops, 1946. 72p.

W

Second Wagner, Philip M. Wine Grapes; Their Selection, Cultivation and Enjoyment. 1st ed. New York; Harcourt, Brace and Co., [1937]. 298p.

Second Wagner, Philip M. A Wine-Grower's Guide. New York; A. A. Knopf, 1945. 230p. (2d ed., rev., 1965. 244p.)

Second Waid, Ernest D. Wheat Culture, Information Concerning Cultivation and Production . . . Pittsburgh, Pa.; Issued by Pennsylvania Lines West of Pittsburgh, Freight Dept., 1911. 12p.

First	Wait, Frona Eunice. Wines and Vines of California. San Francisco; The Bancroft Co., 1889. 215p.
Second	Wallace, Henry A. Clover Culture. Des Moines, Iowa; Homestead Co., 1892. 156p.
Second	Wallace, Henry A. Clover Farming. Des Moines, Iowa; Wallace Publishing Co., 1898. 222p. (Wallaces' Farm Library No. 2) (2d ed., 1900.)
First	Wallace, Henry A., and E. N. Bressman. Corn and Corn-Growing. Des Moines; Wallace Pub. Co., 1923. 253p. (Corn and Corn Growing . . . rev. by J. J. Newlin, Edgar Anderson, and Earl N. Bressman. 5th ed., New York; John Wiley and Sons, Inc., 1949.)
Second	Walter, Jack. The Apple. Gasport, N.Y.; Friend Manuf. Co., 1944. 64p.
First	Ward, H. Marshall. Disease of Plants. London; Society for Promoting Christian Knowledge, New York; E. and J. B. Young, 1896. 196p. (Later ed. titled: Diseases in Plants. New York and London; Macmillan Co., 1901. 309p.)
Second	Ward, H. W. The Book of the Peach; Being a Practical Handbook on the Cultivation of the Peach Under Glass and Out-Of-Doors. London and Newcastle-On-Tyne; The W. Scott Pub. Co., Ltd., 1903. 113p.
Second	Ward, Harding. Potato Culture, with an Improved Method of Cultivation. The Disease, Its Cause and Remedy. London; Simpkin, Marshall, Hamilton, Kent and Co., Ltd., 1892. 22p.
Second	Ward, Henry W. Potato Culture for the Million. London; Eyre and Spottiswoode, 1891. 24p.
First	Warder, John A. American Pomology; Apples. New York; Orange Judd Co., 1867. 744p.
Second	Wardle, Robert A. The Problems of Applied Entomology. New York; McGraw-Hill, 1929. 587p.
Second	Ware, George W., et al. Southern Vegetable Crops. New York; American Book Co., 1937. 467p.
Second	Ware, Lewis S. Beet-Sugar Manufacture and Refining. 1st ed. New York; J. Wiley and Sons, 1905–1907. 2 vols.
Second	Ware, Lewis S. Production, Requirements and Selection of Sugar Beet Seed. Philadelphia; n.d. 86p.
First	Ware, Lewis S. The Sugar Beet, Including a History of the Beet Sugar Industry in Europe. Philadelphia; H. C. Baird and Co., 1880. 323p.
Second	Warner, H. G. Florida Fruits and How to Raise Them. Rev. ed. Louisville, Kentucky; J. P. Morton and Co., 1886. 347p.
First	Washburn, Frederic L. Injurious Insects and Useful Birds; Successful Control of Farm Pests. Philadelphia and London; J. B. Lippincott Co., 1918. 453p. (2d ed., rev., 1925. 453p.)
Second	Watt, George. The Wild and Cultivated Cotton Plants of the World; A Revision of the Genus Gossypium, Framed Primarily with the Object of Aiding Planters and Investigators Who May Contemplate the Systematic Improvement of the Cotton Staple. London, New York, Bombay and Calcutta; Longmans, Green and Co., 1907. 406p.
Second	Watts, R. L. Market Garden Guide. Louisville, Ky.; The Weekly Market Growers Journal, 1912. 123p.
Second	Watts, R. L., and G. S. Watts. The Vegetable Growing Business. New York; Orange Judd Pub. Co., 1945.

First Watts, Ralph L. Vegetable Gardening. New York; Orange Judd Co., 1912. 511p.

Second Waugh, F. A. The American Apple Orchard. New York; Orange Judd Co., 1908. 215p.

Second Waugh, F. A. The American Peach Orchard. New York; Orange Judd Co., 1908. 215p.

Second Waugh, F. A. Beginners' Guide to Fruit Growing. New York; Orange Judd Co., 1912. 120p.

Second Waugh, F. A. Dwarf Fruit Trees. New York; Orange Judd Co., 1906. 125p.

Second Waugh, F. A. Fruit Harvesting, Storing, Marketing. New York; Orange Judd Company, 1901. 224p.

Second Waugh, F. A. Plums and Plum Culture. New York; Orange Judd Co., 1901. 371p.

Second Waugh, F. A. Success with Stone Fruits, Including Chapters on the Planting and Cultivation of Cherries, Peaches, Plums, Apricots, etc. St. Joseph, Mo.; Fruit-Grower Co., 1905. 68p. (Brother Jonathan Series, no. 10)

Second Waugh, F. A. Systematic Pomology. New York; Orange Judd Co., 1903. 288p.

First Weatherwax, Paul. The Story of the Maize Plant. Chicago, Ill.; The University of Chicago Press, 1923. 247p.

Second Weaver, J. E., and W. E. Brunner. Root Development of Vegetable Crops. New York; McGraw-Hill, 1927. 351p.

Second Weaver, John C. American Barley Production: A Study in Agricultural Geography. Minneapolis, Mo.; Burgess Pub. Co., 1950. 115p.

Second Weaver, John E. Root Development of Field Crops. New York; McGraw-Hill Book Co., Inc., 1926. 291p.

First Weber, H. J., and L. D. Batchelor, ed. The Citrus Industry. Volume I. History, Botany and Breeding. 1st ed. University of California Press; Berkeley and Los Angeles, 1943. 1028p.

Second Weed, Clarence M. Farm Friends and Farm Foes: A Text-Book of Agricultural Science. Boston and New York; D. C. Heath and Co., 1910. 334p.

Second Weed, Clarence M. Fungi and Fungicides; A Practical Manual, Concerning the Fungous Diseases of Cultivated Plants and the Means of Preventing Their Ravages. New York; The O. Judd Co., 1894. 228p.

First Weed, Clarence M. Insects and Insecticides; A Practical Manual Concerning Noxious Insects and the Methods of Preventing Their Injuries. Hanover, N.H.; The Author, 1891. 281p. (2d ed., rev., New York; Orange Judd, 1895. 334p.)

First Weed, Clarence M. Spraying Crops; Why, When, and How. New York; Rural Pub. Co., 1892. 110p. (2d ed., rev., New York; 1894. 4th ed., rev., Orange Judd Company, 1903.)

Second Weed, Howard Evarts. Spraying for Profit. Griggin, Ga.; The Horticultural Pub. Co., 1899. 72p. (Rev., "Sprayology" Simplified. Rogers Park, Chicago; The Horticultural Publishing Company, 1906. 61p.)

Second Wehmeyer, L. E. The Genus Diaporthe Nitschke and its Segregates. Ann Arbor; University of Michigan Press, 1933. 349p.

Second Wehmeyer, L. E. A Revision of Melanconis, Pseudovalsava, Prosthecium and Titania. Ann Arbor; University of Michigan Press, 1941. 616p.

Second Weindling, L. Long Vegetable Fibers. New York; Columbia University Press, 1947. 311p.

Second Weldon, George P. Apple Growing in California: A Practical Treatise Designed to Cover Some of the Important Phases of Apple Culture Within the State. Sacramento, Calif.; California State Print. Off., 1914. 124p.

First Went, F. W., and K. V. Thimann. Phytohormones. New York; The Macmillan Co., 1937. 294p.

Second Westcott, Cynthia. The Gardener's Bug Book; 1,000 Insect Pests and Their Control. Garden City, N.Y.; The American Garden Guild, Inc. and Doubleday Doran and Co., 1946. 590p. (Rev. ed., 1956. 579p. 3d ed., 1964. 625p.)

Second Western Adjustment and Inspection Co., Chicago. Losses to Wheat, What to Look For and Where to Find It. Being One of a Series of Articles in Relation to Crops, Their Common Diseases, and Insect Pests to Which They are Subject. Chicago, Ill.; Hail Dept., Western Adjustment and Inspection Co., 1919. 61p.

Second Westgate, John M. Alfalfa Growing for Seed and Hay. South Bend, Ind.; Birdsell Manufacturing Co., 1910. 48p.

Second Weston, W. Diseases of Potatoes, Sugar Beet and Legumes. London; Longmans, Green, 1948. 86p.

First Wetzel, Herbert H., et al. Laboratory Outlines in Plant Pathology. Ithaca, N.Y.; The Authors, 1916. 207p. (2d ed., Philadelphia and London; W. B. Sauders Co., 1925. 231p.)

First Whetzel, Herbert H. An Outline of the History of Phytopathology. Philadelphia and London; W. B. Saunders Co., 1918. 130p.

Second White, F. S. Apple and Peach Growing . . . St. Louis; Issued by the Rock Island-Frisco Lines, n.d. 38p. (Reprinted, St. Louis, 1907. 62p.)

Second White, F. S. Corn and Wheat Culture . . . St. Louis, Mo.; C. P. Curran Printing Co., 1911? 100p. (Issued by the Frisco Line)

Second White, F. S. Potatoes . . . St. Louis; Issued by the Rock Island-Frisco Lines, 1908. 73p.

Second White, Herbert C. The Pecan Tree, How to Plant It, How to Grow it, How to Buy It . . . A Comprehensive Treatise Upon Pecan Culture. Dewitt, Ga.; G. M. Bacon Pecan Company, Inc., 1907. 24p. (G. M. Bacon Pecan Groves, Inc. 6th ed., Atlanta; Foote and Davies Co., 1917. 33p.)

Second White, J. J. Cranberry Culture. New York; Orange Judd Co., 1870. 126p. (New ed., New York; Orange Judd Company, 1885. 131p.)

First White, W. N. Gardening for the South. New York, Athens, Ga.; C. M. Saxton and Co., Wm. N. White, 1856. 402p. (New York, A. O. Moore and Co. 1859)

Second Whitehead, Charles. Hints on Vegetable and Fruit Farming. 4th ed. London; Murray, 1893. 53p.

Second Whitehead, G. E. Plain Fruit Growing for Small Families. Black, 1949.

Second Whitehead, G. E. Tomatoes, Mushrooms and Other Choice Foodcrops. London; Black, 1944. 93p.

Second Whitehead, G. E. Plain Vegetable Growing. London; Black, 1941. 94p.

Second Whitehead, Stanley B. Garden Weeds and Their Control. London; Dent, 1949. 160p.

Second Whitehead, Tatham. The Potato in Health and Disease. 2d ed, rev. and enl. Edinburgh; Oliver and Boyd, 1945. 400p. (3d ed., rev. and enl., 1953. 744p.)

Second Whitner, J. N. Gardening in Florida. Jacksonville, Fla.; C. W. DaCosta, 1885. 146p.

Second Whitner, J. N. A Manual of Gardening in Florida. Fernandina, Fla.; Published by the Florida Mirror, 1881. 73p.

Second Whitten, J. C. Apple Culture, with a Chapter on Pears. St. Joseph, Mo; The Fruit-Grower Co., 1906. 88p. (Brother Jonathan Series no.9)

First Whitten, J. C. Hints on Pruning. St. Joseph, Mo; Fruit-Grower Co., 1906. 73p. (Brother Jonathan Series, no.8)

Second Whitten, J. C. How to Grow Strawberries. St. Joseph, Mo; The Fruit-Grower Co., 1905. 61p. (Brother Jonathan Series, no.3)

First Whyte, R. O. Crop Production and Environment. London; Faber and Faber, 1946. 372p.

First Wickson, E. J. The California Fruits, and How to Grow Them. 1st ed. San Francisco; Dewey and Co., 1889. 575p. (2d ed., rev., 1891. 599p. 7th ed., rev., 1914. 513p.)

First Wickson, E. J. The California Vegetables in Garden and Field. San Francisco; Pacific Rural Press, 1897. 336p. (2d ed., rev., 1910. 367p. 3d ed., rev., 1913. 326p.)

Second Wilder, Gerrit P. Fruits of the Hawaiian Islands. Honolulu; Hawaiian Gazette Co., Ltd., 1907. 77p. (Rev. ed., 1911.)

Second Wilder, M. P. The Horticulture of Boston and Vicinity. Boston; Privately Printed, Tolman and White, Printers, 1881. 85p.

Second Wilkinson, Albert E. The Apple. Boston, Mass.; Ginn and Co., 1915. 492p.

Second Wilkinson, Albert E. The Encyclopedia of Fruits, Berries, and Nuts and How to Grow Them. Philadelphia, Pa.; Blakiston Co., 1945. 271p.

Second Wilkinson, Albert E. Modern Strawberry Growing. Garden City, N.Y.; Doubleday, Page and Co., 1913. 210p.

Second Wilkinson, Frederick. The Story of the Cotton Plant. New York; Appleton, 1899. 191p. (Reprinted 1902, 1904 and 1915.)

First Will, G. F., and G. E. Hyde. Corn Among the Indians of the Upper Missouri. St. Louis; W. H. Miner, 1917.

Second Willcox, John. Peach Culture. A Complete Treatise for the Use of Peach Growers. Bridgeton, N.J.; West Jersey Nursery Co., [1886]. 86p.

Second Willcox, Oswin W. ABC of Agrobiology; The Quantitative Science of Plant Life and Plant Nutrition for Gardeners, Farmers and General Readers. 1st ed. New York; W. W. Norton and Co., Inc., 1937. 323p.

Second Wilson, Andrew. Insects and Their Control. New Brunswick, N.J.; Printed by Thatcher-Anderson Co., 1929. 342p. (Revised Second Printing, 1931. 383p.)

Second Wilson, Archie D., and C. W. Warburton. Field Crops. St. Paul, Minn.; Webb Pub. Co., 1912. 544p. (4th ed., 1912. 544p. Reprinted 1915. Revised ed., 1918, 1919, 1923. 515p. Reprinted 1920.)

Second	Wilson, Charles M., ed. New Crops for the New World. New York; Macmillan Co., 1945. 295p.
Second	Wilson, Harold K. Grain Crops. 1st ed. New York; McGraw-Hill Book Co., 1948. 384p. (2d ed. New York, Toronto, London; McGraw-Hill Book Co., Inc. 1955.)
Second	Wilson, Harold K., and Alvin H. Larson. Identification and Judging; Crops, Weeds, Diseases. St. Paul, Minn.; Midway Book Co., 1940. 65p.
First	Wilson, Perry W. The Biochemistry of Symbiotic Nitrogen Fixation. Madison, Wis.; Univeristy of Wisconsin Press, 1940. 302p.
Second	Wing, Charles B. How to Grow Alfalfa and Other Legumes. Mechanicsburg, Ohio; C. B. Wing, 1915. 28p.
First	Wing, J. E. Alfalfa Farming in America. Chicago; Sanders Pub. Co., 1909. 480p. (Reprinted 1912, 1916.)
Second	Wing, J. E., and Bros. Seed Co. Alfalfa and How to Grow It. Mechanicsville, Ohio; The J. E. Wing and Bros. Seed Co., 1908. 33p.
Second	Winkler, H. G. Vegetable Forcing: Parts I, II, and III. Columbus Ohio; The Winkler Book Concern, 1895. 157p.
Second	Wolf, F. A., and E. T. Wolf. The Fungi. New York, London; Wiley, Chapman and Hall, 1947. 2 vols.
Second	Wood, W. The Publications of the Prince Society, Established May 15th, 1858. Wood's New-England Prospect. Boston; 1865.
Second	Woodward, G. E. and F. W. Woodward's Graperies and Horticultural Buildings. New York; Geo. E. Woodward and Co.; Orange Judd Co., 1865. 139p.
Second	Woodward, R. T. Woodward's Book on Horticulture; The Raising of Large and Small Fruits, The Diseases of the Same, and the Making and Care of Lawns. Boston; 1897. 74p.
Second	Woolverton, L. The Canadian Apple Grower's Guide. Toronto; William Briggs, 1910. 264p.
Second	Work, P. Vegetable Production and Marketing. New York, London; Wiley, Chapman and Hall, 1948.
Second	Wormald, Hugh. Diseases of Fruits and Hops. London; Crosby Lockwood and Son Ltd., 1939. 290p. (3d ed. rev., 1955. 325p.)
Second	Woronin, Michael S. *Plasmodiophora Brassicae, Urheber der Kohlpflanzen-Hernie = Plasmodiophora Brassicae, the Cause of Cabbage Hernia*, English translation by C. Chupp (Ithaca, N.Y.: American Phytopathological Society, 1934 [*Phytopathological Classics* no. 4]). (Originally published in *Jahrbücher für Wissenschaftliche Botanik* 11 (1878): 548–574.)
Second	Wright, Horace J. Onions. London; Agricultural and Horticultural Assoc., Ltd., 1908. 20p.
Second	Wright, Walter P. Everyman's Encyclopaedia of Gardening. Rev. and enl. ed. London; J. M. Dent and Sons, Ltd.; New York; E. P. Dutton and Co. Inc., 1930. 495p. (Later eds. by Stanley B. Whitehead: 1952, 786p.; New title: Encyclopedia of Gardening. Newton Centre, Mass; C. T. Branford Co., 1960. 789p.)
Second	Wright, Walter P. An Illustrated Encyclopaedia of Gardening. London;

J. M. Dent and Sons., Ltd.; New York; E. P. Dutton and Co., 1911. 323p. (Later Publication Date, 1932. 493p.)

Second Wrightson, John. Fallow and Fodder Crops. London, U.K.; Chapman and Hall, 1889. 276p.

Second Wyamn, D. Hedges, Screens and Windbreaks. New York; McGraw-Hill Book Co., Inc., 1939.

Second Wyld, R. The How-to Book on Strawberries. Honeoye Falls, N.Y.; 1953. 112p.

Y

Second Young Men's Christian Association, Portland Ore. Apple Growing in the Pacific Northwest . . . Portland, Oreg.; The Portland, Oregon Young Men's Christian Associaiton, 1911. 215p.

Z

Second Zimmermann, A. Botanical Microtechnique; a Handbook of Methods for the Preparation, Staining, and Microscopical Investigation of Vegetable Structures, translated from the German by James Ellis Humphrey. New York; H. Holt and Co., 1893. 296p. (New York, H. Holt and Co., 1901. 296p.)

First Zirkle, C. The Beginnings of Plant Hybridization. Philadelphia; University of Pennsylvania Press, 1935. 231p.

The Core Agricultural Literature Project asked Kansas State University to check their libraries' holdings of these historical core monographs; only 38% were owned.

D. Popular and Trade Periodicals

Among the most valuable mediums for the dissemination of agricultural information for farmers and orchardists were the early agricultural periodicals. They were clearinghouses of general information and borrowed liberally from each other. Before 1834 farm periodicals presented news without regard for types of plants, animals, or farms. Shortly after that date, however, news began to be organized by subject. Specialized farm literature began to appear with regularity around 1870 in the United States and was significant by 1880. Evans and Salcedo indicate that one-third of all farm periodicals published in 1880 were in specialized subjects, and by 1920 these accounted for 42% of all farm periodicals.[54]

Horticulture is considered to have had the first specialized agricultural

54. James F. Evans and Rodolfo N. Salcedo, *Communications in Agriculture: The American Farm Press* (Ames: Iowa State University Press, 1974).

periodicals in the United States, followed soon by dairying. These popular, general distribution periodicals offer one of the best historical records on American agriculture. Their value is highly rated by crop scientists who need to find historical records, as well as by agricultural historians. For a more extensive discussion of popular periodicals in agriculture see an earlier volume in this series.[55]

It is not possible to identify and determine the historical value of popular periodicals in the usual way, that is, by using source documents to locate citations to the early literature. Literature citations usually relate to scholarly and scientific publications; references to the popular periodical literature are few. Other methods must be used to identify and establish the relative historical merits of this material. Journal lists, library catalogs, and historical writings concerned with agricultural periodicals and history were used. Prime among these was the Stuntz compilation created by the U.S. Department of Agriculture Library in 1941.[56] A master list was prepared from this and other sources and bibliographic information added.

The following criteria were established for retaining and ranking titles in the final list:

(1) Early publishing date of the title
(2) Length of the run of the periodical
(3) National or international influence, as compared to state or local importance
(4) Wide knowledge of the title, its contents, or its editors

The titles have been systematically searched in printed and electronic data sources. Since the intention here is to assist in identifying and ranking crops and gardening periodicals for preservation, those which are known to have been microfilmed have not been included.

It must be noted that the list covers only U.S. and some Canadian publications, and popular rather than scientific or scholarly journals. These more popular materials may contribute to a broader understanding of the culture and the times in which they existed. The numbers in the left-hand column indicate relative rank:

1 = Top-ranked periodical with primary preservation priority
2 = Periodicals of less importance, but worthy of preservation

55. Henry T. Murphy and Dorothy W. Wright, "Primary Historical Literature, 1860–1949," in Wallace C. Olsen, ed., *The Literature of Animal Science and Health* (Ithaca, N.Y.: Cornell University Press, 1993), pp. 310–378.

56. Stephen C. Stuntz, compiler, *List of the Agricultural Periodicals of the United States and Canada Published during the Century July 1810 to July 1910* (Washington, D.C.: U.S. Dept. of Agriculture, 1941 [*USDA Misc. Publication* no. 398]).

The popular and trade periodicals' list has 835 titles of which 495 have first ranking.

Historically Important Popular and Trade Periodicals in Crops, 1850–1950

w = weekly; m = monthly; // = known to have ceased on this date.

1 Acadian orchardist. Wolfville, N. S. (1892–1903), Wolfville and Kentville, Nova Scotia (1903–1910). w. v.1–1892–July 1910//?

1 Agriculturist and floral guide. Mexico, Mo. (1870–1875), St. Louis (1877–1881). m. 1870–81. Seems at some time to have had titles Apiculturist and floral guide and Agriculturist and home circle.

2 The Alfalfa belt. Indianola, Nebr. m. v. 1–6? Jan. 1901–1906.

1 Amateur gardening. Springfield, Mass. m. v. 1–7, no. 5. Jan. 1892–Sept. 1897//

1 American agricultural annual. New York. 1867–? Supersedes in part Rural annual and horticultural directory.

2 American chemical fertilizer and crop journal. Sept. 1886–//

1 American cider and vinegar maker. Established as
(a) American cider maker, v. 1–4, Jan. 1892–1894.
Continued as
(b) Cider maker and fruit packers' journal, v. 5?, 1895). New York. v. 6?–13?, 1896–1903.

1 American cotton grower. Atlanta, Ga. m. v. 1–6. 1935–1940//.

1 American cotton planter and the Soil of the South. Established as
(a) American cotton planter, v. 1–4, Jan. 1853–Dec. 1856. Running title: Cotton planter and soil. Montgomery, Ala. Vol. 5–9 1857–1861//.

1 American elevator and grain trade. United with Grain dealers journal to form Grain and feed journals consolidated. Chicago. v. 1–49 1882–1930//.

1 American farm and horticulturist. Lakewood, Ohio. (Apr. 1889–1890); Richmond, Va. (1890–1893). q. v. 1–5? Apr. 1889–1893//.

2 The American farm and orchard. Mexico, Mo. m. v. 1–5? Sept. 1901–1906.

1 American farmer. Baltimore. Established as
(a) The American farmer. Baltimore. m. (1819, 1832–1834), w. (1820–1831). v. 1–15. 1819–1834.
Continued as
(b) Farmer and gardener and livestock breeder and manager. Baltimore. w. v. 1. 1834–1835.
Continued as
(c) The farmer and gardener. Baltimore. w. v. 2–6. 1835–1839.
Continued as
(d) American farmer and spirit of the agricultural journals of the day. Baltimore. w. (1839–1845) m. (1845–1850). n.s. [3rd. ser.] v. 1–6, 1839–1845. 4th ser. v. 1–6, 1845–1850.
Continued as
(e) American farmer. Baltimore (1850–1861), Baltimore and Richmond (1861). m. 4th ser. v. 7–14. 1850–1859. 5th ser. v. 1–3. 1859–1861//.

1 American farmer and gardener. Washington, D. C. m. v. 1–7, no. 8. 1897 (1900)–Sept. 1903//?

1 American feed and grain dealer. Established as Country grain shipper, 1916–1922. National grain journal, 1922–1946. Superseded by Bulk feed. Minneapolis. 1947–v. 42, 1958//.

1 American florist. Chicago and New York. American Florist Co., v. 1–75 1885–1931//.

2 American fruit and farm. Paonia, Calif. v. 1–? 1908–//?

1 American fruit and nut journal. Established as
(a) The Trade journal. Petersburg, Va. v. 1–Sept. 1904.
Continued as
(b) American nut journal. Petersburg, Va. m. v. 1–3, no. 62. Oct. 1904–June 1906.
Continued as
(c) American fruit and nut journal. Petersburg, Va. m. v. 3, no. 63–v. 5, no. 92. July 1906–Dec. 1908.

2 American fruit and vegetable journal. Chicago. m. 1–3, no. 1. Feb. 1900–Feb 1901//? Absorbed by the Farm, field and fireside, and replaced by the monthly edition of that periodical.

1 American fruit grower. Charlottesville, Va. Established as Virginia fruit grower and farmer, v. no. 1–4, Sept. 1915–Aug. 1917. Absorbed Fruit grower (St. Joseph, Mo.) Sept. 1917. Oct. 1917 united with Green's fruit grower (SEE American fruit grower [Chicago]).

1 American fruit grower. Chicago; Cleveland [etc.]. Established as
(a) Green's fruit grower and home companion, 1889–1909.
Continued as
(b) Green's fruit grower, 1917–1918.
Continued as
(c) Green's American fruit grower, 1918–1921.
Continued as
(d) American fruit grower magazine, 1922–1931. Chicago. American Fruit Grower Company. v. 52– 1932–

1 American fruit grower. Gilman, Ill. Established as
(a) Fruit grower. Gilman, Ill. m. v. 1. 1869–1870.
Continued as
(b) American fruit grower. Gilman, Ill. m. v. 2–3? 1870–1872.

1 American fruit growers' journal. Atlanta, Ga. m v. 1–3? 1897–1900.

1 American fruit growers' union. A magazine. Chicago. m. v. 1– Mar. 25, 1897–1898; v. 1, no. 1, has title American fruit growers' union; a weekly magazine.

2 The American fruiter. New York. w. v. 1, no 1. June 21, 1883//?

2 American fruits. Rochester, N. Y. v. 1–24 April 1904–November 1916//. Continued by
(a) American nurseryman.

1 The American gardener's chronicle. Washington Heights, N. Y. m. (Sept.–Dec.? 1852), w. (Jan.-Aug. 1853). v. 1–2? Sept. 1852–Aug.? 1853.

1 American horticulturist. Established as
(a) Michigan horticulturist. Detroit, Mich. v. 1–2. 1885–1886//. Continued in Popular gardening and fruit growing.

2 American horticulturist. Fowler, Ind. v. 1, no. 1–2. Nov.–Dec. 1911//.

1 American horticulturist. Wichita, Kans. Established as
(a) Smith's small fruit farmer. Lawrence, Kans. q. v. 1–2? 1891–1892? Continued as
(b) Smith's fruit farmer. Lawrence, Kans. (1893), Topeka, Kans. (1894). v. 3?–6? 1893–1896//? lists as absorbed by The American southwest.
1 American nursery trade bulletin. Merged into American nurseryman. Rochester, N. Y. v. 1–22. 1916–1927//.
1 American nurseryman and the national nurseryman. Established as
(a) American fruits, v. 1–24, 1904–1916.
(b) American nurseryman, v. 24–70). Rochester, N. Y. American Fruits Publishing, 1917–
1 American nut culture. Rochester, N. Y. American Fruits Publishing, 1918–//?
1 American nut journal. Merged into National nut news, later Modern school store. Rochester, N. Y. American Fruits Publishing, v. 1–35 1914–1931//.
2 American nut trade bulletin. Supplement to American nut journal. Rochester, N. Y. v. 1–6, 1923–1926//.
1 American nut trade journal. Norfolk, Va. v. 1– 1905–
1 American pecan journal. Waco, Tex. m. v. 1– Oct. 1940–
1 American pomologist. Philadelphia. v. 1. 1851//.
1 American pomologist. Washington, D. C. American Pomological Society, 1–11, 1912–1915//; 1924–1925//.
1 American produce grower. Chicago. m. v. 1–5. Sept. 1926–Apr. 1930//. Absorbed by American fruitgrower magazine, Chicago.
1 American riceman and farmer. Established as
(a) Southwestern oil news and rice paper, v. 1–, 1899–1902?. Houston. v. 4– 1902–//?
1 American sugar industry and beet sugar gazette. Established as
(a) Beet sugar gazette, v. 1–6, 1899–1904. v. 6– 1904–//?
1 American wool and cotton reporter. Established as
(a) American wool, cotton and financial reporter, v. 1–5, 1887–1891. Continued as
(b) Wool and cotton reporter and financial gazette, v. 5–10, 1891–1896. Continued as
(c) Wool and cotton reporter, v. 10–22, 1896–1908, later
(d) America's textile reporter. Boston. v. 22–1908–
2 Appalachian area apple marketer. Martinsburg, W. Va. v. 1, no. 1–10. 1938–1939//.
2 Apple specialist. Merged in Green's fruit grower and home companion. Quincy, Ill. v. 1–4,1903–1908//.
1 The Arkansas cultivator. Little Rock, Ark. s-m. v. 1– 1899–Jan. 15, 1902. Absorbed by the Cotton planters' journal, Jan. 15, 1902?. Combined with the Southern agriculturist, Feb. 1, 1904?
2 Arkansas fruit and farms. Established as
(a) Ozark produce journal, 1–2. Also called Ozark fruit and farms. Fort Smith, Ark. v. 1–8, 1912–1916//.
2 Arkansas fruit grower. Fort Smith, Ark. m. v. 1–3. 1900–Dec. 1902. Merged in the Oklahoma farm journal, Jan. 1903.
1 Arkansas fruit grower. Harrison, Ark. m. v. 1–10? 1892–1898. Merged in the Western fruit grower, Jan. 1899.

2 Arkansas fruits. Fayetteville, Ark. m. Sept. 1912–
2 Arkansas gardener. Batesville, Ark. Arkansas Federation of Garden Clubs, 1941–
2 Arkansas state rice journal. Stuttgart, Ark. m. v. 1–? Apr. 1909–//?
2 Art and gardens. Corvallis, Oreg. v. 1, no. 1. Spring 1948//.
2 Banner of the south and planters' journal. Established as
(a) Banner of the south, not agricultural. Augusta, Ga. 1870–1873//.
1 Bean-bag. Also called Bean-bag and pea journal. St. Louis. East Lansing, Mich. Lightner Publishing, v. 1–23 1918–1940//?
1 Beet sugar enterprise. Grand Island, Nebr. (1890–1891), Lincoln, Nebr. (1891–1892). m. v. 1–2? 1890–1892.
1 Bell's garden review. Windsor, N. Y. q. v. 1– 1886//.Merged in Popular gardening and fruit growing, Nov. 1886.
2 Better crops with plant food. Established as
(a) Better crops, 1923–1927. New York. American Potash Institute, 1927–
1 Better fruit. Hood River, Oreg.; Northwest Fruit Growers' Association. m. v. 1– 1906–
1 Better homes and gardens. Established as
(a) Fruit, garden and home. Des Moines, Iowa. Meredith Pubs. 1922–
1 Black patch journal [Tobacco]. Springfield, Tenn. m. v. 1– May 1907–
2 Blades of grass. New York, N.Y. Golf and Lawn Supply Corp. 1940–
1 Blossom. Keene Valley. Kingston N. Y. v. 1–5?, 1888–1893//.
1 British Columbia united farmer. Established as
(a) Fruit magazine, 1908–1913.
Continued as
(b) Fruit and farm, 1913–1914.
Continued as
(c) British Columbia fruit and farm magazine, 1914–1918.
Continued as
(d) British Columbia farmer, 1919–1920. Vancouver, B.C. Fruit and Farm Co., Ltd., 1921–1924//.
1 Broom and broom corn news. Arcola, Ill. Collins Bros., 1912–
2 Broom corn journal. Charleston, Ill. s-m. v. 1–3? June 1902–1905.
1 Broom corn reporter. Fort Scott, Kans. m. v. 1–2? 1886–1888.
1 Broom corn review. Merged into Broom and broom corn news. Wichita, Kans. v. 1–20 1912–1931//.
1 Burley tobacco grower. Lexington, Ky. Burley Tobacco Growers' Assoc., v. 1–8 1922–1929//.
1 California apiculturist. Oakland; Los Angeles. v. 1, no. 1–12. 1882//.
1 California citrograph. Los Angeles. California Citrograph Publishing, 1915–
1 California culturist. San Francisco, Calif. v. 1–3. June 1858–Dec. 1860//.
1 California farmer and fruit grower. Established as
(a) Farmer and dealer. San Francisco. m. v. 1– 1876 (1880)-1891.
Continued as
(b) California farmer and fruit grower. San Francisco. m. v. 1891.
2 California floriculturist. Los Angeles. no. 1–5. 1901–1902//.
1 California florist and gardener. San Francisco and Los Angeles. m. v. 1–2, no. 4 May 1888–Apr. 1889//. Merged into Pacific rural press June 22, 1889.
1 The California fruit news. Established as
(a) California fruit grower, 1888–April 1906, Oct. 1908–1914. Sacramento. April

1906–Oct. 1908. During 1894–1895 called California fruit grower and fruit review. w. v. 51–1915–
1 California garden. Point Loma, Calif. San Diego Floral Assoc., 1909–
1 The California horticulturist and floral magazine. San Francisco. m. v. 1–10. Nov. 1870–Dec. 1880; merged in Pacific rural press.
1 California olive industry news. San Francisco. v. 1– Jan. 1934–
1 California producer. Los Angeles, Calif. w.? v. 1–3? 1894–1897?
2 California products. Los Angeles, Calif. m. v. 1–2, no. 3. July 1905–Mar. 1906.
1 California sugar beet. San Francisco. m. v. 1, no. 1–5. Sept. 1897–Apr. 1898//?
1 Canadian cigar and tobacco journal. Montreal. Tac Publications. v. 1–62, no. 6. 1896–June 1956. Continued as Tobacco and variety journal.
1 Canadian florist and cottage gardener. Peterborough, Ont. q. v. 1–2? 1885–1887.
1 Canadian florist. Toronto (1905–1909); Peterborough; Streetsville, Ont. s-m. v. 1– 1905–
2 Canadian fruitgrower. Niagara-on-the-Lake, Ont. Scott Publishing Co. Also called Canadian fruitgrower and gardener; Niagara fruitman. m. v. 1– 1944–
1 Canadian gardeners' and florists' exchange. London, Ontario. 1900–//?
2 Canadian grain journal, miller and processor. Established as
(a) Canadian grain journal, 1945–1946.
Continued as
(b) Canadian grain journal and monthly seedsman, 1946–1950. 1950–
1 Canadian grower. Established as
(a) Canadian horticulture and home magazine. Grower's edition. v. 62–70, 1939–1947. Toronto. Fruit Growers' Association of Ontario. 1948–1951//.
1 Canadian homes. Established as
(a) Canadian homes and gardens 1925–Jan. 1960. Toronto. McLean-Hunter Publishers, Feb. 1960–
1 Canadian horticulture and home magazine. Toronto; Peterborough; Oshawa, Ont.; Fruit Growers' Association of Ontario. 1–37. 1877–1914//.
Continued in two editions:
Floral edition (v. 38–70, no. 5, 1915–May 1947) and Fruit edition (v. 62–70, no. 1, 1939–Jan. 1947).
Fruit edition continued as Growers' edition (v. 62–70, no. 1, 1939–Jan. 1947), later continued as Canadian grower.
Floral edition superseded by Your garden and home.
1 The Canadian horticulturist. St. Catherine's, Ont. (1878–1887), Toronto (1887–1888), Grimsby and Toronto (1889–1899), Grimsby, Ont. (1900–1903), Toronto (1904–1908), Petersborough (1909–1910). m. v. 1–56. 1878–1933; v. 1–27 published by the Fruit growers' association of Ontario.
1 Canadian miller and cerealist. Established as
(a) Canadian miller and grain elevator. v. 1–5, no. 3. Apr. 1909–1913. Montreal. 1913–1916. Merged into Journal of commerce.
1 Canadian miller and grain trade review. Toronto. m. v. 1– 1883 (1895)-1895.
1 Canadian milling and feed. Established as
(a) Canadian milling and feed journal, v. 1–25, 1920–1944.
Continued as
(b) Milling and feed, v. 25–28, 1944–1947. Montreal. Wallace Press, v. 28–1947–

1 Canner/packer (began as Canner and dried fruit packer, 1895–1915. Canner, 1916–1955. Canner and freezer, 1955–1958). Louisville; Chicago. 1959–
2 Cannery producer. Waterloo, Wisc. v. 1–4, 1902–1905//.
1 Canning trade, the canned food authority of the world. Established as
(a) Trade, 1878–1904.
Continued as
(b) Canned goods trade, 1904–1912. Baltimore. 1913–
1 Carolina fruit and truckers' journal. Established as
(a) Carolina fruit and truck growers' journal, v. 1–17, 1897–1906. Wilmington, N.C. Eastern Truck and Fruit Growers' Association. American Fruit and produce Travelers' Association, v. 17– 1906–
1 The Carolina planter, a monthly register for the state and local societies. Columbia, S. C. m. v. 1. July 1844–June 1845//.
1 Carolina planter. Columbia, S. C. w. v. 1. Jan. 15, 1840–Jan. 13, 1841//; merged in the Farmers' register.
1 Carolina planter. Florence, S. C. s-m. v. 1–3? 1894–1897.
1 Cash crops. New York, N.Y. m. v. 1, no. 1– Oct. 1921– Not pub. June–Sept.
1 Cereals and feed. Established as
(a) Cereals. Milwaukee. m. v. 1, no. 1–3? May-July? 1900.
Continued as
(b) Cereals and feed. Milwaukee. m. v. 1, no. 4?-v. 4, no. 11. Aug. 1900?-Mar. 1904//. Combined with Flour and feed, Waukegan, Ill., April 1, 1964.
1 Chicago hay journal (began as Hay there, v. 1–2?, 1893–1895). Chicago. v. 3?–7?, 1895–1900//?
2 Cincinnati packer. Cincinnati. v. 1– 1902–//?
1 Citrograph. Redlands, Calif. v. 1–43, no. 17 -Nov. 7, 1908//?
1 Citrus and vegetable magazine. Established as Citrus, 1938–1941. Citrus magazine, 1942–1960). Tampa, Fla. Florida Citrus Exchange, 1961–
1 Citrus grower. Kissimmee. Orlando, Fla. Florida Citrus Growers, v. 1–4 1938–1942//.
1 Citrus industry. Tampa. Bartora, Fla. Association Publications, 1920–
1 Citrus leaves. Redlands, Calif. Mutual Orange Distributors. Pure Gold Association, v. 1–37 1919–1957//.
1 The Clover leaf. South Bend, Ind. v. 1– Mar.-Sept. 1881.
1 Clover leaf. South Bend, Ind. m. v. 1–3? 1892–Aug. 1895; merged in Farm and fireside, Aug. 1895.
1 Colorado fruit grower. Manzanola, Colo. w. v. 1–3? 1894–1896//; absorbed by Field and farm before Nov. 1896.
1 Colorado horticulturist. Greeley, Colo. q. v. 1. 1875//?
1 Coming's garden, poultry, and small fruits. Rockford, Ill. m. v. 1. 1891–1892.
2 Common-tater. Established as Coast grower, v. 1, no. 1–2, 1945. V. 1, no. 3–v. 7. 1945–1952. Vancouver, B.C. Coast Vegetable Marketing Board.
1 The Connecticut farmer and gazette and horticultural repository. Established as
(a) The Farmers' gazette. New Haven. w. Sept. 1840.
Continued as
(b) Connecticut farmer's gazette. New Haven. w. (Sept. 1840–Oct. 1842), s-m. (Oct. 1842–Oct. 1844). v. 1–4? Sept. 1840– Oct. 1844.
Continued as

(c) The Connecticut farmer and gazette, and horticultural repository.
Continued as
(d) Connecticut farmer and horticulturist. New Haven. s-m. n. ser., v. 1, no. 1–4. Nov. l-Dec. 15, 1844//?
Merged in Cultivator.

1 Connecticut Valley farmer; a monthly journal of agriculture, horticulture, and rural economy. Established as
(a) Connecticut Valley farmer and mechanic. v. 1, no. 1–4. Amherst; Springfield, Mass. 1–2. 1853–1854. n.s. v. 1. 1855//? Merged into New England farmer.

2 Connecticut Valley tobacco grower. Hartford. v. 1–4 no. 5. Sept. 10, 1923–Mar. 21, 1927//.

2 Contact. Clarkston, Ga. bi-m. May 1935–

2 The Corn belt. Chicago. m. v. 1–9. Dec. 1895–Dec. 1902//.

1 Corn belt magazine. Decatur, Ill. m. v. 1, no. 1– Dec. 1905–

1 Corn belt farmer. Established as
(a) Corn, v. 1–4, 1912–1915. Merged into Iowa farmer and corn belt farmer. Des Moines, Iowa. Harry B. Clark Co. Farm Publishing Co. [1925]. v. 4–16 1915–1927//.

1 Corn miller, devoted to corn, rye, oat milling etc. Supplement to Millstone, v. 10–16. Indianapolis, Ind. 1–9 1885–1892//?

1 Corn. New York. Corn Industries Research Foundation, 1– 1945–

2 Corn. Waterloo, Iowa. v. 1–3 1912–1914//.

1 Cotton. Atlanta, Ga. m. v. 1– 1901–//? Oct. 1908 absorbed Southern mills; Nov. 1908 absorbed Boston journal of commerce, and assumed the volume numbering of that periodical.

1 Cotton. New York. w. v. 1–3. 1880–1881//?

1 Cotton and farm journal. Established as
(a) Cotton planters' journal, v. 1–5. Memphis, Tenn. v. 1–6. 1897–1903//?

2 The Cotton bale. Devoted to the interests of the southern farmer. Shreveport, La. m. v. 1–2, no. 5. Mar. 1903–Feb. 1904//?

2 The Cotton belt. Memphis, Tenn. s-m. v. 1– Jan. 1882–1887.

2 Cotton co-op of South Carolina. Columbia, S.C. v. 1–6. 1931–1938//?

2 Cotton digest. Houston, Tex. v. 1– 1928–

1 Cotton facts. New York. [American Cotton Association?], 1876–1931//.

1 The Cotton gin and oil mill press. Established as
(a) The Ginner and miller. Dallas, Tex. (1900–Aug. 1903). Memphis, Tenn. (Sept. 1903–), Memphis and Dallas (-June 1901). m. (1900–1903), w. (1904). v. 1–5, no. 26. Jan. 1900–June 27, 1904.
Continued as
(b) The Cotton and cotton oil news and The Ginner and miller. Memphis and Dallas. w. v. 5, no. 27–v. 11, no. July 4, 1904–
Continued as
(c) The Cotton and cotton oil news. Memphis and Dallas. w. v. -11, no. 27. - v. 36, 1935.
Continued as
(d) The Cotton gin and oil mill press. v. 37– 1936–

2 Cotton ginner. Dallas, Tex. Superseded by Cotton ginners' journal. v. 1–3. 1919–1922//.

2 Cotton ginners' journal. Dallas, Tex. v. 1–14. 1929–1942//.
1 Cotton ginners' journal. Waco, Tex. m. v. 1– Apr. 1897–1905; absorbed the Waco weekly tribune.
1 Cotton growers' journal. Atlanta, Ga. m. v. 1– Sept. 1903–
2 The Cotton journal. Atlanta, Ga. w. v. 1–4, no. 27. May 17, 1906–Nov. 13, 1909//.
1 Cotton news. St. Matthews, S.C. American Cotton Association, v. 1–5 1919–1926//?
1 Cotton oil press. Washington, D.C.; National Cottonseed Products Association; American oil chemsists' Society. v. 1–19. 1917–1935//.
1 The Cotton plant. Marion, S. C. (1883–1886), Greenville; S. C. (1887–1890, 1898–1904), Orangeburg C. H., S. C. (1891), Columbia, S. C. (1892–1896), Spartanburg, S. C. (1897). m. v. 1–2, no. 31. 1883–Sept. 23, 1904//.
Combined with
(a) Progressive farmer, Raleigh, N. C. which
Continued as
(b) Progressive farmer and Cotton plant.
2 Cotton planter. Montgomery, Ala. J. B. Stern, 1914–//?
1 Cotton record. (Established as Cotton trade journal, 1907. Cotton trade journal, an illustrated monthly). Savannah, Ga. v. 1–22. 1907–1928//.
1 Cotton review. New Orleans. m. Oct. 1882//?
2 The Cotton seed. Atlanta, Ga. m. v. 1–12. 1906–1912//.
1 Cotton Seed oil magazine. Established as
(a) Cotton. Atlanta, Ga. m. v. 1–3? 1900–May, 1902.
Continued as
(b) Cotton seed oil magazine. Atlanta, Ga. m. v. 3?-58. 1902–1930//.
1 Cotton trade journal. New Orleans. v. 1– 1921–
2 Cotton trade journal. Savannah, Ga. w. v. 1–6. 1900–Aug. 1907//; absorbed Sept. 1, 1907 by Cotton trade journal; an illustrated monthly.
1 The Cotton world. New Orleans. w. v. 1–3, no. 10. Apr. 2, 1887–May 5, 1888//?
2 Country calendar. Harrisburg, Pa., New York N.Y. m. v. 1, no. 1–8. May-Dec. 1905. Merged into Country life in America. Preceeded by a sample number called v. no. 1, issued Dec. 1904.
2 Country elevator operator. Merged into Grain and Feed Dealers Association newsletter. St. Louis. Grain and Feed Dealers National Association, v. 1–3 1946–1949//.
1 County agent's magazine. Established as
(a) Crop improvement bulletin, no. 1. Continued as
(b) Crop improvement page, nos. 2–14.
Continued as
(c) County agent and farm bureau, no. 15– v. 14, no. 7. Chicago. County Agent Publishing, 1916–1926//.
1 County farmer and trucker. Norfolk, Va. w. v. 1–2? 1884.
2 Cranberry grower. Cranmoor, Wis. m. v. 1–3, no. 7. Jan. 1903–July, 1905//.
2 Crop comments. Phoenix, Ariz. Arizona Fertilizers, Inc. 1947–
2 Crop news and views. Mechanicsburg, Ohio. 1946–
1 Crop protection digest. Bulletin series. Washington, D.C. Crop Protection Institute. 1921–Crops. Dallas, Tex. v. 1– Jan. 1907–
1 Currie's monthly [Horticulture]. Milwaukee. v. 1–4. Jan. 1885–Dec. 1888//. Absorbed by Popular gardening and fruit growing, Dec. 1888.

1 Date palm. Indio, Calif. J. Win. Wilson, 1912–//?

1 Desert grapefruit. Phoenix, Ariz.; California Desert Grapefruit Industry Board. v. 1–4. 1944–1948//?

1 Diamond walnut news. Stockton, Calif.; Diamond Walnut Growers. v. 1–60. May 1917–Dec. 1978.
(a) 1917–1922 as Diamond brand news.
Issued 1917–1956 by the association under its earlier name: California Walnut Growers Association.
Merged with Sunsweet standard to form Diamond/Sunsweet news.

1 The Dixie miller. Nashville, Tenn. m. v. 1–62. 1894–1924//.

2 Doings in grain at Milwaukee. Milwaukee, Wisc. 1–15. 1912–1927//.

2 East Texas fruit and truck journal. Jacksonville, Tex. w. v. 1. 1904.

2 Eastern fruit grower. Boyce, Va.; Charles Town, W. Va. Ralph N. Dorsey, m. 1938–

2 Eastern fruit. Philadelphia. m. v. 1, no. 1–6/7. Jan.-June/July 1912//.

1 Eastern New York horticulturist. Chatham, N. Y. q. v. 1–2, no. 4. July 1897–July 1899//.

1 Eastern shore farmer and fruit culturist. Established as
(a) Strawberry culturist. Salisbury, Md. q. v. 1–3. Apr. 1893–Jan. 1896.
Continued as
(b) Strawberry culturist and small fruit grower. Salisbury, Md. m. v. 4–7, no. 6. Dec. 1896–Jan. 1901.
Continued as
(c) Eastern shore farmer and fruit culturist. Salisbury, Md. (1901), Georgeown, Del. (1902). m. v. 8–9, no. 2. Sept. 1901–Feb. 1902//?

2 Eclectic ruralist and nursery exchange journal. Rochester, N.Y. m. v. 1, no. 1 Dec. 1872//.

1 Evergreen grower and horticulturist. Elgin, Ill. m. v. 1–8? 1890–1898.

1 Evergreen and fruit-tree grower. Strugeon Bay, Wis. m. v. 4, no. 2–10. Feb.–Nov. 1874. Began as
(a) Evergreen and forest tree grower. v. 1–4, no. 1. Aug. 1871–Jan. 1874.
Continued as
(b) Evergreen. v. 4 no. 11–v. 5. Dec. 1874–July 1876//.

1 The Family garden. New York. m. v. 1, no. 1–4. April–July 1871//.

1 The Family repository and horticultural cabinet. New London, Conn. m. v. 1– Aug. 1861– Succeeds The Repository.

2 Farm and cotton journal. Memphis, Tenn. m. [n.d.]

2 Farm and floral world. Mobile, Ala. w. v. 1. 1906–1908?

1 Farm and fruit belt. Siloam Springs, Ark. m. v. 1–2? 1899–1902.

1 The Farm and garden. Philadelphia. m. v. 1–7. Sept. 1881–1888; merged in Farm and fireside, Springfield, Ohio, of which it was eastern edition, 1889, and continued as Farm and fireside, eastern edition.

1 Farm and garden and floral guide. Established as
(a) Farm and garden review. Thorn Hill, N. Y. q. v. 1. 1887–1888.
Continued as
(b) Farm and garden and floral guide. Thorn Hill, N. Y. q. v. 2? 1888–1889.

1 Farm and garden. Chicago. bi-m. v. 1– 1877–1881?

2 Farm and garden. Established as Woman's national farm and garden association bulletin. New York. Woman's National Farm and Garden Association, 1914–

1 Farm and garden. New York. m. v. 1, no. 1–9. Jan.-Sept. 1853//
Succeeded The Plow; made up of articles from the weekly New York agricultor.
1 Farm and garden. Tulsa, Okla. v. 1–13. -1946//.
Absorbed by Farm and home.
1 The Farm and the garden. Clinton, S. C. m. v. 1–5? June 1867–1872?
2 Farm and orchard. Keyser, W. Va. m. v. 1–5. 1910–1914//?
1 Farm and orchard. Las Cruces, N. M. m. v. 1– 1894–1899?
2 The Farm and orchard review. St. Joseph, Mich. m. v. 1– Sept 1905–1908.
1 Farm and vineyard and American gardener's assistant. Established as
(a) Farm and vineyard. Erie, Pa. m. v. 1–4, no. 8. May? 1886–Dec. 1890.
Continued as
(b) Farm and vineyard and American gardener's assistant. Erie, Pa. m. (1891–1892), q. (1892). v. 4, no. 9–v. 6. Jan. 1891–1892//?
Combined with Grape belt. Dunkirk, N. Y., 1892?
2 Farm crops. Winnipeg, Manitoba. m. v. 1– 1905–July 1910//; became Bulman's farm and motor magazine Aug. 1910.
2 Farm, field and fireside monthly. Chicago. m. v. 1– Jan. 1900–1903//.
Absorbed in April 1901, The American fruit and vegetable journal.
V. 1–2, 1900–1901 as American fruit and vegetable journal.
Continued as
Farm, field and fireside monthly combined with the American fruit and vegetable journal.
1 Farm, garden and hearth. Keokuk, Iowa. s-m. v. 1. 1882–1883.
2 Farm money maker. Established as
(a) American truck farmer. St. Louis, Mo. m. v. 1–4. Oct. 1903–Dec. 1905.
Continued as
(b) Farm money maker. St. Louis, Cincinnati, and Dallas. m. v. 5–7. Jan. 1906–June 1907//?
1 Farm, orchard and garden. Ingersoll, Ontario. m. v. 1–3? 1893–1896.
1 Farmer and fruit grower and semi-weekly Times-Union. Established as
(a) The Florida dispatch. Jacksonville, Fla. w. v. 1–20? 1869–Jan. 1889.
United with Florida farmer and fruit grower and
Continued as
(b) The Florida dispatch, Farmer and fruit grower. Jacksonville, Fla. w. n. ser., v. 9, no. 4, v. 1, no. 1–v. 5, no. 17. Jan. 28, 1889–Apr. 29, 1893. n. ser., v. 9, no. 4 is Jan. 28, 1889, v. 1, no. 1 is Feb. 7, 1889.
Continued as
(c) The Florida farmer and fruit grower (with which is incorpornted the Florida dispatch). Jacksonville, Fla. w. n. ser., v. 5, no. 18–May 6, 1893–
Continued as
(d) Farmer and fruit grower. Jacksonville, Fla. w. v. 1908.
Continued as
(e) Farmer and fruit grower and semi-weekly Times-Union. Jacksonville, Fla. s-w. v. 1909– . Continued as Semi-weekly Times-Union, not agricultural.
1 Farmer and fruit grower. New Whatcom, Wash. s-m. v. 1–1890 (1896)–1897.
2 Farmer and fruit grower. Carterville, Ill. w. v. 1. 1899.
1 The Farmer and fruit grower of Southern Illinois. Anna, Ill. s-m. (1877), w. (1877–1885). v. 1–8? 1877–1885?

Merged in Prairie farmer, 1885?

1 The Farmer and gardener. Augusta, Ga. s-m. v. 1–2? 1871–Mar. 1873. Suspended but revived almost immediately, Oct. 1872.

1 Farmer and gardener. Lancaster, Pa. m. v. 1– Jan. 1873–

1 Farmer and planter. Pendleton, S.C. v. 1–12, 1850–1861//?

1 Farmer and trucker. Berkley, Va. v. 1, 1895–1896.

2 Farmer and trucker. Hastings, Fla. v. 1, 1904//.

1 The Farmer's companion and horticultural gazette. Detroit, Mich. m. v. 1–4, no. 3. Jan. 1853–Sept. 1854 Merged in Michigan farmer, Oct.? 1854.

2 Farmer's Cotton Union news. Greenville, S.C. Union Printing and Pub. Co. w. 1904–//?

1 Farmer's fruit farmer. Pulaski, N. Y. m. v. 1–3? 1899–1902.

1 Farmer's journal and horticultural magazine. Rutland, Vt. m. v. 1–April–Aug. 1833.

2 Farmer's life and fruit journal. Established as
(a) Farmer's life. Absorbed Fruit journal and Western farmer and assumed its numbering. Merged into Western farm life. Denver. 1. 1913. v. 9, no. 1–5, Mar.–July 1914//.

1 Farmers marketing journal. Dallas, Tex. Farmers Marketing Assoc. of America. 1–5. 1927–1932//?

2 Farm-garden and poultry. Philadelphia (1902), Hommonton, N. J. (1902–). m. v. 1–?. 1902– //?.

2 Feed and feeding digest. St. Louis, Mo. Grain and Feed Dealers National Assoc.

2 Feed bag. Milwaukee. Aug., 1925–

2 Feed knowledge. Kansas City, Mo. v. 1–4. 1909–1911//?

1 Feed stuffs. New York. C. G. Wood, ed., v. 1–4 1922–1926//.

1 Feed trader and retailer. St. Louis, Mo. Grain and Feed Dealers National Assoc. Oct. 1946–

2 Feeding and marketing (began as Stock yards nugget, 1912–1916). Kansas City, Kans. v. 1–12 1912–1919//?

1 Feedingstuffs. Established as
(a) American hay, flour and feed journal. 1901–1914.
Continued as
(b) Flour, hay, grain and feed. 1914–1917. New York. 1917–1922//.
Superseded by Feed stuffs.

1 Feedstuffs. Minneapolis, Minn. May, 1929–

1 Field of horticulture. Little Rock, Ark. v. 1–7?, 1893–1899//?

2 Field's seed sense. Shenandoah, Iowa. v. 1–17. 1913–1929//?

1 Fig and olive journal. Established as
(a) Olive journal. v. 1–2. 1916–1917.
Continued as
(b) Fig and olive journal. Los Angeles. v. 3–6. 1918–1922//.
Merged into Orchard and farm.

2 Floral and family magazine. Mobile, Ala. m. v. 1–4? 1899–1903.

1 Floral and fruit magazine. Washington, D. C. m. v. 1–2? Mar.? 1878–1879.

1 The Floral cabinet. New York. m. v. 1– Jan. 1843–//?

1 The Floral Californian. Petaluma, Calif. q. v. 1–2? Jan. 1, 1878–1879?

1 Floral home. St. Louis, Mo. m. v. 1. 1888–1889.

1 The Floral instructor. Ainsworth, Iowa. v. 1–2. 1880–1882//?
2 Floral life. Philadelphia (1903–1905), Springfield, Ohio (1906–1908). m. v. 1–6, no. 7. Jan. 15, 1903–July 1908//
Succeeded Meehan's monthly; v. 1–2 are called old ser. no. 139–161
Absorbed the Mayfower, Oct. 1906
Combined with Household journal, Springfield, Ohio, Aug. 1908.
1 The Floral magazine and botanical repository. Philadelphia. v. 1, no. 1–5. 1832–1834//.
1 The Floral monthly. Portland, Maine. m. v. 1–2. 1880–1881//?
1 Floral world. Chambersburg, Pa. m. v. 1. 1898–1899.
1 Floral world. Highland Park (Chicago), Ill. v. 1–3. 1883–1886; absorbed by Popular gardening for town and country, Jan. 23, 1886.
1 Floral world. New Brighton, Pa. m. v. 1. 1881.
2 Floral world. Springfield, Ohio. m. v. 1–2, no. 5. Oct. 1901?-Feb. 1903//; merged in Home and flowers, Mar. 1903.
1 Florida farmer and fruit grower, with which is incorporated the Florida dispatch. Jacksonville, Fla. 1–9, 1869–1881. n.s. v. 1–9, 1882–1889. n.s. (s. 3) v. 1–11, 1889–1899//. Also called Florida dispatch and Farmer and fruit grower.
2 Florida fruit and truck grower. Ocala, Fla. (May? 1905–), Como, Fla. (Dec. 1905–). v. 1– 1905–1906.
2 Florida fruit digest. Jacksonville, Fla. 1936–
2 Florida fruit grower and investor. Jacksonville, Fla. (Sept. 1904), Miami, Fla. (Oct. 1904–1905). m. v. 1. Sept. 1904–1905.
1 Florida fruit grower. Highland Park, Fla. w. v. 1–2. 1886–1888.
2 Florida fruit lands review. Kansas City, Mo. m. v. 1– Aug. 1909–//?
2 Florida fruit world. Tampa, Fla. v. 1, no. 1–8. 1928–1929//.
2 Florida fruits and flowers. Bartow, Fla. 1–5. 1924–1926//.
1 Florida grove and garden. Jacksonville, Fla. m. v. 1. 1887.
1 Florida grower. Established as
(a) Florida fruit and produce news, w. v. 1–3, 1908–1911. Jacksonville; Tampa. Florida Growers Publishing, w. v. 4– 1912– .
Also issued as
(b) Florida grower. Monthly edition. Tampa. 1911–1913//.
2 Florida homes and flowers. St. Petersburg, Fla. 1936–
1 Florida horticulturist. v. 1–? -1887//.
1 Florida planter. Ft. Meyers, Fla. All-American Sugar League. 1, 1919–
2 Florida strawberry journal. Lakeland, Fla. m. v. 1–5? April 1902–June 1906.
1 Florist and horticultural journal. Established as
(a) Philadelphia florist and horticultural journal. Philadelphia. m. v. 1, no. 1–12. May 1852–Jan. 1853. No. 10–12 were issued Jan. 1863 without date to bring beginning of v. 2, to Jan. 1853. Continued as
(b) Florist and horticultural journal. Philadelphia. m. v. 2–4, no. 9. Jan. 1853–Dec. 1855//.
1 The Florist's exchange and horticultural trade world. New York. w. v. 1– Dec. 8, 1888–
1 Florist's friend and gardener's manual. Established as
(a) Florist's friend and family visitor. Pleasantville, Pa. m. v. 1. Jan.-Dec. 1874? Continued as
(b) Florist's friend and gardener's manual. Erie, Pa. m. v. 2? 1875//?

2 Florists' news digest. New York. 1938–1939//.
1 Florists' review. Chicago, Ill. w. v. 1– . Dec. 1897– .
(a) 1897–1912 as Weekly florists' review.
2 Flour and feed. Waukegan, Ill. (1900–1904), Milwaukee, Wis. (1904–1910). m. v. 1– Nov. 1900– .
Absorbed Cereals and feed, Milwaukee, April, 1904, and continued at that place.
2 Flour and grain world. Seattle, Wash. v. 1–11. 1916–1926//.
1 Flour trade news. New York. m. v. 1–10, no. 6. June 1902–May 1907//.
Merged in American hay, flour and feed journal, May 1907.
2 Flower and feather. Chattanooga, Tenn. Chattanooga Audubon Society; Elise Chapin Wild Life Sanctuary. Aug. 1945–
2 Flower grower. Established as
(a) Modern gladiolus grower, v. 1–4, 1914–1917. Calcium. Albany, N. Y. Madison Cooper. Williams Press, v. 5– 1918–
2 Flower lore. New York. 1–4. 1916–1920//.
2 Flower talks. New York. N.Y. 1–4, July 1916–June 1920//.
1 Food facts. Chicago. Wheat Flour Institute. Aug. 1930–Food packer. Established as
(a) Canning age. 1922–1943.
Continued as
(b) Food packer and canning age. 1943. Pontiac, Ill. [etc.]. 1920–1958//. Absorbed by Canner/packer.
1 Food, home, and garden. Philadelphia. m. no. 1–9, n. ser. 1–4, no.? Jan. 1889–Jan. 1900// Consolidated with Vegetarian magazine, 1900.
2 Food research. Champaign, Ill. Jan./Feb. 1936–
2 Food topics. Milwaukee, Wis. m. v. 1–3, no. 9. Mar. 1903–Aug. 1905.
2 The Four seasons. Towson, Md. Towson Nurseries, Inc. 1938–
1 Fruit and farm. Oklahoma City, Okla. w. v. 1– 1897–1899.
2 Fruit and farm. Springdale, Ark. m. v. 1. 1900.
1 Fruit and grape grower. Charlottesville, Va. m. v. 1–4, no. 2. Aug. 1886–Sept. 1889//?
2 Fruit and irrigation age. Salinas, Calif. m. v. 1– June? 1903?
2 Fruit and produce dealer. New York. v. 1, no. 1–16. Feb.–July 1920//?
1 Fruit and produce journal. New York. w. v. 1– July 28, 1883–//?
2 Fruit and produce marketer. Portland, Ore. v. 1–7. 1913–1916//.
(a) Fruit and produce distributor, v. 1–5.
2 Fruit and produce. Boston. w. v. 1. 1893–1894.
2 Fruit and vegetable digest. Charles Town, W. Va. July/Aug. 1945–?
Absorbed by the Eastern fruit grower [n.d.].
1 Fruit and vegetable grower. Cheswold, Del. m. v. 1. 1889–1890.
1 Fruit and vegetable grower. Makanda, Ill. m. v. 1–4? 1896–1900.
2 Fruit and vegetable review. Orange, Calif. v. 1– 1939– .
Prior to v. 7 called variously California citrus review and avocado record; California citrus and fruit and vegetable record; California fruit and vegetable review.
2 The Fruit belt. Grand Rapids, Mich. m. v. 1–6, no. 7 (whole no. 54). June 1905–//?
Absorbed The Rural home, Jan. 1906? In 1909 called The Fruit belt and Michigan dairyman.
2 Fruit belt reflex. Makanda, Ill. w. v. 1. 1904.
2 Fruit dealer. Philadelphia. v. 1, no. 1–4. 1930–1931//.
2 Fruit dispatch. New York. Fruit Dispatch Company. v. 1–17. 1916–1931//.

1 Fruit exchange review. Los Angeles, Calif. w. v. 1. 1896–1897.
2 Fruit, farm and flowers. Boonville, Ark. m. v. 1, no. 1–3. Oct. 1–Dec. 1903.
1 Fruit farm. Russellville, Ky. v. 1– Sept. 1879–
1 Fruit farmer. Fruitland, Tex. w. v. 1. 1893.
1 Fruit grower and farmer. Established as
(a) Montana fruit grower. Missoula, Mont. m. (1895–1896), w. (1897–1901). v. 1–9, no. 34. May 1895–Jan. 5, 1900.
Continued as
(b) Edward's fruit grower and farmer. Missoula, Mont. w. v. 9, no. 35. v. 11? Jan. 12, 1900–1901.
Continued as
(c) Fruit grower and farmer. Weekly ed. of Missoulian. Missoula, Mont, w. v. 12?-13? 1902–1903.
2 Fruit grower and gardener. Chicago. m. v. 1–1910.
1 Fruit grower and horticulturist. Lacon, Ill. m. v. 1–2, no. 2. Jan. 1890–Feb. 1891//?
2 Fruit grower. Cove, Ark. m. v. 1–2? 1909–//?
1 Fruit grower. Delta, Colo. m. v. 1–3? 1895–1898?
1 Fruit grower. Lowell, Mass. m. v. 1. 1885–1886.
1 Fruit grower. St. Joseph, Mo. Established as
(a) Western fruitgrower. v. 1–23.
Continued as
(b) Fruit grower and farmer. v. 23–26.
Continued as
(c) Fruit grower. v. 26–28. –1917//.
Merged into American fruit grower.
1 The Fruit growers' journal. Cobden, Ill. (1883–1907), Traynor, Iowa. w. (1883–1886), s-m. (1887–1907), m. (1908). v. 1– 1893–Mar. 1908
Merged in Green's fruit grower, Mar. 20, 1908.
1 Fruit growers' magazine. Harrisburg, Pa. v. 1– Oct. 1904– Eastern edition of Western fruit grower, St. Joseph, Mo.
2 Fruit journal and western farmer. Established as
(a) Intermountain fruit journal. v. 1–7, 1910–June 1913. Grand Junction, Colo. July 1913–Jan. 1914//.
United with Farmer's life to form Farmer's life and fruit journal (SEE).
1 Fruit products journal and American fruit manufacturer. Established as
(a) American vinegar industry, 1921–1922.
Continued as
(b) American vinegar industry and fruit products journal, 1922–1923.
(c) Fruit products journal and American vinegar industry, 1923–1943. New York. Avi Publishing, 1943– v. 29, 1950//.
1 Fruit review. South Pasadena, Calif. m. v. 1. 1895–1896.
1 Fruit trade journal and produce record. Established as
(a) Fruit trade journal. New York. w. v. 1–10, no. 16. 1888–Feb. 10, 1894.
Continued as
(b) Fruit trade journal, dairy and produce record. New York. w. v. 10, no. 17–v. 24, no. 15. Feb. 17, 1894–Jan. 26, 1901.
Continued as
(c) Fruit trade journal and produce record. New York. w. v. 24–87. 1901–1933//.

2 Fruit varieties and horticultural digest. Supersedes Horticultural digest. Wooster, Ohio. American Pomological Society, 1946–
1 Fruit world. Philadelphia. w. v. 1. 1891–1892.
2 Fruit world. South Bend, Ind. v. 1–2 1922–23//.
1 Fruit. Established as
(a) The Grape belt. Dunkirk, N. Y. m. v.1 , 1893.
Also had weekly local edition which later united with Chautauqua farmer to form Grape belt and Chautauqua farmer.
Combined with the Farm and vineyard and American gardener's assistant and continued as
(b) The Grape belt and the Farm and vineyard. Dunkirk, N. Y. m. v. 1894–95.
Continued as
(c) Fruit. Dunkirk, N. Y. m. v. -9, 1895–1900.
2 Fruitage. For fruit men only. Portland, Oreg. m. v. 1–5? 1896–1901.
2 Fruitland. Seattle, Wash. v. 1, no. 1–6, Feb.-Nov. 1922//.
1 The Fruitman and gardener. Established as
(a) The Fruitman. Marcus, Iowa. bi-m. v. 1–5, no. 10. Mar. 1898–Oct. 1902.
Continued as
(b) The Fruitman and garden guest. Mount Vernon, Iowa. m. v. 5, no. 11–v. 8. Nov. 1902–Dec. 1905.
Continued as
(c) The Fruitman and gardener. Mount Vernon, Iowa. m. v. 9–21. 1906–1919//.
1 Fruits and flowers of Oregon and Washington. Portland, Oreg. m. v. 1–2. July 1891–June 1892//.
2 Fruits and flowers. Chautauqua, N. Y. m. v. 1–5? 1906–1910//?
1 Fruits and gardens. Established as
(a) Fruit belt. v. 1–25. 1905–1927. Chicago. American Pomological Society, v. 1–47 1905–1931//.
2 Garden and city. Boston. v. 1, Jan.-Dec. 1875//.
1 Garden and farm. Established as
(a) Success with the garden. Reed City, Mich. m. v. 1– Dec. 1894–1897.
Continued as
(b) Success with the garden and farm. Reed City, Mich. m. v. 1898.
Continued as
(c) Garden and farm. Reed City, Mich. (1898–1899), Springfield, Ohio, and Chicago (1900), Chicago (1901–1902). m. v. – 8, no. 4. –Apr. 1902//.
Merged in Green fruit grower, May 15, 1902.
2 Garden beautiful. Vancouver, B.C. v. 1–8. 1932–1939//.
2 Garden digest. Pleasantville, N.Y. v. 1–14. 1931–1942//.
Merged into Flower grower.
2 Garden, farm and fireside. Syracuse, N. Y. w. v. 1–3? 1900–1902?
2 Garden flowers. Rochester, N.Y. v. 1–2, no. 1. 1928–1929//.
2 Garden forum. Pattonville, Mo. 1– 1939–
2 Garden gossip; recording the activities of gardening in the mid-South. Richmond, Va. June 1925–
2 Garden homes. San Francisco. v. 1, no. 1–3. July-Sept. 1928//. v. 1 no. 1–2 as v. 12 no. 7–8, continuing the numbering of Western homes and gardens. Merged into Better flowers.

1 The Garden magazine. New York, N.Y. m. v.1–39. Feb. 1905–Aug. 1924//. Continued as
(a) Garden magazine and home builder Absorbed Farming.
2 Garden news. Geneva, N.Y. Stern's Nurseries. [n.d.]
1 Garden notes. East Sumner, Maine. m. v. 1–3? 1890–1893.
2 Garden notes. Van Wert, Ohio. no. 1–22. 1920–1926//?
2 Garden, orchard and farm. Vancouver, B.C. 1–11. 1908–1918//? 2 Garden. New Orleans, La. "A magazine of plants and flowers." v. 1, no. 1–2, Mar.–June 1935//.
1 Garden. New York [etc.]. Established as
(a) Gardener's chronicle of America. v. 1–54. 1905–1950.
Continued as
(b) Garden and the gardener's chronicle. v. 54–55. Garden through 1951//.
2 Garden. Seattle, Wash. v. 1, no. 1–27. Oct. 1932–Sept. 1933//.
1 Gardener's monthly and Horticulturist. Ed. by Thomas Meehan. Established as
(a) Gardener's monthly and horticultural advertiser. Philadelphia. m. (Sample no. v. 1, no. 1 issued Oct. 1858), v. 1–17. Jan. 1859–Dec. 1875.
United with the Horticulturist and journal of rural art and rural taste, and continued as
(b) Gardener's monthly and Horticulturist. Philadelphia. m. v. 18–30, no. 1, Jan. 1876–Jan. 1888//.
Merged in American garden, Feb. 1888. 2 The Gardener's magazine and orchard directory. Rochester, N. Y. Prospectus announcing this periodical was published Apr. 1834, but subscription list proved too small, and the matter collected was published in Goodsell's Genesee farmer.
1 The Gardener's magazine. Boston. m. v. 1. Sept. 1854–1855?
2 Gardening magazine. Established as
(a) Flowers and gardens. v. 1, no. 1–8. Jan.–Sept. 1937.
Continued as
(b) Madison Cooper's gardening magazine. v. 1, no. 9–v. 6, no. 10. 1937–1942. Calcium, N.Y. 1942–1943//.
Merged into Gardener's chronicle of America.
1 Gardening. Chicago. s-m. v. 1–33. 1892–1925//.
2 Gardens and modern homes. Established as
(a) Modern homes. v. 1, no. 1–6.
Continued as
(b) Modern homes and gardens. v. 1, no. 7–8. Cleveland, Ohio. v. 1, no. 1–9. 1934–Jan. 1935//.
2 The Ginseng garden. Joplin, Mo. (1903–1905?), Elmhurst, Pa. (1906). q. v. 1–3 no. 2. Mar. 1902–1906//?
2 The Ginseng grower. Wausau, Wis. m. v. 1– Sept. 1904–
2 Ginseng news. Cane Valley, Ky. q. v. 1–2. Mar. 1903–1904.
2 Golden gardens. Los Angeles. California Garden Club Federation. Oct. 1932–
2 Goodfruit grower. Yakima, Wash. Washington State Fruit Commission, 1947–
1 Grain and feed journals consolidated. Established as
(a) Grain dealers' journal, v. 1–65, 1898–1930. Chicago, v. 66– 1931–
1 Grain and feed review. Established as
(a) Cooperative manager and farmer, v. 1–23, 1910–1933. Minneapolis. v. 24– 1934–

2 Grain and hay reporter. Pittsburgh, Pa. s-w. v. 1– 1904–July 1910//?
1 Grain and provision review. Chicago, Ill. 1875–1881.
2 Grain dealers' journal. Winnineg, Manitoba. m. v. 1– June 1905–1907.
2 Grain grower. Westminster, Md. m. v. 1. 1889–1890.
2 Grain growers' journal. Minot, N. Dak. m. v. 1– June 1906–1909.
2 The Grainman's guide. Decatur, Ill. w. v. 1– 1905–July 1906
Combined with The Hay and grain reporter, Chicago, which continued as The National hay and grain reporter.
1 The Grape belt and Chautauqua farmer. Established as
(a) The Grape belt. Brocton, N. Y. (1893–1894), Dunkirk, N. Y. (1895–1901). w. (1893–1894), s-w. (1895–1901). v. 1– 1893–1901.
Continued as
(b) The Grape belt and Chautauqua farmer. Dunkirk, N. Y. s-w. 1902–//?
1 The Grape culturist. St. Louis. v. 1–3, no. 6. Jan. 1869–June 1871//? Ed. by George Husmann.
2 The Grape grower. San Francisco. v. 1– Mar./Apr. 1947–
2 Green gold. Portland, Oreg. v. 1–3. 1943–1946//.
1 Green Mountain Culturist, devoted to agriculture, horticulture, science, education and the mechanic arts. Middlebury, Vt. v. 1, no. 1–5. June-Oct. 1852//.
2 Green thumb. Littleton, Col. Colorado Forestry and Horticulture Assn. Feb. 1944–
2 Grove and garden. Jacksonville, Fla. m. v. 1– 1894–1902.
2 Grower talks. Chicago. George F. Ball, Inc., 1937–
1 Grower. Established as
(a) Canadian grower. Ontario, Ont. Ontario Fruit and Vegetable Growers' Assoc., 1915–
2 Gulf coast citrus fruit grower and southern nurseryman. Established as
(a) Citrus fruit grower and Gulf coast orchardman. Dec. 1910–Feb. 1911. Houston, Tex. v.March 1911–v. 2, no. 2, Feb. 1912//.
2 Gulf coast grower. Foley; Silverhill, Ala. v. 1–2, no. 8. Mar. 1923–Apr. 1925//.
2 Gulf fauna and flora bulletin. Ruston, La. v. 1, no. 1–3. June-Dec. 1899//?
2 Gulf states grower. Established as Gulf coast grower, v. 1–2 no. 4. Mobile, Ala. Gulf Coast Horticultural Society. v. 2 no. 4–6. 1916–1917//.
1 Hammond's floral friend and household journal. Established as
(a) Floral friend and household journal. Sanford, Maine (1884–1885), Banneg-Beg, Maine (1885). m. v. 1–2? 1884–1885.
Continued as
(b) Hammond's floral friend and household journal. Banneg-Beg, Maine (1886), Derry Depot P. O., N. H. (1887–1891), Arlington, R. I. (1892–1894). m. v. 3?-10? 1886–1894.
1 Hawaiian planters' monthly. Honolulu. Hawaiian Sugar Planter's Assoc. 1–28. 1882–1909//. Superseded by Hawaiian planters' record.
1 Hawaiian planters' record. Honolulu. Hawaiian Sugar Planter's Assoc. 1909–
2 Hay and produce trade review. Established as
(a) Hay dealers' review Cohocton, N. Y. m. v. 1. 1901–Mar. 1901.
Combined with the Haymaker, Kansas City, Mo., taking its numbering and continued as
(b) Hay dealers' review and Haymaker. Cohocton, N. Y. m. v. 12, no. 10–v. 13, no.? April. Continued as
(c) Hay and produce trade review. Cobleskill, N. Y. m. v. 13, no.?- 1901.

1 Hay, straw and grain reporter. Sept. 1875.
1 Hay trade journal. Established as
(a) Hay trade journal and directory. Canajoharie, N. Y. w. v. 1–9? 1892–1900. Continued as
(b) The Hay trade journal. Canajoharie, N. Y. w. v. 9–38. 1893–1929//.
1 Haymaker. Established as
(a) Hay-men's gazette. Cherokee, Kans. (1888–1892), Kansas City, Kans. (1893), Kansas City, Mo. (1894) m. v. 1– 1888–1894.
Continued as
(b) Haymaker. Kansas City, Mo. v. -12, no. 9. 1895–Mar. 1901.
Combined with the Hay dealers' review, Cohocton, N. Y., March 1901 to form Hay dealers' review and Haymaker.
2 Herb grower magazine. Established as
(a) American herb grower. Falls Village, Conn. Laurel Hill Farm, 1947–
1 Hoffy's orchardist's companion; or fruits of the United States. Established as
(a) Orchardist's companion. Philadelphia. q. (1841) m. (1841). v. 1. 1841. Continued as
(b) Hoffy's orchardist's companion; or Fruits of the United States. Phildelphia 1. v. 2. 1842–1843//?.
1 Home acres. Established as
(a) Woman's National Agricultural and Horticultural Association Quarterly, 1914–1916. Continued as
(b) Woman's National Farm and Garden Association Quarterly, 1916–1917, 1917–1921, its Bulletin.
Continued as
(c) Farm and garden, 1921–1926.
Continued as
(d) Home acres and countryside magazine, 1931–1933. Grosse Point, Mich. Chicago. Woman's National Farm and Garden Association, v. 21– 1933–
1 Home and flowers, formerly How to grow flowers, consolidated with Success with flowers. Established as
(a) Success with flowers. West Grove, Pa. m. v. 1–14, no. 9. Oct. 1890–June 1904.
Combined with Home and flowers, July 1904, and continued as
(b) Home and flowers, formerly How to grow flowers, consolidated with Success with flowers. West Grove, Pa. m. v. 14, no. 10– v. 16, no. 8. July 1904–May 1906.
Merged in Vick's family magazine, Dansville, N. Y., June 1906.
1 Home and flowers. Established as
(a) How to grow flowers, v. 1–8, 1896–1900. Springfield, Ohio, v. 9–16, 1900–1904//.
1 Home and garden. Bridgeton, N. J. m. v. 1. 1887–1888.
1 Home and garden. Philadelphia, Pa. m. v. 1– June 1901–
1 Home and garden. St. Paul. Established as
(a) American home and garden. Minneapolis. m. v. 1– 1895–Apr. 1897. Continued as
(b) Home and garden. St. Paul (1897–1899, 1901), Minneapolis (1900). m. v. May 1897–1901.
2 Home electra fruit and garden. Oakland, Calif. [1920]-Sept. 1924//.

1 The Home, farm, and orchard. Newburgh, N. Y. w. v. 1–6. 1870–1875//?

1 Home florist. Springfield, Ohio. v. 1–4 no. 4. 1898–1901//.
Merged into Home and flowers (Springfield, Ohio) .

1 Home garden. Concord, N.H. New York. Ameican Garden Guild, v. 1–22, no. 6. Jan. 1943–Dec. 1953//.
Absorbed by Flower grower.

2 Home gardening for the South. Established as Home gardening, 1–3. New Orleans. New Orleans Garden Society, v. 1–12 1940–1952//.

1 Homes and gardens of tomorrow. Established as
(a) Home and garden review. v. 1–13. 1928–1934. Chicago. v. 34–36. 1935–1936//?

2 Homes and grounds. San Francisco. v. 1–2. 1916–1917//.

2 Homes and grounds. Waycross, Ga. v. 1, no. 1–4. 1915–1916//.

2 Homes of the west. Established as
(a) Garden quarterly. v. 1–5 no. 1, 1933–Apr./June 1937.
Continued as
(b) Homes of the west and Garden quarterly. v. 5 no. 2, Sept. 1937. San Francisco. v. 5. no. 3–v. 6 no. 6, Oct. 1937–June 1938//.

1 Hoosier horticulture. Lafayette, Ind. Indiana Horticulture Society, v. 1–35 1919–1953//.

1 Hop circular. 1877–1883.

1 The Hop growers' journal. East Springfield, N. Y., and New York. m. v. 1– Oct.? 1862–1863.

2 Hop growers' reporter. Established as Aurora Borealis, 1900; continued as Hop growers' reporter. Aurora, Oreg. w. 1908.

1 The Hop journal. New York. w. (in summer), m. (in winter). v. 1– Jan. 1863–

1 Horticultural advertiser. Cincinnati. m. v. 1. Jan. 1855–//?
Succeeded Western horticultural advertiser?

1 Horticultural art journal. Rochester, N. Y. m. v. 1–6, no. 11. Jan. 1886–Nov. 1891//.

1 Horticultural digest. Ames, Iowa. American Pomological Society. v. 1–2. 1944–1945//. Superseded by Fruit varieties and horticultural digest.

2 Horticultural edition of Evening observer; Tuesday and Friday issues of each week have same number.

1 Horticultural gleaner. Austin, Tex. m. v. 1–2. 1896–1897//?

1 Horticultural magazine. Montreal. m. v. 1– 1899–//?

1 Horticultural magazine. Rochester, N. Y. m. v. 1– Jan. 1888–//?

2 Horticultural marketplace. Rochester, N.Y. v. 1, no. 1–7. Sept. -Oct. 1910//?

1 Horticultural monthly. Morrisania, N. Y. m. v. 1, no. 1–9? Jan.-Sept. 1859//.

2 Horticultural news. Detroit, Mich. q. v. 1–13. Oct. 1934–winter 1946/47//.

2 Horticultural newsletter. Ardmore, Pa. May 25, 1933–

1 The Horticultural register and gardener's magazine. Boston. m. v. 1–4. Jan. 1835–Dec. 1838//
Merged in New England farmer, which continued as New England farmer and Horticultural register.

1 The Horticultural register. Philadelphia. 1867–1868.

1 The Horticultural reporter. Denison, Tex. June 1888. Special edition of the Denison, Tex., Sunday gazetteer, June 1888, for the Texas state horticultural society meeting, June 27–28, 1888.

1 The Horticultural review and botanical magazine. Established as
(a) Western horticultural review. Cincinnati. m. v. 1–3. Oct. 1850–Sept. 1853. Western horticuitural advertiser published and distributed free with Western horticultural review. Continued as
(b) The Horticultural review and botanical magazine. Cincinnati. m. v. 4 (n. ser., v. 1), no. 1–12. Oct. 1853–Dec. 1854//. Formed by the union of the Western horticultural review and The American botanist, prospectus of which appeared in June/ 1853 but which was never published.
1 Horticultural topics. New Brunswick, N.J. New Jersey Assoc. of Nurserymen. 1– 1937–
1 Horticultural trade journal. Floral Park, N. Y. m. v. 1–2? 1895–1897.
1 Horticultural world. Kansas City, Mo. m. v. 1. 1887.
1 Horticulture. Boston. w. v. 1–37. Dec. 3, 1904–July 25, 1923. New series v. 1– Aug. 1923–
1 Horticulture. Cuyahoga Falls, Ohio. m. v. 1–2? Mar. 1896–1898.
1 The Horticulturist and farm journal. Warren, Ohio. m. v. 1. 1887.
1 The Horticulturist and farmer. Mexico, Mo. (1867), De Soto, Mo. (1868). m. (1867), s-m. (1868). v. 1. 1867–Dec. 1868.
1 Horticulturist. Pilot Point, Tex. m. v. 1–2, no. 6. 1889–Mar. 1891//?
1 House and garden. Cleveland, Ohio. m. v. 1–2? Jan. 1872–1873; made up from pages of weekly Ohio farmer.
1 House and garden. New York. m. v. 1. 1878–1879.
1 House and garden. Philadelphia (1901–1909), New York (1910). m. v. 1–June 1901–
1 House beautiful's practical gardener. New York. 1– 1940–
1 The Household journal with which is incorporated Floral life. Established as
(a) The Household journal. Springfield, Ohio. m. v. 1. January-July 1908. Absorbed Floral life and continued as
(b) The Household journal with which is incorporated Floral life. Springfield, Ohio. m. v. 1–14. Aug. 1908–1917//.
1 Illini horticulture. Quincy, Ill. v. 1– . 1913– .
(a) v. 1–37, Feb. 1948 as Illinois horticulture
(b) April 1948 as Illini horticulture
1 Illinois Farmer. St. Paul, Minn. and Chicago. Established as
(a) Farmer; devoted to the advancement of agriculture in all its brances. v. 1–4 1886–88. Continue as
(b) Orange Judd Farmer. v. 4–72. 1888–1924. Continued as
(c) Orange Judd Illinois Farmer. v. 72–75. 1924–27. Continued as
(d) Illinois farmer. v. 75–78. 1927–30//. Merged into Prairie farmer.
1 Illinois Fruit Exchange news. Carbondale [etc.]. v. 1–5, 1922–1928. n.s. v. 1–11. 1936–1947//.
1 Illinois horticulturist. Shelbyville, Ill. q. v. 1. 1884–1885.
2 Illinois vegetable growers bulletin. [n.p.] Illinois State Vegetable Growers Assoc. May 1941–
1 Illustrated floral work. Rochester, N. Y. q. v. 1–2? Jan.? 1874–1875.
2 Indiana gardens. Winamac [etc.]. Garden Club of Indiana. 1– 1939–
1 Indiana horticulturist. Alert, Ind. m. v. 1. 1891–1892.
2 Iowa gardens. Council Bluffs, Iowa. Federated Garden Clubs of Iowa. m. Sept. 1946–

2 Iowa horticulture; a monthly bulletin of plant life. Des Moines, Iowa. m. v. 1. Dec–Jan. 1908.

2 The Irrigation fruit grower and agriculturist. Established as
(a) The Western slope fruit grower. Paonia, Colo. m. v. 1– Nov. 1905–1906.
Continued as
(b) Colorado fruit grower. Paonia, Colo. (1907–1908), Grand Junction, Colo. (1909–June 1910). m. v. 2?-5, no. 6. Jan. 1907–June 1910.
Continued as
(c) The Irrigation fruit grower and agriculturist. Denver. m. v. 5, no. 7. July 1910//?

1 Journal of agriculture and horticulture. Established as
(a) Illustrated journal of agriculture. v. 1–18. 1879–1897. Montreal. n.s. v. 1–40. 1898–1936//.

2 Kansas gardener. Salina, Kan. 1949–

2 Kansas gardens. McPherson, Kan. Kansas Associated Garden Clubs. v. 1–3. 1927–1929//?

1 Kansas horticulturist. Iola, Kans. -June 1896.

2 Kentucky horticulture. Henderson, Ky. q. v. 1–10. 1924–1934//.

2 The Kernel, incorporating National nut news. Decatur, Ill. v. 1–3. 1938–1940//. Union List of Serials lists as "the business magazine of the pop corn industry." Suspended Nov. 1939–July 1940. Merged into Peanut journal.

2 Kernel . . . (The Wheat kernels). Calgary, Alberta. Jan. 1928–

2 Kernel. Kansas City, Kan. Equity Union Grain Co. 1–4. 1928–1931//.

1 Ladies' floral cabinet and pictorial home companion. New York. m. v. 1–16. no. 1. Jan. 1, 1872–Jan. 1, 1887//
United with American garden, Jan. 1887.

2 Lake district farms and gardens. Duluth, Minn. Oct. 1939–

2 Lake district horticulturist. Established as
(a) Lake district berry culture. v. 1–2. 1937–1938. Duluth, Minn. v. 2 no. 2/3–5, 1938//.
Superseded by Lake district farms and gardens.

1 Landmarks [Horticulture]. (Iona) Peekskill, N. Y. m. v. 1. no. 1–8. Aug. 1862–Mar. 1864//.

2 Landscape gardening. –Oct. 1900.
Combined with Park and cemetery, which continued as Park and cemetery and Landscape gardening, Nov. 1900.

1 Leaf-Chronicle. Established as
(a) Tobacco leaf. Clarksville, Tenn. w. v. 1–1869–1890.
Combined with Chronicle and continued as
(b) Tobacco leaf-Chronicle. Clarksville, Tenn. w. 1891–1899.
Continued as
(c) Leaf-Chronicle. Clarksvllle, Tenn. d. and s-w. 1899–1910//?

2 Leaf tobacco news. Viola, Wis. w. v. 1–1906–1908.

1 Leaflet. [Floriculture]. South Haven, Mich. m. v. 1–4? 1890–1894.

2 Lone star gardener. Tyler, Tex. Texas Garden Clubs. 1949–

1 Lookout daisy [Floriculture]. Lookout Mountain, Tenn. m. v. 1–2? 1897–1899.

2 Los Angeles daily fruit world. Los Angeles, Calif. d. v. 1– June 3, 1903.

1 Louisiana cotton boll. Vermilionvllle, La. w. v. 1–1872–//?

1 Louisiana farmer and rice journal. Crowley, La. s-m. v. 1–3? 1896–1899.

2 M.P.G. news. Presque Isle, Me. Maine Potato Growers, Inc. Aug. 1946–
1 Magazine of horticulture, botany and all useful discoveries and improvements in rural affairs. Established as American gardeners' magazine and register of useful discoveries and improvements in horticulture and rural affairs, v. 1–2, 1835–1836. Boston and New York. v. 3–34, 1837–1868//. Merged into Tilton's journal of horticulture. v. 3–? edited by C. M. Hovey.
1 Manitoba horticulturist. Winnipeg, Manit. Manitoba Horticultural and Forestry Assoc. 1–8. 1914–1921//.
1 Market and farm. Syracuse, N. Y. v. 1–? 1884–1887//?
1 Market garden. Minneapolis. v. 1–? 1893–1903//.
1 Market growers' journal. Established as
(a) Weekly market growers journal. Merged with American vegetable grower. Louisville, Ky. later Akron, Ohio. Babcox Publications, v. 1–86 1907–1957//.
1 Market journal. New York. v. 1–? 1879–1887//.
1 Market review and farm journal. Akron, Ohio. v. 1–? 1894–1899//.
2 Marketmen's journal. Milwaukee. v. 1– 1903–//?
1 Maryland fruit grower. College Park, Md. Feb. 1931–
1 Maryland tobacco grower. Baltimore. Dec. 15, 1928–
1 Mayflower. Floral, N. Y. (1885–1886, 1889–1906), Queens, N. Y. (1887–1888). m. v. 1–21, no. 9? Jan. 1885–Sept. 1906//
Merged in Floral life, Oct. 1906.
1 Meehan's monthly; a magazine of horticulture, botany and knidred subjects. Philadelphia. 1–12, 1891–1902//.
Superseded by Floral life.
2 Meehans' garden bulletin. Germantown, Pa. m. v. 1–5. 1909–1914//.
1 Melon growers monthly. Charleston, Mo. v. 1– Nov. 1899–//?
1 Merchants' and planters' price current. Mobile, Ala. 1861.
1 Miami valley horticulturist. Dayton, Ohio. m. v. 1– 1899–1903.
1 Michigan fruit-grower and practical farmer. Established as Practical farmer and fruit grower, v. 1–4, 1893–1896. Grand Rapids, Mich. v. 5–7? 1897–1899//.
2 Michigan gardener. Ludington, Mich. 1905–//?
2 Michigan gardener. Royal Oak, Mich. Michigan Horticultural Society, v. 1–13 1934–1947//?
1 Michigan gardening and fruit grower. Decatur, Mich. v. 1– 1896–//?
2 Michigan horticulture. Fenville, Mich.; Michigan state horticultural society. v. 1, May 1911–June 1912//?
1 Mid-south cotton news. Memphis, Tenn. Established as
(a) Cotton Association news. 1–10. 1922–1932.
Continued as
(b) Mid-south cotton association news. v. 10–12. 1932–1935. Little Rock; Memphis. v. 12–16. 1935–1939//.
Continued as
(c) Mid-south cotton news. New series, v. 1– 1940–
2 Midwest fruitman. Wathena, Kan. v. 1–12, no. 8. June 1927–Feb. 1939//?
1 Milling and grain news. Established as Spring wheat milling news, v. 1–2, 1902–1903. Omaha and Sioux Falls, S. Dak. v. 2–42 1903–1923//.
1 Minnesota farmer and gardener and education journal. Established as
(a) Minnesota farmer and gardener, v. 1, 1860–1861. St. Paul. v. 2 1862//.
1 The Minnesota horticulturist. St. Paul, Minn.; Minnesota State Horticultural Society. v. 1– 1866–.

2 The Mississippi planter. Gulfport, Miss. w. v. 1– Feb. 6, 1909–

1 The Mississippi planter and mechanic. Grenada, Miss. (1857), Jackson, Miss. (1858). m. v. 1–2? Apr. 1857–March? 1858.

1 The Missouri and Arkansas farmer and fruitman. Established as
(a) Kansas City progress and western farm journal. Kansas City, Mo. w. v. 1– 1888–1894.
Continued as
(b) Missouri and Arkansas farmer and fruitman. Kansas City, Mo. m. 1895– //?

1 Modern farming. Established as
(a) Southern tobacconist and manufacturers' record. Durham, N. C. (1888–1889), Richmond (1890?-1904). w. v. 1–16, no. 30. 1888–1905. Continued as (b) The Southern tobacconist and modern farmer. Richmond. w. v. 16, no. 31–v. 19, no. 9. Nov. 1904–Sept. 1907.
Continued as
(c) Modern farming. Richmond. m. v. 20, no. 10–v. 23, no. 5. Oct. 1907–May 1909//. Consolidated with Progressive farmer, April 23, 1909.

1 Modern farming. Formed by the union of Trucker and farmer, and Modern sugar planter, continuing numbering of the latter. Merged into Southern ruralist. New Orleans. Modern Farming Publishing, George E. Nelson, ed., v. 44–59 1914–1929//.

1 Montana fruit grower. Missoula, Mont. v. 1–9. April 1895–Jan. 1900//.

2 Montana wheat grower. Lewiston, Mont. v. 1–2. 1923–1925//.

1 The Monticello farmer and grape-grower. Charlottesville, Va. m. v. 1–5? 1883–1888.Napa Valley wine press. St. Helena, Calif. 1949–

2 Mountain grower. Martinsburg, W. Va. m. v. 1– 1930–

2 Nation's garden. Wilmington, N.C. m. v. 1 no. 1–10. Apr. 15, 1924–March 1925//?

1 The National farmer. Bay City, Mich. Established as
(a) The Michigan sugar beet. Bay City, Mich. w. v. 1–6, no. 9. Mar. 10, 1899–May 13, 1904. Continued as
(b) The Sugar beet culturist and dairy advocate. Bay City, Mich. w. v. 6, no. 10–v. 8, no. 45. May 20, 1904–Jan. 25, 1907.
Continued as
(c) The National farmer. Bay City. Mich. w. v. 8–18. 1907–1918//.

1 National fruit grower and gardener. St. Joseph, Mich.; Chicago. Established as
(a) Central States fruit grower, v. 1–4, 1896–1899.
Continued as
(b) National fruit grower, v. 5–15, 1899–1910//.

2 National grain grower and Dakota trade journal. Also called National grain grower and Equity farm news. Fargo, N.D. 1–7 1910–1916//?

2 National grain growers' journal. Fremont, Nebr. v. 1– 1904–//?

1 National hay and grain reporter. Established as
(a) Hay and grain reporter, v. 1–? 1901–1906.
Merged into Price current-grain reporter, later Grain world. Chicago. 1907–1914//.

1 National horticultural magazine. Henning, Minn.; Washington, D.C. American Horticultural Society. Aug. 1922–Dec. 1959//.
Continued as American horticultural magazine.

1 National horticulturist. Cambridge, Md. v. 1–3? 1890–1893//.

2 National horticulturist. Council Bluffs, Iowa. v. 1–4 1909–1911//.

1 National nurseryman. Merged into American nurseryman. Rochester, N. Y. Hatboro, Pa. American Association of Nurserymen, v. 1–47 1893–1939//.
1 National plant, flower and fruit guild magazine. New York. National Plant, Flower and Fruit Guild, v. 1–23 1912–1933//.
1 National provisioner. New York and Chicago. American Meat Institute, v. 1– 1891–
2 Nebraska garden news. Scottsbluff, Nebr. 1948–
2 Nebraska horticulture. Lincoln. Nebr. v. 1–10. 1911–1921//.
1 Nebraska horticulturist. Established as Nebraska horticulturist and silk journal, v. 1–2, 1883–1885. Bowers, Nebr. v. 3–? 1886–1893//.
2 Nebraska wheat grower. Hastings, Nebr. s-m. 1924–1927//.
1 New England farmer, and horticultural register. Boston. Battleboro, Vt. v. 1–24, 1822–1846. v. 47–93 1865–1913//.
1 New England florist. Established as
(a) Boston flower market and New England florist. Boston. w. v. 1, no. 1–13. Mar. 7–May 28, 1896. Continued as
(b) New England florist. Boston. w. v. 2–5, no. 13. June 4, 1896–May 25, 1899//.
Has cover title New England florist, but retains title Boston flower market and New England florist as inside title until v. 4, no. 48.
2 The New England tobacco grower. Hartford, Conn. m. v. 1–8, no. 4. Mar. 1902–Dec. 1905//.
2 New Jersey farm and garden. Established as Garden State farmer. v. 1–2 no. 26, 1930–Nov. 1931. Sea Isle City, N.J. v. 2 no. 27, Dec. 1931–
2 New Jersey gardens. Pittsfield, Mass. [etc.] Federated Garden Clubs of New Jersey. v. 1–5. 1930–1932//.
2 New York nursery notes. Geneva, N.Y. irreg. [n.d.]
2 New York State fruit grower. Albion. Medina, N. Y., New York Fruit Grower Publishing, Co., 1918–
2 News for nurserymen. Louisiana, Mo. m. v. 1–8 no. 2 [i.e. no. 3], Aug. 1924–Oct. 1931; Sept. 1935–Sept. 1936//.
1 Niagara farmer and fruit grower. Suspension Bridge, N. Y. w. v. 1. 1877–1878.
2 North American horticulturist. Monroe, Mich. m. v. 1–8. 1895–Aug. 1902//?
1 North Carolina planter. Raleigh, N. C. m. v. 1–4, no. Jan. 1858–1861? v. 4, no. 4 is April 1861.
1 The Northern farmer and practical horticulturist. Newport, N. H. bi-w. v. 1–2. July 7, 1832–Aug. 23,1834//; First three numbers irregular, July 7, Aug. 25, Oct. 6, 1832, after these bi-w.
2 Northern fruit grower. Howard Lake, Minn. m. March 1911–
1 Northwest dairyman and farmer. Established as
(a) Northwest horticulturist, v. 1–3, 1888–1890.
(b) Northwest horticulturist, agriculturist and stockman, v. 4–7, 1891–1894.
(c) Northwest horticulturist, agriculturist, and dairyman, v. 8–31, 1895–1917.
(d) Northwest dairyman and horticulturist. v. 31–35. 1917–1921. Tacoma and Seattle, Wash. v. 35–47. 1921–1932//.
2 Northwest fruit grower. Seattle, Wash. v. 1–4. 1920–1922//.
2 Northwest fruit grower; the authoritative magazine of the deciduous fruit industry. Wenatchee, Wash. Established as Wenatchee fruit. v. 1–11. 1929–1939//?
1 Northwest gardens. Seattle. Mar. 1933–
1 Northwestern farmer and horticultural journal. Dubuque, Iowa. m. v. 1–6. 1857–1862//.

Merged into Iowa homestead.

1 Nurseries and orchards. Louisiana, Mo. Stark Bros. Nurseries. 1894–April 1895.

2 Nurseryman. Des Moines, Iowa. m. v. 1 no. 1–5. Apr. –Sept. 1929//.

2 Nurserymen's news. College Park, Md. Maryland Nurserymen's Assn. Dec. 1942–

1 The Nut-grower; devoted to the interests of the Southern nut-growers' association. Poulan, Ga. m. v. 1–17. 1902–1919//.

2 Nut grower. Downingtown, Pa. Established as Nut tree growers guide, v. 1–2 no. 5. m. v. 1–6 no. 8. Aug. 1924–Apr./May 1930//.

2 Nut kernel. [n.p.] Pennsylvania Nut Growers Assoc. 1948–

2 Ohio cultivator. Columbus, Ohio. v. 1–22. 1845–66//.
Merged into Ohio farmer.

2 Ohio farmer. Cleveland, Ohio. v. 1– . 1852– .

1 Ohio horticulturist. Warren, Ohio. Absorbed by Popular gardening before 1888.

1 Oil mill gazetteer [Cotton seed oil]. Schulenberg, Tex. (1897 (1906)-1906), Brownsville, Tex. (1907–1910). m. v. 1– 1897 (1906)-

2 Oklahoma cotton grower. Oklahoma City, Okla. 1– 1921–

2 Oklahoma gardener. Oklahoma City. Oklahoma Assn. of Garden Clubs. 1– 1933–

1 Ontario fruit grower. Ontario, Calif. w. v. 1. Dec. 1882–1885?

1 Orange blossom. Santa Ana, Calif. m. v. 1. 1891–1892.

1 Orange grower. Rialto, Calif. m. v. 1–1888 (1892)-1892; continued as weekly edition of Orange belt, 1893.

1 Orange growers' journal. Riverside, Calif. June 1884–

1 Orchard and farm. Established as
(a) Orchard and farm. Petaluma, Calif. m. v. 1–5? 1886–1891.
Continued as
(b) California orchard and farm. Petaluma, Calif. m. (1891–May 1896), w. (June 1896–1905?). v. 6?- 1891–1905? Absorbed Western bee journal, Aug. 1905.
Continued as
(c) Orchard and farm. San Francisco, Calif. m. v. –10, no. 5. 1906–May 1910.
No volume numbers. May 1906 printed at Nevada City, Calif.
June 1910 absorbed Irrigation and continued as
(d) Orchard and farm; combined (June: Consolidated) with Irrigation. San Francisco, Calif. m. v. 10–34. 1910–1922//?

1 The Orchard and garden. Little Silver, N. J. m. v. 1–14. Jan. 1879–1892//? absorbed American wine and fruit grower, Nov. 1885
Absorbed American florist and farmer, Dec. 1885.

2 Orchard and garden monthly. Quincy, Ill. m. v. 1. Oct. 15, 1905–1906?

1 Orchard and mill. Mianus, Conn. q. v. 1– 1889–1902.

1 The Orchard and vineyard. Peru, Nebr. (1869–1871), Grand Island, Nebr. (1871–1872). m. v. 1–3? 1869–1872.

1 Orchard fruits. Effingham, Ill. m. v. 1. 1892–1893.

1 Orchard, vineyard, and berry garden. Cawker City, Kans. m. v. 1–3 ? 1885–1888.

1 Orchardist and agricultural advertiser. Tyler, Tex. w. v. 1– 1890 (1895)-//?

2 Oregon apple news. Corvallis, Ore. irreg. v. 1, no. 1– Mar. 1910–July 1910//?

2 Oregon grower. Salem, Ore. 1–5. 1919–1924//.

2 Oregon nurseryman and florist. Corvallis, Oreg. v. 1–6. 1946–1951//.

2 Organic farmer. United with Organic gardening to form Organic gardening and farming. Emmaus, Pa. Rodale Press, v. 1–5 1949–1953//.

2 Organic gardening. United with Organic farmer to form Organic gardening and farming. Emmaus, Pa. Rodale Press, v. 1–21 1942–1953//.

2 Our garden journal. New York. Elinore E. Harde, ed., 1–3 1917–1922//?
1 Our horticultural visitor. Established as
(a) Southern Illinois horticultural visitor. Kinmundy, Ill. q. v. 1, no. 1. May 1895.
Continued as
(b) Our horticultural visitor. Kinmundy, Ill. m. v. 1, no. 2–v. 12. July 1895–1906//.
2 Over the garden wall. Cleveland. Ohio Federation of Negro Garden Clubs, 1944–//?
2 Ozark farm and fruit belt. Siloam Springs, Ark. m. v. 1– 1897 (1907)-1909.
1 Ozark farmer grower. Established as
(a) North Arkansas farmer and fruit grower. Mammoth Spring, Ark. s-m. v. 1– 1896–1899. Continued as
(b) Ozark farmer grower. Mammoth Spring, Ark. s-m. v. 1899.
1 Ozark field and orchard. Cabool, Mo. m. v. 1–2? 1897–1899.
1 Ozark mountain fruit and farm. Rogers, Ark. m. v. 1–2? 1895–1896.
1 Pabor Lake pineapple. Pabor Lake Colony, Avon Park, Fla. 1893–1902//. Incorporated with the South Florida sun 1903.
2 Pacific coast garden. San Francisco. v. 1–5. 1932–1935//.
2 Pacific coast nurseryman. Portland, Oreg. Pacific Coast Assn. of Nurserymen. v. 1–4. 1935–1938//.
2 Pacific Coast nurseryman and garden supply dealer. Established as Pacific coast nurseryman. Arcadia, Cal. California Assoc. of Nurserymen and California Cactus Growers Assoc., 1942–
1 Pacific fruit grower. Los Angeles, Calif. m. v. 1–2, no. 2. Apr. 1887–Feb. 1888//.
1 Pacific fruit world. Established as
(a) Fruit world. Los Angeles. w. v. 1– 1897–Dec. 1899?
Continued as
(b) Pacific coast fruit world and farm home journal. Los Angeles (Los Angeles and San Francisco 1901). w. v. -14, no. 3. Jan. 1900?-June 1903?
Continued as
(c) Pacific fruit world. Los Angeles. w. v. 14, no. 4–v. 32, no. 18. 1903?-July 2, 1910/? v. 27–29 omitted in numbering; 1910 has caption title, Pacific weekly fruit world; has daily editions Los Angeles daily fruit world, 1903–1906?, and San Francisco daily fruit world, 1903–1906? New York daily fruit world, 1903–?
2 Pacific hop grower. Mt. Angel, Ore. Established as Oregon hop grower. v. 1–7. 1933–1940//?
1 Pacific wine, brewing and spirit review. Established as
(a) San Francisco merchant, v. 1–22, 1880–1889.
(b) Merchant and viticulturist, v. 22–23, 1889–1890.
(c) Pacific wine and spirit review, v. 24–55, 1890–1913. San Francisco. suspended 1919–1935 v. 56–61 1913–1935//.
1 Packer. Omaha, Nebr. v. 1–? 1894–1900//?
1 Park's floral magazine (began as Park's floral gazette, 1871–1878). Mt. Vernon, Ohio. Fannettsburg. Libonia. La Park, Pa. 1878–//?
2 The Peach grower, fruit culturist, and truckers magazine. Savannah and Atlanta, Ga. m. v. 1–5, no. 4. Mar. 1903–June 1907//?
1 Peach growers' journal and apple trade review. Sussex, N. J. (1899–1903), Middleport, N. Y. (1903–1909). m. v. 1–5, no. 12. Sept. 1899–Oct. 1904//? v. 5, no. 2–12, also numbered n. ser., v. 1, no. 1–8.

2 Peanut journal and nut world (began as Peanut journal, v. 1–10, 1921–1931). Suffolk, Va. 1931–
1 The Pennsylvania farmer and gardener. Established as
(a) The Farmer and gardener. Philadelphia. m. v. 1–5. Sept. 1859–Dec. 1863. New series began July 1860 with v. 2 (n. ser., v. 1).
Continued as
(b) The Pennsylvania farmer and gardener. Philadelphia. m. v. 6, no. 1–6. Jan.–June 1864//.
2 Pioneer pecan press. Bend; San Saba, Tex. 1–6. 1921–1928//.
2 Plant life. Arcadia. Stanford, Calif. American Plant Life Society, 1945–
2 Plant talk. Boston. m. v. 1– Jan. 1908–1 The Planter. Columbia, S. C. w. v. 1. Jan.-July 1843.
2 Planter and ginner. Waco, Tex. m. v. 1–2? 1898–1900.
1 Planter's guide [Floriculture]. Ainsworth, Iowa. m. v. 1–2? 1886–1888.
2 Planter's magazine. Jackson, Ala. m. v. 1– Sept. 1904–1907.
1 The Planter's journal. Established as
(a) Cotton planter's journal. Memphis, Tenn. m. (1897–1899), s-m. (1899–1902. v. 1–5, no. 12. Nov. 1897–June 15, 1902.
Continued as
(b) Cotton and farm journal. Memphis, Tenn. s-m. v. 5, no. 13–v. 6, no. 5? July 1, 1902–Mar. 15, 1903?
Continued as
(c) The Planter's journal. Memphis, Tenn. m. (July 15, 1903–June 1904), w. (July 13, 1904–May 19, 1905), s-m. (June 1905–July 1910). v. 12. 1903–July 1910//?
Absorbed Tennessee poultry journal, Arkansas cultivator (Jan 15, 1902) and American ginner; incorporated under name Cotton planters' journal, Jan. 1906.
1 The Planters' journal. Vicksburg, Miss. (1880), Vicksburg, New Orleans and Memphis (1881–1884), Vicksburg (1885–1886), Birmingham, Ala. (1887–1888). m. v. 1–18? Jan. 1880–1888.
1 Planters' journal. Ashville, Ala. w. v. 1– 1889 (1893)-1893. Published by the National cotton planters' association of America.
1 The Pleasant Valley fruit and wine reporter. Hammondsport, N. Y. m. 1870–July 1871.
1 The Pomological magazine. Cincinnati, Ohio. bi-m. v. 1. Sept. 1842–June 1843.
2 Potato chipper. Cleveland, Ohio. National Potato Chip Institute. 1941–
2 Potato exchange news. Idaho Falls, Idaho. Idaho Potato Growers Exchange. 1923–//?
2 Potato grower. Established as
(a) Minnesota potato exchange weekly, 1920–1922.
(b) Potato digest, 1922–1923). St. Paul, Minn. Minnesota Potato Growers Exchange, 1923–25//.
2 Potato horizons. Gering, Nebr. Lockwood Grader Corp. [n.d.]
1 Potato magazine. Chicago. Potato Association of America, v. 1–5 1918–1923//.
1 Potato world. Chicago. National Potato Institute, 1932–
1 Potatoes. Washington, D.C. National Potato Council. v. 1– [n.d.]
1 The Practical entomologist. Philadelphia. Entomological Society of Philadelphia. v. 1–2 1865–1867//.
1 Practical fruit grower and farm magazine. Established as

(a) Southwest, v. 1–6, 1894–1900.
(b) Practical fruit grower, v. 7–10, 1900–1903). Springfield, Mo. v. 10–13, 1903–1907//.

1 Practical fruit grower. Springfield, Mass. v. 1–? 1887–1890//?
1 Practical nurseryman and horticultural advertiser. Huntsville, Ala. v. 1–10, 1893–1902//.
1 Practical planter. Memphis, Tenn. v. 1–3 1870–1873//.
1 Produce exchange bulletin. New York. 1878–1890//.
1 Produce exchange reporter. New York. 1854–1900//.
1 Produce journal. Buffalo, N. Y. v. 1–? 1889//?
2 Produce news. Established as
(a) Tuck's reporter and distributor, 1903.
(b) Fruit and produce news, 1903–1907). New York. Chicago. Cincinnati. Dallas. 1908–
1 Producer and progressive farmer. Ames, Iowa. v. 1 1876–1877//.
1 Producer-consumer. Amarillo, Tex. v. 1–20 1935–1952//?
2 Producers review. Established as Texas producers' review, 1904. Dallas. v. 1–7 1904–1906//?
2 Professional gardener. Mineola, N. Y. National Assoc. of Gardeners, 1949–
2 Progressive eastern fruitgrower. Rochester, N.Y. v. 1, no. 1–3. 1911//?
2 Provision news. Chicago. v. 1–? 1900–1905//.
1 Prune packers' guide. San Jose, Calif. v. 1–? 1894–1896//.
1 Pruning knife. Du Quoin, Ill. v. 1–? 1896–//?
1 Purdy's fruit recorder. Established as Purdy's recorder and evaporator, 1891. Palmyra, N. Y. 1892–1894//.
2 Pure food era. Detroit. v. 1–? 1901–//?
2 Pure products. New York. v. 1–17 1905–1921//.
1 Rice industry. Houston. v. 1–13 1898–1911//?
1 Rice journal and southern farmer. Established as Rice journal. Crowley, La. (Signal Printing Co.); Houston, Tex. m. v. 1– 1898–//?
1 Rice planter. Pointe a la Hache, La. v. 1– 1861–//?
2 Rogue River fruit grower. Medford, Oreg. m. v. 1, no. 1–11. April 1909–May 1910//? No issues, Jan.-Feb. 1910.
2 Rogue River Valley pear-o-scope. Medford, Ore. Fruit Growers League. v. 1–10. 1933–1942//.
1 The Rural Californian. Established as
(a) Southern California horticulturist. San Jose, Calif. (1877), Los Angeles (1878–1879). m. v. 1–2. Sept. 1877–Dec 1879.
Continued as
(b) Semi-tropic California. Los Angeles. m. v. 3–6. Jan. 1880–1883.
Continued as
(c) The Rural Californian. Los Angeles. m. v. 7–38. 1883–1914//.
1 Rural Northwest, Pacific fruit grower and dairyman. Portland, Oreg. s-m. v. 1–6, no. 5. Sept. 15, 1891–Nov. 15, 1896
Combined with Oregon agriculturist, and continued as
(a) Oregon agriculturist and the Rural northwest, Dec. 1, 1896.
1 Rural recorder and fruit grower. Altus, Ark. s-m. v. 1–3 ? 1877–1880.
1 Saline Valley farmer and fruit grower. Stonefort, Ill. w. v. 1–3? 1899–1901.

2 San Francisco produce review. San Francisco. d. (July 15, 1901–Oct. 1901) w. (Oct. 1901–Feb. 26, 1902). no. 1–110? July 15, 1901–Feb. 26, 1902?

2 San Francisco trade journal [Canning]. San Francisco. w. v. 1–16. June 3, 1898–1906//?

1 Science and horticulture. Los Angeles and San Diego, Calif. m. v. 1–2. 1891–1892//.

2 Seed and garden merchandising. Established as Southern seedsman, 1939–1958. New Orleans. San Antonio, Tex. 1959–

2 Seed merchant. Chicago. v. 1–3. 1934–1936//.

1 Seed-time and harvest. La Plume, Pa. q. (1880–1882), m. (1882–1892). v. 1–12. Jan.? 1880–Mar. 1892//?
Merged in American farmer and Farm news.

1 Seed-time and harvest. Scranton, Pa. w. v. 1–2? 1897–1898. Date of beginning given as 1879 to connect with paper of this name published at La Plume, Pa., but not a continuation of that paper.

2 Seed-time and harvest. Scranton, Pa. m. v. 1–1905–1908. Not connected with earlier papers of this name at Scranton or La Plume, though giving date of establishment as Jan. 1880.

1 Seed trade news. Chicago. w. 1923–

1 Seed world. Chicago. Seed Trade Reporting Bureau; Seed Trade Publications. s-m. 1915–

1 Semi-tropic culturist and advertiser. San Diego, Calif. w. v. 1–1892 (1899)-1900.

1 Semi-tropic California. Established as Southern California horticulturist, 1–2, 1877–1879//. 3–5. 1880–1882//.
Merged into Rural California.

1 The Semi-tropical. Jacksonville, Fla. m. v. 1–4, no. 4. Sept. 1875–Apr. 1878//.

1 Semi-tropical farmer. Los Angeles. w. v. 1–2? June? 1875–1876.

1 The Semi-tropical planter. Established as
(a) The Semi-tropical planter. San Diego, Calif. m. v. 1, no. 1–8, v. 2, no. 1. May 1887–Jan. 1888. No Feb. 1888 issue
Continued as
(b) The San Diego magazine; California. San Diego, Calif. m. n. ser., v. 1, no. 1–3. Mar.–July 1888.
Continued as
(c) The San Diego magazine and tropical planter. San Diego, Calif. m. v. 2, no. 1–2 (whole no. 13–14). Aug.–Sept. 1888.
Continued as
(d) The Semi-tropical planter. San Diego, Calif. m. n. ser., v. 3–7, no. 38. Oct.? 1888–Oct. 1891//?

1 Sentinel and hop-growers' journal. Hamilton, N. Y. w. v. 1– 1862 (1895)-1895.

1 Sickle and sheaf. Oskaloosa, Kans. w. v. 1–2? 1873–1874.

1 The Soil of the South. Columbus, Ga. m. v. 1–7. Jan. 1851–Jan. 1857//
Absorbed by the American cotton planter, Jan. 1857. In 1853 called Soil of the South and Tropical farmer.

1 Sorghum growers' guide and farm journal. Madison, Ind. m. v. 1– 1883–1890.

1 The Sorgo journal and farm machinist. Cincinnati, Ohio. m. (1863–1867?), q. (1868–1869). v. 1–7, Jan. 1863–Oct. 1869.

2 South Dakota wheat grower. Aberdeen, S.Dak. v. 1–6. 1924–1930//.

1 Southern California horticulturist. Los Angeles. Sept. 1877–
1 Southern farmer and gardener. Aug. 1844.
1 Southern farmer and horticulturist. Newberry, S. C. s-m. v. 1– 1897–1900?
1 Southern florist and gardener. Established asPM(a) Southern florist and gardener. Louisville, Ky. m. v. 1–3. Aug. 1894–July 1897.
Continued as
(b) Florist and gardener. Louisville, Ky. m. v. 4, no. 1–4. Aug.–Nov. 1897.
Continued as
(c) Southern florist and gardener. m. v. 4, no. 5–v. 6, no. 6. Louisville, Ky. Dec. 1897–June 1899//.
Absorbed by Land and a living, July 1899.
1 Southern florist and nurseryman. Fort Worth, Tex. Southern Florist Publishing, 1915–
2 Southern florist. Poulan, Ga. m. v. 1. 1908.
2 Southern food processor. Established as Southern canner and packer. Athens, Ga. 1– 1940–
1 Southern fruit and farm reporter. Established as
(a) The West Tennessee argus. Humboldt, Tenn. w. v. 1–4? 1879 (1882)–1883.
Continued as
(b) Southern fruit and farm reporter. Humboldt, Tenn. w. v. 5? 1884.
1 Southern fruit and produce journal. Established as
(a) Orange grower. New Orleans. s-m. v. 1. 1898–1899.
Continued as
(b) Southern fruit and produce journal. New Orleans. s-m. v.2? 1899.
1 Southern fruit and vegetable reporter. Monticello, Ga. s-m. v. 1–4? 1897–1900.
1 Southern fruit farm. Somerville, Tenn. m. v. 1–2? 1899–1901.
1 The Southern fruit grower. Established as(a) Southern fruit grower. Jersey, Tenn. v. v. 1– Jan. 1896–1898?
Continued as
(b) Southern fruit and truck grower. Chattanooga, Tenn. m. v. 1899?–1901.
Continued as
(c) The Southern fruit grower. Chattanooga, Tenn. m. -v. 25. 1901–1921//.
2 Southern fruit journal. Montezuma, Ga. m. v. 1. 1904.
1 Southern garden. New Orleans. m. v. 1–3? 1894–1896.
2 Southern gardening. Forsyth, Ga. v. 1–2. 1949–1950//.
2 Southern home and garden. Ft. Worth; Dallas, Tex. v. 1–14 no. 2. Nov. 1933–Sept. 1946//. Absorbed by Sun-up.
1 Southern horticultural journal. Denison, Tex. (1888–1890), Weatherford, Tex. (1891). s-m. v. 1–4? 1888–1891.
1 Southern horticulture and market garden. New Orleans. m. v. 1–3? Aug. 1897–1899.
1 Southern horticulturist and farmer. Bryan, Tex. m. v. 1– 1889 (1892)–1892.
1 Southern horticulturist. Canton, Miss. 1–2. 1869–1870//.
Superseded by Swasey's southern gardener.
2 Southern horticulturist. Humboldt, Tenn. m. v. 1–2? 1909–July 1910//? Not connected with earlier paper of same name published at same place.
1 Southern horticulturist. Humboldt, Tenn. s-m. v. 1–2? 1891–1892.
2 Southern life, home and garden magazine. Raleigh, N.C. Established as
(a) Southern home and garden. v. 1 no. 1–2, Apr.–May 1938.

Continued as

(b) Southern garden. v. 1 no. 3–12, June 1938–March 1939.

Continued as (c) Southern life, home and garden magazine. Apr. 1939–V. 4 no. 6, Sept. 1941//.

1 Southern plantation. Montgomery, Ala. Alabama State Grange. w. v. 1–4? 1874–1878.

1 Southern planter and grange. Atlanta. v. 1–? 1878–1879//?

1 The Southern planter. Glasgow, Ky. July 1850–1851?

1 Southern planter. Macon, Ga. s-m. v. 1. Dec. 1842–

1 Southern planter. Monroe, N. C. w. v. 1–2. 1888–1889

1 Southern planter. Natchez, Miss. (Jan.-Aug. 1842), Washington, Miss. (Sept. 1842–). m. v. 1– 1842– Specimen numbers issued in 1841 as Farmer and mechanic, Washington, Miss. May–June, July–Aug., Sept.–Dec., each issued in a combined number.

1 Southern planter. Richmond. m. v. 1–21. Jan. 1841–Dec. 1861; 1846 called n. ser., v. 1. Proposed publishing a weekly edition, Jan. 1846.

1 The Southern planter and farmer. Richmond. Established as

(a) The Southern planter. Richmond. m. n. ser., v. 1. Jan.–Dec. 1867.

United Jan. 1868 with The Farmer and continued as

(b) The Southern planter and Farmer. Richmond. m. (1868–1881), s-m. (Jan.–June 1882), m. (July 1882–). n. ser., v. 2–5 (1868–1871), 9–11 (1872–1874?). n. ser., no vol. no. (1875), v. 37–43 (1876–1882). Jan 1868–Dec. 1882.

1 Southern planters' record. Wilmington, Del. v. 1–2 1871–1872//.

2 Southern seedsman. New Orleans. H. L. Peace Publications. m. 1938–

2 Southern shipper. Houston, Tex. v. 1– 1909–//?

2 Southern Texas truck growers' journal. San Antonio, Tex. 1–3. 1913–1915//?

1 Southern tobacco journal. Danville, Va. (1886–1891), Winston, N. C. (1892–1900), Winston-Salem, N. C. (1900–1910). w. v. 1– 1886– . Annual Christmas number takes place of last number of year. 1906–1908, volumes are numbered 27–32 instead of 37–42.

2 Southern tobacco leaf. Tullahoma, Tenn. m. v. 1, no. 1–6. Feb. 15–July 15, 1910//?

1 Southern vineyard. Los Angeles. w. v. 1? 1861?

1 Southland farmer. Established as

(a) Texas fruits, nuts, berries, and flowers. San Antonio, Tex. m. v. 1. May 1907–June 1908. Continued as

(b) Southern orchards and homes. Houston, Tex. m. v. 2–6. 1908–1911.

(c) Southern orchards and farms. v. 6–10. 1911–1914.

(d) Southland farmer. Houston, Tex. v. 11–41. 1914–1926//.

Absorbed Texas stockman and farmer, Jan. 1916, and adopted its numbering.

1 Southwest wheat grower. Enid, Okla. v. 1– 1922–

2 Southwestern crop and stock. Lubbock, Tex. 1– 1947–

1 Southwestern farm and orchard. Las Cruces, N. Mex. v. 1–7? 1894–1900//.

1 Southwestern farmer and American horticulturist. Established as Southwestern farmer. Wichita, Kans. v. 1–6? 1896–1901//.

1 Southwestern farmer. Established as

(a) National rice and cotton journal, 1904–1905.

(b) National rice and cotton journal, "Southwestern farmer," 1905. Houston. v. 3–12 1906–1912//?

2 Southwestern grain and flour journal. Established as Southwestern grain journal, v. 1–2, 1902–1903. Wichita, Kans. v. 3– 1904–
2 Southwestern horticulturist. Fort Worth, Tex. 1–2. 1912–1913//.
1 Special crops. Skaneateles, N. Y. q. v. 1–2 ? 1890 1891.
1 Special crops. Skaneateles, N. Y. m. n. ser., v. 1–9, no. 95. Apr. 1902–July 1910*; called revival of earlier periodical of same name. Title has been preservation microfilmed by the Albert R. Mann library, Cornell University.
2 St. Louis packer. St. Louis. w. v. 1– 1900–1901.
1 Strawberry specialist. Kittrell, N.C. v. 1–6 1897–1903//.
2 Strawberry. Three Rivers, Mich. v. 1–2 1906–1907//.
2 Subtropical gardening. Established as Subtropical gardening and fruit growing. Brooksville, Fla. Sun Publishing, 1938–
2 Suburban California. Pasadena, Calif. Established as
(a) Pacific garden, 1–9, 1907–1915?
Continued as
(b) Suburban California. v. 10 no. 1–7, –July 1916//.
1 Success with the garden. Rosehill, N. Y. v. 1–4? 1894–1897//.
1 Sugar beet era. Tablerock, Nebr. v. 1–3? 1895–1897//.
2 Sugar beet grower. Denver. v. 1–3, 1900–1902//.
1 Sugar beet grower. Greeley, Colo. Mountain States Beet Grower Assoc. 1–2 1920–1921//.
1 Sugar beet journal. Saginaw, Mich. Farmers and Manufacturers Beet Sugar Assoc., 1–
1 Sugar beet. Geneva, Nebr. v. 1–3? 1895–1897//.
1 Sugar beet. Mason City, Iowa. Denver. Ogden, Utah. v. 1–8 1924–1931//.
1 Sugar beet. Philadelphia. v. 1–32 1880–1911//.
1 Sugar journal. Crowley, La. New Orleans. Pipes Publications, 1938–
1 Sugar planter. West Baton Rouge, La. v. 1– 1852–//?
1 Sugar planters' journal. Established as
(a) Louisiana sugar bowl, 1870–1884.
(b) Louisiana sugar bowl and farm journal, 1885–1893.
(c) Sugar bowl and farm journal, 1893–1894.
Merged into Modern sugar planter. New Orleans. v. 25–41 1894–1910//.
2 Sun-up. San Antonio, Tex. v. 1–5 no. 6. Feb. 1946–Jan. 1950//. Absorbed Southern home and garden.
1 Sunsweet standard. Stockton, Calif.; Sunsweet Growers, v. 1–62; June 1917–Nov. 1978//. Began as
(a) Dry ground, June–July 1917.
(b) Grower's voice, Aug.–Dec. 1917.
Issued 1917–June 1958 by the association under its earlier name: California Prune and Apricot Growers Association.
Merged with Diamanod walnut news to form Diamond/Sunsweet news.
1 Swasey's southern gardener. Established as Southern horticulturist, v. 1–2, 1869–1870. Tangipahoa, La. v. 1 1871//.
2 Sweet potato journal. Shreveport, La. 1– 1946–
2 Tabb potato service. Special bulletin. 1–35. 1926–1927//.
2 Tea and coffee trade journal. Established as
(a) Tea and coffee journal, 1901.
(b) Tea, coffee and sugar, v. 1–4, 1901–1903). New York. v. 4– 1903–

1 Texas farming and citriculture. Established as
(a) Lower Rio Grande Valley magazine, 1924–1929.
(b) Texas citriculture, 1929–1935.
(c) Texas citriculture and farming, 1935–1936. Harlingen, Tex. Watson Publishing, v. 13– 1937–
1 Texas fruit grower. Myrtle Springs. Edgewood, Tex. v. 1–? 1896–1906//.
1 Texas gardener. Beeville, tex. v. 1–2? 1893–1894//.
1 Texas journal of horticulture. Tyler, Tex. v. 1–2? 1888–1889//.
1 Texas planter and breeder. Dallas. v. 1–5? 1896–1900//.
1 Texas planter and farmer. Dallas. v. 1–5 1881–1885//?
1 Texas tobacco plant. Willis, Tex. v. 1–3? 1897–1899//.
2 Texas wheat grower. Amarillo, Tex. Texas Wheat Growers' Assoc. 1–3. 1926–1929//.
2 Thompson's garden consultant. Los Angeles, Calif. v. 1 no. 1–6. –Apr. 1946//.
1 Thru the garden gate. Kalamazoo, Mich. Federated Garden Clubs of Michigan, 1941–//? Tilton's journal of horticulture and florist's companion. Established as (a) American journal of horticulture and florist's companion, v. 1–5, 1867–1869. Boston, v. 6–9, 1869–1871//.
1 Tilton's journal of horticulture and florist's companion. Established as
(a) American journal of horticulture and florist's companion. Boston. m. v. 1–5. Jan. 1867–June 1869.
Combined with The Magazine of horticulture, botany and all useful discoveries and improvements in rural affairs to form
(b) Tilton's journal of horticulture and florist's companion. Boston. m. v. 6–9. July 1869–Dec. 1871//.
Combination effected Jan. 1869 and new name dates from that issue, but volume title-0age retains earlier name through v. 5; cover-title and engraved title pages are Tilton's journal of horticulture and floral magazine.
1 Tobacco. New York. 1– 1886–
2 Tobacco grower. Farmville, Va. v. 1–9. 1932–1941//.
2 Tobacco growers' journal. Tippecanoe City, Ohio. v. 1– Feb. 1917–//?
1 Tobacco herald. Deerfield, Wis. w. v. 1–3, no.? Sept. 1885–Apr. 1888; merged in Lake Mills Leader.
1 Tobacco jobber. East Stroudsburg, Pa.; New York. 1– 1926–
1 Tobacco journal. Lancaster, Pa. w. v. 1–2? 1891–1892.
1 Tobacco leaf and cotton plant. 1873–1875.
1 Tobacco leaf. New York. w. v. 1– 1865– . Separate vol. no. dropped after June 1, 1904.
1 Tobacco news and prices current. Louisville, Ky. v. 1–3. 1877–1880//?
2 Tobacco news. Florence, S.C. v. 1, no. 1–10. 1930–//?
2 Tobacco news. Greenville, N.C. v. 1 no. 1–17. 1938//.
2 Tobacco news. Philadelphia. 1–7. 1910–1918//?
2 Tobacco plant. Cincinnati, Ohio. w. v 1–3? 1899–1901.
1 Tobacco plant. Durham, N. C. w. v. 1–4? 1872–1875.
2 Tobacco planter. Guthrie, Ky. w. v. 1– 1908–July 1910//?
2 Tobacco planter. Louisville, Ky. v. 1–7 1922–1929//.
1 Tobacco record. New York. 1– 1915– . Supersedes Tobacco jobber and retailer.
1 Tobacco reporter. St. Louis. w. v. 1–2? 1887–1888.
1 Tobacco review. Established as

(a) National tobacco review. Chicago. m. v. 1–3? 1885–1887.
Continued as
(b) Tobacco review. Chicago. w. v. 4? 1888.

2 Tobacco tidings. Cincinnati. w. v. 1– 1907–July 1910//?

2 Tobacco tips for the grower, warehouseman, dealer, manufacturer and allied tobacco interests. Manning, S.C. v. 1 no. 1–48. 1932–1933//.

2 Tobacco trade review. Cincinnati. w. v. 1. 1900.

1 Tobacco trades. Boston. m v. 1–4? 1898–1901.

1 Tobacco worker. Louisville, Ky. m. v. 1–28. 1897–1924. n.s. v. 1– 1940–

1 The Tobacco world. Established as
(a) Tobacco age. Philadelphia. w. v. 1– 1883–1894.
(b) The Tobacco world. Philadelphia. s-m. -v. 73, no. 3. 1895–1953//.

1 Tobacconist and farm journal. Calhoun, La. s-m. v. 1–6? 1894–1899.

1 Torch. Toronto, Ont. Canadian Crops Assoc. 1941–

1 Trade bulletin [Produce]. Montreal. w. v. 1– June 1903–

1 Trade journal and international horticulturist. Harrisburg, Pa. and New York, v. 1–3, January 1888–October 1890//?
(a) International horticulturist, v. 1–2.

1 Trade journal and Virginia agriculturist. Lynchburg, Va. w. v. 1–2? 1889–1890.

2 Tree gold. Indianapolis, Ind. 1914–//?

1 The Tri-state farmer and gardener. Established as
(a) The Farm magazine. Knoxville, Tenn. m. v. 1–Jan. 1895–.
Continued as
(b) The Tri-state farmer. Chattanooga, Tenn. m. v. –4. –1898.
Continued as
(c) The Tri-state farmer and gardener. Chattanooga, Tenn. m. v. 5. (n. ser., v.3)–v. 13, no. 2. 1899–Feb. 1909. Volumes are irregularly numbered.

2 Tri-State tobacco grower. Raleigh, N.C. Tobacco Growers Cooperative Assoc., v. 1–6 1921–1926//.

2 Tropical and subtropical America. N. Y. m. v. 1, no. 1–5. Jan.-June 1908//.

1 The Truck farmer of Texas, San Antonio. Tex. (1899–1901), Dallas and San Antonio. Tex. (1902–Mar. 1906), Dallas, Tex. (Apr. 1906–1909). m. v. 1–11, no. 5. July 1899–Nov. 1907//. Vol. 6, no. 4–v. 9, no. 3 (May 1902–Sept. 1905–) have title The Truck farmer. Sept. 1903 has number v. 7, no. 8; Nov. 1903. v. 7, no. 9, but reference is made in the latter to an Oct. issue, no volume 2, so numbered, was ever published, v. 3, no. 1 following immediately after v. 1, no. 12; the publishers probably changed from a 12-number to a 6-number volume in this manner. v. 1. no. 6–v. 4. no. 4 official organ of the Texas truck growers' association.

2 Trucker and farmer. New Iberia; New Orleans, La. Established as Louisiana trucker and farmer. v. 1–9. 1909–1914//. United with modern sugar planter to form Modern farming.

2 Trucker. Chadbourn, N. C. w. v. 1– Jan. 1902–1908.

1 Truckers' and planters' journal. Chadbourn, N. C. w. v. 1– 1896–1900//?

2 U and I cultivator. Salt Lake City. Utah-Idaho Sugar Co. 1941–

2 U and I farm messenger. Salt Lake City. Utah-Idaho Sugar Co. 1930–1937//.

2 Trade journal for canners, packers, grocers, and allied interests. San Francisco. v. 1–5. 1906–1909//?

2 Union gardener and florist. Chicago. v. 1– 1905–//?

2 Union guide. Established as
(a) Texas truck grower and shippers' guide, 1902–1903.
(b) Shippers' guide, 1903.
(c) Texas truck grower and shippers' guide, 1903–1904.
(d) Shippers. guide, 1904. Southern shippers' guide, 1904–1909. Houston, Tex,. 1909–//?
2 United florist news. Toronto. United Florists of Canada, 1945–//?
1 United States miller and Weather and crop journal. Established as
(a) United States milling and manufacturing journal, v. 1–14, 1878–1889.
(b) United States miller and the milling engineer, v. 14–16, 1889–1891.
(c) United States miller, v. 17–19, 1892–1894. Milwaukee. v. 19– 1894– .
1 United States tobacco journal. New York. v. 1– 1874–
1 Utah pomologist and gardener. Established as
(a) Utah pomologist. St. George, Utah. m. v. 1–2? 1870–1871.
Continued as
(b) Utah pomologist and gardener. St. George, Utah. m. v. 3–7? 1872–1876//?
2 Valley farmer and citrus grower. San Benito, Tex. v. 1–4. 1923–1926//.
2 Valley potato grower. Grand Forks, N.D. Red River Valley Potato Growers Assoc. 1– 1946–
2 Vegetable grower; devoted to the growing of vegetables, fruits and flowers. Chicago; Spencer, Ind. 1–8. 1912–1915//?
2 Vegetable growers news. Norfolk, Va. 1– 1946–
2 Vegetable lore. New York. m. v. 1– July 1917–
1 Vermont agriculturist, devoted to agriculture, horticulture, and floriculture. Brandon, Vt. m. v. 1, no. 1–7. June 1877–Apr. 1878//? Feb.–Mar. 1878 not issued.
1 Vick's floral guide. Rochester, N. Y. v. 1–5? 1873?-1877//.
1 Vick's illustrated magazine. Established as
(a) Vick's monthly magazine, v. 1–14, 1878–1891.
(b) Vick's illustrated monthly magazine, v. 15–22, 1891–1899. Rochester, NY v. 23–24, 1899–1900//.
2 Vineyard of the east. New Orleans. v. 1–6. 1920–1925//.
1 Vineyardist. Fruithurst, Ala. w. v. 1–5? 1899–1903.
1 Vineyardist. Penn Yan, N. Y. s-m. v. 1–9? 1886–1894.
2 Washington fruit grower. Established as
(a) Fancy fruit. North Yakima, Wash. m. v. 1–2, no. 21. July 1907–Apr. 1909.
Continued as
(b) Washington fruit grower. North Yakima, Wash. m. v. 3, no. 22–31. May 1909–April 1910//.
2 Weekly florist. Chicago. v. 1–6. 1924–1927//.
2 Weekly fruit and produce news. New York. v. 1–2. 1934–1935//?
1 Western brewer and journal of the barley, malt and hop trades. Chicago and New York. v. 1– 1876–
1 Western brewing and distributing. Established as Western brewing world, 1893–1941. Los Angeles. 1942–
1 Western canner and packer. San Francisco. v. 1–50. 1909–1958//.
Absorbed Pacific canner 1926; Packer and canner 1926.
Absorbed by Canner/packer.
1 Western Colorado, a fruit, farm, and sugar beet journal. Grand Junction, Colo. s-m. v. 1. May 10–Oct. 25, 1899//.

2 Western cotton journal and farm review. Established as
(a) California cotton journal. 1–4, 1925–1927.
(b) Western cotton journal. v. 5, no. 1. San Francisco. v. 5–11. 1928–1930//.
1 Western farmer and gardener, devoted to agriculture, horticulture and rural economy. Cincinnati. v. 1–5. Sept. 1839–July 1845. v. 1. as Western farmer.
1 Western farmer and horticulturist, Established as
(a) Western horticulturist. Ainsworth, Iowa. m. v. 1– 1879–1884.
Continued as
(b) Western farmer and horticulturist. Ainsworth, Iowa, m. v. 1885–1892.
2 Western feed and seed. Established as Western feed and seed dealer. San Francisco. R. M. Beeler, 1945–
2 Western field. San Francisco. m. v. 1–23. 1902–1914//.
1 Western florist and fruit grower. Perry, Okla. s-m. v. 1–7? 1899–1905.
2 Western florist and nurseryman. San Francisco. v. 1– 1947–
2 Western florist and nurseryman. Los Angeles. v. 1–16. 1919–1933//.
2 Western fruit and vegetable grower. Denver. Western Fruit and Vegetable Grower Publishing, 1923–//?
1 The Western fruit-grower. St. Joseph, Mo. m. v. 1–21, no. 7. Jan. 1897–//?
Absorbed American gardening, Dec. 1904 and Arkansas fruit grower, Harrison, Ark., Jan. 1899; Oct. 1904 begins Eastern edition at Harrisburg, Pa. with title Fruit growers' magazine.
2 Western fruit grower. Established as
(a) Grape grower, 1947.
Continued as
(b) California fruit and grape grower, 1948–1950. San Francisco. E. B. Weinard, 1950–
2 Western fruit jobber. Denver. Western Fruit Jobbers Assn. of America. v. 1–18 no. 12. 1914–Apr. 1932//.
1 Western garden and poultry journal. Des Moines, Iowa. 1–10. 1890–1899//?
1 The Western gardener. Leavenworth, Kans. m. v. 1, no. 1–6. Sept. 1870–Mar. 1871//
Merged in Western pomologist, Apr. 27, 1871; no number issued Dec. 1870.
1 Western grower and shipper. Los Angeles. Vegetable and Melon Growers and Shippers, 1929–
2 Western home, fruit and garden. Established as
(a) Berry grower, 1920–1922.
(b) Western fruit, berries and orchard, 1922–1924.
(c) Western fruit, flower and garden, 1924–1926. Seattle. Western Fruit, Inc., 1926–1927//.
2 Western homes and gardens. San Jose, Calif. Established as
(a) Pacific grower. 1–10, 1915–1925?
Continued as
(b) Pacific grower, home and garden. 1926–1927.
Continued as
(c) Pacific flower grower, home and garden. 1927.
(d) Western homes and gardens. 1928–v. 16 no. 6, June 1932//.
1 Western horticultural review; devoted to horticulture, pomology, grape culture, wine manufacture, rural architecture, landscape gardening, meteorology, etc. Cincinnati. v. 1–3. 1851–1853//.

Continued as Horticultural review and botanical magazine.

2 Western horticulture, gardens and homes. Established as
(a) Home designer. 1–8. 1921–1925.
(b) Home designer and garden beautiful. v. 9–13. 1925–1928. Oakland, Cal. Feb.-Sept. 1928//. Merged into Better flowers.

2 Western horticulture. San Francisco. v. 1–2. 1927–1928//.

1 Western horticulturist. Freeport and Chicago, Ill. m. v. 1– 1870–//?

1 Western horticulturist. Pittsburgh. m. v. 1. Jan. 1856–//?

2 The Western New York apple. Barker, N. Y. m. v. 1, no. 1–7. June–Dec. 1908//.

1 Western planter. Kansas City, Mo. v. 1–3. 1871–1873//.

1 The Western pomologist and gardener. Established as
(a) The Western pomologist. Des Moines. m. v. 1–2, no. 6. Jan. 1870–June 1871.
Combined with Western gardener, Leavenworth, Kans., Apr. 27, 1871
Continued as
(b) The Western pomologist and gardener. Des Moines and Leavenworth, Kans. m. v. 2, no. 7–v. 3, no. 6. July 15, 1871– June 15, 1872//.
Combined with the Horticulturist, July 1872. Feb.-Dec. 1871 have caption title The Pomologist.

1 Western tobacco journal. In 1878 called Western tobacco journal and wine and liquor reporter. In 1880 called Western tobacco journal and grocer's review. Cincinnati. v. 1– 1874–

1 Western tree planter and fruit grower. Elgin, Ill. v. 1–2? 1888–1889//.

1 Wheat belt review. Edmonton, Alb. Free Press Pub. Co. w. 1906–?

2 Wheat field. Pendleton, Oreg. [n.d.]

1 Wheat grower. Grand Forks, N.D. North Dakota Wheat Growers Assn. s-m. v. 1–8. 1923–1931//.

2 Wheat growers advocate. Enid, Okla. 1–2. 1931–1933//?

1 Wheat growers journal. Wichita, Kan.; Kansas City, Mo. s-m. v. 1–12. 1920–1932//.

1 Whitlock exposition recorder. Established as
(a) Whitlock's horticultural advertiser, 1867–1868.
(b) Whitlock's horticultural recorder, 1868. New York. 1868–1869//.

2 Wild flower. Washington, D. C. Wild Flower Preservation Society, 1924–

1 The Wine and fruit grower. Established as
(a) American wine and grape grower. New York. m. v. 1–5. July 1878–1883.
Continued as
(b) The Wine and fruit grower. New York. m. v. 6?-7?, v. 8, no. 1– 1884–Nov. 1885, Feb. 1886–//?
Announced as absorbed by Orchard and garden, Nov. 1885, but, as the Orchard and garden failed to consummate its purchase, publication was resumed by its owner, Feb. 1886.

1 Wine and fruit reporter. New York. w. v. 1–12. 1864–1879//?

2 Wine and spirit and brewers' review. Toronto. m. v. 1– 1903–

1 Wine and spirit bulletin. Louisville, Ky. v. 1–32. 1886–1918//?

1 Wine and spirit gazette. New York. v. 1–18. 1887–1905//?

2 Wine merchant. Established as American wine merchant. San Francisco. 1942–

2 The Wine review combined with Wine news. Los Angeles. m. v. 1–18. 1933–1950//. NAL Absorbed by Wines and vines.

1 Wines and vines. Established as
(a) California grape grower. v. 1–10. 1919–1929.
(b) California grower. v. 10–14. 1929–1933.
Changed back to
(c) California grape grower. v. 14–15. 1933–1934.
Continued as
(d) Wines and vines. San Francisco. v. 15– 1934–
1 Wisconsin horticulturist. Baraboo and Madison, Wis. (1896–1902), Sparta and Madison, Wis. (1903). m. v. 1–7. Mar. 1896–Feb. 1903//. Under the management of the Wisconsin state horticultural society.
2 Wisconsin potato journal. Madison, Wis. v. 1– June 1915–
2 The Wisconsin sugar beet. Menomonee Falls, Wis. m. v. 1–9, no. 12. 1901–July 1910//?
1 Wisconsin tobacco leaf. Janesville, Wis. w. v. 1–11? 1889–1899.
1 Wisconsin tobacco reporter. Established as Edgerton independent, Dec. 1874
Continued as
(a) Wisconsin tobacco reporter. Edgerton, Wis. w. v. 3– Apr. 1877–
1 Wisconsin weather and crop journal. Milwaukee. m. v. 1–3? 1892– Oct. 1894
Merged Oct. 1894 in United States miller, which continued as United States miller and weather and crop journal.
2 Wood's crop special. Richmond. m. v. 1–5. 1906–July 1910//?
1 Woodward's architecture, landscape gardening and rural art. New York. no 1–12. 1867–1868.
1 Woodward's record of horticulture. New York. F. W. Woodward. 1–2. 1866–1867//.
2 Your garden and home. Toronto. v. 1–5 no. 6, June 1947–June 1951//.
Supersedes Canadian horticulture and home. Floral edition.
2 Your garden and home. Cleveland. v. 1–16 no. 10, May 1927–Dec. 1942//.
Preceded by introductory number dated April 1927.

E. Scholarly Journals and Experiment State Publications

Early United States scientific serial literature of crops appeared in two primary sources: the general journals of agriculture, botany, and entomology; the reports and bulletins of state agricultural experiment stations and those the U.S. Department of Agriculture. Around 1880 specialized journals devoted to the scientific treatment of crops began to appear.

To identify scholarly crops journals and establish their relative importance, citation analysis of a limited number of documents was executed using some of the data from the original source documents identified in Section B. Details and information about the titles were gathered from these volumes:

(1) U. S. Dept. of Agriculture. *Yearbook of Agriculture, 1936 and 1937*. Thirteen of the twenty-three sections of the 1936 *Yearbook* dealt with improvements in

Table 11.1. Most frequently cited scholarly journals and serials

Journal	Microfilmed[a]
Agricultural Gazette of New South Wales	—
Agricultural Journal of India	—
American Botanist	—
American Breeder's Association Annual Report or Proceedings	—
American Entomologist	UMI
American Journal of Botany	PR
American Midland Naturalist	UMI
American Miller	—
American Naturalist	UMI
American Potato Journal	UMI
Angewandte Botanik	SWETS
Annals of Applied Biology	—
Annals of Botany	PR
Annals of Missouri Botanic Gardens	UMI
Arbeiten, Biologische Zentralanstalt für Land- und Forstwirtschaft	—
Berichte der Deutschen Botanishen Gesellschaft	—
Bibliographia Genetica	—
Botanical Gazette	UMI
Botanical Magazine (Japan)	—
Bulletin of the Torrey Botanical Club (New York)	UMI
California Agricultural Experiment Station Bulletin	SPAULDING
Canadian Journal of Research	UMI
Centralblatt für Bakteriologie und Parasitenkunde	IDC
Colorado Agricultural Experiment Station Bulletin	SPAULDING
Comptes Rendus Académie de Sciences (Paris)	PR
Cornell Agricultural Experiment Station Bulletin	SPAULDING
Cytologica (Japan)	PR
Genetica	UMI
Genetics	UMI
Hereditas	UMI
Hilgardia	UMI
Illinois Agricultural Experiment Station Bulletin	SPAULDING
Indian Journal of Agricultural Sciences	UMI
International Bulletin of Plant Protection	—
Iowa Agricultural Experiment Station Bulletin/Research Bulletin	SPAULDING
Japanese Journal of Botany	PR
Japanese Journal of Genetics	—
Journal of Agricultural Research	UMI
Journal of Agricultural Science (Cambridge)	UMI
Journal of Economic Entomology	UMI
Journal of Genetics	UMI
Journal of Heredity	UMI
Journal of Mycology	—
Journal of Pomology and Horticultural Science	—
Journal of the American Society of Agronomy	IDC
Journal of the Kansas Entomological Society	UMI
Journal of the New York Entomological Society	MANN
Journal of the Royal Horticultural Society, London	UMI
Journal of the Royal Microscopy Society	—
Journal of the Washington Academy of Science	—

Table 11.1. Continued

Journal	Microfilmed[a]
Madras Agricultural Journal	—
Maine Agricultural Experiment Station Bulletin	SPAULDING
Mededeelingen der Landbouwhogeschool (Wageningen)	—
Mycologia	UMI
Nature	UMI
New Jersey Agricultural Experiment Station Bulletin	SPAULDING
New York Agricultural Experiment Station Bulletin	SPAULDING
New York Horticultural Society Memoirs	—
Philosophical Transactions of the Royal Society of London	PR
Phytopathology	UMI
Plant Physiology	UMI
Planta	UMI
Proceedings of the American Society of Horticultural Science	—
Proceedings of the Florida State Horticultural Science Society	—
Proceedings of the National Academy of Science	UMI
Proceedings of the Potato Association of America	—
Proceedings of the Royal Society of London	PR
Proceedings of the Washington State Horticultural Society	—
Science	UMI
Scientific Agriculture	UMI
Scientific Proceedings of the Royal Dublin Society	—
Soil Science	PR
Torreya	MANN
Transactions of the American Entomological Society	UMI
Transactions of the Massachusetts Horticultural Society	—
Transactions of the Royal and Horticultural Societies, London	—
USDA Yearbook	PR
Vermont Agricultural Experiment Station Annual Report	SPAULDING
Virginia Agricultural Experiment Station Bulletin	SPAULDING
Zeitschrift für Pflanzenkrankheiten (Pflanzenpathologie) und Pflanzenschutz	—
Zeitschrift für Pflanzenernährung, Düngung und Bodenkunde	VCH Publishing

[a]UMI = University Microfilm, Inc.; PR = Princeton Microfilm; MANN = Mann Library Microfilm, Cornell University; IDC = Inter Documentation Company.

crops. The 1937 *Yearbook* was considered a continuation from the previous year and had the subtitle "Better Plants and Animals—II." Of the forty-three detailed chapters in the two volumes, thirty-one dealt with plant breeding and crop improvement. These well written thirty-one chapters covered all United States crops or groups of crops. Each volume has 1,200 to 1,500 pages.

(2) Heald, Frederick D. *Manual of Plant Diseases*. New York and London: McGraw-Hill, 1933. 953p.

(3) Dickson, James G. *Outline of Diseases of Cereal and Forage Crop Plants of the Northern Part of the United States*. Minneapolis, Minn.: Burgess Pub., 1939. 259p.

Table 11.1 has the composite titles for the top 83 scholarly journals as cited by authors covering the 1870–1940 literature. The scientific literature of this period clearly showed the value of German and British journals to United States crop science. Those important titles are included as well, although not recommended for preservation.

Additional Sources

Slate, George L. *Bibliography of North American Pomological Literature in History of Fruit Growing and Handling in United States of America and Canada 1860–1872*. 1st ed. University Park, Pa., Regatta City Press and American Pomological Society, 1976. pp. 311–327.

Cowan Memorial Library. Ministry of Agriculture and Fisheries (U.K.) *Hand List of Books in English Printed after 1830* (London, 1939). 35 leaves.

Index

Authors and titles in the core list of monographs (pp. 150–292), the core list of journals (pp. 308–313), the reference titles (pp. 327–382), and the lists of primary historical literature (pp. 399–499) are *not* included in this index.